Pain Therapeutics
Current and Future Treatment Paradigms

RSC Drug Discovery Series

Titles in the Series:

1: Metabolism, Pharmacokinetics and Toxicity of Functional Groups
2: Emerging Drugs and Targets for Alzheimer's Disease; Volume 1
3: Emerging Drugs and Targets for Alzheimer's Disease; Volume 2
4: Accounts in Drug Discovery
5: New Frontiers in Chemical Biology
6: Animal Models for Neurodegenerative Disease
7: Neurodegeneration
8: G Protein-Coupled Receptors
9: Pharmaceutical Process Development
10: Extracellular and Intracellular Signaling
11: New Synthetic Technologies in Medicinal Chemistry
12: New Horizons in Predictive Toxicology
13: Drug Design Strategies: Quantitative Approaches
14: Neglected Diseases and Drug Discovery
15: Biomedical Imaging
16: Pharmaceutical Salts and Cocrystals
17: Polyamine Drug Discovery
18: Proteinases as Drug Targets
19: Kinase Drug Discovery
20: Drug Design Strategies: Computational Techniques and Applications
21: Designing Multi-Target Drugs
22: Nanostructured Biomaterials for Overcoming Biological Barriers
23: Physico-Chemical and Computational Approaches to Drug Discovery
24: Biomarkers for Traumatic Brain Injury
25: Drug Discovery from Natural Products
26: Anti-Inflammatory Drug Discovery
27: New Therapeutic Strategies for Type 2 Diabetes: Small Molecules
28: Drug Discovery for Psychiatric Disorders
29: Organic Chemistry of Drug Degradation
30: Computational Approaches to Nuclear Receptors
31: Traditional Chinese Medicine
32: Successful Strategies for the Discovery of Antiviral Drugs
33: Comprehensive Biomarker Discovery and Validation for Clinical Application
34: Emerging Drugs and Targets for Parkinson's Disease
35: Pain Therapeutics: Current and Future Treatment Paradigms

How to obtain future titles on publication:
A standing order plan is available for this series. A standing order will bring delivery of each new volume immediately on publication.

For further information please contact:
Book Sales Department, Royal Society of Chemistry, Thomas Graham House, Science Park, Milton Road, Cambridge, CB4 0WF, UK
Telephone: +44 (0)1223 420066, Fax: +44 (0)1223 420247,
Email: booksales@rsc.org
Visit our website at www.rsc.org/books

Pain Therapeutics

Current and Future Treatment Paradigms

Edited by

Charlotte Allerton
Pfizer Worldwide Research and Development,
Groton, Connecticut, USA

RSC Publishing

RSC Drug Discovery Series No. 35

ISBN: 978-1-84973-645-9
ISSN: 2041-3203

A catalogue record for this book is available from the British Library

Published by The Royal Society of Chemistry,
Thomas Graham House, Science Park, Milton Road,
Cambridge CB4 0WF, UK

Registered Charity Number 207890

For further information see our web site at www.rsc.org

Preface

Pain continues to be a dominant issue for healthcare worldwide as a key symptom associated with nearly every disease state. Opiates and non-steroidal anti-inflammatory drugs (NSAIDs) remain as the market-leading analgesics, despite leaving large numbers of patients ineffectively treated and exposed to significant safety concerns. Pain research has been active over the last decade in the search for novel medicines to fill this gap, but, disappointingly, this has not translated to the approval of many novel pain analgesics with differentiated efficacy and safety profiles. More recently, there has been a shift away from the research strategies underwritten by preclinical models of efficacy and treating pain as a uniform condition, towards strategies based on human confidence in rationale, underwritten by human translation strategies and an understanding of the role of the target in the pathophysiology of the disease, to enable the selection of the right patient populations for clinical trials. These evolving strategies give us renewed confidence that we will soon see the advancement of differentiated pain medications towards regulatory approval.

This book is a contemporary review of the field of pain therapeutics, including the historical medicines which still dominate standard-of-care treatments, such as opiates and NSAIDs, as well as the new mechanisms and combinations/reformulations that have dominated the regulatory approvals over the last decade, such as Lyrica® and the new formulation of OxyContin®. In addition, several of the chapters cover targets and mechanisms that have been explored as potential treatments of pain with limited success so far, such as delta-opioid ligands, although there are notable exclusions such as the area of purinergic receptors since these have been reviewed extensively elsewhere. Finally coverage is provided of the new emerging mechanisms currently providing excitement in the field of pain research, such as selective sodium channel modulators and biased opiate ligands.

RSC Drug Discovery Series No. 35
Pain Therapeutics: Current and Future Treatment Paradigms
Edited by Charlotte Allerton

Published by the Royal Society of Chemistry, www.rsc.org

Throughout this book the challenges in pain research that have underwritten the lack of success in delivering new analgesics to patients are highlighted, such as the lack of translation seen with pre-clinical models to the current regulatory and commercial challenges, as well as the changes in strategies that are being adopted to ensure greater success in the future. Finally, the book covers case studies of two novel products approved during the last decade, Lyrica and the Butrans® patch, which provide insights into the discovery and development journey for both an oral and non-oral product.

My thanks and appreciation go to all of the authors for the huge effort they have put into providing comprehensive and enlightening chapters in the important areas of pain research and development; as well as to Rosalind Searle and Gwen Jones for their continued guidance on behalf of the Royal Society of Chemistry.

Charlotte Allerton

Contents

RSC Drug Discovery Series No. 35
Pain Therapeutics: Current and Future Treatment Paradigms
Edited by Charlotte Allerton

Published by the Royal Society of Chemistry, www.rsc.org

CHAPTER 1

The Disease of Pain and Current Market Trends

CHARLOTTE M. N. ALLERTON,* ZAHID ALI AND DEVYN M. SMITH

Neusentis, Pfizer, Portway Building, Granta Park, Cambridge, U.K.
*Email: charlotte.allerton@pfizer.com

1.1 Introduction

Pain continues to be a dominant issue for healthcare worldwide, with the current market-leading medications, such as opiates and non-steroidal anti-inflammatory agents (NSAIDs), leaving large numbers of patients ineffectually treated, and exposed to potential safety concerns. This large unmet medical need is not only unacceptable for patients, but also results in significant cost implications for healthcare institutions, healthcare payers and the broader economy through lost work time. Over the last decade very few new pain drugs have been launched, and those that have been tend to be dominated by new formulations designed to target known mechanisms to improve side-effect profiles or enhance patient convenience. The lack of new approaches designed to target novel pathophysiological mechanisms, means that the treatment of patients who respond inadequately to currently available pain medications remains a major unmet medical and socio-economic need. This paucity of new mechanism treatments has been underwritten by significant challenges in preclinical research, a lack of understanding of the complexity of the pathophysiology of pain, a regulatory and payer requirement for any new medication to show differentiation based on efficacy and/or safety over current standard of

RSC Drug Discovery Series No. 35
Pain Therapeutics: Current and Future Treatment Paradigms
Edited by Charlotte Allerton

Published by the Royal Society of Chemistry, www.rsc.org

care, and significant challenges in clinical trial design. This chapter discusses these issues, and the different strategies that are evolving to address them and ensure success in the delivery of novel analgesics in the future, to solve the large unmet medical need in pain.

1.2 The Disease of Pain

The importance of pain in helping to protect against tissue injury is best exemplified by rare genetic disorders such as congenital insensitivity to pain, when severe injury and mutilation of body parts can be sustained due to prolonged contact with injurious stimuli.[1] This protective role of pain is well appreciated, but the chronic pain conditions associated with diseases such as osteoarthritis and neuropathic pain are less well understood. The differences between different types of pain are briefly described in the following section, although a fuller description is provided elsewhere.[2] Migraine is an important pain-related condition, but is beyond the scope of this book.

1.2.1 Acute and Chronic Pain

1.2.1.1 Acute Pain Conditions

Acute pain conditions last approximately 12 weeks or less and are associated with an acute injury or inflammation. Examples include headache, sprains, strains and acute postsurgical pain. The pain is a result of the direct or indirect activation of primary afferent nociceptors in injured tissues which transmit action potentials to the central nervous system (CNS). For this reason these are often referred to as acute nociceptive pain conditions. Although requiring analgesic intervention, these conditions often resolve spontaneously and are not directly associated with long-term health or socio-economic consequences.

1.2.1.2 Chronic Pain Conditions

By contrast, chronic pain can last for years or decades. Although often associated with disease or injury these conditions can sometimes evolve from seemingly inconsequential events and in some instances there is no identifiable injury or definable pathophysiology. Examples of chronic pain conditions are osteoarthritis pain, pain associated with peripheral nerve injury such as painful diabetic neuropathy (DPN), postherpetic neuralgia (PHN) and post-traumatic neuropathic pain (PTNP). In fact, chronic pain is insidious and associated with a host of different conditions such as immune/inflammatory diseases including rheumatoid arthritis and multiple sclerosis; diseases of visceral tissue including the heart, gastrointestinal tract, kidney and reproductive organs; and diseases associated with damage to the CNS such as post-stroke pain. There are multiple ways of classifying pain, but one way of classifying chronic pain is into conditions of chronic nociceptive pain such as osteoarthritis; conditions of chronic neuropathic pain such as DPN, PHN, and post-stroke pain; and conditions of dysfunctional pain such as fibromyalgia.

Chronic pain is often characterised by an ongoing pain that is not directly linked to a noxious stimulus or a magnitude of pain that is not commensurate with the stimulus, such that, for example, a relatively brief movement of the osteoarthritic joint can lead to long-lasting pain. The qualitative nature of the pain can be extremely distressing, taking on the nature of electric shocks, burning, crushing, lancing or throbbing.

Chronic pain conditions may also be associated with unpleasant sensory changes such that innocuous stimuli may be perceived as unpleasant and painful (allodynia), for example when bedclothes are pulled over the skin of some PHN and DPN patients. Many chronic pain patients also have an extreme hypersensitivity to everyday painful stimuli (hyperalgesia) such that a sharp point applied to the skin leads to a much greater pain response when compared to when the same stimulus is applied to non-painful skin. Taken together, the ongoing pain and hypersensitivity symptoms of chronic pain are extremely unpleasant. The relentless nature of the pain leads to extreme suffering and helplessness such that the affective dimension of pain puts profound restrictions on the patient's life.

The psychosocial consequences of chronic pain can lead to feelings of helplessness, anxiety and depression as well as increased disability, absenteeism from work and a breakdown in family and marital relationships. Taking all of these factors into consideration, while chronic pain can be associated with many distinct disease conditions, it also needs to be considered a disease in its own right. The challenge is that pain is the manifestation of multiple complex pathophysiological processes and there is no one common mechanism underlying all chronic pain conditions. The complexity becomes even more challenging in that even in a single disease such as DPN there are likely to be multiple processes at work; although patients may share commonality in suffering a metabolic disease and pain, the mechanisms underlying pain may vary from patient to patient. Understanding different mechanisms underlying the pain is the first step in developing better therapeutic options for the large number of patients who continue to suffer chronic pain conditions.

1.2.2 Mechanisms Underlying Pain

Most pain conditions have both a peripheral (nociceptors) and central (CNS) component with different underlying mechanisms contributing to the pain. Some of these mechanisms are briefly reviewed. A more detailed summary can be found elsewhere.[2,3]

1.2.2.1 Nociceptor Activation and Sensitisation

Nociceptors are small-diameter, specialised, primary afferent nerve fibres innervating almost all peripheral tissues which respond to high-intensity (noxious) stimuli. The loss of nociceptors is associated with a loss of acute pain responses which can contribute to the development of injuries and lesions such as those seen in small-fibre neuropathies, for example those associated with diabetes. Under normal conditions, nociceptors respond to mechanical, thermal and both exogenous and endogenous chemical stimulation. These signals are converted to sodium-channel-mediated action potentials which are transmitted to the CNS where they may be perceived as painful. Injury or inflammation such as is found in nociceptive pain conditions leads to the release of a wide variety of mediators such as prostaglandins, bradykinin, histamine, protons, interleukins and 5-HT from inflammatory and immune cells. These mediators not only activate nociceptors leading to ongoing pain but also sensitise nociceptors to thermal and mechanical stimuli such that a stimulus applied to inflamed or injured skin is far more painful than the same stimulus applied to non-injured skin. This is known as primary hyperalgesia. The opportunity to target inflammatory and peripheral neuronal targets is discussed in subsequent chapters.

1.2.2.2 Spontaneous Activity in Nociceptors

In most cases nociceptors only generate action potentials when activated by endogenous or exogenous stimuli. An important exception to this is in peripheral neuropathic pain conditions where there is evidence for the presence of spontaneous action potentials in nociceptors. It is thought that this barrage of spontaneous activity in nociceptors plays an important role in mediating ongoing pain as well as leading to changes in CNS excitability, connectivity and morphology (see below). Detailed preclinical and clinical investigations indicate a pivotal role for voltage-gated sodium channels in the generation of spontaneous activity. Consistent with this, empirical evidence with non-subtype-selective sodium channel blockers such as local anaesthetics and anticonvulsants are beneficial in some patients. However, a more detailed review has concluded that randomised controlled trials with a number of non-subtype-selective sodium channel blockers show little difference from placebo.[4] This is likely to be at least partly due to the inability to achieve pharmacologically relevant concentrations at the specific subtypes of sodium channels associated with ectopic activity generation due to adverse events associated with blocking sodium channels in the cardiovascular system and CNS. The opportunity for a more targeted approach by the development of subtype-selective sodium channel blockers is discussed in **Chapter 4**.

1.2.2.3 Central Nervous System Modulation and Central Sensitisation

The CNS is not a passive recipient of neuronal activity from nociceptors. Instead, neuronal activity from nociceptors is subject to extensive excitatory and inhibitory modulation during transmission to higher centres. For example, in the dorsal horn of the spinal cord, spinal and supraspinal excitatory and inhibitory processes can modulate the nociceptive signals to a considerable extent.

Neuronal activity of nociceptors can cause profound changes in the excitability of the CNS. This is best exemplified by central sensitisation which is the result of a barrage of activity from directly activated or spontaneously active nociceptors leading to an enhanced excitability of the CNS.[5] This enhanced excitability is a form of use-dependent synaptic plasticity with the initial phase mediated in part by glutaminergic *N*-methyl-D-aspartate (NMDA) and calcium-permeable α-amino-3-hydroxy-5-methyl-4-isoxazolepropionic acid (AMPA) receptors as well as voltage-gated calcium channels. This activation leads to a cascade of downstream effects which are detailed elsewhere.[6] Central sensitisation is a key mechanism that underlies hyperalgesia as well as allodynia. One important difference in the hyperalgesia generated by central sensitisation when compared to the primary hyperalgesia discussed above is that the hyperalgesia is not restricted to the injury site. Instead, this hyperalgesia can cover large areas that surround the injury site in otherwise normal skin and for this reason it is called secondary hyperalgesia.

1.2.2.4 Additional Central and Peripheral Nervous Mechanisms Leading to Pain

Preclinical evidence suggests that following nerve injury, low-threshold Aβ fibres start to contribute to the nociceptive response.[5] In addition, a number of mechanisms may contribute to produce a disinhibition of central structures such as the spinal cord. These mechanisms include structural changes which may be driven by the loss of peripheral and central neurons following peripheral nerve injury, as well as changes in connectivity between injured peripheral nerves and their targets during the regeneration process. Evidence from imaging studies indicates that marked neurodegeneration can occur in a number of chronic pain conditions as evidenced by decreases in grey matter volume and density.[7] One mechanism of increasing importance is the role of neuroimmune interactions. Immune interactions in both the peripheral nervous system mediated by macrophages and T lymphocytes, as well as in the CNS mediated by microglia and astrocytes, are likely to play an important role in the genesis and maintenance of chronic pain.[8]

1.3 Recent Trends in Pain Research

1.3.1 Pain Research over the Last Decade

In order to address the high unmet medical need still prevalent in pain populations, considerable research efforts have occurred in this area over the last decade. Interest in pain research, as measured through publication output, has been at an all-time high with pain-related publications over the last decade nearly matching the number released during the previous 40 years.[9] Despite this intense focus on the research of potential pain therapies, new medical entities based on novel mechanisms of action account for only approximately 5% of the pain drug launches during this time period.[10] The reasons for this low return on investment, which has led to an exit from pain research by several major pharmaceutical companies, warrants discussion.

Much research into analgesics was initially based on the assumption that all pain was potentially treatable by one single drug entity, without full recognition of the pathophysiology of differing pain states underwritten by complex molecular action at multiple and different central and peripheral sites of action.[4] The search for a broadly applicable analgesic often led researchers to focus on central mechanisms likely to achieve efficacy through suppression of pain sensation symptoms, rather than through treatment of the underlying cause of pain. Examples of mechanisms pursued in this category include the substance P receptors NK1 and NK-2, as well as the glutamate receptors, NMDA, metabotropic glutamate receptor 5 (mGluR5), AMPA, among others. A focus on these mechanisms brings challenges regarding safety and side-effect profile, which is already a challenge for the existing dominant opiate analgesics. Any future therapies would need to show at least equivalent efficacy to opiates with a significantly improved side-effect profile to meet the requirements of regulators and payers. An equivalent side-effect profile would likely only be tolerated by these bodies if there were significant efficacy advantages, which, with a central mechanism, would be hard to achieve over opiates in broad patient populations. An additional perceived advantage of these central mechanisms was that they often provided familiar G-protein coupled receptor (GPCR) targets, where medicinal chemistry and early screening experience enabled rapid delivery of suitable clinical compounds. However, speed of delivery is only valuable when combined with the right molecule for the right target.

Investigation of the validation of target selection during the last decade also exposes further challenges faced by researchers of novel pain analgesics. Following identification of the biological target of choice, and discovery of a selective ligand, further target validation research to strengthen the confidence in the rationale of that approach to have broad treatment for pain usually took place in preclinical models of pain. A range of models have been adopted, predominantly in rats and mice, focused on replicating the effects of acute, neuropathic, inflammatory and other pain diseases (such as for cancer pain).[11] These models have been largely unsuccessful in predicting efficacy in the clinic for novel pain medicines. Many potential reasons for this lack of translation have been investigated, including the poorly controlled behavioural nature of the models; the mismatch between the evoked pain stimulus in the preclinical model and the pathophysiology of the pain stimulus in the human disease population; bias in outcome due to lack of blinding in preclinical models; an over-reliance on hypersensitivity endpoints in preclinical models rather than the more prevalent spontaneous and continuous pain types seen in human patient populations; and a focus on acute medium-throughput preclinical models to enable rapid progress in discovery programmes but with little relevance to the unmet medical need in chronic pain. One example of this lack of translation between clinical outcome and efficacy in non-clinical models was seen with the fatty acid amide hydrolase (FAAH) inhibitor PF-04457845. This compound had shown anti-nociceptive effects in rodent models where pain sensitivity is enhanced to external stimuli.[12] At doses that inhibited plasma and brain FAAH, PF-04457845 reduced mechanical hyperalgesia and allodynia in the Freund adjuvant model through the expected cannabinoid-mediated mechanism with equivalent efficacy to the positive control naproxen. Similarly, in a non-inflammatory model of pain, it reduced the joint compression threshold in an equivalent manner to the positive control celecoxib. Such behavioural outcomes in these preclinical models were believed to be indicative of potential therapeutic benefit for the treatment of pain in osteoarthritis patients. In the clinic, despite decreasing plasma FAAH activity by greater than 96% leading to elevation of endogenous substrates known to reduce pain sensation through a central cannabinoid-receptor-mediated pathway, PF-04456845 showed no change in Pain WOMAC score compared to placebo, while the positive control naproxen behaved as expected.[13] There are a number of possible explanations for this lack of translation, such as differing FAAH inhibitory activity centrally compared to peripherally, compensatory pathways, pathway differences between rodent and humans; as well as the general issues with preclinical models described earlier. Discovering in a proof of concept trial that a mechanism does not have the horsepower required to give a meaningful pain response is an expensive and unsustainable way to learn, and makes earlier investment in gaining a deep understanding of the target pathway, pathophysiology of disease and translation to humans more compelling. There are exceptions to this lack of translation, for example with tanezumab, where efficacy in preclinical models broadly aligned with efficacy in associated human pain populations.[14] This translation is more likely to occur in examples where the molecular understanding of the pain mechanism is understood, allowing selection of appropriate models.

The final area of challenge in pain research is clinical trial design and selecting the right dose, which are challenges across many different disease areas. Clinical trials for novel pain mechanisms are conducted in pain populations where the chosen mechanism is believed to have the greatest chance of efficacy, either based on preclinical models of particular pain states or an understanding of the target's involvement in the treatment of specific pain types. However, there is enormous heterogeneity within pain disease populations, such as the osteoarthritis patients discussed in the FAAH example, and despite strict exclusion criteria to try to impose greater homogeneity, it is still a challenge for a single mechanism to show efficacy in this population type. We know already that our most common pain analgesics are only responsive in some pain populations with limited success. Increasingly, as the industry grows its understanding of the mechanisms underlying pain, clinical trials will be designed utilising this knowledge to select patients through phenotyping or genotyping strategies, which in turn will lead to a greater efficacy signal. A recent example of this strategy was the selection of erythromelalgia patients, who have a $Na_V1.7$ gain-of-function mutation, for participation in a trial aimed at testing the efficacy of a $Na_V1.7$ blocker in modulating pain.[15] These patient stratification approaches could then continue after launch by enabling physicians to select pain treatments based on diagnostics, leading to a greater success in treatment and a lower payer burden through avoiding experimentation with numerous different treatment paradigms.

As well as the challenges in selecting patients in whom to test a novel pain mechanism, there have also been considerable challenges in selecting the dose or doses at which to run the trial. In order to show efficacy in patients it is unquestionably important to ensure that the chemical or biological drug entity has exposure at the site of action, is engaging the target and has the desired downstream pharmacological outcome in humans.[16] The same limited understanding of the underlying pathways of some of the new mechanisms tested in pain during the last decade that caused the issues described above, also makes understanding exposure, occupancy and biomarker response at the site of action a considerable challenge. There have been numerous failed clinical trials in pain, as well as other diseases, run with novel mechanisms where it is uncertain whether the negative outcome was due to a poor choice of target or a poor choice of dose. When the wrong dose is selected, then the utility of that mechanism in the treatment of pain is never tested, which can lead to researchers incorrectly halting research into a particular mechanism as well as a complete lack of learning around the mechanism from that clinical experiment. An example from the obesity disease area occurred with the MC4 agonist MK-0493, where the compound displayed only a small, marginally significant effect on 24 hour energy intake, insignificant impact on weight loss, no meaningful information from exploratory biomarkers and no effect on blood pressure, which contrasted the dog pharmacology. The paper concluded that it was plausible that clinically relevant concentrations had not been achieved in this study, meaning that little was learnt about the mechanism in that disease state.[17] There are similar examples from the pain field, many unpublished, and to turn the tide of success in drug discovery, programmes need to ensure that the fundamentals of target engagement are understood.

While the challenges faced in pain research over the last decade have been significant, there are already many strategies emerging to ensure the mistakes are learnt from and there is greater success in the future. These strategies will be addressed in Section 1.4.

1.3.2 Dominant Pain Treatments Over the Last Decade

Considering the challenges encountered by pain researchers over the last decade, it is not surprising that there has been a paucity of new mechanisms launched into the market place. Current pain treatment continues to be dominated by opioid analgesics and NSAIDs, with over 50% of the current market for chronic pain treated through these mechanisms and a further approximately 20% being treated with combinations of these mechanisms or the more selective non-steroidal anti-inflammatory COX-2 inhibitors.[18] Alternative modes of action for the treatment of chronic pain are provided by antiepileptics such as the alpha2delta ($\alpha_2\delta$) ligand pregabalin (Lyrica®) and antidepressants such as the serotonin norepinephrine reuptake inhibitor (SNRI) duloxetine (Cymbalta®), as well as local anaesthetics.[19] For neuropathic pain, approximately 50% of the market share is covered by antiepiletics such as pregabalin as well as opioid analgesics. This is an interesting outcome considering the research that has been reported showing that the number of patients that need to be treated in the primary care setting to achieve a 50% reduction in neuropathic pain for one patient is four.[20] The share of the global pain market currently attributed to these well-known drug classes in 2010 is shown in Figure 1.1.[21]

The launch of new pain medications over the last decade has been dominated by reformulations and combinations, with only approximately 5% of launches accounted for by new medical entities with novel mechanisms of action.[22] The new medical entities launched over the last decade are shown in Table 1.1, along with previously launched medications which represent novel mechanisms in the context of approval for use in pain treatment during the same time period.

The new product launches summarised in Table 1.1 were responsible for around 30% of the global pain sales for 2010 (not including gabapentin enacarbil, which was not launched).[21,23] Pregabalin, duloxetine and tapentadol are prescribed in the primary care setting, while ziconotide is reserved for a specialised hospital setting. Gabapentin enacarbil is an extended-release formulation of a gabapentin prodrug. While these medications class as new pain treatments delivered over the last decade, with the exception of ziconotide most of the compounds work through known mechanisms of action with differentiation being achieved either through novel dual pharmacology, improved pharmacology or being first in class in new pain populations. This reiterates again the challenges of the last decade in identifying truly novel, new

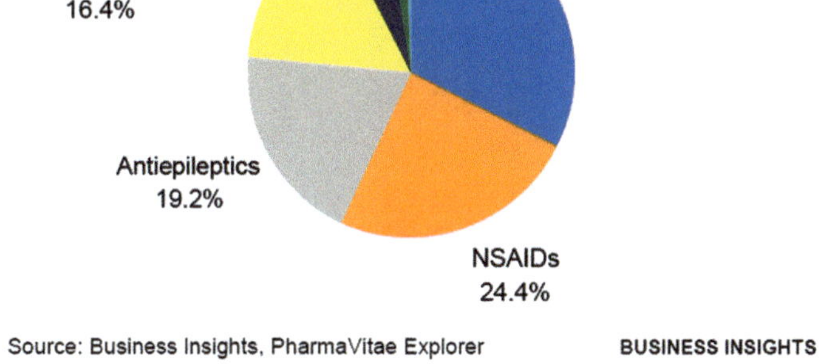

Figure 1.1 Global pain market share by drug class, 2010.

Table 1.1 New pain treatments launched over the last decade.

Generic name	*Proprietary name*	*NME*	*Novel target*	*Target class*	*Company*	*FDA approval*
Gabapentin enacarbil	Horizant®	Y	N	$\alpha_2\delta$ gabapentin prodrug	GSK	2012 for PHN
Tapentadol	Nucynta®	Y	Y	Opiate/NRI	Grunenthal	2009 for moderate/severe acute pain
Pregabalin	Lyrica®	Y	N	$\alpha_2\delta$	Pfizer	2004 for NeP 2007 for fibromyalgia
Duloxetine	Cymbalta®	N	N	SNRI	Eli Lilly	2004 for DPN 2008 for fibromyalgia 2010 for musculoskeletal pain
Ziconotide	Prialt®	Y	Y	N-type Ca channel blocker	Elan	2004 for intrathecal delivery for severe chronic pain

DPN, diabetic neuropathy; NeP, neuropthaic pain; NRI, norepinephrine reuptake inhibitor; PHN, postherpatic neuralgia; SNRI, serotonin and norepinephrine reuptake inhibitor.
Source: www.fda.gov/ and company websites.[84]

classes of pain therapeutics and the need for continued, more successful research in the area to meet the still high unmet medical need. These pain therapeutics are discussed in more detail later in the book, in the chapter relevant to their underlying mechanism of action.

Given the effectiveness of current therapies, combined with the paucity of new mechanisms, it is not surprising that reformulations and combinations have achieved such great market success. The reformulations of known mechanisms are either focused on reducing substance abuse in the treatment of chronic pain through various tamper-resistant oral technologies and/or achieving more convenient dosing through extended-release technologies or local administration. In addition, formulation changes can have an advantageous safety profile through either reducing systemic exposure to drug through local administration, or through providing less variation in systemic exposure through a slower release of active ingredient orally. The three top-selling reformulations over the last decade are shown in Table 1.2, and are all based on known mechanisms.

Purdue's OxyContin® (extended-release oxycodone) provides convenient dosing of the active mu opioid agonist oxycodone ingredient for moderate to severe chronic pain patients and is widely prescribed. This drug, as with many opiates, has been a target for drug abusers, and hence Purdue has switched to a tamper-resistant reformulation of OxyContin®, which was approved in 2010, where extended-release oxycodone has been reformulated in a polymer matrix that makes crushing difficult and forms a viscous gel with water to prohibit injection.[24] The reformulation of opiates to target reducing abuse, misuse and addiction has been an active strategy in the industry (*e.g.* Pfizer's Remoxy®), and is addressing a significant healthcare concern, since a 2007 US government survey reported that 13.2% of individuals misuse pain relievers at some point in their lives.[25] Tamper-resistant reformulations have focused on physical barriers to prevent the crushing, chewing or dissolution into liquids that enable abuse through injection or snorting which gives the rapid exposure required to

Table 1.2 The three top-selling pain reformulations in 2010.

Drug name	*Description*	*Target class*	*Company*	*FDA approval*	*2010 sales ($m)*
OxyContin®	Oral, ER oxycodone	Mu agonist	Purdue	1995 for chronic pain 2010 tamper-resistant formulation	3881
Lidoderm®	Local anaesthetic patch	Ion channel modulator	Endo	1999 for PHN	1083
Duragesic®	Fentanyl transdermal patch	Mu agonist	Johnson & Johnson	1991 for chronic pain	748

ER, extended-release; PHN, postherpatic neuralgia.
Source: Company 10K reports from www.sec.gov/ and www.fda.gov/

achieve the euphoric effects. Increasingly we now are seeing additional technologies enabling sequestered opioid antagonists to counteract any opioid effects if the tablets are crushed, or a sequestered aversive ingredient designed to create adverse events if the tablets are administered following crushing or dissolution. For all of these reformulations the companies face challenges in showing the safety and reproducibility of any formulation, and regulatory challenges related to proof of impact on opioid abuse. Opiates are discussed in greater detail in **Chapter 2**.

The Lidoderm® (lidocaine) patch provides an example of topical administration of a local anaesthetic, which provides pain relief while avoiding the toxic side effects seen with systemic exposure to this non-selective ion channel modulator. The Lidoderm® patch is approved for neuropathic pain resulting from PHN, although there has been literature evidence for its efficacy in the treatment of other pain conditions, for example carpal tunnel syndrome.[26] The Duragesic® fentanyl patch offers its differentiation through transdermal rather than topical delivery of the active ingredient. In this case the potent opioid analgesic is delivered transdermally, which provides advantages for patients in terms of duration of efficacy, ease of administration, particularly for patients struggling to swallow, and avoidance of some systemic side effects such as gastrointestinal constipation. The Duragesic® patch was the first fentanyl patch to the market, although generic versions are now available. In addition alternative opioid transdermal patches, such as Purdue's buprenorphine patch Butrans®, are increasing their market dominance following a later launch date (see **Chapter 13**). Topical and transdermal pain medications account for around 7% of worldwide chronic pain sales in 2010, and are particularly favoured in Japan (which accounts for 7.9% of the global market) where this route of administration is very popular because there is a lower acceptance level for strong oral painkillers.[22]

Fixed-dose combinations of pain medications are targeted at combining the efficacy of differentiated pain mechanisms, which can be safely co-administered, in one tablet for patient convenience. Johnson and Johnson's Ultracet®, which is a fixed-dose combination of tramadol and paracetamol (acetaminophen), has achieved success in the acute pain market with 2010 sales of around $134 million. Fixed-dose combinations can provide real therapeutic benefit. For example, 1000 mg of paracetamol is associated with approximately 1 in 3.8 patients achieving more than 50% pain relief.[27] Similarly, a low (60 mg) dose of codeine is associated with 1 in 16.7 patients achieving more than 50% pain relief. However, a combination of the two agents results in 1 in 2.2 patients achieving more than 50% pain relief,[28] thereby keeping doses of both drugs to safe and tolerated levels while increasing the overall analgesic efficacy. These assets often provide an opportunity to increase a pain medication's market share, although the technical challenges of combining two active ingredients into one tablet from a pharmaceutical sciences and drug metabolism perspective cannot be ignored. In many cases it may be more advantageous to pursue dual pharmacology in one molecule, such as with tapentadol. The ultimate challenge is to pursue combinations with synergistic mechanisms of actions, leading to enhanced efficacy compared to the additive impact of their individual efficacy.

An additional change over the last decade in pain management has been the increased patient use of non-drug approaches. These include everything from the application of heat patches, such as Pfizer's ThermaCare® heatwrap range, to exercise routines, to physiotherapy, to approaches such as acupuncture.[29] While these non-drug therapies are not the focus of this book, their impact on the pain market should be considered and many clinics are now offering access to these approaches alongside the more traditional drug therapy.

1.3.3 Recent Trends in the Regulatory, Payer and Commercial Environment

A key to understanding the size of the unmet need in pain is to assess the overall costs to society of inadequately treated pain. In 2011 the Institute of Medicine (IOM) released a report, '*Relieving Pain in America: A Blueprint for Transforming Prevention, Care, Education, and Research*', which estimated the annual cost of chronic pain in the United States at $560–635 billion, including healthcare expenses, medication, and lost productivity (*e.g.* inability to work).[30] The IOM report utilised the 2008 Medical Expenditure Panel Survey (MEPS) data to evaluate the economic burden of pain in the United States.[31] The direct healthcare-related costs of pain were estimated to be $260–300 billion, with indirect (productivity) costs of $300–330 billion. The authors estimated that over 40% of the direct costs of pain was borne by private insurers, with Medicare paying 25%, Medicaid paying 8%, and individuals paying 17% of the costs. This puts a tremendous burden on all players involved in paying for treating pain. This suggests that there is indeed still significant unmet need in the pain space and new treatments could certainly drive down the overall costs of pain to society. Interestingly enough, according to the National Institutes of Health (NIH), these estimates for pain are higher than those for cardiovascular disease, metabolic disease, and cancer.[32]

While the overall projected costs for the cost of inadequately treated pain are staggering, the total cost of prescription medications for treating pain is relatively modest. According to a study by Decision Resources, the total dollars spent on prescription medication for pain worldwide was $20.6 billion in 2010 and is forecast to remain flat with spend of $20.3 billion in 2020.[18] The US market is forecast to shrink from $13.6 billion in 2010 to $11.9 billion in 2020, due to the number of currently branded pain medications forecast to lose patent exclusivity in the next 6 years.[18] Because of this, generic medications comprised nearly 25% of the dollar market in 2010, and are forecast to grow to over one-third of the dollar market in 2020, while growing to over 75% of the total prescriptions written. The 2010 spend of $13.6 billion in the United States is just over 2% of the total direct healthcare costs from pain, indicating a tremendous amount of spend on other pain-related healthcare costs such as productivity loss, hospitalization, physician, *etc.*

Given the tremendous unmet need in patients with pain highlighted in this chapter, there have been several organizations that have focused on calling for change. First, the IOM report referred to above called for the establishment of a comprehensive national strategy for treating pain in patients, citing it as a national public health issue. It recommended a cultural transformation in how pain is judged and perceived by patients, providers and payers to improve overall treatment and management of pain. The draft elements of the plan that the IOM recommended were:

- Increase awareness of pain and its downstream health consequences.
- Increase emphasis on pain prevention.
- Improve delivery and financing of pain assessments, treatment and management led by the federal government programmes.
- Leverage public health communication strategies to educate patients on self-management approaches to manage their own pain.
- Address disparities in the assessment and treatment of pain across subgroups of Americans.

In addition to the IOM call for reform, other groups such as the Mayday Fund, a family foundation dedicated to reducing suffering caused by pain, have been actively involved in calling for, and implementing, changes to pain treatment.[33] In addition, the Mayday Fund provides grants to further research in improving pain treatment. Finally, others have called for more investment from the NIH in pain, and more philanthropy directed to pain research, to further help meet the challenge of poorly treated pain.[34] Given that the various reports mentioned above were published in the last 2–3 years, the overall impact of the recommendations has yet to be seen.

A key question is what impact regulatory and payer constraints have had on the pain market. In the regulatory space, one key problematic aspect with pain medications is opioid abuse. A description of the new technologies being deployed was given earlier in this chapter. In addition to these efforts by companies, the US Food and Drug Administration (FDA) is playing an active role in trying to mitigate the amount of opioid abuse that is occurring. To combat this abuse problem, in February 2009, the FDA announced that all extended-release opioid analgesics will be subjected to Risk Evaluation and Mitigation Strategies (REMS) to minimise the risk of abuse. In July 2012, the FDA formally approved a REMS programme for extended-release (ER) and long-acting (LA) opioid medications.[35] The REMS requires companies marketing ER/LA opioid medications to have training available for relevant prescribing healthcare professionals on the appropriate practices as well as having educational materials available for both prescribers and patients on the safe use of opioid medications.[36] It will be several years before the impacts of the REMS is seen in the US marketplace.

A recent study has highlighted how large of a problem the abuse of prescription pain medications in the US is, and why the FDA is proactively working with pharmaceutical companies to combat this. This study used data from the National Survey on Drug Use and Health from 1985 through 2009. The study claimed that abuse of prescription pain medications is the second most common form of illegal drug use after marijuana in the United States.

It also claimed that there was a 500% increase in the number of people seeking treatment for dependency to prescription opioids from 1997 to 2007 and that there was a 129% increase in visits to hospital emergency departments due to non-medical use of prescription pain medications between 2004 and 2009. This study shows that there is certainly a medical need for tamper-resistant opioid medications.[37] In addition, over 15 000 Americans die each year from overdosing on prescription painkillers—more than die from cocaine, heroin and other illegal drugs. In fact, drug overdoses are now the largest cause of accidental death in the United States, with over 39100 deaths in 2009—more than were killed in motor vehicle accidents.[38] This rise in deaths from accidental overdose has created much greater media attention to the problem,[39] as well as an increase in prosecution of unscrupulous physicians[40] and clinics.[41]

Another important regulatory consideration for pain medications is the opportunity for switching a prescription-only (Rx) medication to an over-the-counter (OTC) medication—usually once the patent on the medication has expired.[42] Of the 106 products that have successfully transitioned from Rx to OTC since 1976, only 6 were pain medications.[43] Of these six, there were only three novel MOAs—ibuprofen, dyclonine and naproxen (Note: aspirin and paracetamol are not on the list given they were already OTC before 1976). The six with their Rx-to-OTC approval dates and initial brand name are listed in Table 1.3. This list of Rx-to-OTC switches is quite limited, with the last successful switch over 10 years ago. It is unlikely that any opioid-based medications will ever transition to OTC due to safety profile; thus, the opportunity in this space for currently marketed prescription pain medications is likely limited.

Finally, what is the impact of recent US healthcare legislation on patients being treated for pain? While it is still early in the implementation of the Patient Protection and Affordable Care Act (which is being phased in gradually with much of the law not taking effect until 2014), there have been some impacts on access to pain medications for patients. One aspect of the new law was focused on patients with flexible spending accounts (FSA). These are accounts that allow patients to use tax-free dollars to buy non-prescription medications, insurance co-payments, or other out-of-pocket health costs, although these FSAs are 'use it or lose it' accounts which may end up in wastage at the end of the year. The Patient Protection and Affordable Care Act forces consumers who want to spend their FSA account on OTC medications to get a doctor's

Table 1.3 Pain medications transitioned from Rx to OTC since 1976.

Generic name	*Transition date*	*Proprietary name*
Dyclonine hydrochloride	1982	Sucrets
Ibuprofen	1984	Advil
Ibuprofen/pseudoephedrine	1989	Advil Cold & Sinus
Naproxen	1994	Aleve
Naproxen/pseudoephedrine	1999	Aleve Cold & Sinus
Ibuprofen/pseudoephedrine pediatric	2002	Children's Advil Cold

Source: www.fda.gov/

prescription first. However, the unintended consequences are that many patients are not able or willing to get a prescription for an OTC medication and either go without or pay out of pocket without using the FSA monies. In addition, it has put an additional burden on healthcare providers to spend time writing prescriptions for OTC medications. This was clearly not the intention of the law, but it has certainly had these unintended consequences.[44] In addition, a recent review focused on the potential impacts of the Patient Protection and Affordable Care Act on pain specialists. Two key impacts mentioned in the review are a reduction in reimbursement and a reduction in access to interventional pain management specialists by patients. Overall, the article paints a very dismal future for patients who wish to access pain specialists.[45]

In summary, pain is an area of tremendous unmet need characterised by extremely high costs to the healthcare system. There has been significant focus from many organisations on highlighting the need for a comprehensive change to the management and treatment of pain, and there should be some positive impacts of this over the next few years. Abuse potential remains a key risk for pain medications and both companies and regulators are focused on minimising the impact of abuse. Finally, the big unknown in the current treatment of pain is the impact of the Patient Protection and Affordable Care Act on patient access to providers and the ability to pay for treatment. With the current dynamic state of play in the commercial, regulatory and reimbursement fields for pain, the next few years will bear close watching to see how these fields evolve.

1.4 Future Trends in Pain Research

1.4.1 New Trends in Preclinical Research Strategies

Based on the learnings described in Section 1.3, progress in the molecular and mechanistic understanding of pain, and a focus on achieving greater return on investment in research and development, pain research is slowly changing to follow a fundamentally different strategy. The strategy used to select and validate targets for the effective treatment of pain is shifting away from identifying potentially broad mechanisms underwritten by preclinical disease models, to a more robust approach driven by human data coupled with human *in vitro* data from patient-specific cells, coupled with an investment in objective translatable biomarkers to enable preclinical dose predictions as well as dose-setting in the clinic. This more robust investment ahead of a drugs journey into the clinic should lead to greater clinical success in Phase 2 and 3 trials through the stratification of patients based on their phenotype or genotype to enable a greater efficacy signal. This vision for the future of analgesic research and development has been elegantly summarised in graphic form (Figure 1.2).[4]

Human genetics provides a powerful tool to validate a target's utility in pain. Examples of targets currently of interest in pain research underwritten by human genetics are $Na_V1.7$ and TrpA1. In the case of $Na_V1.7$, there are loss-of-function mutations leading to patients suffering from congenital

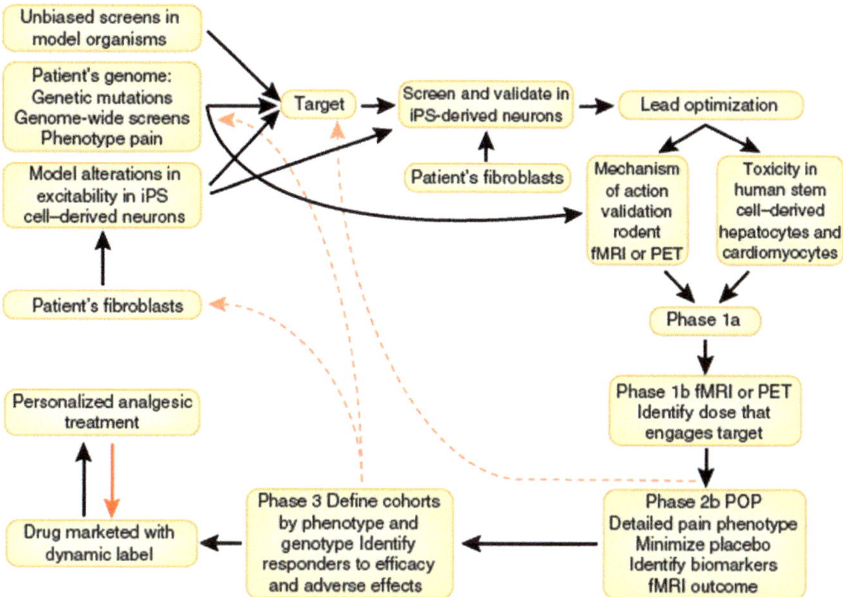

Figure 1.2 A proposed new analgesic drug development pathway. Adapted by permission from Ref. 5 Macmillan Publishers Ltd: Nature Medicine, **16**(11), 1241–1247, Copyright 2010. http://www.nature.com/nm/journal/v16/n11/abs/nm.2230.html

insensitivity to pain and gain-of-function mutations leading to erythromelagia patients who suffer, among other symptoms, debilitating episodes of burning pain and eythema.[1] In the case of TrpA1, a gain-of-function mutation is associated with oversensitivity of the channel to its pharmacological triggers leading to familial episodic pain syndrome for the affected individuals.[46] There is currently significant investment across the pharmaceutical industry building the understanding of the genotypes and phenotypes of different patient populations, to enable more precise treatment paradigms and to enable understanding of the underlying biology (see **Chapter 11**).

While human genetics provides a good starting point for target identification and validation, further understanding of the mechanism and pathway is required to underwrite the safety of the approach as well as its part in treating the pathophysiology of pain, which is an important determinant in selecting biomarkers and patients. Human *in vitro* tools can provide this further depth of understanding either to build on the genetic evidence or to build rationale in the absence of genetic data, but access to human cells such as dorsal root ganglia has been restricted. There has therefore been substantial stem cell research focused on the production of human *in vitro* cell systems.

Significant advances in stem cell biology over the last few years have enabled the generation of specialised cell types, including neurons, from human pluripotent stem cells facilitated by extrinsic factors.[47] These human *in vitro* systems enable assessment of different mechanisms abilities to modulate neuronal excitability, in dorsal root ganglia derived from both healthy and patient tissue. In addition to assessing the efficacy, and potentially also safety, of new mechanisms, these human cells also enable a more phenotypic screening strategy approach to the identification of new targets. In this case a range of chemotypes, with diversity of underlying target activity, would be screened for their impact on the disease-relevant outcome of decreasing neuronal excitability. If a potent hit is identified, then further work would elucidate the mechanism of activity and enable a resulting discovery programme.

As described earlier, building the link between engagement of the target and the downstream response is important, both in understanding the role of the target in the pathophysiology of disease, and hence patient selection, but also in identifying biomarkers that can be measured in the clinic to assess target engagement and enable success in dose-setting. Ideally these biomarkers should be measurable both preclinically and clinically to enable the understanding of human translation. Increasingly we are seeing investments in more objective measures of target occupancy, which can act as surrogates to true biomarkers, such as positron emission tomography (PET) imaging to assess receptor occupancy, or functional magnetic resonance imaging (fMRI) to assess blood oxygenation and flow to functioning areas of the brain. Additionally more mechanistic biomarkers can be selected, such as a capsaicin flare model in preclinical as well as clinical studies involving TrpV1 modulators, which is the pathway through which capsaicin elicits its pain response.[48] These approaches offer significant advances over the pain animal models described previously, as they enable an objective assessment of *in vivo* pharmacology preclinically and clinically to underwrite dose selection and an understanding of whether the target is truly being modulated at the concentrations tested in the clinic.

The combination of an investment in human biology, understanding of the pathophysiology of pain, human-relevant *in vitro* screening tools and translatable biomarkers will build a much stronger foundation for success in progressing a compound into the clinic. Building on the knowledge of the target in the design and patient selection for clinical trials will also be key for success, and this will be discussed further in **Chapter 11**.

1.4.2 Summary of Emerging Mechanisms

Any review of potential targets in pain will quickly come to the conclusion that the area of novel analgesic research is rich with opportunity. This is partly due to the host of peripheral and central mechanisms that contribute to the development of chronic pain conditions. In addition, pain is associated with many physiological and psychological responses such that, for example, the aversive aspect of pain is closely linked to anxiety, stress and depression. Taken together, this results in a long list of GPCRs, ion channels and enzymes that can serve as potential targets for novel analgesics. Despite this promise, as detailed in Section 1.3.1, there has been little return on more than a decade of intense investigation. In short, the targets, molecules and/or development strategies that have been the focus of pain research and development have not been right.

As detailed elsewhere in this chapter, efforts are under way to address these and to hopefully reduce the failure rate in the future.

Potential analgesic targets are best assessed in terms of their potential ability to modulate pathophysiological mechanisms underlying chronic pain conditions. One obvious approach is to separate targets into those that target peripheral mechanisms and those that target the CNS.

1.4.2.1 Peripheral Mechanisms

Inflammatory diseases are often associated with pain. In recent years, monoclonal antibodies (MABs) have been used to treat the disease underlying a number of painful inflammatory conditions. Results from clinical studies suggest that mechanisms targeted by the antibody may also be exploited for their analgesic potential.[49] These include the IL-6 receptor targeted antibody tocilizumab which is approved for use in rheumatoid arthritis (RA). There is a strong rationale for this as IL-6 levels are known to be elevated in painful RA as well as in low back pain, and treatment with tocilizumab is associated with a reduction in the pain of RA. In a similar way, MABs that target TNFα such as etanercept were developed as disease-modifying treatments for RA and are associated with reductions in the pain of RA. Although it is unclear whether any of the existing treatments targeting IL-6 or TNFα will be developed specifically as analgesics, they do, however, provide human evidence for analgesia associated with reducing the effects of inflammatory mediators. What remains unclear is whether blockade of these mediators in conditions such as osteoarthritis (OA) pain, where there is far less overt inflammation, is sufficient to confer analgesia. As well as MABs, small molecules such as inhibitors of JAK, which is a key component of the cytokine signalling mechanism,[50] and the recombinant IL-1 antagonist anakinra may also have analgesic effects, although evidence suggests that anakinra is ineffective in OA pain.[51]

An alternative peripheral approach is to target mechanisms that directly act on the peripheral nociceptors. In this respect, perhaps one of the biggest recent breakthroughs in analgesic research occurred with the discovery that the anti-NGF MAB tanezumab had markedly superior analgesic efficacy in OA pain when compared to NSAIDs. The development of anti-NGF treatments was initially paused as some OA patients participating in clinical trials of anti-NGF MABs developed rapidly progressing OA. Efforts are currently under way to better understand the benefit–risk profile of this class of treatment. Independent of the development of anti-NGF MABs for pain, there is interest in the general approach of targeting NGF pathways as well as the broader neurotrophin pathways as a novel approach for developing analgesic drugs. For example, an alternative approach is to block the tyrosine kinase A (TRK A) receptor which is one of the receptors through which NGF signals.

Voltage-gated sodium and potassium channels play a critical role in action potential transduction and transmission in peripheral nerves. The observation that some sodium channel subtypes are predominately located on nociceptive neurons has led to the possibility of developing blockers that are free of the

CNS and cardiovascular adverse events associated with non-selective sodium channel blockers. Chief among these sodium channel subtypes are the $Na_V1.7$, $Na_V1.8$ and $Na_V1.9$ channels, which are predominately located on nociceptors, as well as the embryonic sodium channel $Na_V1.3$, which is upregulated in nociceptors following nerve injury.[52] There has been considerable excitement about $Na_V1.7$ as an analgesic target on the basis of the human genetic gain-of-function and loss-of-function data and this has been discussed elsewhere in this volume. There is not the same strength of genetic evidence supporting $Na_V1.8$, but the location of these channels on nociceptors and the biophysical properties of the channel, taken together with preclinical data, make this an intriguing target. Clinical efficacy results from the first selective $Na_V1.7$ and $Na_V1.8$ blockers are likely to be available in the near future and will undoubtedly determine the future direction of research in the area of sodium channels and pain. The clinical development of selective $Na_V1.9$, $Na_V1.3$ and subtype-selective potassium channel modulators is important to better understand the utility of peripherally acting voltage-gated ion channel blockers for the treatment of pain.

The peripheral terminals of primary afferents express a host of GPCR and ligand-gated ion channels. These include members of the TRP superfamily of ligand-gated ion channels including TRPV1 which is activated by protons, heat and capsaicin; TRPA1 which is activated by cold and a range of chemical stimuli including cinnamaldehyde, methyl salicylate and isothyocynates; and TRPM8 which is activated by cold and menthol.[53] While a number of selective TRPV1 antagonists have progressed into clinical trials, efficacy has undoubtedly been limited by the presence of hypothermia. Whether this is an insurmountable obstacle remains to be seen. The relative importance of TRPM8 and TRPA1 is yet to be elucidated. However, the recent association between a gain-of-function TRPA1 mutation and familial episodic pain syndrome has suggested that TRPA1 might be an important therapeutic target.

1.4.2.2 Central Mechanisms

As with the peripheral nervous system, there are multiple potential targets and mechanisms in the CNS that are thought to contribute to chronic pain. These include mechanisms associated with the release and modulation of neurotransmission from the central terminals of primary afferents, through to mechanisms associated with modulation of the cognitive responses to nociceptive inputs. The remainder of this section highlights just a few of the potential opportunities.

A powerful way of modulating nociceptive transmission is to directly or indirectly activate endogenous inhibitory mechanisms within the CNS. Indeed, this is at least partly the mechanism by which opioids mediate their analgesic activity. A key central inhibitory system is the GABA system which is thought to play an important role in nociceptive transmission and processing in the CNS. The challenge is to develop GABA modulators with an acceptable benefit–toleration profile. This may be achieved by selectively targeting the

$GABA_A$ subtype and by developing allosteric modulators as opposed to full agonists. Other inhibitory mechanisms that have shown early promise but have failed to translate to efficacy in the clinic are CB2 agonists and FAAH inhibitors, as discussed above. Other inhibitory mechanisms for which there is a strong rationale but a difficult development path is the development of glycine agonists as well as approaches involving the activation of opioid receptors in a way that improves upon the benefit–toleration profile. Overall the difficulty has been to engage these mechanisms in a way that maintains analgesic efficacy while reducing the CNS adverse events associated with modulating mechanisms that are not only involved in pain but also in other functions.

An alternative way to modulate nociceptive processing in the CNS is to prevent excitation and sensitisation. One approach is to block N-type calcium channels (Cav2.2), which play a key role in neurotransmission in the CNS. The importance of targeting N-type calcium channels is supported by evidence of the therapeutic benefit of the ω-conotoxin ziconotide in patients with moderate to severe pain who are refractory to opioids. In addition, N-type calcium channels may play a role in microglia-mediated cytokine release in the CNS.[54] The challenge is to develop treatments that target this channel in pain-related mechanisms and can be administered orally, without having the serious unwanted effects associated with block of N-type calcium channels in the CNS.

Both metaboropic and ionic glutamate receptors have long been associated with the modulation of nociceptive processing in the CNS.[55] Importantly, the NMDA receptor has been seen as a key target, particularly with respect to the potential to reverse central sensitisation mechanisms with NMDA receptor blockers. Unfortunately, as discussed earlier, blockade of NMDA receptors is associated with multiple central adverse events and to date all approaches aimed at improving the therapeutic index of NMDA blockers have met with little success. Other novel approaches aimed at targeting central sensitisation mechanisms include targeting the brain-specific atypical protein kinase C (PKC) isoform PKMζ, which is implicated in long-term potention mechanisms.[56] Preclinically, PKMζ levels in the spinal cord are increased following peripheral painful stimulation, and nociceptive responses which are dependent upon central sensitisation mechanisms are reversed by PKMζ inhibitors.

A number of mediators are implicated in pain-related neuroimmune interactions in the CNS. In addition to N-type calcium channels discussed above, the purinergic system is thought to play an important role. Indeed, there is evidence for a role for the purinoreceptors P2X4, P2X7, P2Y6 and P2Y12 which are all expressed in microglia.[57] Although it is not yet clear which of these receptors is most important for targeting in terms of optimising any analgesic benefit, one advantage of targeting purinorectors in the CNS is that they appear to be expressed at low levels under normal conditions and are upregulated following injury. This suggests that targeting these purinorecptors may not result in significant CNS adverse events as they may not play a very important role in normal function. However, this is yet to be determined.

A number of additional peripheral and central targets have not been discussed here, but also show promise such as resolvins, ASICs, CX3CR1 and

Toll-like receptors.[58] In addition, as human genetic evidence becomes increasingly available, new targets will be identified and the importance of previously identified targets will be refined.

1.4.3 Optimising Early Analgesic Clinical Trials

The primary aim of analgesic clinical trials in human patients is to help to define the benefit risk profile of the compound as well as to better understand the physiological consequences of targeting a particular mechanism. In most cases, proof of concept (POC) studies are the first opportunity to explore analgesic efficacy for novel compounds. As a consequence, carefully designed POC studies and clinical programmes allow for early go/no go decisions which can help minimise future unnecessary patient exposure to an ineffective or potentially unsafe or poorly tolerated compound. The financial consequences of early decision-making in analgesic drug development are important, as the direct and indirect costs of even the simplest POC study are excess of a million dollars. This is compounded further in Phase 2b and 3 trials as the costs and consequences of progressing a compound with a less than ideal profile from POC into late-stage development is much greater.

Major concerns with the way in which analgesic trials have been conducted to date are:

- There has been a lack of optimisation and standardisation in the way in which studies have been designed, conducted, analysed and interpreted which makes it difficult to understand the benefit–risk profile of the agents being investigated as well as the clinical meaningfulness of the results.
- As there is less confidence in the translatability of preclinical data to the human patient, an inevitable consequence is for novel mechanisms that the first true exploration of analgesic efficacy is made in the human patient. In lieu of better translatable methodologies, decision-making in human has to be made faster and far less expensive.
- The consistency and reliability of efficacy results for an established analgesic varies from trial to trial, with some clinical studies even demonstrating a complete lack of efficacy although the compound has previously demonstrated robust efficacy.[3]
- Using a specific analgesic to treat a specific indication has resulted in variability in analgesic response which cannot be solely accounted for by safety/tolerability of pharmacokinetic considerations.

The way in which each of these issues is currently being addressed is summarised below.

1.4.3.1 Optimisation and Standardisation

The Initiative on Methods, Measurement, and Pain Assessment in Clinical Trials (IMMPACT) is comprised of invited participants from academia,

Table 1.4 IMMPACT recommendations: core domains for clinical trials of chronic pain treatment efficacy and effectiveness.

Pain
Physical functioning
Emotional functioning
Participant ratings of global improvement
Symptoms and adverse events
Participant disposition (including adherence to the treatment regimen and reasons for premature withdrawal from the trial)

This table has been reproduced with permission of the International Association for the Study of Pain® (IASP).[84]

regulatory agencies (FDA, European Medicines Agency), NIH, US Veterans Administration, consumer support and advocacy groups, and industry. The mission of IMMPACT is to develop consensus reviews and recommendations for improving the design, execution and interpretation of clinical trials of treatments for pain. Since the inception of IMMPACT in November 2002, thirteen meetings have focused on chronic pain, one on acute pain and one on acute and chronic pain in adolescents. A full exposition of all of the activities of IMMPACT is beyond the scope of this discussion. However, important recommendations of the initiative have included the identification of core outcome domains (Table 1.4). The aim of identifying these domains and associated outcome measures has been to ensure a more complete investigation and reporting of study results as well as allowing a standardised way in which clinical trial data can be collated, compared and interpreted. Other important discussions have included an assessment of patient-reported outcomes, clinician-reported outcomes and laboratory-reported outcomes, and the complex interplay between these measures. In addition, the difficulty of assessing the clinical meaningfulness of a statistically significant reduction in pain intensity has been addressed. This included comparisons to the arthritis field where 'low disease activity state' and 'patient acceptable symptom state' are being considered to better understand what the patient would consider a substantial response given the current treatment options. The relevance to pain is, for example, that a meaningful response is one that reduces pain to a score of 4 or less on the 11-point NRS where it is known that there is less interference with physical and emotional functioning when compared to higher ratings. Other IMMPACT meetings have focused on statistical considerations regarding multiple endpoints, individual *versus* group responses to analgesic treatments and optimising the design of randomised clinical trials. A fuller understanding of the discussions and recommendations of IMMPACT may be gained by referring to the IMMPACT website and associated publications.[59]

1.4.3.2 Translatability of Preclinical Data to the Human Patient

Efforts have been underway to obtain evidence of pharmacodynamic activity of novel putative analgesic agents in humans as early and as efficiently possible. Despite this, it is unlikely that analgesic research will be in the position of some other indications where evidence of human pharmacodynamic activity can be obtained in relatively few patients in Phase 1 studies. There is, however, considerable scope for reducing the size, cost and time associated with conventional POC studies from more than 100 subjects to studies of less than half that size. Key to achieving this is to better utilise crossover design where sample sizes are typically a lot smaller, meaning that studies can be both faster and less expensive to run. However, one concern with crossover studies has been the challenge of retaining subjects in the study and the observation of substantial period effects. One way of overcoming this has been to limit the crossover to no more than two periods.[12] Where the study involves the investigation of more than two treatments, two parallel crossover cohorts may be employed.[12]

It is also possible to take this one step further and to try to obtain evidence of analgesic activity in single-dosing studies prior to proceeding to multiple-dosing Phase 1 studies, which may be useful for high-risk targets where investment needs to be staged. The classical example is that of dental pain which is sensitive to a wide range of analgesic intervention related to nociceptive pain: NSAIDs, paracetamol and opioids. There has been some concern that this study is conducted in otherwise healthy volunteers and is a pain that originates within the trigeminal system and may therefore lead to false positives and false negatives for novel mechanisms. Although these concerns are largely unsubstantiated, a single dose equivalent of the standard OA pain model which utilises functional endpoints has been developed and demonstrates the analgesic efficacy of naproxen within 6 hours of a single dose administration (Ali *et al.* manuscript in preparation). Similarly, in a fMRI study, evidence of pharmacodynamic activity of a single dose of naproxen in response to pressure pain applied to the OA knee of less than 20 subjects has been demonstrated, suggesting this may be a sensitive way of getting early evidence of analgesic activity in small numbers of subjects.[60]

A final approach is to maximise the opportunity to obtain evidence of pharmacodynamic activity in healthy human volunteers ahead of conducting studies in patients. An example of this is the utilisation of biomarkers of pharmacodynamic activity such as the inhibition of IL-1β release by a P2X7 antagonist.[61] However, the limitation of this particular approach is that it requires exogenous stimulation of IL-1β release and, based upon clinical results from other P2X7 receptor antagonists, does not predict analgesic efficacy. An alternative approach using healthy volunteers is to utilise human experimental models of pain and hyperalgesia to help predict analgesic efficacy in patients. One example of this is the utilisation of capsaicin-induced flare and ultraviolet-induced irradiation models to provide evidence of pharmacodynamic activity of a selective TRPV1 antagonist in healthy human volunteers.[48] However, although some of the measures of pharmacodynamic activity demonstrated positive results, other measures did not. Subsequent studies in dental pain[62] and migraine[63] did not find evidence of analgesic activity with this compound, suggesting that healthy human volunteer markers of pharmacodynamic activity need to be chosen with care. A meta-analysis of analgesic results from healthy human volunteer models of pain is currently under way with the aim of identifying and recommending healthy human volunteer standards for conducting studies into pharmacodynamic activity of novel analgesics. This is being conducted with the EuroPain project which is part of the Innovative Medicines Initiative.[64]

1.4.3.3 Consistency and Reliability of Efficacy Results

A key factor in helping to ensure assay sensitivity of analgesic clinical trials is to reduce the magnitude of the placebo effect, as well as to reliably predict it. Results from an analysis of 17 PHN and 55 painful DPN trials indicated a greater improvement in pain scores in the placebo group of DPN when compared to the placebo group of PHN.[65] This suggests that the actual pain state in which the trial is conducted may lead to differences in the magnitude of the placebo response. The manner in which the trial is conducted is also likely to have an important impact on the magnitude of the placebo response. For example, in a meta-analysis of three neuropathic pain studies, the higher-recruiting centres were found to lead to the greatest improvement in pain of the placebo group.[66] Furthermore, as part of the EuroPain project, patient experiences—including relationship with the investigator and verbal suggestions—were found to play a central role in determining the magnitude of placebo response in patients in clinical trials.[67] As part of the EuroPain project further prospective work is currently under way to better understand the placebo and nocebo effect, including hidden *versus* open paradigms as well as a meta-analysis of placebo data from industry clinical studies in OA. In addition, prospective clinical studies which form part of EuroPain benefit from the inclusion of questionnaires and detailed phenotyping of patients, which may help further an understanding into factors that contribute to the placebo effect.

A number of approaches have been taken to reduce the magnitude of the placebo effect. The first of these has been employed in crossover design studies when subjects are blinded as to when a treatment period is about to start or end by dosing them with matching placebo during the pretreatment washout period as well as between dosing periods. This ensures that the subject is unaware of when a dosing period is about to start or finish, and thereby reducing any expectation associated with the start of a new dosing regimen. It also ensures that the subject is temporally unable to identify when a dosing period, be it placebo or the active treatment, starts and finishes. This has helped to successfully reduce the magnitude of the placebo effect in the few studies in which it has been employed.[12] Other approaches have included employing placebo run-in periods with the aim of excluding placebo responders, although it is not clear if these studies have achieved the aim of increasing assay sensitivity.

As well as reducing the magnitude of the placebo response, efforts have been made to optimise the magnitude of the active treatment effect. The best example of this has been the employment of a 'flare' design in studies of OA pain which relies on pain getting worse once the subject has their active treatment, typically an NSAID, withdrawn. Only those subjects showing this 'flare' response go on to receive another active treatment, which is typically another NSAID or COX-2 inhibitor. This has the benefit of excluding subjects who, for whatever reason, are not sensitive to analgesic approaches that target prostanoid mechanisms. A natural variation of this clinical trial design is to employ an open-label active run-in period when all subjects receive active treatment prior to randomisation and only those who show a response to the active treatment go on to be randomised to either active or placebo. Despite the rational approach taken with these designs, it is not clear whether these have delivered the level of assay sensitivity that was initially hoped for.

The opportunity to evolve the area of analgesic clinical trial design has been greatly enhanced by the FDA recently launching a public–private partnership, the Analgesics Clinical Trial Innovations Opportunities Network (ACTION).[68] This initiative will include a detailed programme of activities aimed at better understanding the placebo effect as well as identifying approaches to develop an evidence-based approach to analgesic clinical trial design.

1.4.3.4 Variability in Analgesic Response

Chronic pain patients sharing the same disease state such as DPN can present with very different symptoms including the quality, intensity and pattern of ongoing pain as well as any disease-associated sensory changes such as hyperalgesia, allodynia and hypoalgesia. This has led to the possibility that different symptoms may reflect differences in underlying mechanisms associated within the disease population. It has also been hypothesised that this may at least partly account for differences in a patient's ability to respond to one analgesic mechanism over a completely different mechanism. In addition, it has been observed that similar clusters of symptoms are present in different pain states such as in DPN and PHN where, despite different aetiologies, groups of patients may share similarities in, for example, the degree of central sensitisation and/or peripheral denervation and/or peripheral nerve hypersensitivity.[69] This raises the possibility that future treatments can be targeted towards the underlying mechanism as opposed to defining the chronic pain patient by the different disease aetiologies. The most comprehensive approach has been led by the German pain network on neuropathic pain where the methodology for determining the sensory profile of patients is determined by the use of a comprehensive battery of 7 tests measuring 13 parameters.[70] These are roughly grouped into thermal detection, thermal pain, mechanical detection and mechanical pain. Currently, as part of the IMI EuroPain project, a study is underway to determine if patients with any type of peripheral nerve injury with different sensory profiles respond differently to different analgesics. A complementary approach has been undertaken using a validated questionnaire, Pain*DETECT*, which consist of nine questions regarding the severity, course, quality and nature of the patients pain and specific neuropathic pain symptoms.[71] A survey of 2100 PHN and DPN patients revealed 5 distinct subgroups of patients showing a characteristic sensory profile. All subgroups were present in the DPN and PHN population, but the frequency of occurrence of the subgroup differed between the two disease states. Results from future investigations will reveal if such a mechanism-based approach is the optimal way of both developing analgesic drugs as well as providing treatments for chronic pain patients.

Another way of identifying potential groups of responders to novel treatments may be through the broader use of genotyping. In the simplest interpretation, it is known that diseases such as inherited erythromylalgia are driven by gain-of-function mutations in the gene associated with the $Na_V1.7$ sodium channel. Inherited erythomylalgia is an extremely rare disease, but if the predisposition for developing chronic pain conditions can be associated to specific genes, this may provide a way of stratifying patients into those who are more or less likely to respond to an analgesic treatment.

1.4.4 Emerging Trends in the Pain Market

There are some key trends emerging in the pain market which are highlighted below, including an increase in the number of novel mechanisms in the development pipeline, a continued focus on reformulations and precedented mechanisms, pharmaceutical and biotech companies with a clear focus in the pain space based on their pipelines, and the emergence of multiple precompetitive consortia focused on enabling the development and commercialisation of new pain therapies.

There have been limited approvals in the pain space over the last decade. This is not due to the current regulatory environment but to the lack of new therapies making it successfully through clinical development. The pain market continues to evolve as new targets and new therapeutic modalities have been discovered. We can see this through a review of the current pipeline. There are a large number of assets in preclinical and clinical development in the pain space (Figure 1.3).

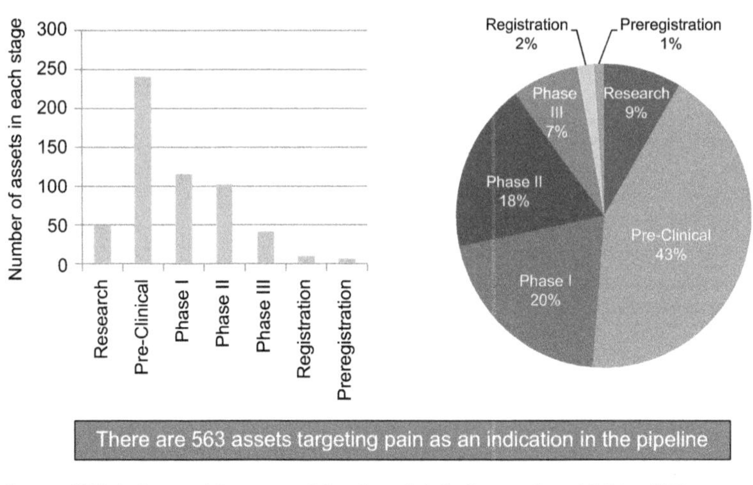

Source: ADIS pipeline search for compounds targeting pain indications –performed October, 2012, www.adis.com

Figure 1.3 The pain pipeline by stage.

An analysis performed in October 2012 found that there were over 550 assets in development for a pain indication and some trends emerged from this analysis. Interestingly, over half were preclinical assets, and just under half were in the clinic. Further analysis (Figure 1.4) indicates that the 25 largest pharmaceutical companies have a total of 133 assets in their pipelines focused on an indication that is pain related. However, over half of these assets are in the pipelines of six companies—Pfizer, Abbott, Astellas, Johnson & Johnson, Merck and AstraZeneca. These six companies have clearly maintained an investment in the pain space, as evidenced by their large pipelines.

If we look at the small companies for those with the largest number of assets in development, seven companies have four or more assets in development (primarily reformulations). These companies are PharmacoFore, DURECT, SK Pharmaceuticals, Glenmark, Nektar, Omeros, and Theraquest. This suggests that there are clear large-company and small-company leaders in the development of pain-related assets, although there are many companies developing assets in pain.

Analysis was also performed on the assets to understand the trends in route of administration and drug class. Nearly 400 (Table 1.5) of the assets have a

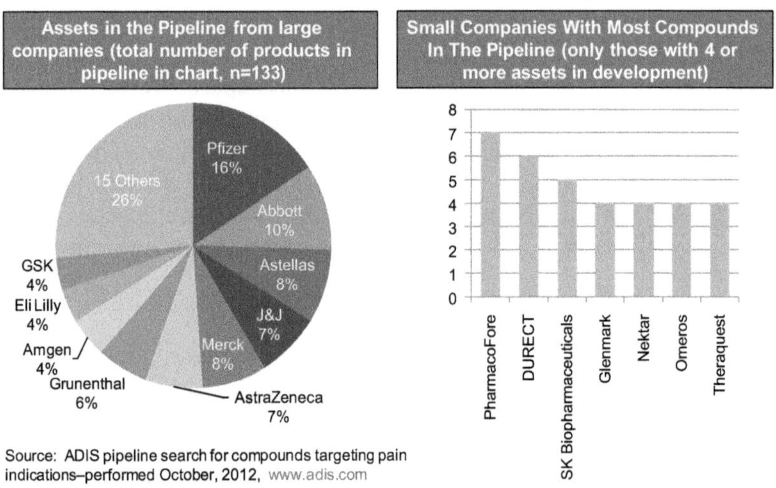

Figure 1.4 The pain pipeline by company.

Table 1.5 Pain pipeline by route of administration.

Administration route	*No. of assets*	*% of total*
Topical	43	11.1
Oral	251	64.5
Injection	35	9.0
Intravenous	60	15.4
Total	389	

Source: ADIS pipeline search for compounds targeting pain indications, performed October 2012, www.adis.com.

route of administration (ROA) listed. Of these, nearly two-thirds are oral, with the remaining mixed between parenteral/intravenous, topical formulations, and various injection modalities (*e.g.* intramuscular, intrathecal, subcutaneous, *etc.*). This shift to more products with non-oral deliveries reflects the shifting marketplace for pain products to those with more complex delivery methods (*e.g.* biologics) and reformulation of old mechanisms (primarily opioids). Commercial approval of biologic therapies for pain could have substantial impact on the treatment paradigm for pain as primary care physicians have traditionally not used biologic therapies, which have usually been used by specialist physicians. However, given the higher costs of biologic therapies *versus* small-molecule therapies, the bar for reimbursement by payers will be significant, requiring new biologics for pain to be differentiated from existing treatments to provide an attractive cost–benefit ratio. Time will tell whether this shift to more biologic-based mechanisms to treat pain will push primary care physicians to become more comfortable in prescribing biologic therapies for their patients. Alternatively, biologics could remain in the realm of the specialists, in this case, pain specialists.

In addition (Table 1.6), we can look at the mechanisms of action for the assets in development, (where these have been publicly stated). As one can see, over half of the assets are either an opioid or some form of a COX targeting asset, both of which have been on the market for many years. Beyond these two mechanisms, other targets of interest in the pipeline include calcium channels, sodium channels, and GABA modulators. These data suggests that there are a considerable number of 'me-too's' or reformulations of existing products in the

Table 1.6 Pain pipeline by mechanism of action.

Mechanism[a]	*No. of assets*	*% of total*
Opioid	246	36.0
COX	121	17.7
Calcium channel	60	8.8
Sodium channel	42	6.1
Adrenergic	41	6.0
TRPV	40	5.8
Cannabinoids	34	5.0
GABA	16	2.3
Nicotinic	14	2.0
NGF	13	1.9
GPCR	10	1.5
Potassium channel	10	1.5
Other[b]	37	5.4

COX, cyclooxygenase; GABA, gamma-aminobutyric acid; GPCR, G-protein coupled receptor; NGF, nerve growth factor; TRPV, transient receptor potential vanilloid (receptor).

[a]Some assets have more than one mechanism of action listed, hence the number in this table is larger than actual assets in development.

[b]'Other' includes mechanisms of action such as cell replacement therapies, TRPV,TRPM, TRPA, other ion channel targets and histamine.

Source: ADIS pipeline search for compounds targeting pain indications, performed October 2012, www.adis.com.

pipeline, given that over half of the assets are in mechanistic areas with launched compounds. However, the data also suggest that there are a significant number of novel mechanisms being pursued, some of which may be significant pain treatments of the future, many of which are discussed in this book. For example, stem-cell based therapies[72,73] from companies like Mesoblast and Celgene have promise for treating pain and the underlying disorders that cause pain, but have not been proven in large, placebo-controlled, randomised Phase 2 studies. Mesoblast, for example, is running a 100-patient Phase 2 trial[74] to test the safety and efficacy of using mesenchymal stem cells to treat patients with chronic lower back pain, with the study apparently over 50% enrolled by mid-2012.[75] The acid-sensing ion channels (ASICS) are another area that has received a lot of press, with a recent *Nature* article about a natural venom that binds to these receptors, and this is another emerging area, albeit high risk.[76] Given the large number of novel mechanisms being studied, from ion channel targets to stem cells, it should prove to be an exciting few years as the data from these clinical studies reads out. Further, there is also significant activity focused on better diagnostic and surgical techniques for treating pain in areas like back pain and OA.[77,78]

Finally, given some of the challenges set out above in developing novel pain assets, there has been significant work in precompetitive consortia focused on developing tools to better enable drug discovery in the pain space. Two large examples include the Innovative Medicines Initiative Funded EuroPain[64] and the London Pain Consortia.[79] Others consortia include the NIH Pain Consortia,[80] the German Research Network on Neuropathic Pain[81] and ACTION.[82]

EuroPain is a precompetitive consortium made up of an international team of researchers and clinicians from both academia and industry whose focus is to increase the understanding of chronic pain, develop new treatments for pain and identify novel biomarkers for different types of pain. The consortium is co-funded by the IMI and the European Federation of Pharmaceutical Industries and Associations (EFPIA). The project is forecast to last 5 years and will receive €6 million from the IMI as well as €12.5 million in-kind contribution from the EFPIA participating members over 5 years. The consortium includes scientists from 12 European universities: King's College London, University College London, Imperial College London, University of Oxford, Christian-Albrechts-University of Kiel, Medical Faculty Mannheim/Heidelberg University, Technische Universität München, Goethe University of Frankfurt, BG University Hospital Bergmannsheil/Ruhr University Bochum, University Hospitals of Aarhus, Rigshospitalet Copenhagen, University of Southern Denmark, and SME Neuroscience Technologies from Barcelona. There is also significant involvement of pharmaceutical companies, including AstraZeneca, Boehringer-Ingelheim, Eli Lilly, Esteve, Grünenthal, Pfizer, Sanofi-Aventis, Abbott, Astellas and UCB.

Unique aspects of the EuroPain consortium include undertaking a large number of preclinical and clinical studies to both find new drugs for treating pain as well as identify better biomarkers for different pain aetiologies. Key to the EuroPain consortium is that delivery of the goals will be through collaboration between all members of the EuroPain network who will share resources and data from their work. Results will be made public during and after the project, which will ensure that the benefits are quickly translated into drug discovery and development efforts.

The London Pain Consortium was funded by a grant from the Wellcome Trust in 2003, which was renewed in 2008 for another 5 years and £5 million. This consortium is made up of researchers from King's College London, Imperial College London, University College London, and the University of Oxford. The member institutions are contributing an additional £1 million. Companies can partner with the consortium members to gain access to the network of leading researchers. For example, Toray, a Japanese company, has partnered with a consortium member and has then been able to leverage the entire consortium's network to enable its research.[83] The consortium has been successful, with over 280 scientific papers published, development of new animal and human models for pain, creation of a publicly available Pain Database and training of more than 30 PhD students and postdoctoral scientists.

In summary, while there has been limited success in bringing pain drugs with novel mechanisms of action to the market, the understanding of the underlying molecular mechanisms of pain has resulted in a strong pipeline of diversified assets. It will be exciting to observe the clinical data from these many novel mechanisms as they move towards commercialization, as patients should have access to even better pain medications in the coming years. In addition, various consortia are focused on both improving early discovery efforts and/or clinical development efforts to enable more pain therapies coming to market.

1.5 Conclusions

From the work described above, it is clear that commitment to the research and delivery of novel pain medicines to address the huge unmet medical need worldwide is still high in academia, in industry and through precompetitive consortia. There is an increased focus on understanding the disease of pain, investigating human evidence for the role of specific targets in pain pathways through genetics analyses and using this depth of knowledge to stratify patients for inclusion in clinical trials to show enhanced efficacy signals. In addition to the focus on new mechanisms, about 50% of the current development investment in pain assets remains in the well-known analgesic mechanisms, in particular opiates and COX programmes, looking to enhance the efficacy without the unwanted side effects. These approaches will also benefit from an increased understanding of these mechanisms of action, for example investigating the differing downstream signalling cascades of mu opioid receptors, as well as enhanced technology solutions. Many of the trends in pain research and development described in this chapter will be covered in more depth in latter chapters. Collectively these chapters will strengthen our understanding of the disease of pain, the best mechanisms to treat the differing subpopulations and the emerging research and development that will increase our success in delivering novel pain medicines to patients.

References

1. J. J. Cox, F. Reimann, A. K. Nicholas, G. Thornton, E. Roberts, K. Springell, G. Karbani, H. Jafri, J. Mannan, Y. Raashid, L. Al-Gazali, H. Hamamy, E. M. Valente, S. Gorman, R. Williams, D. P. McHale, J. N. Wood, F. M. Gribble and C. G. Woods, *Nature*, 2006, **444**, 894–898.
2. *Wall and Melzack's Textbook of Pain*, ed. S. McMahon and M. Koltzenburg, Elsevier/Churchill Livingstone, Philadelphia, 5th edn, 2006.
3. R. S. Griffin and C. J. Woolf, in *Principles of Pharmacology: The Pathophysiological Basis of Drug Therapy*, ed. D. E. Golan, A. H. Tashjian, E. Armstrong and A. W. Armstrong, Lippincott, Williams and Wilkins, Baltimore, MD, 2nd edn, 2007, pp. 263–282.
4. Z. Ali, J. E. Palmer and V. Goli, in *Textbook of Pain*, Elsevier/Churchill Livingstone, Philadelphia, 6th edition, in press.
5. C. J. Woolf, *Nat. Med.*, 2010, **16**, 1241–1247.
6. M. Costigan, J. Scholz and C. J. Woolf, *Annu. Rev. Neurosci.*, 2009, **32**, 1–32.
7. A. V. Apkarian, Y. Sosa, S. Sonty, R. M. Levy, R. N. Harden, T. B. Parrish and D. R. Gitelman, *J. Neurosci.*, 2004, **24**, 10410–10415.
8. P. J Austin and G. J. Moalem-Taylor, *Neuroimmunology*, 2010, **229**, 26–50.
9. I. Kissin, *Anesth. Analg.*, 2010, **110**, 780–789.
10. www.centerwatch.com/drug-information/fda-approvals/.
11. J. S. Mogil, *Nat. Rev. Neurosci.*, 2009, **10**, 283–294.
12. K. Ahn, S. E. Smith, M. B. Liimatta, D. Beidler, N. Sadagopan, D. T. Dudley, T. Young, P. Wren, Y. Zhang, S. Swaney, K. Van Becelaere, J. L. Blankman, D. K. Nomura, S. N. Bhattachar, C. Stiff, T. K. Nomanbhoy, E. Weerapana, D. S. Johnson and B. F. Cravatt, *J. Pharmacol. Exp. Ther.*, 2011, **338**, 114–24.
13. J. P. Huggins, T. S. Smart, S. Langman, L. Taylor and T. Young, *Pain*, 2012, **153**, 1837–1846.
14. P. W. Mantyh, M. Koltzenburg, L. M. Mendell, L. Tive and D. L. Shelton, *Anesthesiology*, 2011, **115**, 189–204.
15. Y. P. Goldberg, N. Price, R. Namdari, C. J. Cohen, M. H. Lamers, C. Winters, J. Price, C. E. Young, H. Verschoof, R. Sherrington, S. N. Pimstone and M. R. Hayden, *Pain*, 2012, **153**, 80–85.
16. P. Morgan, P. H. Van Der Graaf, J. Arrowsmith, D. E. Feltner, K. S. Drummond, C. D. Wegner and S. D. A. Street, *Drug Discov. Today*, 2012, **17**, 419–424.
17. R. Krishna, B. Gumbiner, C. Stevens, B. Musser, M. Mallick, S. Suryawanshi, L. Maganti, H. Zhu, T. H. Han, L. Scherer, B. Simpson, D. Cosgrove, K. Gottesdiener, J. Amatruda, B. J. Rolls, J. Blundell, G. A. Bray, K. Fujioka, S. B. Heymsfield, J. A. Wagner and G. A. Herman, *Clin. Pharmacol. Therapeut.*, 2009, **86**, 659–666.
18. Decision Resources. Pain Management Study—Chronic Pain, November, 2011.

19. Decision resources. Pain Management Study—Neuropathic Pain, November 2011.
20. N. B. Finnerup, M. Otto, T. S. Jensen and S. H. Sindrup, *Med. Gen. Med.*, 2007, **9**, 36.
21. Scripps Business Insights™. The Pain Market Outlook to 2016.
22. www.centerwatch.com/drug-information/fda-approvals/ Accessed 9 October 2012.
23. Company 10Ks and annual reports.
24. S. P. Stanos, P. Bruckenthal and R. L. Barkin, *Mayo Clin. Proc.*, 2012, **87**, 683–694.
25. C. C. Upshur, R. S. Luckmann and J. A. Savageau, *J. Gen. Intern. Med.*, 2006, **21**, 652–655.
26. S. Nalamachu, R. S. Crockett, A. R. Gammaitoni and E. M. Gould, *Med. Gen. Med.*, 2006, **8**, 33.
27. www.medicine.ox.ac.uk/bandolier/booth/painpag/Acutrev/Analgesics/Leagtab.html Accessed 12 February 2013.
28. www.medicine.ox.ac.uk/bandolier/booth/painpag/Acutrev/Analgesics/Acutepain2007combo.pdf Accessed 12 February 2013.
29. X. G. Tao and E. J. Bernacki, *J. Occup. Environ. Med.*, 2005, **47**, 1298–1306.
30. Institute of Medicine. *Relieving Pain in America: A Blueprint for Transforming Prevention, Care, Education, and Research*. The National Academies Press, Washington, DC, 2011.
31. D. J. Gaskin and P. Richard, *J. Pain*, 2012, **13**, 715–724.
32. *National Heart, Lung, and Blood Institute Fact Book Fiscal Year 2010*, U.S. Dept. of Health and Human Services, National Heart, Lung, and Blood Institute, Bethesda, MD 2011.
33. Mayday Fund. Call to Revolutionize Chronic Pain Care in America—November, 2009. www.maydayfund.org/index.html. Accessed 31 December 2012.
34. D. Borsook, *Cerebrum*, 2012, May/June, 1–16. http://dana.org/news/cerebrum/detail.aspx?id = 39160. Accessed 31 December 2012.
35. U.S. Food and Drug Administration. www.fda.gov/drugs/drugsafety/informationbydrugclass/ucm163647.htm. Accessed 31 December 2012.
36. U.S. Food and Drug Administration. www.fda.gov/Drugs/DrugSafety/InformationbyDrugClass/ucm309742.htm. Accessed 31 December 2012.
37. R. Miech, A. Bohnert, K. Heard and J. Boardman, *J. Adolesc. Health*, 201210.1016/j.jadohealth.2012.07.01.
38. K. D. Kochanek, J. Xu, S. L. Murphy, A. M. Minino and H.-C. Kung, *Deaths: Final Data for 2009, by Division of Vital Statistics*. Centers for Disease Control Prevention, National Center Statistics, Vital System United States Department of Health and Human Services, 2011, **60**(3), p. 11.
39. T. Catan, D. Barrett and T. W. Martin, *Wall Street Journal*, 5 October, 2012. http://online.wsj.com/article/SB10000872396390444223104578036933277566700.html?KEYWORDS = pain + clinic. Accessed 31 December 2012.

40. Associated Press, *Wall Street Journal*, 6 December 2012. http://online.wsj.com/article/AP58ce2ea78af64546999bf01fa3b9fc2e.html?KEYWORDS = pain + clinic. Accessed 31 December 2012.
41. T. W. Martin, *Wall Street Journal*, 25 December 2012. http://online.wsj.com/article/SB10000872396390444223104578036933277566700.html?KEYWORDS = pain + clinic. Accessed 31 December 2012.
42. Consumer Healthcare Products Association, www.chpa-info.org/media/resources/r_8041.pdf. Accessed 31 December 2012.
43. Consumer Healthcare Products Association, www.chpa-info.org/media/resources/r_4620.pdf. Accessed 31 December 2012.
44. J. Adamy, *Wall Street Journal*, 9 March 2011 http://online.wsj.com/article/SB10001424052748704692904576166554110739560.html?mod = djemHL_t Accessed 31 December 2012.
45. L. Manchikanti and J. A. Hirsch., *Pain Physician*, 2012, **15**, E629–E640.
46. B. Kremeyer, F. Lopera, J. J. Cox, A. Momin, F. Rugiero, S. Marsh, C. G. Woods, N. G. Jones, K. J. Paterson, F. R. Fricker, A. Villegas, N. Acosta, N. G. Pineda-Trujillo, J. D. Ramírez, J. Zea, M.-W. Burley, G. Bedoya, D. L. H. Bennett, J. N. Wood and A. Ruiz-Linares, *Neuron*, 2010, **66**(5), 671–680.
47. S. M. Chambers, Y. Qi, Y. Mica, G. Lee, X-J Zhang, L. Niu, J. Bilsland, L. Cao, E. Stevens, P. Whiting, S.-H. Shi and L. Studer, *Nat. Biotechnol.*, 2012, **30**(7), 715–720.
48. B. A. Chizh, M. B. O'Donnell, A. Napolitano, J. Wang, A. C. Brooke, M. C. Aylott, J. N. Bullman, E. J. Gray, R. Y. Lai, P. M. Williams and J. M. Appleby, *Pain*, 2007, **132**(1–2), 132–141.
49. I. P. Chessell, A. Dudley and A. Billinton, *Drug Discov Today*, 2012, **17-**(15–16), 875–879.
50. J. H. Coombs, B. J. Bloom, F. C. Breedveld, M. P. Fletcher, D. Gruben, D. J. M. Kremer, R. Burgos-Vargas, B. Wilkinson, C. A. Zerbini and S. H. Zwillich, *Ann. Rheum. Dis.*, 2010, **69**, 413–416.
51. X. Chevalier, P. Goupille, A. D. Beaulieu, F. X. Burch, W. G. Bensen, T. Conrozier, D. Loeuille, A. J. Kivitz, D. Silver and B. E. Appleton, *Arthritis Rheum.*, 2009, **61**, 344–352.
52. B. T. Priest, *Curr. Opin. Drug Discov. Devel.*, 2009, **12**(5), 682–692.
53. L. Vay, C. Gu and P. A. McNaughton, *Br. J. Pharmacol.*, 2012, **165**(4), 787–801.
54. N. Tokuhara, K. Namiki, M. Uesugi, C. Miyamoto, M. Ohgoh, K. Ido, T. Yoshinaga, T. T. Yamauchi, J. Kuromitsu, S. Kimura, N. Miyamoto and Y. Kasuya, *J. Biol. Chem.*, 2010, **285**(43), 33294–33306.
55. K. M. Wozniak, C. Rojas, Y. Wu and B. S. Slusher, *Curr. Med. Chem.*, 2012, **19**(9), 1323–1334.
56. A. Latremoliere and C. J. Woolf, *J. Pain*, 2009, **10**(9), 895–926.
57. M. Tsuda, H. Tozaki-Saitoh and K. Inoue, *Curr. Opin. Pharmacol.*, 2012, **12**(1), 74–79.

58. J. V. Bergera, L. Knaepena, S. P. M. Janssena, R. J. P. Jakena, M. A. E. Marcusa, E. A. J. Joostena and R. Deumensa, *Brain Res. Rev.*, 2010, **67**, 282–310.
59. www.immpact.org/. Accessed 1 February 2012.
60. M. Giménez, J. Pujol, Z. Ali, M. López-Solà, O. Contreras-Rodríguez, J. Deus, H. Ortiz, C. Soriano-Mas and J. Monfort, Submitted to *Arthritis Rheum.* 2013.
61. Z. Ali, B. Laurijssens, T. Ostenfeld, S. McHugh, A. Stylianou, N. Garman, P. Scott-Stevens, L. Hosking, O. Dewit, J. Richardson and C. Chen, *Br. J. Clin. Pharmacol.*, 2012, **75**(1), 197–207.
62. J. Palmer, R. Lai, N. Thomas, J. Bullman, Z. Ali, A. Baines and J. Appleby, *Eur. J. Pain*, 2009, **13**(S1), S202.
63. B. Chizh, J. Palmer, R. Lai, F. Guillard, J. Bullman, A. Baines, A. Napolitano and J. Appleby, *Eur. J. Pain*, 2009, **13**(S1), S202.
64. www.imieuropain.org/. Accessed 1 December 2012.
65. R. H. Dworkin, D. C. Turk, S. Peirce-Sandner, M. P. McDermott, J. T. Farrar, S. Hertz, N. P. Katz, S. N. Raja and B. A. Rappaport, *Pain*, 2010, **150**(1), 12–16.
66. M. C. Irizarry, D. J. Webb, Z. Ali, B. A. Chizh, M. Gold, F. J. Kinrade, P. D. Meisner, D. Blum, M. T. Silver and J. G. Weil, *Clin. J. Pain*, 2009, **25**(6), 469–476.
67. L. Vase, K. N. Nørskov, G. L. Petersen and D. D. Price, *Philos Trans R Soc London, Ser. B*, 2011, **366**(1572), 1913–1921.
68. www.actionppp.org/. Accessed 1 February 2013.
69. C. Maier, R. Baron, T. R. Tölle, A. Binder, N. Birbaumer, F. Birklein, J. Gierthmühlen, H. Flor, C. Geber, V. Huge, E. K. Krumova, G. B. Landwehrmeyer, W. Magerl, C. Maihöfner, H. Richter, R. Rolke, A. Scherens, A. Schwarz, C. Sommer, V. Tronnier, N. Uçeyler, M. Valet, G. Wasner and R. D. Treede, *Pain*, 2010, **150**(3), 439–450.
70. R. Rolke, R. Baron, C. Maier, T. R. Tölle, R. D. Treede, A. Beyer, A. Binder, N. Birbaumer, F. Birklein, I. C. Bötefür, S. Braune, H. Flor, V. Huge, R. Klug, G. B. Landwehrmeyer, W. Magerl, C. Maihöfner, C. Rolko, C. Schaub, A. Scherens, T. Sprenger, M. Valet and B. Wasserka, *Pain*, 2006, **123**(3), 231–243. Erratum in: *Pain*, 2006, **125**(1–2), 197.
71. R. Baron, T. R. Tölle, U. Gockel, M. Brosz and R. Freynhagen, *Pain*, 2009, **146**(1–2), 34–40.
72. J. A. Singh, *BMC Med.*, 2012, **10**, 44.
73. P. Sacerdote, S. Niada, S. Franchi, E. Arrigoni, A. Rossi, V. Yenagi, L. de Girolamo, A. E. Panerai and A. T. Brini, *Stem Cells Dev.*, 2013, **22**(8), 1252–1263.
74. www.clinicaltrials.gov/ct2/show/NCT01290367?term = mesoblast&rank = 1 Accessed 1 December 2012.
75. www.lifescientist.com.au/article/421875/mesoblast_hits_50_enrolment_back_pain_trial/. Accessed 1 December 2012.

76. S. Diochot, A. Baron, M. Salinas, D. Douguet, S. Scarzello, A. S. Dabert-Gay, D. Debayle, V. Friend, A. Alloui, M. Lazdunski and E. Lingueglia, *Nature*, 2012, **490**(7421), 552–555.
77. S. P. Hughes, A. J. Freemont, D. W. Hukins, A. H. McGregor and S. Roberts., *J. Bone Joint Surg. Br.*, 2012, **94**(10), 1298–1304.
78. J. C. Lotz, V. Haughton, S. D. Boden, H. S. An, J. D. Kang, K. Masuda, A. Freemont, S. Berven, D. K. Sengupta, L. Tanenbaum, P. Maurer, A. Ranganathan, A. Alavi and N. L. Marinelli, *Radiology*, 2012, **264**(1), 6–19.
79. www.lpc.ac.uk/html. Accessed 1 December 2012.
80. http://painconsortium.nih.gov/purpose.html. Accessed 31 December 2012.
81. www.neuro.med.tu-muenchen.de/dfns/e_index.html. Accessed 1 December 2012.
82. www.actionppp.org/. Accessed 1 December 2012.
83. www3.imperial.ac.uk/corporatepartnerships/partners/successes/accesstoconsortium. Accessed 1 December 2012.
84. D. C. Turk, R. H. Dworkin, R. R. Allen, N. Bellamy, N. Brandenburg, D. B. Carr, C. Cleeland, R. Dionne, J. T. Farrar, B. S. Galer, D. J. Hewitt, A. R. Jadad, N. P. Katz, L. D. Kramer, D. C. Manning, C. G. McCormick, M. P. McDermott, P. McGrath, S. Quessy, B. A. Rappaport, J. P. Robinson, M. A. Royal, L. Simon, J. W. Stauffer, W. Stein, J. Tollett and J. Witter, *PAIN*, 2003, **106**(3), 337–345.

CHAPTER 2

Oral Opioids

VEERAINDAR GOLI,* DAVID PRYDE AND KIYO OMOTO

Pfizer Inc., 4000 Centregreen Way, Suite 260, Cary, NC 27513, USA
*Email: veeru.goli@pfizer.com

2.1 Introduction

Opium is produced by drying the latex of opium poppy (*Papaver somniferum*). The cultivation and use of opium as a painkiller, for food and for rituals probably precedes written records. Writings from many cultures around the world, including early Sumerian, Babylonian and Egyptian texts, mention the value of opium as an analgesic. An ancient Indian text, the Rig Veda, mentions the use of Soma for medicinal purposes and surgery. However, the use of opioid analgesics has been a double-edged sword. Despite the uncontested benefits, abuse and dependence has been a perpetual problem in many cultures and was even responsible for the two Opium Wars between England and China between 1839 and 1860.[1]

Suffice it to say that the use and misuse of legal and illegal opium alkaloids has been a global phenomenon since ancient times and continues to be an issue today. Prescription drug abuse was declared a national crisis in the United States in 2011.[2] For prescription opioid use, the pendulum, at least in the United States, continues to swing from one extreme to the other. Townsend Harris's quotation over a century ago still holds true: 'The expense of a war could be paid in time; but the expense of opium, when once the habit is formed, will only increase with time.'

Unfortunately, despite thousands of years of experience, we have yet to learn how to use these extremely effective and valuable alkaloids and modify them to

RSC Drug Discovery Series No. 35
Pain Therapeutics: Current and Future Treatment Paradigms
Edited by Charlotte Allerton

Published by the Royal Society of Chemistry, www.rsc.org

preserve the beneficial/analgesic effects without the (undesirable) abuse and dependence. Two hundred years have elapsed since the isolation of the morphine structure in 1803 by the German pharmacist Serturner, but only in the last 40 years have we molecularly cloned the three classic opioid receptors (mu, delta, kappa) and have a better understanding of mechanism of action. Even today, despite multiple concerns about the use and abuse of opioids, this class of drugs still holds a promising repertoire of new molecules that have yet to find a match in the treatment of acute analgesia and other analgesic indications

We have been reasonably successful in developing semisynthetic and synthetic opioids with modest improvement in efficacy and safety profiles. The next step and challenge for the medicinal chemists is to develop novel and potent oral opioids that lack the typical mu-opioid-mediated side effects. The manipulation of the opioid pharmacology to develop drugs with targeted analgesic effects while blocking the euphoric effects and addiction potential has a long and frustrating history which led Schaumann (1956) to believe that analgesic action and addiction were inseparable. But the development of agonist-antagonists with lesser addiction potential than pure agonists was a step in the right direction.

Another area of research which holds promise is the naturally occurring opioid peptides as opposed to the plant-based alkaloids. Although inherently more difficult to formulate, one advantage of endogenous opioids is that on metabolic degradation they break down to amino acids (unlike opioids) which are non-toxic and do not cause organ damage.

The Holy Grail is to develop a potent opiate that acts selectively at the receptors to produce superior analgesia and none of the untoward short-term effects like respiratory depression, constipation and sedation or long-term effects such as tolerance and dependence. Although research in this area is exploding in the preclinical space, it has yet to become a clinical reality. This chapter focuses on the advances in medicinal chemistry, use in clinical practice and future direction of oral opioids.

2.2 Opiate Pharmacology and Medicinal Chemistry

2.2.1 Opioids and Opiates

Opium comprises a mixture of alkaloids and it is approximately 200 years since the primary active ingredient of opium was discovered to be morphine.[3] In 1973, radiolabelled morphine was found to locate to specific regions in the brain, thereby identifying the site of action of morphine to be at brain-centred opiate receptors, or in modern terms, the opioid receptors.[4] The naturally occurring alkaloids that act through opioid receptors are termed opiates, while the term opioid is used broadly to describe all agents that act through these receptors. In the 1960s and 1970s Martin and co-workers proposed several receptor subtypes, termed mu (μ), kappa (κ) and delta (δ).[5] Later, in the 1990s, a fourth receptor subtype was identified (the 'opioid-receptor like' or ORL_1

receptor).[6] Most commonly, these four subtypes are referred to by their IUPHAR terminology as the mu-opioid (MOP), kappa-opioid (KOP), delta-opioid (DOP) and nociceptin-opioid (NOP) receptors. Other receptors have also been proposed as possible opioid receptors but remain poorly characterised and will not be covered any further here.

The endogenous ligands for opioid receptors are peptides, initially identified by Hughes and Kosterlitz in 1975 as 'endogenous morphines' or endorphins.[7] Endorphins are derived from pro-opiomelanocortin (POMC) and preferentially bind to the MOP receptor. Enkephalins were also discovered as endogenous opioid peptides, are derived from pro-enkephalin and are selective for the DOP receptor. Dynorphins are derived from pro-dynorphins and are KOP receptor selective, while nociceptin is a 17-amino-acid peptide that is selective for the NOP receptor.[8] All endogenous opioids are located in the central nervous system (CNS) where they function as neurotransmitters. The primary endogenous ligands acting at the opioid receptor subtypes are shown in Table 2.1.

2.2.2 Opioid Receptors and Their Pharmacology

Opioid receptors are seven-transmembrane Gi/Go-protein coupled receptors that function as neurotransmitter modulators that are found throughout the body, but their primary distribution is in the CNS.[9] They are localised in somatodendritic membranes and in axonal fibres, and are found throughout cell types that are associated with pain pathways including dorsal root ganglia, spinal cord and brainstem.[10] Much of the early research of opioid pharmacology was based on the MOP receptor, which is a 400-amino-acid protein comprising a substantial intracellular C-terminal domain.[11] The G-proteins that couple with opioid receptors contain G-αi/o isoforms that are sensitive to pertussis toxin.[12] Receptor activation in spinal dorsal horn or in neurons in the thalamus releases G-protein α and $\beta\gamma$ subunits which inhibit adenylate cyclase and activate potassium and voltage-gated calcium channels resulting in the inhibition of presynaptic neurotransmitter release and reduction of spontaneous neuronal activity. MOP receptor agonists inhibit the release of

Table 2.1 Endogenous opioid peptides that target the receptor subtypes.

Endogenous opioid peptide	*Sequence*	*Preferred receptor subtype*
Endomorphin 1	H-Tyr-Pro-Trp-Phe-NH_2	MOP
Endomorphin 2	H-Tyr-Pro-Phe-Phe-NH_2	MOP
Leu-enkephalin	H-Tyr-Gly-Gly-Phe-Leu-NH_2	DOP
Met-enkephalin	H-Tyr-Gly-Gly-Phe-Met-NH_2	DOP
Dynorphin A	H-Tyr-Gly-Gly-Phe-Leu-Arg-Arg-Ile-Arg-Pro-Lys-Leu-Lys-NH_2	KOP
Dynorphin B	H-Tyr-Gly-Gly-Phe-Leu-Arg-Arg-Gln-Phe-Lys-Val-Val-Thr- NH_2	KOP
Nociceptin	H-Phe-Gly-Gly-Phe-Thr-Gly-Ala-Arg-Lys-Ser-Ala-Arg-Lys-Leu-Ala-Asn-Gln-NH_2	NOP

serotonin, noradrenaline (norepinephrine) and gamma-aminobutyric acid (GABA) and indirectly increase the firing activity of dopaminergic cells and dopamine release, which is intensely pleasurable and a common characteristic of many drug-dependence pathways. MOP receptor agonists have a significant neuroendocrine effect, increasing the release of prolactin and growth hormones while decreasing the release of leutinising hormone.[13,14]

Before molecular cloning of the opioid receptor genes, pharmacological studies had identified two forms of the MOP receptor, two forms of the DOP receptor and three forms of the KOP receptor. However, cloning studies revealed a single gene encoding each receptor subtype (*OPrm1*, *OPrd1*, *OPrk1*, *OPrl1*), and it became clear that the previously characterised receptors are not different receptor subtypes. There are a number of explanations that reconcile the discrepancy between the pharmacological studies and the existence of single genes, including differential gene splicing, heterodimerisation of receptors and receptor bias due to differential intracellular signalling components.

2.2.3 Opioid Receptor Mediated Side Effects

Opioid agonists have a broad range of undesirable side effects including sedation, nausea and vomiting, pruritus and myoclonus, but the two problematic ones are respiratory depression and reduced gastric motility.[15] Chronic opioid use also leads to tolerance and dependence. Many but not all of the side effects diminish as tolerance develops. Opioids that act upon MOP receptors in the bowel inhibit gastric emptying and reduce propulsive peristalsis of the intestine to produce a constipating effect. Opioids cause respiration to slow and become irregular through engaging opioid receptors in the brainstem[16] leading to hypercapnia and hypoxia, which appears to correlate with opioid receptor occupancy, such that rapid binding agents in particular are high risk for respiratory depression. Several schools of thought exist around opioid tolerance, including down-regulation of receptors, receptor phosphorylation to reduce compound binding affinity, and functional decoupling of receptors from G-proteins.[17] Tolerance to analgesia is common but not to opioid-induced constipation. Opioid dependence also results from a complex combination of neurochemical, environmental and genetic factors. Many opioids are non-selective across the opioid subtypes, and many also act upon other receptors and channels; for example, some agents are known to engage *N*-methyl-D-aspartate (NMDA) receptors, resulting in neuropathic pain and tolerance. Different compounds may have different modalities against the opioids that they have affinity for. For example, some drugs are agonists at MOP, DOP and KOP receptors, while others may be partially agonistic at MOP receptors and antagonistic at KOP receptors. Opioid research is therefore a highly complex field of balancing selectivity, complex pharmacologies and differential engagement of several other signalling pathways; identifying a drug that can effectively navigate through this highly complicated landscape is an enormous challenge. Table 2.2 summarises the primary effects of engaging the opioid receptor subtypes, their endogenous ligands and the most common side effects attributed to each subtype.

Table 2.2 Summary of the four major opioid receptor subtypes and their distribution, function and endogenous ligands.

	MOP	*DOP*	*KOP*	*NOP*
Primary receptor-mediated physiology	Analgesia, sedation, euphoria	Analgesia, antidepressant	Analgesia, sedation	Nociception, GI motility, locomotor activity, analgesia
Primary safety concerns	Respiratory depression, physical dependence, GI motility	Respiratory depression, seizures	Respiratory depression, dysphoria	Anxiety, depression
Receptor distribution	Brain, spinal cord, sensory neurons, GI tract	Brain, sensory neurons	Brain, spinal cord, sensory neurons	Brain, kidney, spinal cord
Endogenous peptides				
Enkephalins	Weak agonist	Agonist	Weak agonist	Inactive
Endorphins	Agonist	Very weak agonist	Very weak agonist	Inactive
Dynorphin A	Weak agonist	Weak agonist	Agonist	Inactive
Nociceptin	Very weak agonist	Very weak agonist	Very weak agonist	Agonist

2.2.3.1 Opiate Structures

Morphine is the most abundant alkaloid found in the opium poppy, approximately 10%, with codeine being the next most abundant at approximately 3%.[18] There are typically some 50 other alkaloid derivatives contained in opium, the most common of which are shown in Figure 2.1.

The most famous and most studied opiate is morphine which takes its name from the Greek god of dreams, Morpheus. It is a potent analgesic through acting at primarily the MOP receptor, but it is also active at both the KOP and DOP receptors. Morphine is usually administered intravenously and orally despite poor bioavailability. Codeine is the 3-methyl version of morphine and is somewhat weaker at MOP, KOP and DOP receptors than morphine. Codeine is the major opium-derived opiate in therapeutic use today. Thebaine is the 3,6-dimethyl version of morphine while the alkaloids noscapine and papaverine are structurally somewhat different and do not act through opioid receptors.

2.2.4 MOP Receptor Ligands

2.2.4.1 Semisynthetic Opioids

Following the discovery and isolation of morphine in the early 1800s, a large number of semisynthetic derivatives have been made, characterised and in many cases launched as opioid receptor targeting drugs in the search for

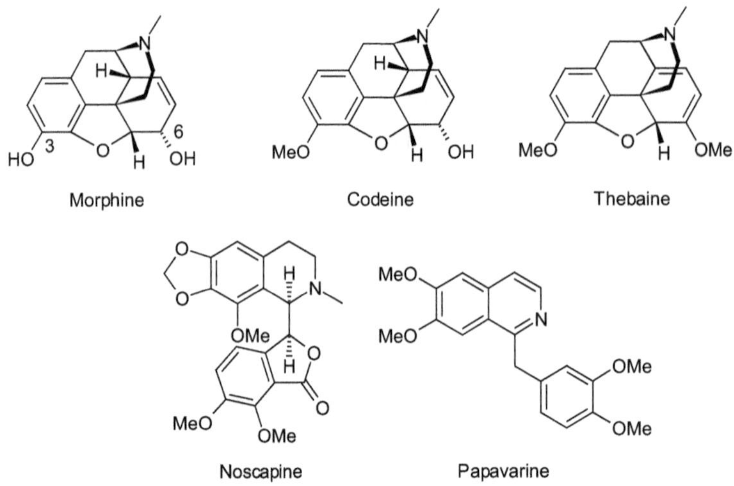

Figure 2.1 Naturally occurring opioids.

analogues that retained the analgesic properties of morphine with fewer side effects.[19] Oxycodone was made in 1916 from thebaine and is one of the world's most prescribed analgesics. Hydrocodone, oxymorphone, levorphanol, hydromorphone and dihydrocodeine were all synthesised shortly afterwards and show similar analgesic properties to that of morphine itself. Diamorphine, more commonly known as heroin, was synthesised in the late 1800s as a more rapid-acting diacetylated prodrug of morphine itself. Diamorphine is prescribed in the UK for pain, although it comes under class A of the Misuse of Drugs Act of 1971. In the United States it is not prescribed for pain and falls under Schedule 1 of the Controlled Substances Act of 1970. Many of these agents have been scheduled as narcotics, with some still in clinical use today for the treatment of pain and as antitussives (Figure 2.2).

Several other semisynthetic agents have been identified as possessing an agonist-antagonist pharmacological profile. For example butorphanol, nalorphine and nalbuphine are agonistic at KOP receptors, but antagonistic at MOP receptors. A larger, lipophilic *N*-alkyl substituent is a common theme for MOP receptor antagonism, and nalmefene, naltrexone and naloxone are all MOP receptor antagonists that have seen utility in the management of alcohol dependence, opioid dependence and opioid overdose.

Buprenorphine is a more recent semisynthetic partial agonist agent made from thebaine, with high affinity but low intrinsic agonist activity for the MOP receptor. Buprenorphine produces typical MOP agonist effects including side effects such as respiratory depression[20] and gut motility but limited by a ceiling effect. Buprenorphine also shows partial agonistic activity at NOP receptors which are postulated to induce a pronociceptive effect and antagonistic at both DOP and KOP receptors which appear to be responsible for less sedation and psychotomimetic effects than morphine.[21] Buprenorphine achieves effective analgesia at relatively low receptor occupancy (5–10%) such that analgesia is

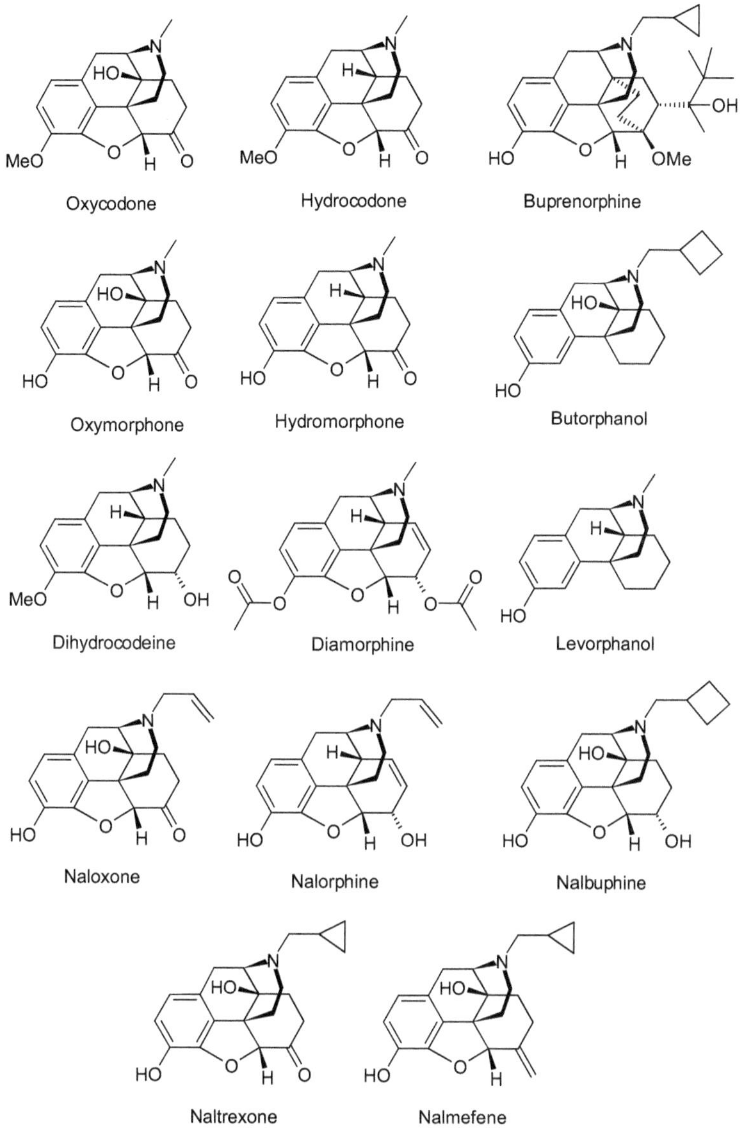

Figure 2.2 Semisynthetic opioids.

largely unaffected by its slow-onset characteristics, while its slow dissociation from the MOP receptor results in a long duration of effect.[22]

2.2.4.2 Synthetic Opioids

Agonists. Substantial effort has been directed at making synthetically simpler opioid versions of the traditional morphine-based skeleton to retain the analgesic properties of the natural substance while reducing some of the

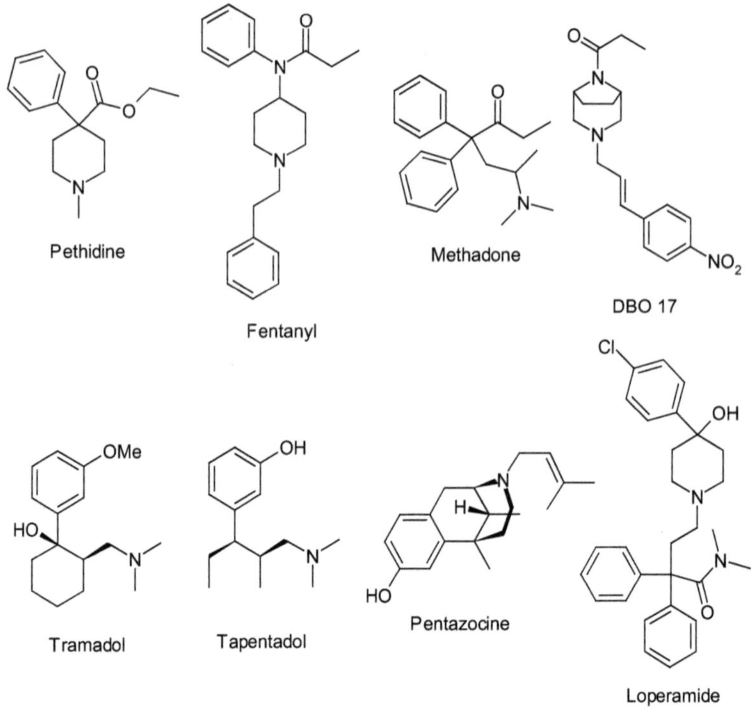

Figure 2.3 Designed MOP receptor agonists.

side effects, some of which are depicted in Figure 2.3. Pentazocine is a synthetic derivative of the morphinan skeleton which has an agonist-antagonist pharmacological profile similar to that of nalorphine and nalbuphine.[23] Pethidine was first made in the 1930s based on a simple piperidine skeleton, but was found to be less potent than morphine, to have a shorter duration of action, and to have a number of neurological side effects.[24] Methadone was also developed in the 1930s as a synthetic opioid, and in most countries is sold as the racemate. Methadone is a potent MOP receptor agonist, and has antagonist activity at both NMDA and nAChR. Its typical elimination half-life is longer than that of morphine at approximately 24 hours. Methadone is used as an analgesic but most commonly is prescribed to manage opioid dependence through mitigating opioid withdrawal symptoms.[25] Fentanyl, an aminopiperidine derivative that is a highly potent MOP receptor agonist, was synthesised as a designed analogue of pethidine[26] that retains many of the same side effects as morphine derivatives. Fentanyl has been variously formulated as an oral agent, sublingual spray, buccal tablets and transdermal patches. Loperamide is a peripherally restricted MOP receptor agonist that increases the tone of intestinal smooth muscle and effectively treats diarrhoea.[27]

Tramadol is another simplified synthetic analogue with an active phase-1 metabolite (M1) *O*-desmethyltramadol that is a very weak agonist of MOP receptors. Tramadol possesses several other pharmacological actions, including noradrenaline (norepinephrine) reuptake inhibition, NMDA receptor antagonism, potent 5-HT_{2c} agonism and muscarinic receptor antagonism, which undoubtedly contributes to its analgesic properties.[28] The structurally related tapentadol was launched by Grünenthal in 2009 for the treatment of pain and is the most recently approved opioid agent in more than two decades.[29] Tapentadol is also a combined weak MOP receptor agonist and a noradrenaline (norepinephrine) reuptake inhibitor and reportedly shows fewer side effects than morphine and oxycodone. Other more recent MOR receptor ligands include the diazabicyclooctane DBO 17[30] which shows good selectivity over the KOP and DOP receptors.

Antagonists. Acylated or amino derivatives of the morphinan skeleton have been synthesised as MOP receptor antagonists (Figure 2.4) including clocinnamox derivatives,[31] amino-imidazole linked[32] and acylated[33] naltrexone derivatives (IBNTxA) and disulfide-linked morphinans such as TAMO.[34] These agents all retain similar levels of MOP receptor affinity to naltrexone, with in some cases greater levels of subtype selectivity.

Several researchers have targeted peripherally restricted versions of MOP receptor antagonists primarily to target MOP receptors in the GI tract as treatment for opioid-induced constipation without compromising centrally mediated analgesia. Low-dose oral naloxone has been used for this purpose, but despite its low oral bioavailability, is freely brain penetrant.

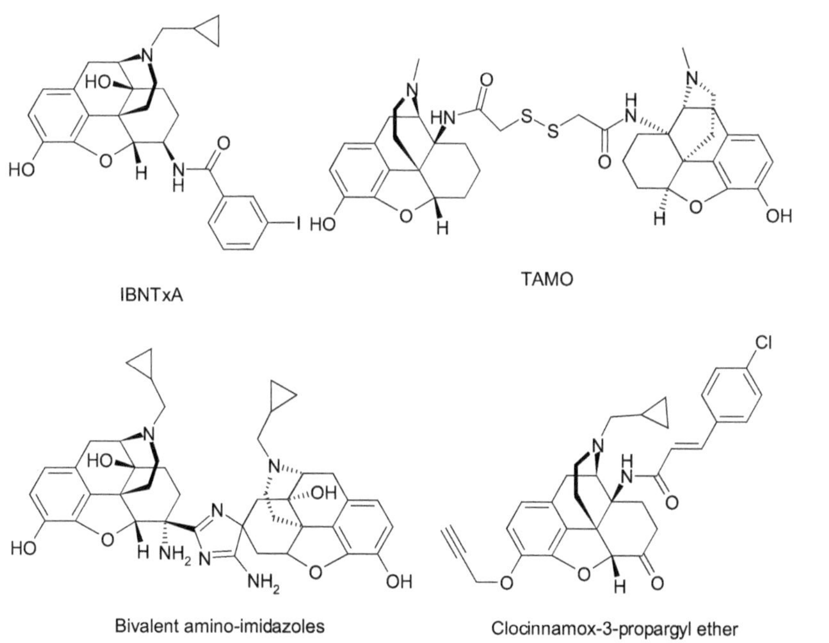

Figure 2.4 Designed MOP receptor antagonists.

N-methylnaltrexone (proprietary name Relistor), a quaternised version of naltrexone developed by Progenics, is not brain penetrant and has been approved as a treatment for opioid-induced constipation.[35] Alvimopan (proprietary name Entereg) is a zwitterionic compound marketed by Adolor with low oral bioavailability that does not cross the blood–brain barrier. It has been approved for the treatment of patients to avoid postoperative ileus following large or small bowel resection.[36] A common metabolite of alvimopan (ADL 08-0011) in which the glycinyl unit has been hydrolysed retains many of the pharmacological and distribution properties of the parent compound (Figure 2.5).

TD-1211 from Theravance is in Phase 2 trials for the treatment of opioid-induced constipation (structure not disclosed). ADL-5945 and ADL-7445 (structures not disclosed) are both MOP receptor antagonists that were being developed by Adolor for opioid-induced bowel dysfunction. In late 2011 Adolor was acquired by Cubist Pharmaceuticals, who have taken ADL-5945 forward into Phase 3 trials.

2.2.4.3 MOP Receptor Pharmacophores

In surveying all of the compounds, synthetic and natural-product inspired, that show affinity for the MOP receptor, there are typically five major pharmacophores, as shown in Figure 2.6. The majority of clinical agents are based on the morphine structure, while several others are based on the morphinan skeleton that lacks the tetrahydrofuran ring of morphine. Sequential simplification of this skeleton has provided various clinical agents that feature the

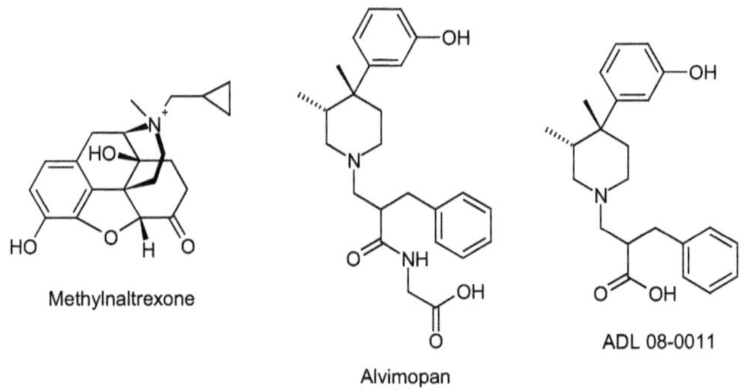

Figure 2.5 Peripherally restricted MOP receptor antagonists.

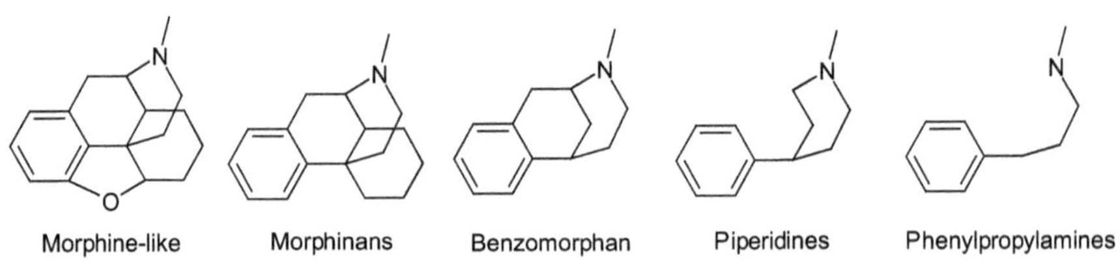

Figure 2.6 The evolution in MOP receptor pharmacophores.

benzomorphan, piperidine or phenpropylamine molecular template. All active compounds within any of these structural types possess a full basic N atom, and as the substituent on this N atom increases in size, compounds tend to shift towards MOP receptor antagonism. This is particularly true in the morphine-like or morphinan skeletons. As indicated above, quaternisation of this basic centre is an effective means of limiting CNS exposure of the ligand.

2.2.5 DOP Receptor Ligands

2.2.5.1 Agonists

While there are currently no DOP receptor agonists on the market, several compounds have been evaluated in clinical trials (Figure 2.7). SNC-80 was the first non-peptide, highly selective DOP receptor agonist to be identified.[37] It has analgesic, antidepressant and anxiolytic effects[38] but caused convulsions at high doses, preventing its further clinical evaluation.[39] JNJ20788560 is a tropane-based molecule that has shown antinociceptive behaviour in several models of inflammatory pain with no evidence of any proconvulsant behaviour or physical dependence.[40] ADL5859 was identified by development of a high-throughput screening hit through to a DOP receptor selective, orally bioavailable compound that was well tolerated in Phase 1 trials,[41] as was its more potent analogue ADL5747.[42] Both compounds had been progressed into a Phase 2 osteoarthritis pain trial but both failed their efficacy endpoints. The isoquinoline compound TAN67 is reportedly still undergoing development.[43]

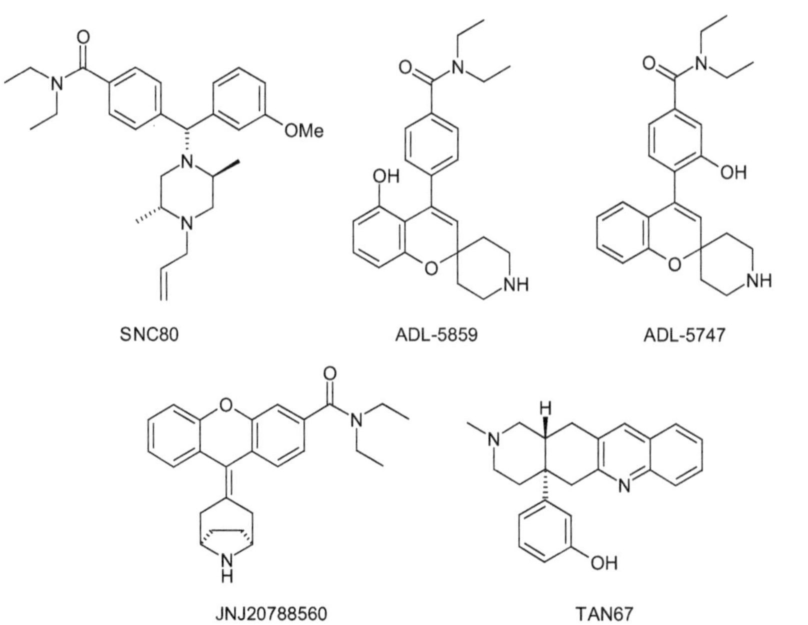

Figure 2.7 DOP receptor agonists.

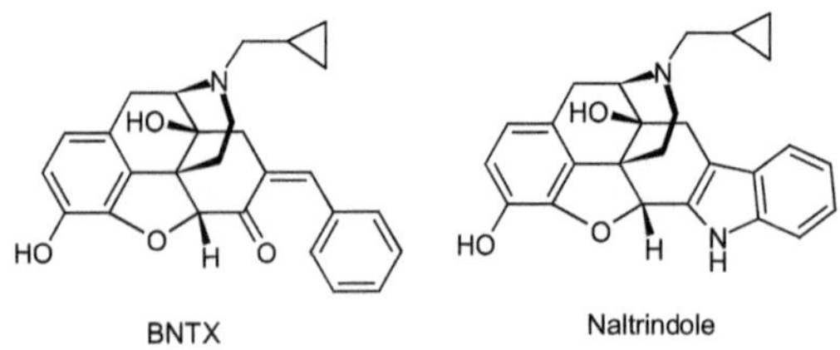

Figure 2.8 DOP receptor antagonists.

2.2.5.2 *Antagonists*

Structural modification of the naltrexone skeleton has produced compounds that are antagonists at the DOP receptor, including 7-benzylidenenaltrexone[44] (BNTX) and naltrindole (NTI) (Figure 2.8).[45]

2.2.6 KOP Receptor Ligands

2.2.6.1 *Agonists*

Ketazocine is an analogue of pentazocine that shows greater selectivity for the KOP receptor and is the prototypical ligand that was used to characterise the KOP receptor (Figure 2.9). Salvinorin A is a terpenoid derived from a Mexican plant that is both a potent KOP receptor agonist and a potent partial agonist at D_2 receptors and notably does not possess the basic N atom that all other opioid ligands do. This compound produces perceptual distortions, but there is no evidence that it is addictive.[46] Spiradoline is a potent synthetic agonist of KOP receptors that was being developed for the treatment of pain but was abandoned due to a high incidence of dysphoria and hallucinations.[47] A very similar compound, U69593, and the structurally related U-50488H are also potent KOP agonists[48] and have been used extensively as pharmacological tools.[49] It has been noted that this compound is somewhat related to the structure of lidocaine and also possesses blocking activity at voltage-gated sodium channels.[50] Interestingly, while this compound potently activates and internalises the human KOP receptor, it does not do so at the rat KOP receptor,[51] indicating that for at least some ligands, significant species differences do exist. Nalfurafine has been approved as an antipruritic for patients undergoing dialysis.

Due to the dysphoria and psychotomimetic side effects observed with KOP receptor agonists, attempts have been made to peripherally restrict ligands to capitalise on KOP receptors found in peripheral sensory afferents but avoid neurotoxicity. Asimadoline is a peripherally restricted full KOP receptor agonist that lacked CNS toxicity in clinical trials but also suffered from poor analgesic efficacy. LPK-26 is a poorly brain-penetrant diamine derivative that is antinociceptive with low substance abuse potential (Figure 2.10).[52]

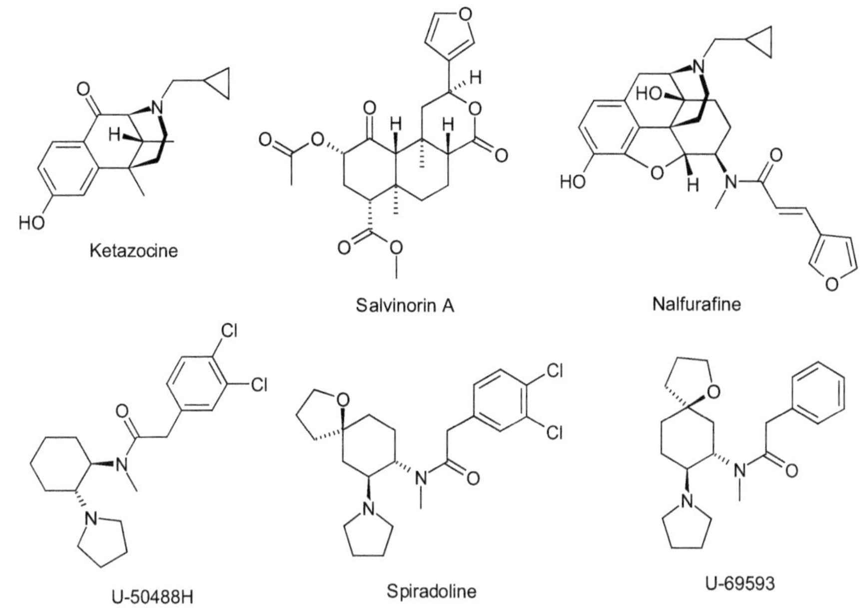

Figure 2.9 KOP receptor agonists.

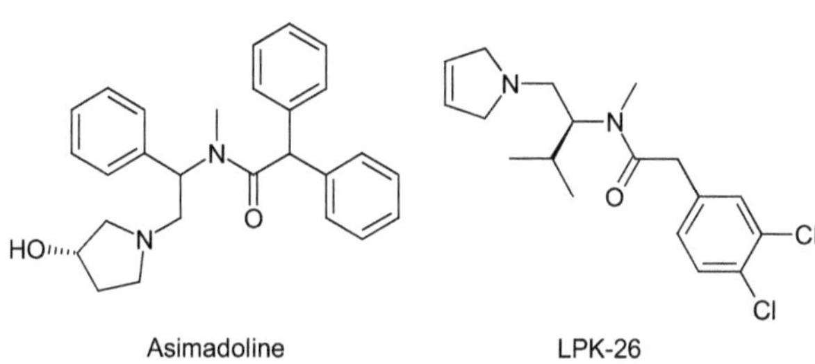

Figure 2.10 Peripherally restricted KOP receptor agonists.

2.2.6.2 Antagonists

As highlighted above, several MOP receptor ligands based on the morphine skeleton also possess KOP receptor antagonist properties, for example buprenorphine which reportedly has antidepressant effects in humans.[53] Medicinal chemistry designs of KOP receptor antagonists based on the morphinans have identified several agents with KOP receptor antagonist selectivity, most notably from the Portoghese group who developed the morphinan skeleton into nor-BNI and 5′-GNTI.[54,55] The 4-phenylpiperidine JDTic was reported by Carrol *et al.*,[56] as KOP receptor selective through growing the *N*-alkyl side chain of the phenyl-piperidine skeleton. All of these agents have a slow onset of action and very long-lasting effects *in vivo*,[57] possibly due to functional changes in *c*-Jun N-terminal kinase activity (Figure 2.11).[58]

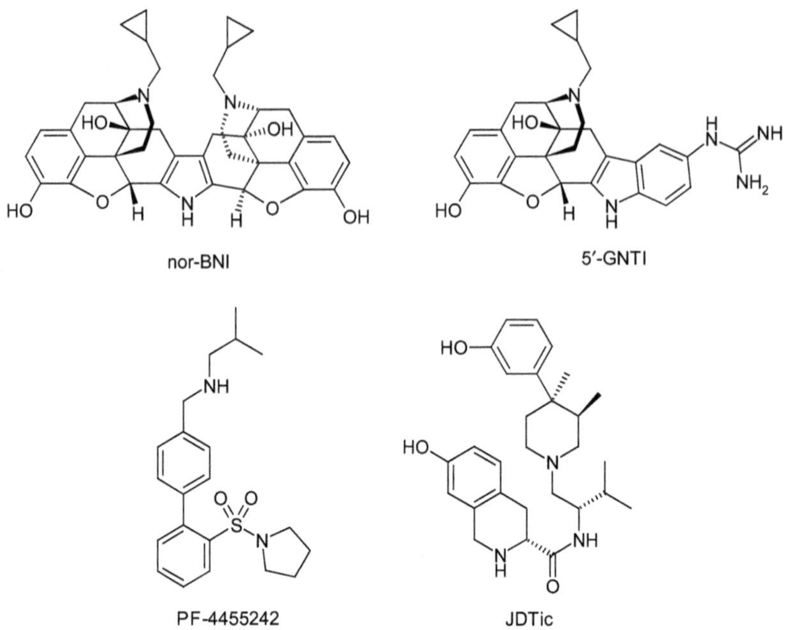

Figure 2.11 KOP receptor antagonists.

PF-4455242 is a KOP receptor antagonist that has been reported by the Pfizer group, with approximately 20-fold selectivity over the MOP receptor, that has been taken into clinical trials for the treatment of depression and substance abuse.[59]

2.2.7 NOP Receptor Ligands

The opioid-like receptor 1 (ORL1) was initially classified as an orphan receptor,[60] but later found to be a member of the opioid receptor family as it had approximately 50% overall sequence similarity (70% similarity for transmembrane domains) to classical opioid receptors (MOP, KOP and DOP receptors) and was eventually named the NOP receptor (nociceptin receptor). However, the receptor does not show a strong response to classical opioid ligands such as morphine, naloxone and endogenous peptides such as dynorphin A,[61] and had been considered an orphan receptor until 1995 when Meunier successfully isolated a dynorphin A-like heptadecapeptide from rat brain.[59,62] It was found to inhibit adenylate cyclase which is coupled with the NOP receptor and provoked hyperalgesia when administered to mice and was therefore named nociceptin. Concurrently, Civelli and co-workers identified the same peptide in porcine brain and observed that it also induced hyperalgesia.[63]

The NOP receptor is widely distributed in both the CNS and periphery. Within the CNS, the receptor has an especially high distribution in the dorsal horn of the spinal cord and is believed to play a significant role in pain transmission. Peripherally, distribution in the kidney has triggered interest from some pharmaceutical companies for urology indications.[64] The *in vitro* pharmacology of the NOP receptor has been investigated intensively, including studies for its role in the inhibition of adenylate cyclase, modulation of ion channels and modulation of MAP kinase.[65] However, a physiological role of the receptor has not been clearly elucidated although it has been proposed to influence pain, anxiety, tolerance, feeding and renal pathways based on studies with peptide ligands.[65,66] It is likely that small-molecule non-peptide NOP receptor agonists or antagonists will help expedite further discoveries of the physiology of this receptor.

2.2.7.1 Agonists

As outlined in Figure 2.12, Roche have described a NOP receptor agonist programme that started in the 1990s from a high-throughput screening hit, **1**.[67] The initial hit was found to be non-selective across opioid receptors, but optimisation led to improved compounds. For example, **2**, in which the 2-tetralinyl group is substituted by 1-indanyl, showed 8-fold higher affinity than **1** (700 pM). Additional modifications to the ring (*e.g.* addition of a methyl group on the 5-position) converted the lead molecule to a derivative with 20-fold and 30-fold selectivity against MOP and KOP respectively (**3**). Further modification of the *N*-8 substituent led the team to discover an analogue with higher selectivity (100-fold selectivity for MOP and 200-fold selectivity for KOP) and very high affinity ($pK_i = 9.41$).[68] The compound was named Ro-64-6198 and was tested for anxiolytic effects in several validated anxiety models of distinctive types of anxiety states and proved to be efficacious at low doses (0.3–3 mg kg^{-1}),

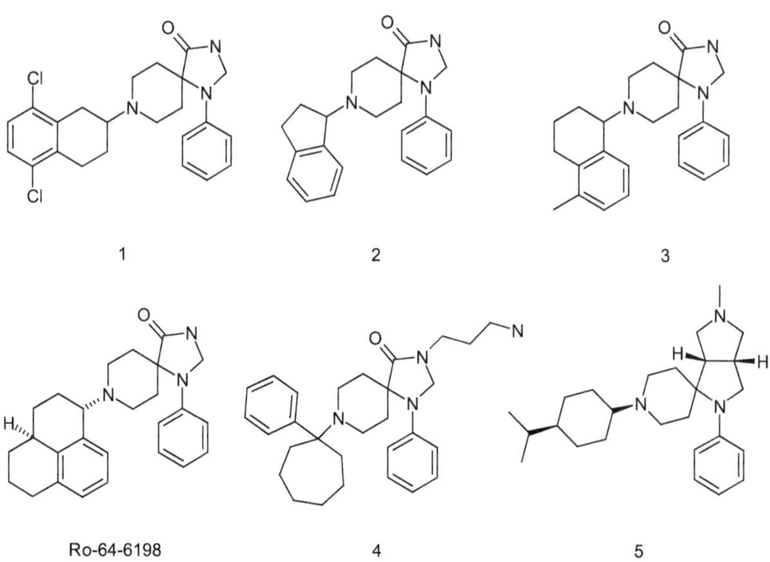

Figure 2.12 NOP receptor agonists.

which supported a role of the receptor in adaptive behavioural fear responses to stress. Another study suggested clinical potential of this compound as an anxiolytic.[69]

Concurrently, groups from Pfizer have also patented a series of triazaspirodecanones as agonists at the NOP receptor, as exemplified by **4**.[70]

In an effort to improve selectivity against other opioid receptors, Roche scientists found a new series of agonists with greater selectivity. Within a similar pharmacophore, a compound **5** with affinity of 0.5 nM and with a selectivity of 1000-fold against other opioid receptors was found, suggesting that right-hand side variations in particular had significant impact on selectivity against other opioid receptors. While it showed a good *in vitro* profile, no anxiolytic effect was observed, attributed to low bioavailabilty, a high volume of distribution and unspecific tissue binding of the molecule. No further development of this series has been reported.[71]

A different series of NOP receptor agonists have been described by Schering, as noted in Figure 2.13. Once again starting from a high-throughput screen of the Schering compound collection, an initial hit **6** was identified ($K_i = 613$ nM).[72] Again, this compound shared the same pharmacophore as Ro-64-6198, with hydrophobic cores of both sides of the terminals and piperidine positive nitrogen on the linker. Initial SAR development focused on substituents of the piperidine nitrogen, from which a benzyl group was found to be the most impactful on activity. Modifications to the benzyl group led to **7** with K_i of 13 nM against the NOP receptor and high intrinsic efficacy (107% GTP-γs stimulation at 100 nM) that showed high selectivity over other opioid receptors (130-fold, 30-fold and 20-fold selectivity over DOP, KOP and MOP, respectively). Piperidine ring modifications resulted in SCH-221510, which has been found to be efficacious in a variety of anxiety models.[73] In the Vogel conflict model, the effect of SCH-221510 was attenuated by a NOP receptor

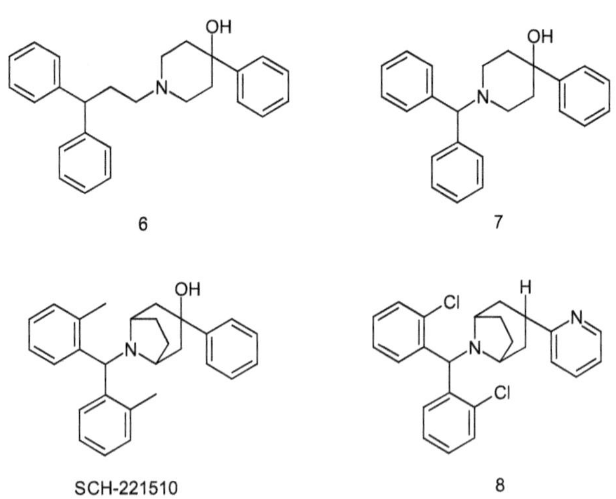

Figure 2.13 NOP receptor agonists.

antagonist, but not by the MOP receptor antagonist naltrexone, suggesting that the compound selectively interacts with the NOP receptor. Further optimisation led to **8**, which was found to have the best overall profile, with NOP receptor activity of 28 nM (GTP-γs EC_{50}), selective over the MOP receptor (80-fold EC_{50}), and with very low affinity for hERG, hPXR, and P450 enzymes. It displayed significant efficacy in guinea-pig models of anxiolysis and cough following oral administration with no CNS side effects.[74]

2.2.7.2 Antagonists

The research and development of NOP receptor antagonists has been led by two Japanese companies, Banyu and Japan Tobacco. The first NOP antagonist, outlined in Figure 2.14, was reported by Banyu who found a lead compound with moderate affinity for NOP (IC_{50} 200 nM) but poor selectivity over MOP and KOP receptors and no functional activity (GTP-γs assay) as an antagonist **9**.[75] SAR optimisation resulted in **10** which demonstrated full antagonist activity in the GTP-γs assay with IC_{50} 270 nM. Most interestingly, a small change of the benzyl group to a cyclooctyl ring switched functional activity from agonist to antagonist. It is likely that the benzyl group has a higher affinity to a binding site within the active state than to an inactive state, while the cyclooctyl group favours the latter. A variety of benzimidazolidinone substituents were introduced to the nitrogen of the lead structures to provide an ethyl analogue **11** that was found to have excellent antagonist activity (IC_{50} 20 nM) and moderate selectivity (>50-fold). Finally, through introduction of a hydroxymethyl group on the piperidine ring, J-113397 was identified, which has antagonist activity of 5.6 nM with high selectivity (1000-fold). This compound provided a pharmacological tool with which *in vivo* function of the receptor could be delineated and was later further optimised to a clinical candidate **12** with *in vivo* efficacy following oral dosing and good brain penetration.[76] So far, no further details have been disclosed for its clinical development.

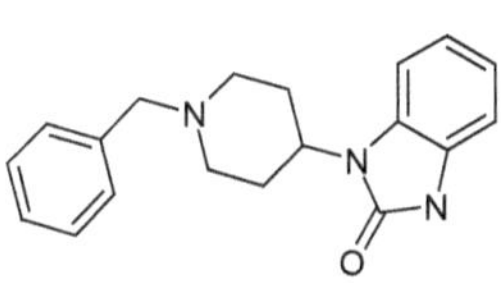

9

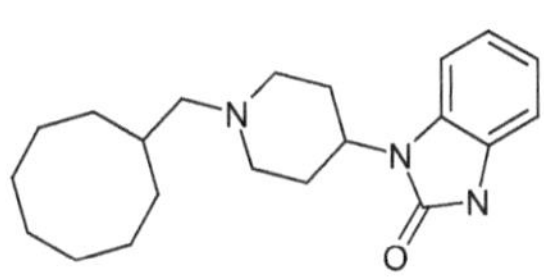

10

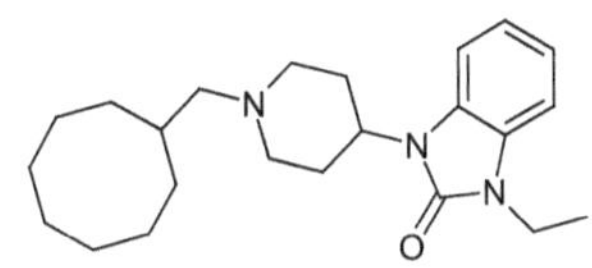

11

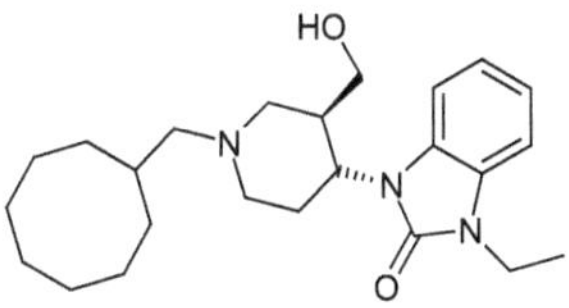

J-113397

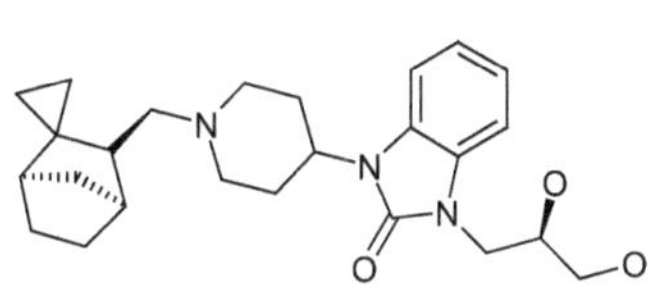

12

Figure 2.14 NOP receptor antagonists.

A second class of NOP receptor antagonists from Banyu is the spiropiperidine series, which showed moderate NOP antagonist activity (GTP-γs IC_{50} 8.3 nM). Through a combinatorial chemistry approach, in which a variety of substituents were introduced to the piperidine nitrogen, they identified **13** (named **C-24**) with sub nM antagonist activity (IC_{50} 0.27 nM) and complete selectivity over other opioid receptors (>10 000-fold).This is illustrated in Figure 2.15. The compound was tested by an *in vivo* model and proved to antagonise locomotor activity triggered by a NOP agonist effectively. Though few reports have been published on this compound, it appears to have high therapeutic potential. Indeed, it was tested in several pain models and proved to suppress the antinociceptive effect of 1 nM orphanin.[77] Finally, a recent publication of the crystal structure of the NOP receptor unveiled the binding mode of this compound to the receptor, with which the essential pharmacophore of other NOP ligands can be associated.

It should be noted that hERG affinity of NOP receptor ligands tend to be high as the pharmacophore of the compounds is very similar to that of hERG inhibitors. Banyu have reported on the high hERG affinity of a NOP antagonist from GSK (Figure 2.16, **SB-612111**,[78] K_i 86 nM), but Banyu were able to optimise away from hERG affinity (**14**, IC_{50} >50 μM) without compromising NOP receptor activity.[79]

A distinct NOP antagonist series has been reported by Japan Tobacco.[80] SAR development started with a random screening hit (51% inhibition at 10 μM)

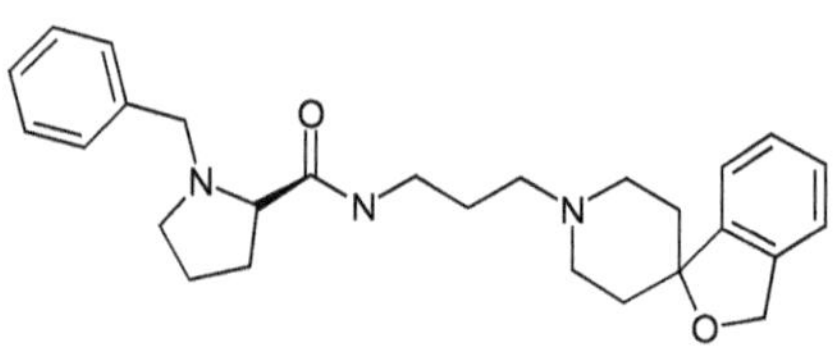

13

Figure 2.15 NOP receptor antagonists.

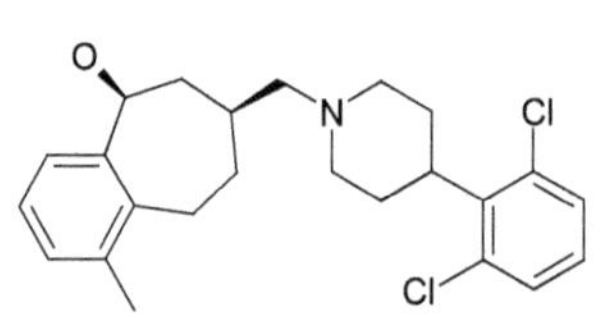

SB-612111

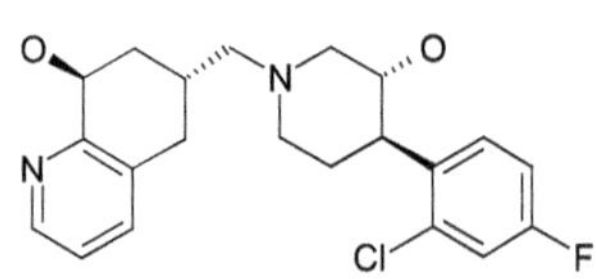

14

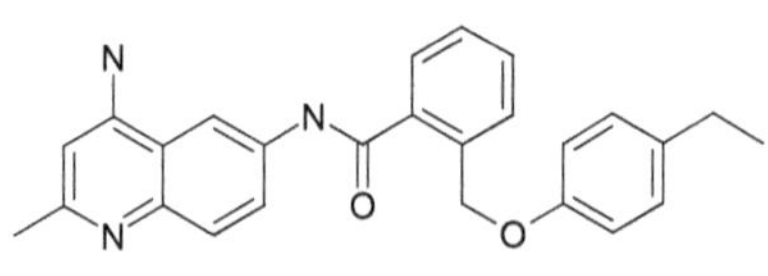

JTC-801

Figure 2.16 NOP receptor antagonists.

from which **JTC-801** was identified for further *in vivo* studies as it showed the highest bioavailability. The compound was tested in a pain model of allodynia induced by intrathecal injection of orphanin and was shown to antagonise the effect *in vivo*. As the analgesic action was not compromised by naloxone, they concluded that the compound did not act via the MOP receptor and showed its effect through a novel pain pathway.

2.2.8 Peptide Ligands

A temptingly simple solution to the search for selective therapeutic ligands for opioid receptors is synthetic versions of their endogenous peptide ligands, as listed in Table 2.1. Despite the drawbacks of this approach which include low oral bioavailability, low CNS penetration and low stability in the systemic circulation, several peptides with a useful *in vivo* profile have been identified.

An early report described analogues of the MOP receptor ligand tetrapeptide dermorphin, including the guanidine derivative ADAMB (Figure 2.17), that was a more potent analgesic than morphine following oral or subcutaneous administration.[81] The 2′,6′-dimethyl-tyrosine derivative DmtDALDA is a long-acting analgesic when administered subcutaneously, and has an unusual pharmacology that involves indirect release of the KOP receptor ligands dynorphins.[82] Guanidinylation of endomorphins has resulted in more brain-penetrant analogues, including guanidino-endomorphin. DAMGO is a common MOP-selective glycinol peptide based on the enkephalin skeleton.[83]

The DOP receptor selective peptidomimetic agonist UFP-512 (Figure 2.18) exhibited antidepressant and anxiolytic activity after intraperitoneal administration.[84] In contrast, the lysine peptidomimetic MZ-2 is a DOP receptor antagonist that shows oral activity in mice.[85]

The dynorphin A analogue, E2078 (Figure 2.19), is a potent KOP receptor agonist that shows analgesia in humans[86] with analogues of this structure showing some activity after inhaled delivery[87] and cyclised versions displaying KOP receptor antagonism.

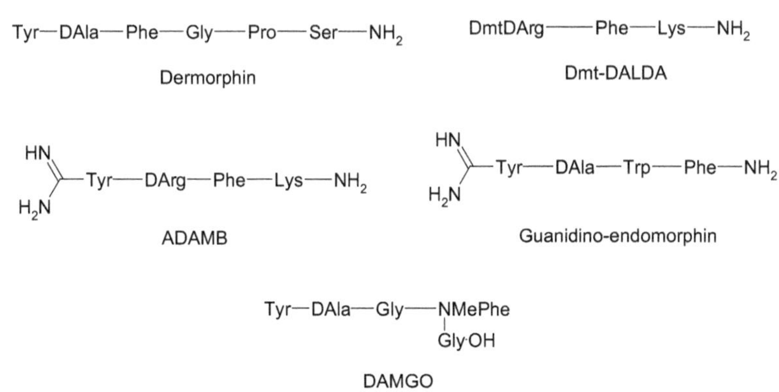

Figure 2.17 MOP-selective peptide analogues.

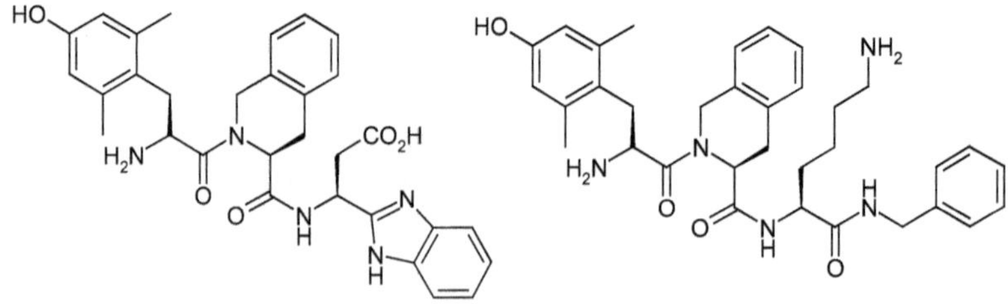

UFP-512 MZ-2

Figure 2.18 DOP-selective peptidomimetic analogues.

NMeTyr—Gly—Gly—Phe—Leu—Arg—NMeArg—DLeu—NHEt

E2078

Figure 2.19 KOP-selective peptide analogue.

Cara Therapeutics has disclosed several all D-tetrapeptide agonists of the KOP receptor. The most advanced of these is CR845 (structure not disclosed), a subnanomolar and highly selective agonist (human MOP and DOP $EC_{50} > 10\ \mu M$) that is peripherally restricted and is currently in Phase 2 clinical trials for the treatment of acute postoperative pain as an intravenous formulation. The peptide agent was specifically designed to be hydrophilic and charged to stay out of the CNS, and has been shown to be safe and well tolerated in several human trials. Female patients undergoing a surgical laparoscopic hysterectomy were administered 0.04 mg kg^{-1} CR845 before surgery and/or after surgery by request. Intravenous morphine was administered to patients who requested it after surgery. CR845 produced a significant decrease in pain intensity and increase in pain relief, particularly when dosed both before and after surgery, and resulted in significantly decreased morphine consumption.[88] Cara has also reported the development of an oral formulation of CR845, in which the active principle was formulated in an enteric-coated capsule with excipients to encourage its oral absorption.[89]

2.2.9 Opioid Receptor Structural Biology

One of the major challenges in SAR development for G-protein coupled receptor (GPCR) programmes is that there is no structural knowledge for the interaction between ligands and receptors. This has made it difficult for medicinal chemists to identify molecules with clinical potential in a short period of time. A significant number of molecules should be designed and tested to understand ligand binding sites, *i.e.* relative locations of hydrogen bond donor/acceptor residues or hydrophobic residues. In 2012 crystal structures of all four opioid receptors were published, providing information about the binding sites of opioids (MOP,[90] DOP,[91] KOP,[92] and NOP[93]). In this respect, the year 2012 was a notable one for medicinal chemistry of the opioid receptors as well as that of general GPCRs.[94] Brief summaries of the crystal structures are given below, along with an outlook on how they could impact on opioid drug discovery.

2.2.9.1 General Features

All four opioid crystal structures showed a typical GPCR structure with seven-transmembrane helices, which has been predicted from the sequence similarity of the opioid receptors to rhodopsin and a sequence alignment that suggested the location of the seven helices. The upper half of the seven helices shapes the orthosteric binding site of the receptors. Comparison of the four structures showed a small plasticity of this domain. Despite high sequence similarity between them, relative location of the seven helices in the upper half is not perfectly conserved. Notably, TM5 (transmembrane 5; fifth helix in sequence) of the NOP receptor shifted more than 4 Å as compared with the MOP receptor and the KOP receptor structures, indicating why classical opioids are not active against the NOP receptor. The core moieties of the ligands are located at the bottom of the binding pockets, which are buried by ECL2 (second extracellular loop) to a small extent. The pockets are open to the extracellular solvent, which may explain why such a diverse series of ligands are recognised by the opioid receptors, from small-molecule modulators to peptides. Interestingly, in the publication of the MOP crystal structure, it was compared with that of the M3 receptor in which the binding site is closed with a layer of tyrosines. In fact, some muscarinic antagonists show very slow kinetics. For example, the dissociation half-life of the clinically used drug tiotropium at the M3 receptor is 34.7 hours.[95] In contrast, opioid ligands show much shorter dissociation half-lives, no more than 1 hour.[96] The other interesting feature is that the binding sites can be divided into a lower part which is highly conserved and an upper part which is less conserved. Ligand engagement of opioid receptors has been described in terms of the 'message–address' concept, in which the ligands have two moieties, one of which is responsible for essential activity while the other controls selectivity. This pharmacological phenomenon can be understood as a direct consequence of structure. In the crystal structures, the core portions of the ligands reside in the lower part of the binding site. For instance, the typical morphinan scaffold sits in the lower half of the MOP and DOP receptors. On the other hand, the upper part accommodates a diverse range of structures, being rich in selectivity determinants. Finally, the receptors are crystallised as a parallel or antiparallel dimer, for which Kobilka and co-workers made some discussion of the relevance to the pharmacology of the MOP receptor. They suggested two interfaces are involved in receptor dimerisation, one consisting of TM5 and TM6 and the other of TM1, TM2, and an intracellular helix which follows TM7. Indeed, there is evidence for dimers of the MOP receptor *in vivo*. It has been suggested that opioid agonists such as DAMGO and methadone reduce tolerance to morphine by facilitating morphine-induced endocytosis through MOP oligomerisation. However, of note is that the dimer structure results from the process of the crystallisation and might not have any physiological relevance.

2.2.9.2 *MOP Mode of Ligand Recognition*[90]

A morphinan derivative, β-FNA, was co-crystallised with MOP, in which the Michael acceptor of the ligand forms a covalent bond with Lys5.39 (the number indicates Ballesteros–Winstein nomenclature). There are 14 residues which shape the binding site. Nine of them, located at the lower part, are the same in the KOP and DOP receptors, indicating why the morphinan core supports activity across opioid receptors. Asp3.32, which is conserved across aminergic and some peptidic GPCRs, is involved in the formation of the salt bridge with the positive nitrogen of the ligand. The phenol ring of the ligand, a typical pharmacophore for opioids, interacts with His6.52 through intervention of two water molecules. The structure also highlights the three non-conserved residues in the upper part of the binding site, which are responsible for interaction with endomorphins. They are just above the β-FNA binding sites, explaining the higher selectivity of the peptides compared with morphine analogues.

2.2.9.3 *DOP Mode of Ligand Recognition*[91]

Naltrindole, a DOP-selective ligand, was utilised to provide the DOP crystal structure. The ligand shares the morphinan core with β-FNA. The core interacts with the DOP receptor in a similar way as in MOP, in that the protonated nitrogen forms a salt bridge interaction with Asp3.32 while His6.52 fixes the phenol ring to the location through a two-water hydrogen-bond bridge, which is also observed in the MOP and KOP crystal structures. It seems that the two waters linking His6.52 and the ligands are essential factors for successful structure-based drug design of opioids. Besides the interactions between the core and the receptor, the upper part of the molecule, *i.e.* the indole ring, interacts with Leu7.35, cognate residues of which are Trp7.35 and Tyr7.35 in MOP and KOP, respectively. The difference in the size of the residues could explain why naltrindole is selective for KOP.[97] It is likely that Leu7.35 can provide sufficient space for the indole ring whereas the bulky residues of Trp7.35 and Tyr7.35 collide with it. Interestingly, the interaction mode underpins the 'message–address' concept of the receptor, as conserved interactions are seen in the bottom of the binding site, the 'message' region, while non-conserved residues are involved in the interaction in the upper part, the 'address' region.

2.2.9.4 *KOP Mode of Ligand Recognition*[92]

JDTic, a KOP ligand in the early stages of clinical development, was selected for the co-crystallisation process. As seen in the MOP crystal structure, the ligand fits tightly in the bottom of the binding site. The two protonated amines and the linker amide nitrogen made tridentate interactions with the Asp3.32. The phenol ring of the isoquinoline ring is associated with His6.52 through mediation of two water molecules, in the same manner as the phenol ring of the β-FNA interacts with His6.52 in the MOP receptor. Interestingly, the other terminal phenol is also trapped by ECL2 with a two-water bridge, suggesting a pivotal role of water in the ligand–protein interactions of opioid receptors. The bioactive structure of the ligand in the binding site adopts a V-shaped conformation, which is also seen in the small-molecule X-ray crystal structure of the ligand. This suggests that the internal strain energy in the bioactive pose would be small, which is consistent with very high activity of JDTic. In addition to the conserved interactions among opioid receptors (Asp3.32 and His6.52), the ligand interacts with the four non-conserved residues (Val2.53, Val2.63, Ile6.55 and Tyr7.35), which are thought to contribute to the exceptionally high selectivity over other opioid receptors (>1000-fold). Homology modelling of the MOP and the DOP receptors based on the KOP structure suggests the four residue mutations weaken interactions or provoke unfavourable interactions. The structure also provided insights for why the ligand shows antagonist activity. At the bottom of the binding site, the cyclopropyl group of the ligand forms hydrophobic interaction with Trp6.48, which is thought to be a key part of the activation mechanism of the GPCRs. It has been suggested that an interaction with the residue blocks the conversion to an active conformation of the receptor.[98]

2.2.9.5 *NOP Mode of Ligand Recognition*[93]

C-24, an NOP selective antagonist from Banyu,[99] was selected for co-crystallisation as it showed high activity against NOP and excellent selectivity over other opioids (>1000-fold). Like MOP, KOP and DOP (Figure 2.20), the residue of Asp3.32 forms the salt bridge interaction with the protonated nitrogen of the benzofuran/piperidine ring. The residue at the 6.52 position is not conserved in NOP. His6.52 in MOP, KOP and DOP is mutated to Gln6.52 in NOP. It seems that this residue difference has an effect on the low activity of classical opioids against NOP. A comparison of the C-24 binding mode with that of JDTic in KOP suggests that His6.52 prevents the active site from creating a sufficient space to accommodate the benzofuran core that is seen in NOP. In addition, the structure shed light on the roles of the other non-conserved residue, Ala5.39, which is Lys5.39 in the classical opioid receptors. The residue change was demonstrated to be involved in a large-scale deformation of the binding pocket. In detail, Lys5.39 in KOP, and potentially in MOP and DOP, forms salt bridges with neighbouring residues (Asp5.35 and Glu6.58). The disruption of this interaction in NOP resulted in an outward shift of the extracellular half of TM5 and an inward shift of TM6, which reshapes the entrance to the pocket. This probably gives rise to distinct features of the NOP receptor.

2.2.9.6 *New Opportunities for Structure-Based Drug Design of Opioids*

The crystal structures have provided deeper understanding of the interaction of the opioid receptors and the ligands, *e.g.* the residues which are responsible for the high affinity and selectivity of the ligands. Commonly, the bottom part of the binding site, which is highly conserved, accommodates core moieties of

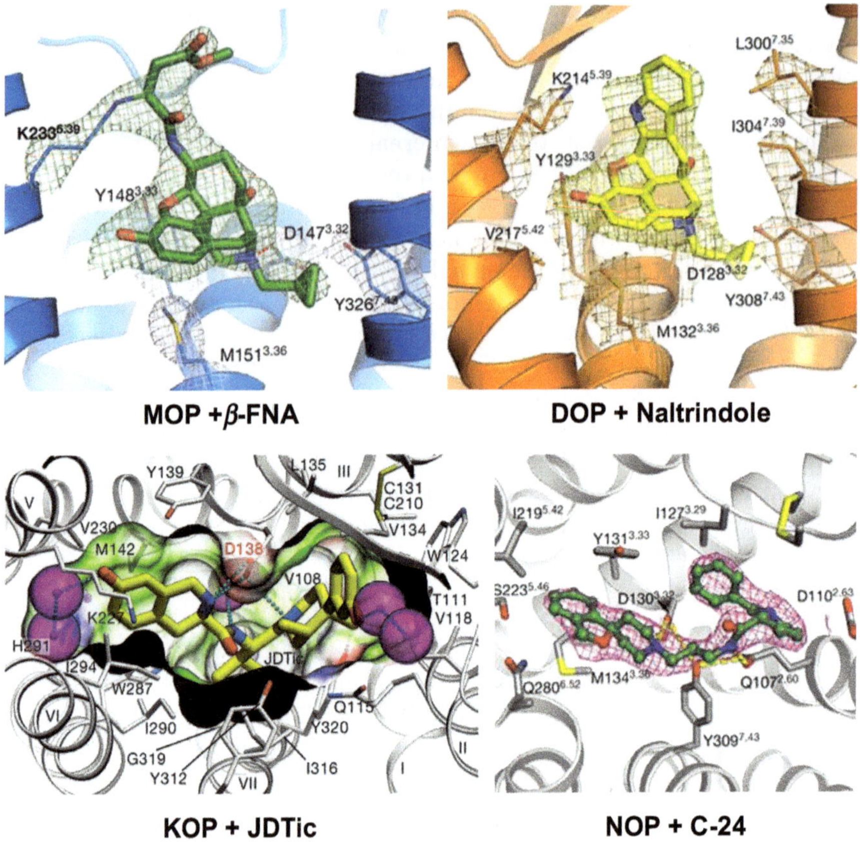

Figure 2.20 Crystal structures of opioid receptors.

opioids while the upper entrance part of the binding site, which is less conserved, recognises variable substituents conferring the ligands' high selectivity profiles. Docking calculations of new hits onto the crystal structures will suggest which portions of a new ligand resides in the bottom part or the upper part.[100] This will guide medicinal chemists to create different strategies for molecule design aiming for high activity and improvement in selectivity over that achieved to date through more classical semisynthetic medicinal chemistry approaches. However, a significant additional enabling breakthrough will be to obtain active state structures of all four opioid receptors as most clinical opioids are agonists.

2.3 Clinical Overview

2.3.1 Rationale

The International Association for the Study of Pain (IASP), the World Health Organization (WHO) and the European Federation of the IASP promoted a

'Global day against pain' in 2004. Brennan *et al.*[101] termed this as a 'point of inflection' in which unreasonable failure to treat pain is viewed worldwide as poor medicine, unethical practice and an abrogation of a fundamental human right.[101] The IASP defines pain as 'an unpleasant sensory and emotional experience associated with actual or potential tissue damage or described in terms of such damage.' Pain is classified as nociceptive (somatic and visceral), or neuropathic. Acute pain originates from the specialised nociceptive nerve endings, and is protective by warning of potential or actual injury. Chronic pain is defined as pain that persists for 1 month beyond the normal time of healing. Chronic pain can be categorised as cancer, chronic non-cancer pain (CNCP), or neuropathic (either malignant or non-malignant). Pharmacological and non-pharmacological approaches are recommended for optimal relief. It is universally acknowledged that pain, on a global scale, remains inadequately treated because of cultural, attitudinal, educational, legal and system-related reasons.[101] Due to the potential for misuse, abuse and diversion, opiophobia remains a major hurdle; patients who clearly need opioid drugs are denied access to them. To this day, no other class of drugs offers the versatility of pain relief provided by opioids. Unfortunately the limited formulary of effective drugs further magnifies the problem.

The clinician faces several challenges. There is no clear consensus on the classification of pain syndromes and the mechanisms involved. There is significant variability in the individual's response to expression of pain and their response to treatment. The clinician also has to deal with political, regulatory and societal biases. Despite these issues, the clinician is obligated to relieve the patient's pain and suffering. Once a clinician willing to navigate these hurdles has been able to identify the right opioid for the right condition and the right patient, constant vigilance is needed to prevent it getting into the wrong hands. This is counterintuitive to a clinician who now has to assume the schizophrenic role of doctor/policeman to relieve the patient's pain and suffering. Better drugs, better tools and better regulations are needed.

2.3.2 Guidelines

Multiple evidence-based guidelines exist in North America and Europe for acute, chronic and cancer pain. All of the guidelines essentially endorse the legitimate use of opioids as part of the treatment, after appropriate diagnosis. A WHO Delphi study conducted in 2007 recognises that there are numerous (at least 23) guidelines in existence globally. It lists controversies, barriers and undesirable practices in pain management, the most prominent of which is the resistance of healthcare practitioners to use opioids when indicated. The report also recommends that WHO needs to take the lead in developing distinct guidelines on three topics: acute pain, chronic malignant pain and CNCP.[102]

2.3.2.1 Cancer Pain Guidelines

In 1986 WHO developed a conceptually simple three-step approach to cancer and palliative pain management that has been shown to be effective in relieving

Table 2.3 Pharmacological approaches to cancer pain: WHO three-step ladder.

Step 3: Severe pain	Morphine, hydromorphone, methadone, leorphanol, fenanyl, oxycodone + non-opioid analgesics ± adjuvants
Step 2: Moderate pain	Paracetamol or aspirin + codeine, hydrocodone, oxycodone, dihydrocodeine, tramadol ± adjuvants
Step 1: Mild pain	Aspirin, paracetamol, non-steroidal anti-inflammatory drugs ± adjuvants

Table 2.4 Summary of WHO recommended 5-step approach to administration of opioids for the treatment of cancer pain.

By mouth	*Oral route is preferred for all opioids for ease of use*
By the clock	Opioids should be prescribed at regular intervals and not 'as needed'
By the ladder	Opioids to be added only when non-opioids are inadequate to control pain in steps 2 and 3 of the WHO ladder
For the individual	Right dose for the patient is the dose that relieves the pain and may vary with the individual
With attention to detail	Ongoing monitoring of patient's response

pain in 90% of patients with cancer (Table 2.3).[103] The basic principles behind the three steps of the ladder include selecting the appropriate analgesic for the pain intensity and individualising the dose by titration of opioid analgesics in step 2 and 3 of the ladder. WHO published comprehensive guidelines for assessment and treatment of cancer pain. When an opioid is indicated, WHO recommends a simple five-step approach which is summarised in Table 2.4.

After more than two decades since the introduction of the WHO step ladder for cancer and palliative pain, all three steps of the WHO ladder are noted to be effective, but it was also noted that potent opioids can be used as a first step for severe pain.[103]

2.3.2.2 Acute Pain Guidelines

Several guidelines exist for acute pain, which mostly focus on the relief of the symptom rather than the cause. An example is the Institute for Clinical Systems Improvement (ICSI) evidence-based guidelines for acute pain published in 2008 (Table 2.4). Similar to the cancer pain guidelines, these guidelines walk the clinician through a proper evaluation and potential evidence-based therapies.[104]

2.3.2.3 Chronic Pain Guidelines

In contrast, the evidence for use in CNCP is limited. The American Pain Society (APS) guidelines published in 2009 conclude that chronic opioid therapy can be effective for carefully selected and monitored patients with CNCP.[105]

We have yet to understand why certain individuals and certain chronic pain conditions respond better than others. Prescribing opioids for CNCP has outpaced advances in scientific evidence. The current literature consists largely of efficacy trials which are not generalisable, yet the guidelines treat these studies as if they are effectiveness studies, which hampers practice.[106] Despite major strides in pain medicine, significant gaps remain in our basic understanding of normal and abnormal pain mechanisms, the chronification of pain, and the classification of the chronic pain syndromes on one hand and the pharmacokinetic and pharmacodynamic responses to drugs on the other. Pain genetics and environment play an important but yet to be determined role.[107]

2.3.3 Clinical Pharmacology

At an individual patient level, there is compelling evidence for the use of chronic opioids. However, clinicians are not yet in a position to predict which patient will respond to various opioids. Many factors have been proposed to contribute to these differences in opioid responsiveness including environmental, psychological and genetic. Empirical trials of several opioids are needed to find the one that may be appropriate for the individual. Patient characteristics and structural differences between opioids contribute to differences in opioid metabolism and hence influence the variability in efficacy, safety and tolerability of specific opioids in diverse populations.[108] Pharmacokinetic and pharmacodynamic variability affects the opioid concentrations and response. The allelic variants in the genes involving the opioid (*UGT2B7*, *OPRM1*, and *ABCB1* genes) and non-opioid systems such as catechol-*O*-methyl transferase (COMT), the enzyme which metabolises dopamine and adrenaline (epinephrine), may affect the efficacy of pain. The regional mu-opioid system response to pain is diminished in healthy volunteers with single nucleotide polymorphisms (SNP) in the *COMT* gene. Some drug-metabolising enzymes and transporters, including cytochrome P450 (CYP), uridine 5′- diphosphate (UDP)-glucuronosyltransferases (UGT), and adenosine triphosphate (ATP)-binding cassette (ABC) transporters, may play a significant role in opioid metabolism as well as influencing the flux of opioids into and out of various opioid target sites (OTS). These processes appear to have the potential to significantly affect interindividual differences in opioid concentrations in the human body and brain.[109] In the future, pharmacogenetic mapping may allow us to predict which opioid will be best suited to a particular individual.[110]

2.3.3.1 Pharmacokinetic Variability

In the liver, opioids undergo phase 1 metabolism by the CYP2D6 and CYP3A4 pathways which have the potential for significant drug–drug interactions. Phase 2 metabolism is by conjugation and has relatively less interaction potential.

Morphine, hydromorphone and oxymorphone undergo phase 2 metabolism; all other opioids are metabolised by CYP-mediated oxidation. Since chronic pain patients are usually on a complex regimen of medications, they are prone to substantial drug interactions especially when it involves the CYP system.

Age, sex, ethnicity and medical conditions, especially hepatic and renal impairment, can have significant implications for the clearance of opioids and thereby their efficacy. Reduced clearance of morphine, codeine, fentanyl and oxymorphone has been noted in older patients. For example, steady state concentrations of oxymorphone may be 40% higher in patients aged over 65 years. Women have 25% higher concentration than men of oxycodone and hydromorphone. There are ethnic differences in the frequency of variant alleles to CYP2D6. Approximately 5–10% of white Europeans are poor metabolisers (PM), whereas a significant number of Chinese and African populations may show CYP2D6 allelic variations that alter metabolism of codeine and hydrocodone. Based on the genetic polymorphisms of the 2D6, patients may be classified as PM, intermediate metabolisers (IM), extensive metabolisers (EM) and ultrarapid metabolisers (UM). This can modify or trigger drug interactions and clinical response.[108]

CYP2D6 is responsible for the metabolism of hydrocodone, codeine, and dihydrocodeine and has interaction potential with selective serotonin reuptake inhibitors, tricyclics, beta-blockers and antiarrythmics, and other drug substrates, inducers and inhibitors of CYP2D6.

CYP3A4 metabolises fentanyl, oxycodone and tramadol is metabolised by both CYP3A4 and CYP2D6. Grapefruit juice (bergamottin) and other CYP3A4 inducers, inhibitors and substrates interact with them.

The liver is the major site for metabolism of almost all opioids and the metabolites are primarily eliminated by the kidneys. Patients with moderate liver or kidney disease need differential dose adjustments depending on the opioid used. Peak plasma levels of oxycodone and its metabolite noroxycodone are increased 50% and 20% respectively in liver and kidney disease, with potential for adverse side effects.[111] On the other hand, the pharmacokinetics of fentanyl and methadone are not affected by hepatic and renal impairment and these drugs may be preferred in end-stage renal disease (Table 2.5).[112]

Morphine, oxymorphone and hydromorphone are metabolised by phase 2 glucuronidation and have little potential for metabolically based interactions but may have additive pharmacodynamic interactions. The activity of UGT2B7 enzyme which is involved in glucuronidation shows significant interpatient variability and may be linked to accumulation of morphine in those individuals.

Opioid transporters are important proteins that influence the absorption, distribution and elimination of opioids. The two major families of drug transporters of relevance to opioid pharmacokinetics are the ABC superfamily of efflux transporters, and the solute carrier (SLC) superfamily of influx transporters.[113] These have the ability to influence gastrointestinal (GI) absorption, facilitate presystemic biliary elimination and transporter expression at the blood–brain barrier and influence clinical efficacy and safety.

Table 2.5 Factors influencing opioid metabolism.

Opioid	*Age*	*Sex*	*Ethnicity*	*Hepatic impairment*	*Renal impairment*
Morphine	Reduced clearance	No effect	Chinese have higher clearance	Dose adjustment recommended	Dose adjustment recommended
Codeine	Caution in older adults	No effect	CYP2D6 allelic variants alter metabolism in Asian and African descent	Dose adjustment recommended	Dose adjustment recommended
Hydrocodone	Caution in older adults	No effect	CYP2D6 allelic variants alter metabolism in Asian and African descent	In combination with paracetamol (acetaminophen). Dose adjustment and liver monitoring	In combination with paracetamol (acetaminophen). Dose adjustment and monitoring
Oxycodone	Concentration higher in elderly	25% higher conc.in women	No effect	Dose adjustment recommended	Dose adjustment recommended
Methadone	Dose adjustment	No effect	No effect	Well tolerated and preferred but dose adjustment may be needed in severe failure	Well tolerated but dose adjustment may be needed in severe failure
Fentanyl	Reduced clearance	No effect	No effect	Well tolerated and preferred and dose adjustment not needed	Well tolerated and preferred and dose adjustment not needed
Hydromorphone	No effect	C_{max} 25% higher in women	No effect	Dose adjustment recommended	Dose adjustment recommended
Oxymorphone	40% higher concentration	No difference	No effect	Contraindicated	Dose adjustment recommended
Tramadol	Not studied	No effect	No effect	Pharmacokinetics significantly altered in liver failure	Pharmacokinetics significantly altered in kidney failure
Tapentadol	Not studied	No effect	No effect	No dose adjustment needed in mild to moderate failure	No dose adjustment in mild to moderate failure

Adapted from H. S. Smith, Opioid metabolism, *Mayo Clin. Proc.*, 2009, **84**(7), 613–624.

2.3.3.2 Pharmacodynamic Variability

MOP, KOP and DOP may have polymorphisms leading to interindividual variability of clinical effects of opioids. There are at least 100 polymorphisms in the human MOP receptor gene (*OPRM1*) with some variants showing altered binding affinities to different opioids.[114] Ross and colleagues compared 'opioid switchers' who did not tolerate morphine with 'controls' who responded to morphine, and showed significant differences in the genotype of the signal transducer and activator of the transcription 6 (*STAT6*) gene between 'switchers' and 'controls.'[115] Opioid receptor heterodimerisation may lead to an increase in functional variety of the GPCRs which may be unique to a tissue. The MOP receptor may form heterodimers with KOP and MOP leading to altered pharmacological properties.[116] Mu–delta heterodimers may lead to increased antinociceptive responses.[117] Other factors that contribute to variability in opioid responses include supraspinal, spinal and peripheral variability of opioid analgesic action.[109]

Although the above factors contribute to variability, the lack of validated tools that can predict who responds to an opioid leaves the clinician to resort to empirical trials with significant dropouts due to adverse events or lack of efficacy. Future knowledge based on disease mechanisms, animal data from knockout mice, pharmacogenetic and gene polymorphisms will better guide physician decisions.[109]

2.3.4 Clinical Considerations

Opioids used in clinical practice may be classified as weak or strong opioids but this may be a misnomer since the activity and affinity differ in individuals. A more palatable classification may be based on their predominant mechanism of action on the specific opioid receptor. The common classifications that have been accepted in the literature are summarised in Tables 2.6 and 2.7.

The solid oral route of administration is preferred in most clinical situations and offers the most practical and convenient form of drug delivery due to the ease of access and ability to self-administer in most conscious patients who can swallow. In others, including children below age 7 years, a liquid oral form may be preferable. Oral formulations also allow inexpensive manufacture and until recently were technologically less challenging. However, the current trend is to manufacture proprietary and complex abuse-resistant oral formulations that make it difficult to manipulate the product to release large amounts of a potentially lethal opioid. Most oral forms of opioids are available in

Table 2.6 Common classifications of opioids accepted in literature.

Traditional	*Based on potency as weak, intermediate and strong opioids*
Origin of the drug	Naturally occurring, semisynthetic and synthetic and endogenous opioids
Function	Pure agonists, agonist-antagonist, partial agonists and antagonists

Table 2.7 Classification of opioids.

Traditional	*Origin of the drug*	*Predominant mode of action*
Strong	*Naturally occurring alkaloids*	*Pure agonists*
Morphine	Morphine	Morphine
Pethidine	Codeine	Fentanyl
Fentanyl	Papaverine	Oxycodone
Oxycodone	Thebaine	Hydrocodone
Oxymorphone	Noscapine	Methadone
Hydromorphone	*Semisynthetic*	Partial agonist
Hydrocodone	Hydromorphone, oxycodone	Bupenorphine
Intermediate	bupenorphine, dihydrocodeine	*Agonist-antagonists*
Bupenorphine	naloxone, naltrexone	Pentazocine
Pentazocaine	*Synthetic*	Nalbuphine
Butorphanol	Methadone, nalbuphine	Nalorphine
Weak	levorphanol, butorphanol	*Pure antagonists*
Codeine	fentanyl, tramadol, tapentadol	Naloxone
Tramadol	*Endogenous opioid peptides*	Naltrexone
Tapentadol	Endorphin	Nalmafene
	Enkephalin	
	Dynorphin	
	Endomorphins	
	Nociceptin	

immediate-release and sustained-release formulations. Opiates can be pharmacologically long-acting by virtue of their half-life or pharmaceutically long-acting by virtue of the formulation.

The oral bioavailability of most common opioids is highly variable, depending on the opiate. Opioids are weak bases and dissociate into ionised and unionised fractions; the unionised fraction is more diffusible than the ionised form. In the acidic environment of the stomach, opioids are highly ionised and therefore poorly absorbed. Conversely, in the alkaline small intestine, they are predominantly unionised and are readily absorbed. However, many opioids then undergo extensive first-pass metabolism in the intestinal wall and liver, resulting in low oral bioavailability. Extended-release formulations may have longer time to peak plasma drug concentrations(T_{max}) and lower peak plasma concentrations (C_{max}) than their immediate-release counterparts. High lipid solubility facilitates opioid transport and confers a more rapid onset and short duration of action.

Repeated short-term use of opioids in an individual does not lose its 'potency' as long as a sufficient time has elapsed between treatments. However, chronic and continuous use of an opioid leads to a need to increase the dose for maintaining the response, a phenomenon known as tolerance which appears to be mediated by NMDA receptors and nitric oxide synthetase and possibly other unknown mechanisms.[118–123] In clinical practice, there is incomplete cross-tolerance between different opioids; hence, if a patient become tolerant to one opioid, they can be converted to another class of opioid (opioid rotation).[124]

The key feature of opioid therapy is individualisation. There is no optimal dose of an opioid and individualisation is the key feature to opioid therapy due to significant patient variability. Pharmacogenomics may unravel who benefits from opioid therapy without the hit-and-miss treatment practised currently. In a study on various strains of mice with fixed-dose morphine, the response ranged from 0% to 80%.[125] Similarly the response in humans varies widely. The choice of an opioid is still empirical and a trained clinician has to use the 'art of medicine' combined with 'state of the art' knowledge of pharmacology to find the optimal drug, dose and response, especially in chronic pain conditions.

In a meta-analysis of opioids in CNCP,[126] both weak and strong opioids outperformed placebo for pain and function in all types of CNCP, including nociceptive and neuropathic pain syndromes. The strong opioids were superior to naproxen and TCA for pain relief but not for function. The most common side effects were constipation and nausea.

Opioids also work fairly well for CNCP of 90 days or less, but the efficacy of opioids in long-term use beyond 90 days is quite controversial. In multiple randomised clinical trials for various pain conditions including noiciceptive and neuropathic pain for short-term use, opioids were effective, but in three systematic reviews evaluating the long-term effectiveness of opioids in CNCP, the evidence was unclear and weak. Many patients withdrew due to adverse events or insufficient pain relief and for those who were able to continue opioids for more than 6 months, the evidence was weak or limited.[127]

Short-term studies provide only equivocal evidence regarding the efficacy of opioids in reducing the intensity of neuropathic pain, whereas intermediate-term studies demonstrate significant efficacy of opioids over placebo, which is likely to be clinically important. Reported adverse events of opioids are common but not life-threatening. Further randomised controlled trials are needed to establish long-term efficacy, safety (including addiction potential), and effects on quality of life.[128] Two recent population studies in Scandinavian countries indicated that chronic pain persisted despite opioid treatment in a large Swedish and Danish population. Almost 73% of patients on long-acting opioids took themselves off the opioids despite persistent pain after 3 years. It was felt that there may be significant comorbid conditions in individuals who stay on opioids, and both studies indicate that opioids alone may not be the solution.[129–131]

Irrespective of which guideline is used, when an opioid is indicated, there are some general principles that are generalisable across all disease states.[132] The strategies and guidelines are beyond the scope of this chapter. All guidelines endorse the proper assessment of the patient, using standardised tools and measures. All guidelines endorse the use of a multidisciplinary approach including non-pharmacological and pharmacological therapies. All endorse the use of opioids in moderate to severe pain. Confusing arrays of resources are available for the clinician who boldly steps up to treat chronic pain. The APS chronic pain guidelines are an example of guidelines for clinicians who wish to practise responsible pain management.[132]

2.3.5 Pure Agonist Opioids

All opioid agonists bind to the opioid receptors, stimulating physiological activity. A drug may stimulate activity in MOP, KOP and DOP receptors even if it preferentially binds strongly to one of the receptors.

2.3.5.1 Mu-Opiate Drugs

Almost all opioids that are used in clinical practice are preferentially active at the MOP, although they are also active at KOP and DOP. Pure opioid agonists have no real ceiling effect and increased dose increases pain relief until intolerable side effects limit further escalation. The classic MOP agonist analgesics include codeine, morphine, oxycodone and hydrocodone, fentanyl, methadone, oxymorphone and hydromorphone. Clinically all mu opioids are equally effective and there is no clear advantage from an analgesic perspective, and there is no evidence for a differential indication of the available opioids, yet traditionally certain opioids have been used for specific conditions.

Codeine. Codeine is used to treat mild to moderate pain but is more commonly known as a cough suppressant. Codeine is considered to be a prodrug which is metabolised to morphine to exert analgesic effects. PM and RM of CYP2D6 do not respond well to codeine. These patients will experience adverse effects more than the analgesic effects. About 11% of codeine is also metabolised by unknown mechanisms to hydrocodone.

Common side effects include drowsiness, constipation, and other typical opioid-related side effects. Respiratory depression is a serious side effect which is potentially lethal to neonates when codeine is metabolised to morphine and is passed through breast milk. The FDA issued a warning in August 2012 about unexpected deaths in children under the age of 6 years who receive codeine with paracetamol (acetaminophen) after tonsillectomy or adenoidectomy.[133] Codeine products are rarely used in chronic pain management. Codeine is marketed as a single ingredient or in combination preparations with paracetamol, NSAIDs or muscle relaxants.

Morphine. Worldwide, morphine remains the archetypal MOP agonist against which all others are compared. It is considered to be a 'strong' opioid, although the potency of several MOP agonists is much more than that of morphine. It is recommended as step 3 of the WHO pain ladder. Morphine is indicated for use in postoperative and chronic malignant pain. It is the preferred opioid in renal colic and angina pectoris. Its use in children is also fairly well characterised. Morphine is not commonly used in chronic pain states, although there is no clear reason not to use it other than its side effects. In most cases, patients who require morphine can be titrated to an oral dose of morphine in a long-acting formulation with immediate-release morphine for breakthrough pain. Oral morphine is poorly absorbed, irrespective of the formulation.

Morphine is metabolised by glucuronidation to two major metabolites, an inactive morphine-3-glucuronide (M3G) and an active morphine-6-glucuronide (M6G) which is analgesic but contributes more to the respiratory depression and other side effects. M3G has no analgesic effect but contributes to neurotoxic side effects such as confusion. A small, clinically insignificant, amount of morphine is metabolised to hydromorphone. Morphine accumulation has been reported in liver disease which can result from decreased plasma clearance and/or increased elimination half-life of the parent drug. In patients with early liver disease, initial lower doses should be used, but at normal dosing intervals. However, as the disease progresses to advanced hepatic failure, longer dosing intervals may be necessary. Morphine is available in multiple generic and branded versions of short-acting and extended-release oral formulations.

Hydromorphone. Hydromorphone is a potent semisynthetic agonist opioid that behaves like other strong opioids.[134] Evidence to date does not support its superiority over morphine for the management of moderate to severe pain, although on a weight for weight basis it is more potent than morphine. It is highly lipophilic and has a more rapid onset of action. Hydromorphone is glucuronidated to the primary metabolite hydromorphone-3-glucuronide which has no analgesic properties but can be neurotoxic. A once-a-day extended-release formulation, Palladone, was approved in 2004 but was voluntarily recalled in July 2005 due to safety concerns about fatal dose dumping when used with alcohol. Hydromorphone is not commonly used in general practice.

Oxymorphone. Oxymorphone has been available in oral formulations since 2006. Oxymorphone is a semisynthetic mu-opioid receptor agonist which is indicated for the relief of moderate to severe pain. Oxymorphone is highly lipophilic and more potent than morphine, but due to its very low oral bioavailability (~10%) the equianalgesic ratio of oral morphine to oral oxymorphone is only 3 : 1.[135] Oxymorphone is metabolised in the liver to 6-hydroxy oxymorphone and oxymorphone-3-glucuronide. Both oxymorphone and its metabolites accumulate in renal failure; oxymorphone is removed during haemodialysis.[136] Drug interactions for oxymorphone have not been well defined; it does not affect the CYP2C9 or CYP3A4 pathways.[137] Oxymorphone IR or ER has a significant food effect and can increase serum levels by up to 50%.[138,139] Co-ingestion of alcohol should be avoided with all opioids; oxymorphone ER levels rise significantly and in an unpredictable manner.[138] Oral oxymorphone has been studied most extensively in patients with CNCP, in industry-funded research. No head-to-head comparisons with morphine have been undertaken. All studies suggest oxymorphone's side effects are similar in frequency and magnitude to those of other oral opioids.

Oxymorphone ER has shown similar efficacy to oxycodone ER and superiority to placebo in randomised, blinded comparisons in patients with chronic low back pain and osteoarthritis pain.[136,140–143] Oxymorphone IR has been shown to be safe and effective for acute postsurgical pain.[144,145]

Two controlled studies in cancer pain[146] involved transitioning cancer patients from morphine or oxycodone ER to oxymorphone ER; analgesia and side effects remained stable on the oxymorphone ER.

Oxymorphone is not approved for use in children; no data exist on paediatric dosing. Due to its long half-life, oxymorphone IR should be dosed every 6 hours. It comes as 5 and 10 mg tabs. Oxymorphone ER can be dosed every 12 hours and comes in 5, 10, 20, and 40 mg tabs. Equianalgesic conversion data range from: 1.2–2 : 1 for oral oxycodone: oral oxymorphone and 1.8–3 : 1 for oral morphine: oral oxymorphone (*i.e.* 18–30 mg oral morphine = 10 mg oxymorphone).[141,142]

Hydrocodone. Hydrocodone is another potent semisynthetic opioid whose use appears to be unique to the United States; almost 99% of the world's production of hydrocodone is prescribed in the United States.[147] It is currently available only in a combination form, with paracetamol (acetaminophen) or NSAIDs in multiple generic versions for pain. The Drug Enforcement Administration (DEA) has classified hydrocodone alone as a Schedule II narcotic but products currently sold in the United States combine the hydrocodone with other non-opioid drugs, and are classified as Schedule III by the Controlled Substances Act (CSA), although there is no scientific basis for the change. This favourable scheduling has an impact on the prescribing practices of primary care doctors, and hydrocodone is associated with more drug abuse and diversion than any other opioid in the United States. Recognising this concern, in January 2013 the FDA Drug Safety and Risk Management Committee has recommended reclassifying all hydrocodone combination products as Schedule II. If this recommendation is accepted by the United States Attorney General, it will have a significant impact on the prescribing patterns and monitoring of these combinations in the United States. Hydrocodone combinations are frequently prescribed by dentists, primary care physicians and emergency department physicians for short-term pain relief of multiple aetiologies. One study of 469 chronic pain patients found hydrocodone with ibuprofen was superior to a codeine with paracetamol (acetaminophen) combination.[148] On 13 January 2011, the FDA asked drug manufacturers to limit the strength of paracetamol (acetaminophen) in prescription drug products, including combination paracetamol (acetaminophen) and opioid products, to no more than 325 mg per dosage unit. Limiting the amount of paracetamol (acetaminophen) per dosage unit in prescription products may reduce the risk of severe liver injury from paracetamol (acetaminophen) overdosing. Majority of the currently available hydrocodone products are not subject to the risk evaluation and mitigation strategy (REMS) since they are short-acting formulations and none are currently available in tamper-resistant formulations.

Hydrocodone is highly bioavailable and has a half-life of 2.5–4 hours. Structurally similar to codeine, it undergoes demethylation to the more potent hydromorphone. Although some have proposed it to be a prodrug, the studies do not support this so far, since the parent compound is also active at the MOP receptor.

A unique side effect of chronic use of hydrocodone can be associated with progressive sensorineural hearing loss. Genetic polymorphisms of drug-metabolising enzymes as well as associated comorbidities such as hepatitis C infection may be significant in the development of hydrocodone ototoxicity, though additional investigations are necessary.[149]

Although it is commonly prescribed, very few studies have evaluated the effectiveness of hydrocodone in chronic pain. An ER version of hydrocodone bitartrate without paracetamol for CNCP appears to be effective and provides an alternative to other potent opioids. The NDA submission is based on data from over 1100 patients with chronic pain participating in the pivotal Phase 3 efficacy study (Study 801), and an open-label Phase 3 safety study (Study 802) of Zohydro ER. Study 801 successfully met its primary efficacy endpoint, demonstrating that Zohydro ER resulted in significantly ($p = 0.008$) improved chronic pain relief compared to placebo.[133] In December 2012, the FDA's Anesthetic and Analgesic Drug Products Advisory Committee (AADPAC) voted against the approval of Zohydro ER in its current form due to concerns about the potential for prescription abuse. It is likely that the company may reformulate the product as a tamper-resistant formulation (TRF).

Oxycodone. Oxycodone is a phenanthrene class of semisynthetic Schedule II opioid which is available in both an IR and a modified-release (8–12 hour duration) preparation. The IR formulation is available as a single entity and in combination with acetaminophen (paracetamol) or aspirin. Lower doses of oxycodone (*e.g.* 2.5 mg, 5 mg, 7.5 mg, 10 mg) in combination with a non-opioid co-analgesic are frequently used for management of acute pain in patients with limited prior opioid exposure. A modified-release formulation of oxycodone is now widely used for management of chronic pain. Oxycodone has liposolubility similar to morphine but has a higher oral bioavailability of 60%. Women clear oxycodone 25% more slowly than men. As oxycodone is metabolised in the liver by *O*-demethylation to form the active oxymorphone in a reaction catalysed by the enzyme CYP2D6 and noroxycodone by the enzyme CYP3A4, PM accumulate the metabolite noroxycodone with a decrease in efficacy. Drug–drug interactions are common with 2D6 inhibitors. Fluoxetine, a potent inhibitor of CYP2D6, increases the oxycodone requirement in a PM. Serotonin syndrome has been reported with the concomitant use of sertraline. Oxycodone has a more favourable pharmacokinetic profile than morphine. Its oral bioavailability is significantly higher and therefore the interindividual variation in bioavailability and expected plasma concentrations is less. Oxycodone has the typical opioid-related adverse effects, but causes fewer hallucinations, releases less histamine and causes less itching.

A recent FDA warning stated that the concomitant use of oxycodone hydrochloride controlled-release tablets with all CYP3A4 inhibitors such as macrolide antibiotics (*e.g.* erythromycin), azole-antifungal agents (*e.g.* ketoconazole) and protease inhibitors (*e.g.* ritonavir) may result in an increase in oxycodone plasma concentrations and may cause potentially fatal respiratory depression. This is more of a risk when an ER formulation is tampered with. Patients receiving oxycodone controlled-release tablets and a CYP3A4 inhibitor should be carefully monitored for an extended period of time, and dose adjustment should be made if warranted.[150]

The effectiveness of oxycodone has been evaluated in multiple studies. Long-term studies from 8 months to 3 years demonstrated effectiveness but had high dropout rates and adverse events.[151–154]

2.3.5.2 Delta Opiate Drugs

Delta-opioid agonists potentially offer a variety of beneficial pharmacological effects including antidepressant effects, analgesia, neuroprotection and in impulse control disorders, possibly without the side effects like constipation and dependency and respiratory depression, although this has yet to be quantified. DOP receptors have been relatively understudied compared to MOP. However, the most promising advance in preclinical research is that DOP agonists may be more effective in chronic persistent pain, as opposed to the MOP agonists which so far have had mixed results. The debate continues as to whether MOP agonists are in fact ineffective in long-term use for chronic pain. DOP agonists weakly modulate acute nociception[155] but appear to be effective in persistent pain,[156–158] stress or chronic morphine treatment[159]

Data suggests that DOP activity is effective in neuropathic, inflammatory pain and appears to reduce chronic pain downstream of aminergic systems.[160–162] This area of research is the most promising for chronic pain.

Currently, there are no delta-opioid pharmaceuticals on the market but the pharmacology has progressed to the point that there are several delta-opioid receptor (DOR) agonists with potent *in vivo* activity.

The GSK product SNC-80 is the leading DOR agonist in development for pain and urinary incontinence. Adolors ADL5747 and ADL 5859 have completed Phase 2 clinical trials for the treatment of pain. Another lead DOR, TAN-67 (Toray), is also in development This area is exciting and if proven to be effective can be a significant advance in modern analgesia.[161]

2.3.5.3 Kappa Opiate Drugs

KOP agonists offer the potential for potent antinociceptive effects without the accompanying undesirable side effects of morphine-like drugs. The first highly selective KOP agonist U-50,488H (Upjohn in 1982) and subsequent modifications of this structure all were highly selective but with significant side effects including psychotomimetic effects and dysphoria, dissimilar to morphine.[163]

TRK-820 (nalfurafine hydrochloride, Toray) may be a more clinically viable compound. Nalfurafine hydrochloride is a novel derivative of the opioid receptor antagonist naltrexone. In fact it is the first non-addictive opioid drug approved in Japan in 2009 as an antipruritic for severe itching associated with haemodialysis (Remitch 2.5 μg capsules). Initial clinical trials for postoperative pain led to a serendipitous finding that patients on TRK-820 did not have any pruritus. The shift in focus in clinical development to pruritus led to the successful launch of Remitch in Japan. So far there has been no evidence of addiction or psychological dependence.[164,165] Nalfurafine hydrochloride showed ameliorating effects on pain, neuropathic pain, drug dependence, schizophrenia and dyskinesia in non-clinical studies and may be useful in these conditions in humans.[166]

CR845 is a highly selective, peptide-based, peripherally restricted KOP agonist currently in development for the treatment of acute and chronic pain. Cara Therapeutics is developing an intravenous formulation of CR845 in the United States for the treatment of acute postoperative pain. The company completed a 200-patient Phase 2 trial in Q2, 2012 but data is unavailable. Earlier in that year, Cara successfully completed a Phase 1 trial with an oral capsule formulation of CR845 which demonstrated a mean oral bioavailability of 16% across all patient groups and peripheral KOR activation at all doses tested as measured by a standard biomarker.[133]

2.3.6 Partial Agonists

Partial agonists have a ceiling effect to pharmacological actions including analgesia. The potential advantage is that they may also have a ceiling effect to adverse events such as respiratory depression. The best-known of the partial agonists is buprenorphine.

2.3.6.1 Buprenorphine

Buprenorphine is a semisynthetic derivative of thebaine which was initially approved for the treatment of opioid addiction. Recently it was approved by the FDA for the treatment of moderate to severe chronic pain as a transdermal patch, and sublingual formulations are also available. Buprenorphine is a partial agonist at the MOP that can also bind to KOP and DOP as a weak antagonist. It has extensive first-pass metabolism and its oral bioavailability is too low for a meaningful clinical effect. Sublingual and transdermal routes bypass the first-pass hepatic metabolism and are clinically effective. The bioavailability is 30%through the sublingual route as opposed to 15% via the transdermal route, and is absorbed rapidly in less than 5 minutes. It is very lipophilic with a very high affinity to the MOP receptor and tends to dissociate very slowly, resulting in a long duration of clinical effect. Buprenorphine has an active metabolite norbuprenorphine which prolongs its action. The elimination half-life is bipasic. Sublingual buprenorphine formulations, which are FDA approved for the treatment of opioid addiction, have been prescribed off-label for the management of chronic pain since they became available in 2002. A pure buprenorphine formulation (Subutex®) and a buprenorphine-plus-naloxone combination (Suboxone®) are available. If a patient is already on a pure opioid agonist, the introduction or 'induction' of buprenorphine may precipitate withdrawals.

2.3.6.2 Meptazinol

Meptazinol was developed under the trade name Meptid by Wyeth in the early 1970s. Almirall Ltd manufactures it in the UK. It was indicated for the treatment of moderate to severe pain in obstetrics and in kidney stones. The usual dose is 200 mg every 4 hours, due to its short duration of action. Meptazinol is a partial agonist at the MOP similar to buprenorphine. It is metabolised to the inactive glucuronide and excreted in the urine. It is not used for chronic pain, given its short duration of action and the availability and advantages of buprenorphine.

2.3.7 Agonist-Antagonists

These drugs have a ceiling effect for analgesia either by activation of mu or kappa receptors and an antagonist effect at the other receptors. The most commonly known agonist-antagonist drugs are nalorphine, pentazocine, nalbuphine, butorphanol and dezocine. These are used rarely if at all in chronic pain settings.

2.3.7.1 Pentazocine

Pentazocine is the prototypical mixed agonist-antagonist which is marketed under several brand names. It is a weak antagonist of the MOP receptor. Talwin NX with naloxone replaced Talwin in the United States, but is sold as Talwin PX without naloxone in Canada. Although it can be used to treat moderate to severe pain, similar to morphine, it may have a ceiling effect to analgesia. It is also more likely to cause psychomimetic effects and hallucinations. There have been reports of visual hallucinations and confusion with therapeutic doses of Talwin, which clear after discontinuation of the drug.[208] It is also not recommended for cardiac pain. Interestingly, it is a Schedule IV drug in the United States.

2.3.7.2 Butarphanol

Butarphanol is a synthetic opioid which is structurally related to levorphanol. It exhibits partial agonist and antagonist activity at MOP and KOP. It is known by the popular brand name Stadol, which was manufactured by Bristol-Myers who recently discontinued its manufacture. Several generic brands are available as a nasal spray for migraine headaches and moderate to severe pain. Studies with butorphanol nasal spray have been performed in postoperative (*i.e.* general, orthopaedic, oral, cesarean section) pain, in postepisiotomy pain, in pain of musculoskeletal origin, and in migraine headache pain. There is no clear rationale for its use in chronic pain conditions. Its short-acting nature encourages overuse and development of tolerance. Severe hypertension has been reported with butarphanol therapy.

2.3.7.3 Nalbuphine

Nalbuphine hydrochloride is a synthetic partial opioid agonist-antagonist which belongs to the phenathrene family, not available in an oral formulation. It has similar potency to morphine.

2.3.8 Antagonists

Relatively minor structural changes convert an opioid agonist drug to an antagonist which blocks activation of opioid receptors and may precipitate withdrawals in opioid-tolerant individuals. Oxymorphone is transformed into naloxone or naltrexone, the two commonly used antagonists. They are FDA approved for the treatment of alcoholism or opioid addiction (naltrexone; *e.g.* Trexan®, Revia®, Depade®) or opioid overdose (naloxone; *e.g.* Narcan®). A long-acting, high-dose depot form of naltrexone (Vivitrol®) for intramuscular injection is also approved for addiction therapy. They are also used in TRFs to deter abuse and misuse. There are several potential uses of an antagonist in chronic pain management, where further research is warranted:

- brief detoxification via IV naloxone for opioid-tolerant patients or those having opioid-induced hyperalgesia
- ultra-low-dose naltrexone monotherapy for Crohn's disease and fibromyalgia or for treatment of irritable bowel syndrome
- ultra-low-dose naloxone in combination with an agonist for acute and chronic pain, to mitigate some of the side effects or prevent overdose.

Oxytrex® is a combination product by Pain Therapeutics, Inc. with oxycodone and 0.002 mg naltrexone. It was in clinical trials with preliminary findings of efficacy and improved side-effect profile. The current status is unknown. This table summarises the available current research on opioid antagonists (Table 2.8).

2.3.9 Dual-Pharmacology Opiate Drugs

Almost all of the natural and derived opiates of interest to analgesia act on the GPCRs. Although one of the original goals of medicinal chemists was synthesis of highly selective and potent opiates, it has become increasingly apparent that in the treatment of chronic pain, multiple mechanisms are involved and drugs that act at more than one target may be more efficacious. However, until recently methadone, which is a synthetic opioid, was the only drug that was commonly known to have an effect on the NMDA receptor in addition to being an MOP agonist. Subsequently tramadol was first introduced in Europe as a dual-acting weak opioid which induces serotonin release and inhibits reupake

Table 2.8 Summary of opioid antagonists for pain in human trials/cases.

Application	*Antagonist dose*	*Clinical outcome*	*Reference*
Naloxone (IV, intrathecal, or oral—as specified)			
Brief opioid detoxification for intractable pain	0.6–1.2 mg IV (ca. 20 minutes)	Significant pain relief at lower opioid doses. Requires close monitoring	Loitman 2006
+ Buprenorphine for postop breakthrough pain	0.08–0.4 mg oral	Effective, long-lasting analgesia	Schmidt *et al.* 1985; La Vincente *et al.* 2008
Monotherapy for acute postop dental pain	0.4 or 1.0 mg oral	Significant analgesia compared with placebo	Levine *et al.* 1988
+ Pentazocine (60 mg) for dental pain	0.4 mg oral	Greater analgesia than monotherapy with pentazocine or 15 mg morphine	Levine *et al.* 1988
+ PCA morphine postop	0.00025 or 0.001 mg/kg/h steady infusion IV	Reduced side effects and lower morphine consumption versus placebo	Gan *et al.* 1997
+ PCA morphine postop	0.8 mg dose/IV bolus (= 0.38 mg/kg/h on average)	No benefit	Sartain *et al.* 2003
+ Nalbuphine (5 mg IV) postop	0.2 mg IV	Improved analgesia	Gordon *et al.* 2007
+ Intrathecal morphine (5 mg) for intractable chronic pain	0.000050 mg/day intrathecal during follow-up	Significantly improved ongoing pain relief	Hamann *et al.* 2008
+ Oxycodone (chronic non-cancer pain; chronic back pain)	10, 20, 40 mg/d oral	No loss of analgesia, improved bowel function at 2 : 1 oxycodone/naloxone ratio	Meissner *et al.* 2009; Simpson *et al.* 2008; Vondrackova *et al.* 2008
Naltrexone (oral)			
Crohn's disease	4.5 mg/day	Favourable response in 89% of patients	Smith *et al.* 2007
Irritable bowel syndrome (IBS)	0.5 mg/day	Pain/symptom relief during first 2 months	Kariv *et al.* 2006
Fibromyalgia	4.5 mg/day	Unknown, but anecdotal reports favourable	SNAPL 2008
Neuropathic pain (combined with methadone)	0.002 mg/day	Improved analgesia at lower methadone dose	Cruciani *et al.* 2003
+ Oxycodone (Oxytrex®) for osteoarthritis or chronic back pain	0.002 mg/day	Reduced opioid consumption and side effects, including resolution of opioid-induced constipation	Chindalore *et al.* 2005; Webster *et al.* 2006
+ Hydrocodone (experimental)	0.25, 0.5, or 1.0 mg	No beneficial effect	Cantilena *et al.* 2004
+ Intrathecal morphine for chronic non-cancer pain	0.2 mg/day	Reduced pain intensity, improved pain relief	Hamann and Sloan 2007
Opioid (methadone) tapering	0.125 or 0.250 mg in multiple doses/day	Reduced withdrawal symptoms and opioid craving compared with placebo	Mannelli *et al.* 2006, 2008

From S. B. Leavitt, Opioid antagonists, naloxone & naltrexone—aids for pain management. An overview of clinical evidence, *Pain Treatment Topics*, 2009, http://pain-topics.org/pdf/OpioidAntagonistsForPain.pdf.

of noradrenaline (norepinephrine). Subsequently tapentadol, a more potent version which has affinity to MOP and is highly selective to norepinephrine reuptake was introduced in 2001. Dual-pharmacology opioids discussed here include tramadol, tapentadol, methadone and levorphanol.

2.3.9.1 Methadone

The use of methadone as a potent opioid agonist has gained renewed interest due to its low cost and potential activity in neuropathic pain syndromes. Methadone is a racemic mixture; one (*S*) stereoisomer acts as a NMDA receptor antagonist, the other (*R*) is a MOP receptor agonist. The NMDA mechanism plays an important role in the prevention of opioid tolerance, potentiation of opioid effects, and possibly efficacy for neuropathic pain. Methadone is highly lipophilic, with rapid GI absorption and onset of action. It has a large initial volume of distribution with slow tissue release. Oral bioavailability is around 80%. It has no active metabolites. The major route of metabolism is hepatic, with significant faecal excretion; renal excretion can be enhanced by urine acidification (pH <6.0). Unlike morphine, no dose adjustment is needed in patients with renal failure since there are no active metabolites.

Methadone is available in tablet, liquid and injectable forms; oral preparations can be used rectally. In the United States, any clinician with a Schedule II DEA licence can prescribe methadone for pain; a special licence is only required to prescribe methadone for the treatment of addiction.

Unlike morphine, hydromorphone, or oxycodone, methadone has an extended terminal half-life of up to 190 hours. This half-life does not match the observed duration of analgesia (6–12 hours) after steady state is reached. A highly linear relationship between plasma AUC and dose exists for both (*R*)-methadone and (*S*)-methadone. This long half-life can lead to increased risk for sedation and respiratory depression, especially in older people or with rapid dose adjustments. Rapid titration guidelines for other opioids do not apply to methadone. ECG monitoring is appropriate due to the risk of development of QT interval prolongation and torsades de pointes.[167] Several studies have demonstrated the efficacy of methadone in chronic pain syndromes but, due to the complexity of dosing and potential for cardiac adverse effects, the use of this opiate should be reserved for experienced practitioners.

2.3.9.2 Levorphanol

This is a neglected potent opioid agonist which has some very unique attributes. It is an agonist and an NMDA antagonist as well as a monoamine reuptake inhibitor of serotonin and noradrenaline (norepinephrine). Similar to methadone, levorphanol has a long half-life, and an elimination half-life which is longer than its analgesic duration of action. Levorphanol's analgesic half-life is comparable to that of methadone (6–8 hours), but its elimination half-life of 11 hours is more predictable than that of methadone. Plasma concentrations increase with chronic use, leading to accumulation and toxicity. It is about 4–8 times more potent than morphine and the usual recommended dose is 2 mg every 6–8 hours. Levorphanol lacks the stigma associated with methadone and its use in addiction medicine. Levorphanol is a strong NMDA receptor antagonist which has generated interest as a treatment for neuropathic pain. Limited research has supported its role as an effective treatment for neuropathic pain, allodynia and hyperalgesia.[168–171] Rowbotham demonstrated a dose–response curve with oral levorphanol for patients with neuropathic pain; 9 mg daily was more effective than 3 mg.[168] As with methadone, levorphanol has not been directly compared with other opioids or adjuvant analgesics for neuropathic pain. McNulty showed in a recent case series of 31 patients (including hospice patients and CNCP patients) that 74% of patients had improved pain relief when switching to levorphanol in the setting of inadequately controlled pain on other opioids.[171] Levorphanol has also been noted to have a positive effect on mood.[172]

Parenteral levorphanol is twice as potent as the oral formulation. Published oral morphine:oral levorphanol equianalgesic ratios range from 30 : 4 to 12 : 1.[170,173] The most recent case series looking at switching from other opioids to levorphanol used a staggered morphine:levorphanol ratio,[173] similar in concept to switching to methadone. Available data indicate these ratios are reasonably safe and effective. The medication is dosed every 6–12 hours depending on an individual patient's duration of analgesia. Opioid-naive patients can start with 6 mg orally a day, divided. Levorphanol is available in 2 mg tablets and 2 mg mL^{-1} or 2 mg 10 mL^{-1} parenteral formulations.

2.3.9.3 Tramadol

Tramadol is a centrally acting, synthetic analgesic with a dual mechanism of action. It was initially introduced in Europe over 30 years ago as a weak opioid. Tramadol is highly bioavailable, up to 75%, and has about 20% of the potency of morphine. It is available in IR and ER versions and in combination with paracetamol (acetaminophen). It is available in multiple generic versions worldwide.

Tramadol has a weak affinity for the MOP receptor and inhibits the reuptake of serotonin and noradrenaline (norepinephrine) at the level of the dorsal horn.[174] Naloxone only reverses about 30% of its analgesic action. Tramadol is extensively metabolised via several pathways including CYP3A4 and CYP2D6. The biologically active *o*-desmethyl metabolite (M1) is responsible for the opioid effects.[175]

About 30% of the dose is excreted unchanged in the urine, and 60% of the dose is excreted as metabolites. In Phase I studies of healthy subjects, concentrations of tramadol were approximately 20% higher in PM compared with EM, and M1 concentrations were 40% lower. Concomitant administration fluoxetine and quinidine could result in increases in tramadol concentrations and decreased concentrations of M1. The pharmacological impact of these alterations on efficacy or safety is unknown.[175]

Concomitant use of SSRIs or monoamine oxidase inhibitors may increase the risk of adverse events, including seizure.

As opposed to other opioids, tramadol is not scheduled and carries a very minimal risk of respiratory depression with overdose and has a very low risk of physical dependence and abuse. Tramadol does carry a small risk of seizures especially with high doses and in individuals with a lowered seizure threshold.[174,176]

It is effective in a variety of chronic pain states including fibromyalgia, diabetic neuropathy, osteoarthritis, and back pain.[177] Five randomised controlled trials involving a total of 374 participants meeting the inclusion criteria compared tramadol to placebo. Evidence from these trials showed that 100–400 mg of tramadol is an effective symptomatic treatment for peripheral neuropathic pain in a review of the literature.[178]

For osteoarthritis pain, tramadol or tramadol/paracetamol decreases pain intensity, produces symptom relief, and improves function, but these benefits are small. Adverse events, although reversible and not life-threatening, often cause participants to stop taking the medication and could limit the usefulness of tramadol or tramadol plus paracetamol.[179]

Tramadol with paracetamol (acetaminophen) was introduced as a combination analgesic (Ultracet) containing 37.5 mg of tramadol and 325 mg of paracetamol (acetaminophen) and approved by the FDA for short-term management of acute pain in August 2001. Since 2004, generic versions have been available. Ultracet was evaluated in three single-dose studies of acute pain arising from third molar extractions and was found to be more effective than placebo and paracetamol (acetaminophen) with an onset of action of 17 minutes and duration of 5 hours.[180] Ultracet was also found to be significantly more effective than placebo in osteoarthritis flares[181] and in chronic pain conditions including chronic low back pain, compared to placebo[182] and codeine,[183] in fibromyalgia.[184] The efficacy of Ultracet did not appear to diminish over a 2 year period and there was no indication of tolerance.[185,186]

The recommended dosage of tramadol/paracetamol(acetaminophen) combination is two tablets every 4–6 hours as needed for pain up to a maximum of eight tablets. The combination does not appear to alter the safety of either drug in combination. CNS effects, hepatotoxicity, and potential for abuse are concerns. The most common side effects include constipation, somnolence, sweating, dizziness and nausea.

2.3.9.4 Tapentadol

Tapentadol (Nucynta®) IR formulation is available in 50 mg, 75 mg and 100 mg tablets and was approved in the United States in 2008 for relief of moderate to severe acute pain in patients 18 years and older.

Tapentadol is a centrally acting synthetic analgesic. Analgesic efficacy is thought to be due to mu-opioid agonist activity and the inhibition of noradrenaline (norepinephrine) reuptake. Oral bioavailability is about 40% with a half-life of 4.5 hours. It is 18 times less potent than morphine in binding to the human mu-opioid receptor and is 2–3 times less potent in producing analgesia in animal models. Tapentadol exerts its analgesic effects without a pharmacologically active metabolite.

The efficacy and safety of tapentadol in the treatment of moderate to severe acute pain has been established in two randomised, double-blind, placebo- and active-controlled studies of moderate to severe pain from first metatarsal bunionectomy and end-stage degenerative joint disease. (package insert).

Tapentadol is mainly metabolised by glucuronidation and, in contrast to tramadol, it does not inhibit or induce P450 enzymes. Based on data from nine Phase 2/3 studies that administered multiple doses (seven placebo- and/or active-controlled, one non-controlled and one Phase 3 active-controlled safety study) the most common adverse events (reported by ≥10% in any tapentadol dose group) were nausea, dizziness, vomiting, and somnolence. Tapentadol is available in IR and ER formulations.

2.3.10 Oral Reformulations and Tamper-Resistant Formulations

Over the last few decades there has been a shift in the prescription patterns of opioids for a broad array of painful conditions with a concomitant increase in misuse, abuse and diversion.[187] Abusers typically tamper with the product in an attempt to get high. Tampering involves manipulating a dosage form to change its drug delivery characteristics and includes crushing or chewing and then swallowing; crushing and snorting; and crushing, then injecting. Although prescription opioids may be formulated for oral use, they are often taken intravenously or intranasally when abused.[188,189] As tolerance to the psychoactive effects of opioids increases with use over time, the user often progresses from the oral route to the more lethal intranasal or intravenous (IV) route to attain greater opioid effects and more rapid onset of action[188,190,191] With the introduction of controlled-release formulations, specifically the original OxyContin in 1995, this abuse took a more lethal turn, since a potentially large lethal dose of pharmaceutical-grade opioid could be released by just tampering with an ER tablet or capsule.

In the United States in 2007, more than one-third (36%) of all poisoning deaths involved opioid analgesics.[192] Since 1999, poisoning deaths in the United States involving opioid analgesics have more than tripled[192,193]; deaths from opioid analgesics have surpassed those from heroin and cocaine.[194,195] Over the 5-year period from 2004 to 2008, the estimated number of emergency department visits owing to non-medical use of pain relievers has increased 111%, from 144 644 in 2004 to 305 885 in 2008.[196] According to the Centers for Disease Control and Prevention, 15 597 Americans died from overdoses involving opioids in 2009, an increase of 109% from 2002.

This phenomenon may appear unique to the United States and may also be a function of the relatively high use of opioids there compared to the rest of the world. Americans constitute 4.6% of the world's population, but consume 80% of the global supply of opioids as well as two-thirds of the world's illegal drugs.[147] Recognising the seriousness of the problem, on 9 July 2012, the FDA approved a REMS for ER and long-acting (LA) opioid medications.[197] The REMS introduces new safety measures to reduce risks and improve safe use of ER/LA opioids while continuing to provide access to these medications for patients in pain. This alone does not appear to be enough to abate the epidemic and the FDA also encourages pharmaceutical manufacturers to develop modified opioid formulations that employ unique and proprietary formulations to deter abuse. However, the FDA will not allow any label claims of abuse deterrence until post-marketing epidemiological data is available to support the claim.

Since prescription opioids are essential for the management of pain, it is important to balance patient's legitimate needs for analgesia with the urgent public need to reduce drug abuse, misuse, and diversion. Reformulations of existing products and newer TRFs have emerged to curtail the misuse epidemic. Although TRFs may reduce abuse of a drug, from a public health perspective, it is analogous to squeezing a balloon: the problem just shifts to another drug. National and global awareness and a multipronged approach are needed to curb the public health epidemic. There are four main strategies to deter opioids abuse (Table 2.9):

- physical barriers to crushing, chewing and extraction
- opioids with aversive agents

Table 2.9 Tamper-resistant formulations and reformulations.

Technology	*Product examples*	*Advantages*	*Disadvantages*
Physical barrier	OxyContin, Opana ER, Remoxy XRT, Col-003, Egalet, Exalgo (hydromorphone ER with OROS technology)	Resists tampering. Effective in preventing intranasal and intravenous routes	Does not prevent overconsumption. Shift of abuse to other products
Sequestered antagonists	Embeda®, ALO-02	Effective in preventing chewing, snorting or injecting. May prevent fatal respiratory depression in overdose situations	May precipitate withdrawals in opioid-dependent individuals
Sequestered aversive components	Oxecta®	Prevents intranasal, intravenous and oral chewing	Does not prevent overconsumption. Too many competing generic oxycodone products
Technology to prevent opioid overconsumption	Lysine-modified prodrug of hydrocodone, NRP 290 (New River Pharmaceuticals)	May address the majority of patients who abuse oral medications	May limit the dose of an opioid used for severe pain

- opioids with sequestered antagonists
- technologies that deter overconsumption of intact drug.

Currently, there are no technologies that can prevent overconsumption of intact drug. Although this constitutes the largest group of abusers, intranasal and IV routes are equally important since they are the most lethal.

Reformulations of existing products involve the addition of excipients or antagonists to render them difficult to tamper with while retaining the pharmacokinetics of the original drug.

2.3.10.1 *Talwin NX (Sanofi-Aventis, U.S., LLC, Bridgewater, NJ)*

A case study of a successful adaptation is pentazocine (Talwin) introduced in 1967, which resulted in a steep increase in IV abuse of crushed Talwin. There was an epidemic of abuse of 'Ts and blues' in which pentazocine and the antihistamine tripelennamine (known as 'blues') were crushed together and injected IV. Talwin was replaced with a reformulated Talwin NX (pentazocine and naloxone hydrochlorides, USP) in 1982. Naloxone when administered orally at 0.5 mg has no pharmacological activity, but when administered parentally at the same dose it is an effective antagonist to pentazocine and a pure antagonist to narcotic analgesics. The original Talwin was withdrawn from the market in 1983 and in the 2 years following this there was a 70% decrease in the rates of mentions by emergency department and medical examiner according to the Drug Abuse Warning Network (DAWN) data.[198] This indicated a successful introduction of a reformulated product.

2.3.10.2 *Reformulated OxyContin (Purdue Pharma LP, Stamford, CT)*

The original formulation of OxyContin was approved in December 1995, and by November 2000 it was recognised that there was a growing problem of misuse and abuse directly attributed to OxyContin. In April 2010 the reformulated OxyContin (ORF) was approved with two post-marketing requirements, a REMS and extensive post-marketing epidemiological studies. The controlled-release (CR) oxycodone was reformulated in a polymer matrix that makes tampering by crushing or chewing difficult,[199] and when immersed in water it forms a viscous gel that resists oxycodone extraction for injection.[199] The reformulation has bioequivalence to the original CR oxycodone. The FDA has instructed the manufacturer to conduct extensive post-marketing surveillance to gauge the effects of the new formulation on abuse. The original OxyContin was withdrawn from the US market. Three separate studies funded by Purdue Pharma presented at an APS meeting suggest that ORF may be fulfilling its intended purpose of reducing misuse and abuse of ORF by non-oral routes only,[200] but the abuse may be shifting to alternative opioids and heroin.[201] On 19 November 2012, Health Canada approved six versions of the easily abusable generic oxycodone.

2.3.10.3 Reformulated Opana ER (Endo Pharmaceuticals, Inc., Chadds Ford, PA)

An ER formulation of oxymorphone (Opana ER) using the polysaccharide hydrogel (PSH) was initially approved in 2006 and was prescribed about 10-fold less frequently than CR OxyContin for many reasons, including lack of familiarity. However, after the new tamper-resistant OxyContin was introduced, there was a rise in the abuse of oxymorphone ER-PSH.

The reformulated oxymorphone ER with a polyethylene oxide (PEO) matrix (INTAC™; Grünenthal GmbH, Aachen, Germany) designed to resist crushing, oxymorphone ER-PEO. ER-PEO was shown to be bioequivalent to the previously marketed oxymorphone ER-PSH.[199] In three open-label studies, 5–40 mg of oxymorphone ER-PEO was shown to be bioequivalent to the previously marketed oxymorphone ER-PSH.[199] This was approved in December 2011 and Endo ceased the production of the original formulation in February 2012.

2.3.10.4 Tamper-Resistant Formulations (TRFs)

Physical Barriers. These formulations resist crushing, chewing, dissolution or chemical extraction, reducing the risk of a rapid release of opioids. They may also be formulated to resist chemical extractions, reducing the risk of dose dumping. Some examples of the technologies in a rapidly crowded space include:

- **COL-003 (Collegium Pharmaceutical, Inc., Cumberland, RI).** Collegium Pharmaceuticals has a patented DETERx™ formulation which consists of small spherical beads that are filled into a capsule. The beads are prepared using a standard spray congealing process. They contain a solid dispersion of the active drug in an ER matrix. All of the formulation excipients used are generally recognised as safe (GRAS), NF-grade, and listed in the FDA's Inactive Ingredient Database for oral administration (company website). The first product to enter Phase 3 trials is COL-003, an ER formulation of oxycodone embedded in a multiparticulate matrix containing oxycodone in a waxy excipient base. When opened, crushed, or chewed, the particles maintain their ER properties. *In vitro*, COL-003 was more resistant than currently available formulations of oxycodone to dissolution in solvents. In healthy volunteers, the pharmacokinetic profile of COL-003 was similar when chewed compared with when it was taken as intended.[202] The oxycodone DETERx formulation can be effectively administered using alternative oral routes of administration such as nasogastric tube and sprinkling onto soft food. (Chewing does not increase blood levels of oxycodone.)[203]

- **Remoxy XRT™ (Pfizer, Inc., New York, USA).** Remoxy XRT™ is a novel formulation of an ER oxycodone which uses a proprietary ORADUR technology. The ORADUR is a novel formulation comprised of a viscoelastic fluid matrix that is filled into a capsule. This matrix resists abuse by crushing and dissolution in water or alcohol.[204] A Phase 3 programme demonstrating safety and efficacy was completed. A study in osteoarthritis demonstrated efficacy with adverse events similar to other opioids,[205] but the initial new drug application for the formulation was not accepted by the FDA and Pfizer continues to address the issues raised by the FDA complete response letter.
- **Egalet Morphine (Egalet Ltd, Værløse, Denmark).** The formulation is an injection-moulded polymer system consisting of an erodible matrix in which morphine sulfate is dispersed. The matrix is partly covered with a water-impermeable, non-erodible shell which leaves both ends of the cylindrical tablet exposed to erosion by the GI fluid. The well-defined, fixed surface erosion area allows a tightly controlled, extended release of morphine over a period of 12 hours. Egalet morphine is designed to maintain its ER properties across a wide range of solvents including ethanol, and is resistant to physical and chemical attempts to alter the slow release of the drug substance contained in the matrix.[206] In a 2-week, double-blind, randomised, exploratory crossover study, once-daily Egalet morphine was as effective as twice-daily CR morphine in adult patients with cancer.[207] Oxycodone and hydrocodone formulations using Egalet technology are also being developed.

Sequestered Opioid Antagonists. Combining an agonist with an antagonist is not a novel technology, and Talwin NX is an example of an older technology that was effective. However, another novel approach is to use an opioid with a sequestered antagonist core that is intended to be released to neutralise the effects of the opioid when tampered for the purpose of abuse. However, the risks of tampering are the same for opioid-dependent non-abusers who may on the one hand precipitate withdrawals when they unintentionally crush the product, but on the other hand it may prevent life-threatening respiratory depression in the patient who ingests a potentially lethal dose of the opioid.

- **Naltrexone Platform (NT) (Pfizer, Inc., New York, USA).** King pharmaceuticals, a wholly owned subsidiary of Pfizer, is developing a wide range of products that utilise its proprietary NT platform. Embeda® was successfully launched but recalled later due to manufacturing issues. Others in the pipeline include ALO-02 (oxycodone and naltrexone) and ALO-04 (hydrocodone and naltrexone)
- **Embeda (morphine sulfate with sequestered naltrexone 100 mg) (Pfizer, Inc., New York, USA).** Embeda is a unique and proprietary pellet formulation

which has a sequestered naltrexone core surrounded by an ER formulation of morphine sulfate in each pellet; it was approved by the FDA in August 2009. This is the first morphine formulation to be FDA approved with this technology. It was voluntarily recalled by the manufacturer because of some problems with the formulation that are yet to be resolved. Compared with ER morphine, Embeda taken intact or chewed was associated with fewer positive subjective effects, including euphoria, and participants indicated they would be willing to pay less for it. During clinical trials, Embeda has not been associated with severe withdrawal symptoms when taken as intended or after chewing.
- **ALO-02 (Pfizer, Inc., New York, USA).** ALO-02 (oxycodone hydrochloride and naltrexone hydrochloride ER capsules) utilises a technology similar to Embeda consisting of pellets with sequestered naltrexone in a fixed ratio to the opioid. Topline results of a Phase 3 safety study were released in October 2012. A registration Phase 3 trial is ongoing.

Aversion Technology. Aversive agents are sequestered and are only released when crushed, chewed or chemically tampered with, causing a dislike of the abused substance and reducing the attractiveness of the formulation to the abuser.

- **Acurox (Acura Pharmaceuticals/Pfizer, Inc., New York, USA).** This was an oxycodone IR product with niacin, developed by Acura Pharmaceuticals. When taken in excessive doses it causes a classic niacin reaction with flushing which is very uncomfortable to the individual. However, niacin's effects can be dulled by food. Analgesic efficacy and abuse liability trials were completed. After a negative determination by the FDA advisory panel in April 2010, the company removed the niacin and reformulated the drug to the currently approved Oxecta® (see below).
- **Oxecta (Pfizer, Inc., New York, USA).** Approved on 17 June 2011, this is a short-acting oxycodone formulation designed to discourage common methods of tampering associated with opioid abuse and misuse. Oxecta is formulated using a proprietary aversion technology (Acura Pharmaceuticals, Inc., Palatine, IL) that incorporates commonly used pharmaceutical ingredients that can be irritating if ingested or inhaled.[196] These include colloidal silicon dioxide, crospovidone, magnesium stearate, microcrystalline cellulose, polyethylene oxide, and sodium lauryl sulfate. Oxecta® utilises both a physical barrier and an aversion technology. This is the only short-acting opioid currently available with a TRF.

Technologies To Deter Overconsumption. A prodrug has little or no pharmacological effect until it is metabolised to an active form, which may be self-limiting after a certain dose. This is an interesting concept which also deters IV and intranasal abuse. A lysine-modified opioid prodrug NRP-290 was initially proposed by New River Pharmaceuticals, but no further information about its development is available after it was acquired by Shire Pharmaceuticals.[208]

2.4 New Opiate Research

2.4.1 Functionally Biased Ligands

There are differences in the ways peptide and non-peptide ligands induce receptor internalisation at opioid receptors which could be explained by biased agonism, in which different ligands that act at the same receptor elicit different responses.[209] Etorphine desensitises the MOP receptor, through high levels of phosphorylation and arrestin coupling, whereas morphine produces little phosphorylation.[210] These data have been corroborated by both compounds behaving similarly in MOP receptor mutants designed to block phosphorylation.[211]

Trevena focuses their research on biased ligands,[212] and have successfully completed a Phase 1 trial of a biased MOP receptor ligand TRV-130,[213] which is being developed for the treatment of postoperative pain. While the structure of TRV-130 has not been disclosed, recent patent publications suggest it could be related to the spirofused tetrahydropyran system shown in Figure 2.21.[214]

2.4.2 Splice Variants, Heterodimers and SNPs

The human MOP gene (*OPRM1*) spans over 200 kb and contains 11 exons that yield some 33 splice variants, many of which have altered intracellular C-terminals, but retain high affinity for opioid ligands.[215] However, several of these splice variants respond differently to GTPγs stimulation, and thereby respond functionally very differently to activation. In some cases, splice variants are insensitive to traditional opioid ligands, but remain sensitive to related peptide agonists.[82]

Despite the MOP receptor being encoded by a single gene, *OPRM1*,[216] over 700 SNPs have been identified in this gene primarily located at exons 1–4, corresponding to the genomic region for the MOP receptor mRNA, and an increasing body of data suggest these SNPs influence receptor function and

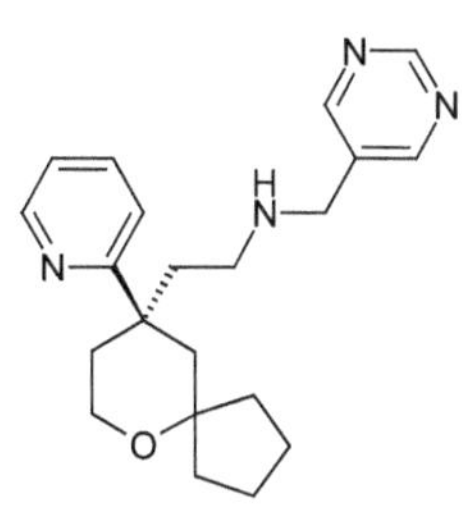

Trevena ligand

Figure 2.21 Trevena MOP receptor-biased ligand.

thereby pain sensitivity. Studies of the A118G SNP have provided the most compelling association with pain-related traits. G-allele carriers of this SNP showed lower sensitivity to mechanical stimulation than AA subjects. This same SNP has also been shown to be associated with opioid sensitivity in which opioid consumption, requirement and 50% effective concentration were all higher than in the corresponding AA subjects.[217]

Significant research has focused on the elucidation of heteromeric opioid receptors, in which two different opioid receptors dimerise. There has been evidence of DOP/KOP, MOP/DOP, MOP/KOP, and NOP/MOP dimers [218] that demonstrate a different distribution pattern, trafficking behaviour and signalling.

All of this research has highlighted a further complexity in the opioid receptor family that raises the possibility of novel ligands achieving a pharmacology and side-effect profile that is distinct from that of the more traditional opioid ligands.

2.4.3 Conjugated Analogues for Peripheral Restriction

Several conjugated analogues have been reported recently that seek to use conjugative strategies to exclude opioids from the CNS. NKTR-118 has been licensed by Nektar Therapeutics to AstraZeneca as a pegylated version of the morphine-based opioid antagonist naloxol that increases the bioavailability of the parent compound when given orally through lowering first-pass clearance, and limits uptake into the CNS.[219] NKTR-118 is currently in Phase 3 trials for the treatment of constipation.[220] NKTR-181 is another polymer-based conjugate from Nektar that has been specifically designed to reduce the rate of passage into the CNS (but not completely exclude the active principle from the CNS) to reduce euphoria and respiratory depression, but not the analgesia provided by the opioid constituent. NKTR-181 is currently in a Phase 2 trial in osteoarthritis patients.[221] A novel version of dynorphin A, CJC-1008, has been found to display analgesia in a Phase 2 trial in patients with PHN. CJC-1008 contains a conjugated maleimido group to promote covalent attachment to serum albumin and increase pharmacokinetic stability.[222] Other peptide conjugation strategies that have served to increase stability include glycosylation and pegylation.[223]

A recent report describes conjugating an early iteration of a Cara D-tetrapeptide[224] with an antibody to fully exclude the peptide from the CNS and endow the peptide cargo with antibody-like pharmacokinetic properties. The resulting conjugate retained agonist potency at the KOP receptor and apparently greatly extended its terminal elimination half-life.[225]

2.4.4 Opioid Activation of the Immune Pathway

Recent reports have highlighted that the activation of opioid receptors leads to the stimulation of the pattern recognition receptor TLR4, and its MyD88-dependent signalling pathway. This ancillary action of opioid stimulation amplifies the opioid-induced elevations in dopamine levels and boosts the reward pathway. Further studies defining the potential of TLR4 as a potential therapeutic target for addiction are anticipated.[226]

2.5 Conclusions

The development of effective medications for pain is a major focus of clinicians and pharmaceutical researchers. Clinicians and medicinal chemists face several challenges in their quest for better drugs to treat pain and suffering. Despite the centuries–long availability of opioids, the manipulation of the naturally occurring alkaloids to improve on the efficacy and safety is fairly new. The medicinal chemist faces the complex challenge of navigating the complicated pharmacologies and differential engagement of several signalling pathways to identify a drug that has the efficacy without any deleterious effects typically seen with opioids.

A relatively 'low-hanging fruit' may be to manipulate the MOP receptor agonists to maintain or improve analgesia and eliminate deleterious effects. Additional advances in KOP, DOP and NOP pharmacology may enable us to develop medicines active at all four known receptors into clinically meaningful analgesics.

Endogenous peptides are tempting due to the lack of side effects seen in the plant-based alkaloids but face challenges of their own with limited blood–brain permeation and low biostability. Hopefully newer generation peptides/peptidomimetics may be able to provide analgesia without addiction and GI side effects.

Clinician researchers need to have a better understanding of the pathophysiology of the diseases they treat and the pain mechanisms they target. Pain geneticists will have to answer the puzzle of the individual variances in pain expression and the variable analgesic response to opioids. Epigenetic mechanisms may provide novel insights into pain chronicity. Better understanding of the pharmacokinetic and pharmacodynamic variability of opioids will help select better drugs for the right patient.

Clinicians need tools to phenotype individuals who respond to their analgesic regimens to choose the right drug for the right patient to achieve the best outcome without the trial and errors of current practices.

A successful outcome requires a fresh, patient-centred approach and a seamless integration of research in medicinal chemistry, molecular modelling, clinical pharmacology and clinical research with the actual clinical practice.

In the meantime, when oral opioids are indicated, all stakeholders have the responsibility of ensuring that the clinician's rod of Asclepius does not become the staff of Caduceus for patients and drug diverters.

References

1. T. W. Wallbank and A. M. Taylor, A short history of the opium wars, in *Civilization: Past and Present*, Harpercollins, New York, 7th edn, 1992.

2. *Epidemic: Responding to America's Prescription Drug Abuse Crisis*. White paper from Office of National Drug Control Policy www.whitehouse.gov/sites/default/files/ondcp/policy-and-research/rx_abuse_plan.pdf.
3. B. L. Kieffer, Opioids: first lessons from knockout mice, *Trends Pharmacol. Sci.*, 1999, **20**(1), 19–26.
4. C. B. Pert and S. H. Snyder, Opiate receptor: demonstration in nervous tissue, *Science*, 1973, **179**(4077), 1011–1014.
5. W. R. Martin, History and development of mixed opioid agonists, partial agonists and antagonists, *Br. J. Clin. Pharmacol.*, 1979, **7**(S3), 273S–279S.
6. J.-C. Meunier, *et al.*, Isolation and structure of the endogenous agonist of opioid receptor-like ORL1 receptor, *Nature*, 1995, **377**(6549), 532–535.
7. J. Hughes, *et al.*, Identification of two related pentapeptides from the brain with potent opiate agonist activity, *Nature*, 1975, **258**(5536), 577–579.
8. A. D. Corbett, *et al.*, 75 years of opioid research: the exciting but vain quest for the Holy Grail, *Br. J. Pharmacol.*, 2006, **147**(S1), S153–S162.
9. D. Peckys and G. B. Landwehrmeyer, Expression of mu, kappa, and delta opioid receptor messenger RNA in the human CNS: a 33P *in situ* hybridization study, *Neuroscience*, 1999, **88**(4), 1093–1135.
10. A. Mansour, *et al.*, Immunohistochemical localization of the cloned μ opioid receptor in the rat CNS, *J. Chem. Neuroanat.*, 1995, **8**(4), 283–305.
11. U. Gether, Uncovering molecular mechanisms involved in activation of G protein-coupled receptors, *Endocrine Rev.*, 2000, **21**(1), 90–113.
12. G. W. Pasternak, Book review: *The Pharmacology of Mu Analgesics: From Patients to Genes*, *Neuroscientist*, 2001, **7**(3), 220–231.
13. E. Chevlen, Opioids: A review, *Curr. Pain Headache Rep.*, 2003, **7**(1), 15–23.
14. A. M. Trescot, S. Datta, M. Lee and H. Hansen, Opioid pharmacology, *Pain Physician*, 2008, **11**, S133–S153.
15. J. A. Kiritsy-Roy, L. Marson and G. R. Van Loon, Sympathoadrenal, cardiovascular and blood gas responses to highly selective mu and delta opioid peptides, *J. Pharmacol. Exp. Therapeut.*, 1989, **251**(3), 1096–1103.
16. K. T. Pattinson, Opioids and the control of respiration, *Br. J. Anaesth.*, 2008, **100**(6), 747–758.
17. T. Koch and V. Höllt, Role of receptor internalization in opioid tolerance and dependence, *Pharmacol. Therapeut.*, 2008, **117**(2), 199–206.
18. H. Kalant, Opium revisited: a brief review of its nature, composition, non-medical use and relative risks1, *Addiction*, 1997, **92**(3), 267–277.
19. M. J. Bogusz, Opiate agonists, *Handbook of Analytical Separations*, 2000, **2**, 3–65.
20. A. Dahan, *et al.*, Comparison of the respiratory effects of intravenous buprenorphine and fentanyl in humans and rats, *Br. J. Anaesth.*, 2005, **94**(6), 825–834.
21. J. D. Leander, Buprenorphine is a potent κ-opioid receptor antagonist in pigeons and mice, *Eur. J. Pharmacol.*, 1988, **151**(3), 457–461.

22. W. K. Bickel, *et al.*, Buprenorphine: dose-related blockade of opioid challenge effects in opioid dependent humans, *J. Pharmacol. Exp. Therapeut.*, 1988, **247**(1), 47–53.
23. A. S. Keats and J. Telford, Studies of analgesic drugs. VIII. a narcotic antagonist analgesic without psychotomimetic effects, *J. Pharmacol. Exp. Therapeut.*, 1964, **143**(2), 157–164.
24. S. Izenwasser, *et al.*, The cocaine-like behavioral effects of meperidine are mediated by activity at the dopamine transporter, *Eur. J. Pharmacol.*, 1996, **297**(1–2), 9–17.
25. M. Bockmuhl and G. Ehrhart, Über eine neue Klasse von analgetisch wirkenden Verbindungen, *Justus Liebigs Ann. Chem.*, 1948, **561**, 52–85.
26. T. H. Stanley, The history and development of the fentanyl series, *J. Pain Symptom Manage.*, 1992, **7**(3), S3–S7.
27. R. A. Stokbroekx, *et al.*, Synthetic antidiarrheal agents. 2,2-Diphenyl-4-(4′-aryl-4′-hydroxypiperidino)butyramides, *J. Med. Chem.*, 1973, **16**(7), 782–786.
28. S. Grond and A. Sablotzki, Clinical pharmacology of tramadol, *Clin. Pharmacokinet.*, 2004, **43**(13), 879–923.
29. C. T. Hartrick, Tapentadol immediate release for the relief of moderate-to-severe acute pain, *Exp. Opin. Pharmacother.*, 2009, **10**(16), 2687–2696.
30. P. Fadda, *et al.*, Antinociceptive action of DBO 17 and DBO 11 in mice: two 3,8 diazabicyclo (3.2.1.) octane derivates with selective μ opioid receptor affinity, *Naunyn-Schmiedeberg's Arch. Pharmacol.*, 1997, **356**(5), 596–602.
31. S. M. Husbands, *et al.*, 3-Alkyl ethers of clocinnamox: delayed long-term μ-antagonists with variable μ efficacy, *J. Med. Chem.*, 1998, **41**(18), 3493–3498.
32. M. Bolognesi, *et al.*, Opioid antagonist activity of naltrexone-derived bivalent ligands: importance of a properly oriented molecular scaffold to guide 'address' recognition at κ opioid receptors, *J. Med. Chem.*, 1996, **39**(9), 1816–1822.
33. S. Majumdar, *et al.*, Generation of novel radiolabeled opiates through site-selective iodination, *Bioorg. Med. Chem. Lett.*, 2011, **21**(13), 4001–4004.
34. S. Archer, *et al.*, 14.alpha.,14′.beta.-[dithiobis[(2-oxo-2,1-ethanediyl)imino]]-bis(7,8-dihydromorphinone) and 14.alpha.,14′.beta.-[Dithiobis[(2-oxo-2,1-ethanediyl)imino]]bis-7,8-dihydro-*N*-(cyclopropyl- methyl)normorphinone: chemistry and opioid binding properties, *J. Med. Chem.*, 1994, **37**(11), 1578–1585.
35. S. B. Sawh, *et al.*, Use of methylnaltrexone for the treatment of opioid-induced constipation in critical care patients, *Mayo Clin. Proc.*, 2012, **87**(3), 255–259.
36. J. Thomas, Opioid-induced bowel dysfunction, *J. Pain Symptom Manage*, 2008, **35**(1), 103–113.
37. S. N. Calderon, *et al.*, Probes for narcotic receptor mediated phenomena. 19. Synthesis of (+)-4-[(.alpha.R)-.alpha.-((2S,5R)-4-Allyl-2,5-dimethyl-1-piperazinyl)-3-methoxybenzyl]-*N*,*N*-diethylbenzamide (SNC 80): A highly

selective, nonpeptide, delta opioid receptor agonist, *J. Med. Chem.*, 1994, **37**(14), 2125–2128.
38. E. J. Bilsky, *et al.*, SNC 80, a selective, nonpeptidic and systemically active opioid delta agonist, *J. Pharmacol. Exp. Therapeut.*, 1995, **273**(1), 359–366.
39. E. Jutkiewicz, *et al.*, Separation of the convulsions and antidepressant-like effects produced by the delta-opioid agonist SNC80 in rats, *Psychopharmacology.*, 2005, **182**(4), 588–596.
40. E. E. Codd, *et al.*, JNJ-20788560 [9-(8-Azabicyclo[3.2.1]oct-3-ylidene)-9H-xanthene-3-carboxylic acid diethylamide], a selective delta opioid receptor agonist, is a potent and efficacious antihyperalgesic agent that does not produce respiratory depression, pharmacologic tolerance, or physical dependence, *J. Pharmacol. Exp. Therapeut.*, 2009, **329**(1), 241–251.
41. B. L. Bourdonnec, *et al.*, Potent, orally bioavailable delta opioid receptor agonists for the treatment of pain: discovery of *N*,*N*-Diethyl-4-(5-hydroxyspiro[chromene-2,4′-piperidine]-4-yl)benzamide (ADL5859), *J. Med. Chem.*, 2008, **51**(19), 5893–5896.
42. B. Le Bourdonnec, *et al.*, Spirocyclic delta opioid receptor agonists for the treatment of pain: discovery of *N*,*N*-diethyl-3-hydroxy-4-(spiro[chromene-2,4′-piperidine]-4-yl) benzamide (ADL5747), *J. Med. Chem.*, 2009, **52**(18), 5685–5702.
43. J. Kamei, *et al.*, Antinociceptive effects of the selective non-peptidic δ-opioid receptor agonist TAN-67 in diabetic mice, *Eur. J. Pharmacol.*, 1995, **276**(1–2), 131–135.
44. P. S. Portoghese, *et al.*, Application of the message-address concept in the design of highly potent and selective non-peptide.delta. opioid receptor antagonists, *J. Med. Chem.*, 1988, **31**(2), 281–282.
45. T. Suzuki, *et al.*, Effect of naltrindole on the development of physical dependence on morphine in mice: A behavioral and biochemical study, *Life Sci.*, 1995, **57**(17), PL247–PL252.
46. F. Yan, B. L. Roth and A Salvinorin., A novel and highly selective κ-opioid receptor agonist, *Life Sci.*, 2004, **75**(22), 2615–2619.
47. M. L. G. Wadenberg, A review of the properties of spiradoline: a potent and selective k-opioid receptor agonist, *CNS Drug Rev.*, 2003, **9**(2), 187–198.
48. R. S. Zukin, *et al.*, Characterization and visualization of rat and guinea pig brain kappa opioid receptors: evidence for kappa 1 and kappa 2 opioid receptors, *Proc. Natl. Acad. Sci. U. S. A.*, 1988, **85**(11), 4061–4065.
49. K. Dosaka-Akita, *et al.*, The kappa opioid agonist U-50,488H antagonizes respiratory effects of mu opioid receptor agonists in conscious rats, *J. Pharmacol. Exp. Therapeut.*, 1993, **264**(2), 631–637.
50. M. K. Pugsley, E. J. Yu and A. L. Goldin, U-50,488H, a kappa opioid receptor agonist, is a more potent blocker of cardiac sodium channels than lidocaine, *Proc. Western Pharmacol. Soc.*, 2000, **43**, 47–50.
51. F. Zhang, *et al.*, (−)U50,488H [(trans)-3,4-dichloro-*N*-methyl-*N*-[2-(1-pyrrolidinyl)-cyclohexyl]benzeneacetamide] induces internalization and

down-regulation of the human, but not the rat, κ-opioid receptor: structural basis for the differential regulation, *J. Pharmacol. Exp. Therapeut.*, 2002, **302**(3), 1184–1192.

52. Y.-M. Tao, *et al.*, LPK-26, a novel κ-opioid receptor agonist with potent antinociceptive effects and low dependence potential, *Eur. J. Pharmacol.*, 2008, **584**(2–3), 306–311.
53. J. A. Bodkin, *et al.*, Buprenorphine treatment of refractory depression, *J. Clin. Psychopharmacol.*, 1995, **15**(1), 49–57.
54. P. S. Portoghese, A. W. Lipkowski and A. E. Takemori, Binaltorphimine and nor-binaltorphimine, potent and selective k-opioid receptor antagonists, *Life Sci.*, 1987, **40**(13), 1287–1292.
55. R. M. Jones and P. S. Portoghese, 5′-Guanidinonaltrindole, a highly selective and potent κ-opioid receptor antagonist, *Eur. J. Pharmacol.*, 2000, **396**(1), 49–52.
56. J. B. Thomas, *et al.*, Identification of the first trans-(3R,4R)-dimethyl-4-(3-hydroxyphenyl)piperidine derivative to possess highly potent and selective opioid κ receptor antagonist activity, *J. Med. Chem.*, 2001, **44**(17), 2687–2690.
57. M. D. Metcalf and A. Coop, Kappa opioid antagonists: successes and future prospects, *AAPS J.*, 2005, **7**(3), E704–E722.
58. M. R. Bruchas, *et al.*, Long-acting κ opioid antagonists disrupt receptor signaling and produce noncompetitive effects by activating c-Jun N-terminal kinase, *J. Biol. Chem.*, 2007, **282**(41), 29803–29811.
59. P. R. Verhoest, *et al.*, Design and discovery of a selective small molecule κ opioid antagonist (2-methyl-*N*-((2′-(pyrrolidin-1-ylsulfonyl)biphenyl-4-yl)methyl)propan-1-amine, PF-4455242), *J. Med. Chem.*, 2011, **54**(16), 5868–5877.
60. C. Mollereau, *et al.*, ORL1, a novel member of the opioid receptor family. Cloning, functional expression and localization, *FEBS Lett.*, 1994, **341**(1), 33–38.
61. J. L. Butour, *et al.*, Recognition and activation of the opioid receptor-like ORL 1 receptor by nociceptin, nociceptin analogs and opioids, *Eur. J. Pharmacol.*, 1997, **321**(1), 97–103.
62. J. C. Meunier, *et al.*, Isolation and structure of the endogenous agonist of opioid receptor-like ORL1 receptor, *Nature.*, 1995, **377**(6549), 532–535.
63. R. K. Reinscheid, *et al.*, Orphanin FQ: a neuropeptide that activates an opioidlike G protein-coupled receptor, *Science*, 1995, **270**(5237), 792–794.
64. J. S. Mogil and G. W. Pasternak, The molecular and behavioral pharmacology of the orphanin FQ/nociceptin peptide and receptor family, *Pharmacol. Rev.*, 2001, **53**(3), 381–415.
65. J. C. Meunier and FQ Nociceptin/orphanin, and the opioid receptor-like ORL1 receptor, *Eur. J. Pharmacol.*, 1997, **340**(1), 1–15.
66. H. Mizoguchi, *et al.*, Dermorphin tetrapeptide analogs as potent and long-lasting analgesics with pharmacological profiles distinct from morphine, *Peptides*, 2011, **32**(2), 421–427.

67. S. Rover, *et al.*, High-affinity, non-peptide agonists for the ORL1 (orphanin FQ/nociceptin) receptor, *J. Med. Chem.*, 2000, **43**(7), 1329–1338.
68. J. Wichmann, *et al.*, Synthesis of (1S,3aS)-8-(2,3,3a,4,5, 6-hexahydro-1H-phenalen-1-yl)-1-phenyl-1,3,8-triaza-spiro[4.5]decan-4-one, a potent and selective orphanin FQ (OFQ) receptor agonist with anxiolytic-like properties, *Eur. J. Med. Chem.*, 2000, **35**(9), 839–851.
69. F. Jenck, *et al.*, A synthetic agonist at the orphanin FQ/nociceptin receptor ORL1: anxiolytic profile in the rat, *Proc. Natl. Acad. Sci. U. S. A.*, 2000, **97**(9), 4938–4943.
70. F. Ito and Y. Ohashi, Preparation of 1,3,8-triazaspiro[4.5]decanones as ORL1 receptor agonists. *Eur. Pat. Appl.*, 2000, *EP 997464 A1 20000503.*
71. S. Kolczewski, *et al.*, Novel hexahydrospiro[piperidine-4,1′-pyrrolo[3,4-c]pyrroles]: highly selective small-molecule nociceptin/orphanin FQ receptor agonists, *J. Med. Chem.*, 2003, **46**(2), 255–64.
72. G. D. Ho, *et al.*, Synthesis and structure-activity relationships of 4-hydroxy-4-phenylpiperidines as nociceptin receptor ligands: Part 2, *Bioorg. Med. Chem. Lett.*, 2007, **17**(11), 3028–3033.
73. G. B. Varty, *et al.*, The anxiolytic-like effects of the novel, orally active nociceptin opioid receptor agonist 8-[bis(2-methylphenyl)methyl]-3-phenyl-8-azabicyclo[3.2.1]octan-3-ol (SCH 221510), *J. Pharmacol. Exp. Ther.*, 2008, **326**(2), 672–682.
74. S. W. Yang, *et al.*, Identification of 3-substituted *N*-benzhydryl-nortropane analogs as nociceptin receptor ligands for the management of cough and anxiety, *Bioorg. Med. Chem. Lett.*, 2009, **19**(9), 2482–2486.
75. H. Kawamoto, *et al.*, Discovery of the first potent and selective small molecule opioid receptor-like (ORL1) antagonist: 1-[(3*R*,4*R*)-1-cyclooctylmethyl-3-hydroxymethyl-4-piperidyl]-3-ethyl-1,3-dihydro-2H-benzimidazol-2-one (J-113397), *J. Med. Chem.*, 1999, **42**(25), 5061–5063.
76. A. Satoh, *et al.*, Identification of an orally active opioid receptor-like 1 (ORL1) receptor antagonist 4-{3-[(2R)-2,3-dihydroxypropyl]-2-oxo-2,3-dihydro-1H-benzimidazol-1-yl}-1-[(1*S*,3*S*,4*R*)-spiro[bicyclo[2.2.1]heptane-2,1′-cyclopropan]-3-ylmethyl]piperidine as clinical candidate, *J. Med. Chem.*, 2009, **52**(14), 4091–4094.
77. C. Fischetti, *et al.*, Pharmacological characterization of the nociceptin/orphanin FQ receptor non peptide antagonist Compound 24, *Eur. J. Pharmacol.*, 2009, **614**(1–3), 50–57.
78. P. F. Zaratin, *et al.*, Modification of nociception and morphine tolerance by the selective opiate receptor-like orphan receptor antagonist (-)-*cis*-1-methyl-7-[[4-(2,6-dichlorophenyl)piperidin-1-yl]methyl]-6,7,8,9-tetrahy dro-5H-benzocyclohepten-5-ol (SB-612111), *J. Pharmacol. Exp. Ther.*, 2004, **308**(2), 454–461.
79. T. Yoshizumi, *et al.*, A novel class of cycloalkano[b]pyridines as potent and orally active opioid receptor-like 1 antagonists with minimal binding affinity to the hERG K+ channel, *J. Med. Chem.*, 2008, **51**(13), 4021–4029.

80. H. Shinkai, *et al.*, 4-Aminoquinolines: novel nociceptin antagonists with analgesic activity, *J. Med. Chem.*, 2000, **43**(24), 4667–4677.
81. T. Ogawa, *et al.*, Synthesis and structure – activity relationships of an orally available and long-acting analgesic peptide, *N*-α-amidino-Tyr-d-Arg-Phe-MeβAla-OH (ADAMB), *J. Med. Chem.*, 2002, **45**(23), 5081–5089.
82. H. Mizoguchi, *et al.*, Involvement of mouse μ-opioid receptor splice variants in the spinal antinociception induced by the dermorphin tetra-peptide analog amidino-TAPA, *Eur. J. Pharmacol.*, 2011, **651**(1–3), 66–72.
83. B. K Handa., *et al.*, Analogues of β-LPH61–64 posessing selective agonist activity at μ-opiate receptors, *Eur. J. Pharmacol.*, 1981, **70**(4), 531–540.
84. R. Vergura, *et al.*, Anxiolytic- and antidepressant-like activities of H-Dmt-Tic-NH-CH(CH2-COOH)-Bid (UFP-512), a novel selective delta opioid receptor agonist, *Peptides*, 2008, **29**(1), 93–103.
85. E. D. Marczak, *et al.*, Orally administered H-Dmt-Tic-Lys-NH-CH2-Ph (MZ-2), a potent μ-/δ-opioid receptor antagonist, regulates obese-related factors in mice, *Eur. J. Pharmacol.*, 2009, **616**(1–3), 115–121.
86. J. Aldrich and J. McLaughlin, Peptide kappa opioid receptor ligands: potential for drug development, *AAPS J.*, 2009, **11**(2), 312–322.
87. B. Brugos, V. Arya and G. Hochhaus, Stabilized dynorphin derivatives for modulating antinociceptive activity in morphine tolerant rats: effect of different routes of administration, *AAPS J.*, 2004, **6**(4), e36.
88. J. B. Jones, F. Menzhagi, S. J. O'Connor, R. H. Spencer and D. T. Chalmers, Analgesic and morphine-sparing effects of the peripherally restricted kappa opioid agonist CR845 after intravenous administration in women undergoing a laparoscopic hysterectomy, in *IASP Milan 2012*, Milan, Italy.
89. S. J. O'Connor, R. H. Spencer, J. B. Jones, D. T. Chalmers and F. Menzhagi, A first-in-man study of the safety, pharmacokinetic and pharmacodynamic profile of orally administered CR845, a novel peptidergic kappa opioid agonist for the treatment of acute and chronic pain and pruritus, in *IASP Milan 2012*, Milan, Italy.
90. A. Manglik, *et al.*, Crystal structure of the micro-opioid receptor bound to a morphinan antagonist, *Nature*, 2012, **485**(7398), 321–326.
91. S. Granier, *et al.*, Structure of the delta-opioid receptor bound to naltrindole, *Nature*, 2012, **485**(7398), 400–404.
92. H. Wu, *et al.*, Structure of the human kappa-opioid receptor in complex with JDTic, *Nature*, 2012, **485**(7398), 327–332.
93. A. A. Thompson, *et al.*, Structure of the nociceptin/orphanin FQ receptor in complex with a peptide mimetic, *Nature*, 2012, **485**(7398), 395–399.
94. R. Van Noorden, Nobel work boosts drug development, *Nature*, 2012, **490**(7420), 320.
95. B. Disse, *et al.*, Ba 679 BR, a novel long-acting anticholinergic bronchodilator, *Life Sci.*, 1993, **52**(5–6), 537–544.
96. M. Kurowski, *et al.*, [3H]-etorphine and [3H]-diprenorphine receptor binding *in vitro* and *in vivo*: differential effect of Na^+ and guanylyl imidodiphosphate, *Brain Res.*, 1982, **249**(2), 345–352.

97. G. Bonner, F. Meng and H. Akil, Selectivity of mu-opioid receptor determined by interfacial residues near third extracellular loop, *Eur. J. Pharmacol.*, 2000, **403**(1–2), 37–44.
98. J. Standfuss, *et al.*, The structural basis of agonist-induced activation in constitutively active rhodopsin, *Nature*, 2011, **471**(7340), 656–660.
99. Y. Goto, *et al.*, Identification of a novel spiropiperidine opioid receptor-like 1 antagonist class by a focused library approach featuring 3D-pharmacophore similarity, *J. Med. Chem.*, 2006, **49**(3), 847–849.
100. T. Beuming and W. Sherman, Current assessment of docking into GPCR crystal structures and homology models: successes, challenges, and guidelines, *J. Chem. Inf. Model.*, 2012, **52**(12), 3263–3277.
101. F. Brennan, *et al.*, Pain management: a fundamental human right, *Anesth Analg.*, 2007, **2007**(105), 205–221.
102. N. Kumar, WHO Normative Guidelines on Pain Management—Report of a Delphi Study to determine the need for guidelines and to identify the number and topics of guidelines that should be developed by WHO, WHO, Geneva, Switzerland, 2007.
103. World Health Organization, *Cancer Pain Relief*, WHO, Geneva, Switzerland, 1996.
104. Institute for Clinical Systems Improvement (ICSI), *Assessment and Management of Acute Pain*, 2011. https://www.icsi.org/_asset/bw798b/ChronicPain.pdf.
105. R. Chou, G. J. Fanciullo, P. G. Fine, *et al.*, Clinical guidelines for the use of chronic opioid therapy in chronic noncancer pain, *J. Pain*, 2009, **10**(2), 113–130.
106. C. R. Chapman, D. L. Lipschitz, M. S. Angst, *et al.*, Opioid pharmacotherapy for chronic non-cancer pain in the United States: a research guideline for developing an evidence-base, *J. Pain*, 2010, **11**(9), 807–829.
107. J. S. Mogil, Pain genetics: past, present and future, *Trends Genet.*, 2012, **28**(6), 258–266.
108. H. S. Smith, Opioid metabolism, *Mayo Clin. Proc.*, 2009, **84**(7), 613–624.
109. H. S. Smith, Variations in opioid responsiveness, *Pain Physician*, 2008, **11**, 237–248.
110. A. Roses, Pharmacogenetics and future drug development and delivery, *Lancet*, 2000, **355**, 1358–1361.
111. OxyContin (oxycodone HCl controlled-release tablets), package insert, Purdue Pharma LP, Stamford, CT, 2007.
112. F. E. Murtagh, M. O. Chai, P. Donohoe, P. M. Edmonds and I. J. Higginson, The use of opioid analgesia in end-stage renal disease patients managed without dialysis: recommendations for practice, *J. Pain Palliat. Care Pharmacother.*, 2007, **21**(2), 5–16.
113. A. A. Somogyi, D. T. Barratt and J. K. Coller, Pharmacogenetics of opioids, *Clin. Pharmacol. Ther.*, 2007, **81**, 429–444.
114. J. R. Ross, J. Riley, C. Quigley and K. I. Welsh, Clinical pharmacology and pharmacotherapy of opioid switching in cancer patients, *Oncologist*, 2006, **11**, 765–773.

115. J. R. Ross, D. Rutter, K. Welsh, *et al.*, Clinical response to morphine in cancer patients and genetic variation in candidate genes, *Pharmacogenomics*, 2005, **5**, 324–336.
116. B. A. Jordan and L. A. Devi, G-protein coupled receptor heterodimerization modulates receptor function, *Nature*, 1999, **399**, 697–700.
117. I. Gomes, A. Gupta, J. Filipovska, H. H. Szeto, J. E. Pintar and L. A. Devi, A role for heterodimerization of μ and δ opiate receptors in enhancing morphine analgesia, *Proc. Natl. Acad. Sci. U. S. A.*, 2004, **101**, 5135–5139.
118. K. Elliott, N. Minami, Y. A. Kolesnikov, *et al.*, The NMDA receptor antagonists, LY274614 and MK-801, and the nitric oxide synthase inhibitor, NG-nitro-L-arginine, attenuate analgesic tolerance to the muopioidmorphine but not to kappa opioids, *Pain Physician*, 1994, **56**, 69–75.
119. G. W. Pasternak, Y. A. Kolesnikov and A. M. Babey, Perspectives on the *N*-methyl-D-aspartate nitric oxide cascade and opioid tolerance, *Neuropsychopharmacology*, 1995, **13**, 309–313.
120. K. A. Trujillo and H. Akil, Inhibition of morphine tolerance and dependence by the NMDA receptor anagonist MK-801, *Science*, 1991, **251**, 85–87.
121. S. Ben-Eliyahu, P. Marek, A. L. Vaccarino, *et al.*, The NMDA receptor antagonist MK-801 prevents longlasting non-associative morphine tolerance in the rat, *Brain Res.*, 1992, **575**, 304–308.
122. A. M. Babey, Y. Kolesnikov, J. Cheng, *et al.*, Nitric oxide and opioid tolerance, *Neuropharmacology*, 1994, **33**, 1463.
123. Y. Kolesnikov, C. G. Pick and G. W. Pasternak, NG-nitro-L-arginine prevents morphine tolerance, *Eur. J. Pharmacol.*, 1992, **221**, 339–340.
124. N. Cherny, C. Ripamonti, J. Pereira, *et al.*, Strategies to manage the adverse effects of oral morphine: an evidence-based report, *J. Clin. Oncol.*, 2001, **19**, 2542–2554.
125. C. G. Pick, J. Cheng, D. Paul, *et al.*, Genetic influences in opioid analgesic sensitivity in mice, *Brain Res.*, 1991, **556**, 295–298.
126. A. D. Furlan, J. A Sandoval, A. Mailis-Gagnon and E. Tunks, *CMAJ*, **174**(11), 1589–1594.
127. Opioids Special Issue. *Pain Physician*, 2008, **11**, S181–S200.
128. E. Eisenberg, E. McNicol and D. B. Carr, Opioids for neuropathic pain, *Cochrane Database Syst. Rev.*, 2006, **3**, CD006146.
129. G. P. Kurita, P. Sjøgren, K. Juel, J. Højsted and O Ekholm, The burden of chronic pain: A cross-sectional survey focusing on diseases, immigration, and opioid use, *Pain Physician*, 2012, **153**(12), 2332–2338.
130. A. B. Gustavsson, J. Bjorkman, C. Ljungcrantz, *et al.*, Pharmaceutical treatment patterns for patients with a diagnosis related to chronic pain initiating a slow-release strong opioid treatment in Sweden, *Pain Physician*, 2012, **153**(12), 2325–2331.
131. J. C. Ballantyne, Chronic opioid therapy and its utility in different populations, *Pain*, 2012, **153**(12), 2303–2304.

132. R. Chou, G. J. Fanciullo, P. G. Fine, *et al.*, Clinical guidelines for the use of chronic opioid therapy in chronic noncancer pain, *J. Pain*, 2009, **10**(2), 113.
133. http://ir.zogenix.com/phoenix.zhtml?c=220862&p=irol-newsArticle&id=1597368.
134. C. Quigley, Hydromorphone for acute and chronic pain, *Cochrane Database Syst. Rev.*, 2002, **1**, CD003447.
135. E. Prommer, Oxymorphone: a review, *Support Care Cancer*, 2006, **14**, 109–115.
136. M. A. Lee, M. F. Leng and R. M. Cooper, Measurements of plasma oxycodone, noroxycodone and oxymorphone levels in a patient with bilateral nephrectomy who is undergoing haemodialysis, *Palliat. Med.*, 2005, **19**, 259–260.
137. M. P. Adams and H. Adieh, Single- and multiple-dose pharmacokinetic and dose-proportionality study of oxymorphone immediate-release tablets, *Drugs Res. Dev.*, 2005, **6**, 91–99.
138. Oxymorphone ER *(Opana ER), Prescribing information*, Endo Pharmaceuticals.
139. Oxymorphone IR *(Opana), Prescribing information*, Endo Pharmaceuticals.
140. M. E. Hale, H. Ahdieh, T. Ma and R. Rauck, Efficacy and safety of OPANA ER (oxymorphone extended release) for relief of moderate to severe chronic low back pain in opioid-experienced patients: a 12-week, randomized, double-blind, placebo-controlled study, *J. Pain*, 2007, **8**, 175–184.
141. M. E. Hale, C. Dvergsten and J. Gimbel, Efficacy and safety of oxymorphone extended release in chronic low back pain: results of a randomized, double-blind, placebo- and active-controlled phase III study, *J Pain*, 2005, **6**, 21–28.
142. A. Kivitz, C. Ma, H. Ahdieh and B. S. Galer, A 2-week, multicenter, randomized, double-blind, placebo-controlled, dose-ranging, phase III trial comparing the efficacy of oxymorphone extended release and placebo in adults with pain associated with osteoarthritis of the hip or knee, *Clin. Ther.*, 2006, **28**, 352–364.
143. A. K. Matsumoto, N. Babu and H. Ahdieh, Oxymorphone extended-release tablets relieve moderate to severe pain and improve physicial function in osteoarthritis: results of a randomized, double-blind, placebo- and active-controlled phase III trial, *Pain Med.*, 2005, **6**, 357–366.
144. J. Gimbel and H. Ahdieh, The efficacy and safety of oral immediate-release oxymorphone for postsurgical pain, *Anesth Analg.*, 2004, **99**, 1472–1477.
145. J. S. Gimbel, D. Walker, T. Ma and H. Ahdieh, Efficacy and safety of oxymorphone immediate release for the treatment of mild to moderate pain after ambulatory orthopedic surgery: results of a randomized, double-blind, placebo-controlled trial, *Arch. Phys. Med. Rehab.*, 2005, **86**, 2284–2289.

146. P. Sloan, N. Slatkin and H. Ahdieh, Effectiveness and safety of oral extended-release oxymorphone for the treatment of cancer pain: a pilot study, *Support Care Cancer*, 2005, **13**, 57–65.
147. L. M. Manchikanti, B. Fellows, H. Ailinani and V. Pampati, Therapeutic use, abuse, and nonmedical use of opioids: a ten-year perspective, *Pain Physician*, 2010, **13**, 401–435.
148. M. Palangio, M. J. Damask, E. Morris, R. T. Doyle, J. G. Jiang, C. J. Landau and A. de Padova, Combination hydrocodone and ibuprofen versus combination codeine and acetaminophen for the treatment of chronic pain, *Clin. Therapeut.*, 2000, **22**(7), 879–892.
149. T. M. Ho, J. T Vrabec and A. W. Burton, Hydrocodone use and sensorineural hearing loss, *Pain Physician*, 2007, **10**, 467–472.
150. Oxycontin package insert, Purdue Pharma L.P., Stamford, CT.
151. R. L. Rauck, S. A. Bookbinder, T. R. Bunker and C. D. Alftine, A randomized, open-label, multicenter trial comparing once-aday Avinza (morphine sulfate extended- release capsules) versus twice-a day OxyContin (oxycodone hydrochloride controlled release tablets) for the treatment of chronic, moderate to severe low back pain: Improved physical functioning in the ACTION trial, *J. Opioid Manage.*, 2007, **3**, 35–43.
152. J. A. Hermos, M. M. Young, D. R. Gagnon and L. D. Fiore, Characterizations of long-term oxycodone/acetaminophen prescriptions in veteran patients, *Arch. Intern. Med.*, 2004, **164**(2), 361–366.
153. R. K. Portenoy, J. T. Farrar, M. M. Backonja, C. S. Cleeland, K. Yang, M. Friedman, S. V. Coluci and P. Richards, Long-term use of controlled-release oxycodone for noncancer pain: Results of a 3-year registry study, *Clin. J. Pain*, 2007, **23**, 287–299.
154. S. H. Roth, R. M. Fleischmann, F. X. Burch, F. Dietz, B. Bockow, R. J. Rapoport, J. Rutstein and P. G. Lacouture, Around-the-clock, controlled- release oxycodone therapy for osteoarthritis-related pain: Placebo controlled trial and long-term evaluation, *Arch. Intern. Med.*, 2000, **160**, 853–860.
155. T. W. Vanderah, Delta and kappa opioid receptors as suitable drug targets for pain, *Clin. J. Pain*, 2010, **26**(Suppl. 10), S10–S15.
156. C. M. Cahill, S. V. Holdridge and A. Morinville, Trafficking of delta-opioid receptors and other G-protein-coupled receptors: implications for pain and analgesia, *Trends Pharmacol. Sci.*, 2007, **28**, 23–31.
157. N. A. C. Kabli and C. M. Cahill, Anti-allodynic effects of peripheral delta opioid receptors in neuropathic pain, *Pain*, 2007, **127**, 84–93.
158. B. Bie and Z. Z. Pan, Trafficking of central opioid receptors and descending pain inhibition, *Mol. Pain*, 2007, **3**, 37.
159. C. L. Schramm and C. N. Honda, Co-administration of delta and mu-opioid receptor agonists promotes peripheral opioid receptor function, *Pain*, 2010, **151**, 763–770.
160. M. Benbouzid, C. Gavériaux-Ruff, I. Yalcin, *et al.*, Delta-opioid receptors are critical for tricyclic antidepressant treatment of neuropathic allodynia, *Biol. Psychiatr.*, 2008, **63**, 633–636.

161. H. Nagase and H. Fujii, Hideaki opioids in preclinical and clinical trials, *Top. Curr. Chem.*, 2011, **299**, 29–62.
162. A. A. Pradhan, K. Befort, C. Nozaki, C. Gavériaux-Ruff and B. L. Kieffer, The delta opioid receptor: an evolving target for the treatment of brain disorders, *Trends Pharmacol. Sci.*, 2011, **32**(10), 581–590.
163. M. Piercey, RA Lahti, LA Schroeder, FJ Einspahr and C Barsuhn, *Life Sci.*, 1982, **31**, 1197.
164. H. Kumagai, T. Ebata, K. Takamori, T. Muramatsu, H. Nakamoto and H. Suzuki, Effect of a novel kappa-receptor agonist, nalfurafine hydrochloride, on severe itch in 337 haemodialysis patients: a Phase III, randomized, double-blind, placebo-controlled study, *Nephrol. Dial. Transplant.*, 2010, **4**, 1251–1257.
165. K. Nakao and H. Mochizuki, Nalfurafine hydrochloride: a new drug for the treatment of uremic pruritus in hemodialysis patients, *Drugs Today (Barc).*, 2009, **45**(5), 323–329.
166. K. Nakao, K. Hasebe, S. Yoshikawa, *et al.*, [Pharmacological effects of nalfurafine hydrochloride, a kappa-opioid receptor agonist.], *Nihon Shinkei Seishin Yakurigaku Zasshi*, 2010, **5–6**, 185–191.
167. M. J. Krantz, L. Lewkowiez, H. Hays, *et al.*, Torsade de pointes associated with very-high-dose methadone, *Ann. Intern. Med.*, 2002, **137**, 501–504.
168. M. C. Rowbotham, L. Twilling, P. S. Davies, L. Reisner, K. Taylor and D. Mohr, Oral opioid therapy for chronic peripheral and central neuropathic pain, *N. Engl. J. Med.*, 2003, **348**, 1223–1232.
169. K. M. Foley, Opioids and chronic neuropathic pain, *N. Engl. J. Med.*, 2003, **348**, 1279–1281.
170. P. R. Hutson and K. J. Williams, Methadone, buprenorphine and levorphanol: alternatives for treatment of chronic pain a comparison of the unique properties of each drug and clinical advantages and disadvantages of their use. *JPSW*, 2009, Jan/Feb.
171. J. Schofferman and D. Mazanec, Evidence-informed management of chronic low back pain with opioid analgesics, *Spine J*, 2008, **8**, 185–194.
172. F. S. Labella, C. Pinsky and V. Havlicek, Morphine derivatives with diminished opiate receptor potency show enhanced central excitatory activity, *Brain Res.*, 1979, **174**, 263–271.
173. J. McNulty, Can levorphanol be used like methadone for intractable refractory pain?, *J. Palliat. Med.*, 2007, **10**(2), 293–296.
174. Ultram (package insert), Ortho-McNeil Pharmaceutical, Inc., Raritan, NJ, 1998.
175. Ultracet (package insert), Ortho-McNeil Pharmaceutical, Inc., Raritan, NJ.
176. J. S. Gardner, D. Blough, C. R. Drinkard, *et al.*, Tramadol and seizures: a surveillance study in a managed care population, *Pharmacotherapy*, 2000, **20**, 1423–1431.
177. L. Moreland, E. W. St. Clair, G. Biasi, S. Manca, S. Manganelli, *et al.*, The use of analgesics in the management of pain in rheumatic diseases, *Rheum. Dis. Clin. North. Am.*, 1999, **25**, 153–159.

178. J. Hollingshead, R. M. Dühmke and D. R. Cornblath, Tramadol for neuropathic pain, *Cochrane Database Syst Rev.*, 2006, **3**, CD003726.
179. M. S. Cepeda, F. Camargo, C. Zea and L. Valencia, Tramadol for osteoarthritis, *Cochrane Database Syst. Rev.*, 2006, **3**, CD005522.
180. R. A. Medve, J. Wang and R. Karim, Tramadol and acetaminophen tablets for dental pain, *Anesth. Progr.*, 2001, **48**, 79–81.
181. JC Silverfield, K. M., N Rosenthal, *et al.*, Efficacy and safetyof combination tramadol and acetaminophen tablets (Ultracet™) for the treatment of pain associated with osteoarthritis flare: a multicenter, outpatient, randomized, double- blind, placebo-controlled study, in *American College of Rheumatology 64th Annual Scientific Meeting*, November 2001, San Francisco, CA.
182. D. G. Borenstein, M. Kamin, N. Rosenthal, *et al.*, Combination tramadol and acetaminophen (Ultracet™) for the treatment of chronic lower back pain: a multicenter, outpatient, randomized, double-blind, placebo-controlled study, in *American College of Rheumatology 64th Annual Scientific Meeting*, November 2001, San Francisco, CA.
183. W. S. Mullican, J. R Lacy, *et al.*, Tramadol/acetaminophen combination tablets and codeine/acetaminophen combination capsules for the management of chronic pain: a comparative trial, *Clin. Ther.*, 2001, **23**, 1429–1445.
184. R. Bennett, M. Kamin, N Rosenthal, *et al.*, Efficacy and safety of a tramadol/acetaminophen combination (Ultracet™) in the management of fibromyalgia related pain: a multicenter, outpatient, randomized, double-blind, placebo-controlled study, in *American College of Rheumatology 64th Annual Scientific Meeting*, November 2001, San Francisco, CA.
185. L. Alwine, Long-term (2-year) analgesic efficacy of tramadol/ acetaminophen tablets. Poster 301, in *European League Against Rheumatism (EULAR) Congress*, 2000, Nice, France.
186. U.S. Food and Drug Administration, Center for Drug Evaluation and Research, *Review and Evaluation of Clinical Data. Tramadol Hydrochloride (37.5 mg)/Acetaminophen Tablets (325 mg)*, www.fda.gov/cder/foi/nda/2001/21123_Ultracet.htm, cited 6 December 2001.
187. A. Rosenblum, L. A. Marsch, H. Joseph, *et al.*, Opioids and the treatment of chronic pain: controversies, current status, and future directions, *Exp. Clin. Psychopharmacol.*, 2008, **16**(5), 405–416.
188. L. Hays, K. L. Kirsh and S. D. Passik, Seeking drug treatment for OxyContin abuse: a chart review of consecutive admissions to a sustance abuse treatment facility in Kentucky, *J. Natl. Compr. Canc. Netw.*, 2003, **1**, 423–428.
189. S. E. McCabe, J. A. Cranford, C. J. Boyd and C. J. Teter, Motives, diversion and routes of administration associated with nonmedical use of prescription opioids, *Addict Behav.*, 2007, **32**, 562–575.
190. W. W. Stoops, K. W. Hatton, M. R. Lofwall, P. A. Nuzzo and S. L. Walsh, *Intravenous oxycodone, hydrocodone, and morphine in*

recreational opioid users: abuse potential and relative potencies. Psychopharmacology (Berl.), 2010, **212**, 193–203.

191. J. R. Havens, R. Walker and C. G. Leukefeld, Prevalence of opioid analgesic injection among rural nonmedical opioid analgesic users, *Drug Alcohol Depend.*, 2007, **87**, 98–102.
192. Centers for Disease Control and Prevention, QuickStats: number of poisoning deaths involving opioid analgesics and other drugs or substances—United States, 1999–2007, *Morb. Mortal. Wkly. Rep. (MMWR)*, 2010, **59**, 1026.
193. M. Warner, L. H. Chen and D. M. Makuc, Increase in fatal poisonings involving opioid analgesics in the United States, 1999–2006, in: *National Center for Health Statistics (NCHS) Data Brief*, National Center for Health Statistics, Hyattsville, MD, 2009.
194. L. J. Paulozzi, R. H. Weisler and A. A. Patkar, A national epidemic of unintentional prescription opioid overdose deaths: how physicians can help control it, *J. Clin. Psychiatr.*, 2011, **72**, 589–592.
195. L. J. Paulozzi, Opioid analgesic involvement in drug abuse deaths in American metropolitan areas, *Am. J. Publ. Health*, 2006, **96**, 1755–1757.
196. Drug Abuse Warning Network, *The DAWN Report: Trends in Emergency Department Visits Involving Nonmedical Use of Narcotic Pain Relievers*, Office of Applied Studies, Substance Abuse and Mental Health Services Administration, Rockville, MD, 2010.
197. U.S. Food and Drug Administration, *Extended-Release (ER) and Long-Acting (LA) Opioid Analgesics Risk Evaluation and Mitigation Strategy (REMS)*. Initial REMS Approval. 08/2012; 07/2012.
198. C. Baum, J. Hsu and R. C. Nelson, The impact of the addition of naloxone on the use and abuse of pentazocine, *Publ. Health Rep.*, 1987, **102**(4), 429.
199. I. H. Benedek, J. Jobes, Q. Xiang and W. D. Fiske, Bioequivalence of oxymorphone extended release and crush-resistant oxymorphone extended release, *Drug. Des. Devel. Ther.*, 2011, **5**, 455–463.
200. S. F. Butler, H. Chilcoat, T. A. Cassidy, R. A. Black, C. Landau, S. H. Budman and P. Coplan, Initial findings on abuse rates and routes of administration among individuals assessed for substance use treatment following introduction of reformulated OxyContin®, in *APS annual meeting*.
201. T. J. Cicero and M.S. Ellis, Letter to the editor., *N. Engl. J. Med.*, 2012, **367**(2), 187–189.
202. M. Schiller, K. A. Johnston, A. Bush and S. Saim,Tamper-resistant properties of a novel,multiparticulate,extended-release formulation of oxycodone for the treatment of moderate to severe pain, in *American Academy of Pain Medicine 26th Annual Meeting*, San Antonio, TX.
203. E. A. P. Kopecky, R. K. Varanasi, A. B. Fleming, Said Saim and S. P. Mayock, Alternative methods of oxycodone DETERx administration, in *American Academy of Pain Medicine 26th Annual Meeting*, San Antonio, TX.

204. M. C. Zamloot, W. Ling-Ling Kang, J. Ross and R. Fu, Remoxy®: A novel formulation of extended release oxycodone developed using the ORADUR technology, *J. Appl. Res.*, 2010, **10**, 3.
205. N. Friedmann, V. Klutzaritz and L. Webster, Efficacy and safety of an extended-release oxycodone (Remoxy) formulation in patients with moderate to severe osteoarthritic pain, *J. Opioid Manage.*, 2011, **7**(3), 193–202.
206. A.-M. Haahr, L. Lyhne-Iversen, S. Ridderberg, M. Nilsson and P. Hemmingsen, Drug abuse resistant,controlled release,using Egalet® dosage units. Poster no 680, in *Proceedings of the 34th Annual Meeting and Exposition of the Controlled Release Society*, Curran Associates, Inc., New York.
207. D. Ridgway, M. Sopata, A. Burneckis, L. Jespersen and C. Andersen, Clinical, efficacy and safety of once-daily dosing of a novel,prolonged-release oral morphine tablet compared with twice-daily dosing of a standard controlled-release morphine tablet in patients with cancer pain: a randomized, double-blind, exploratory crossover study, *J. Pain Symptom Manage.*, 2010, **39**(4), 712–720.
208. N. Katz, Abuse-deterrent opioid formulations: are they a pipe dream?, *Curr. Rheumatol. Rep.*, 2008, **10**, 11–18.
209. T. Kenakin, Functional selectivity through protean and biased agonism: who steers the ship?, *Mol. Pharmacol.*, 2007, **72**(6), 1393–1401.
210. J. L Whistler. and M. von Zastrow, Morphine-activated opioid receptors elude desensitization by β-arrestin, *Proc. Natl. Acad. Sci. U. S. A.*, 1998, **95**(17), 9914–9919.
211. H. Zheng, *et al.*, Modulating μ-opioid receptor phosphorylation switches agonist-dependent signaling as reflected in PKCϵ activation and dendritic spine stability, *J. Biol. Chem.*, 2011, **286**(14), 12724–12733.
212. S. Rajagopal, *et al.*, Quantifying ligand bias at seven-transmembrane receptors, *Mol. Pharmacol.*, 2011, **80**(3), 367–377.
213. *Trevena Completes FTIH Study of TRV130, an Injectable Mu-Opioid Biased Ligand for Acute Pain.* www. trevenainc. com/ news- details. php?id = 44. Accessed 13 December 2012.
214. D. Yamashita, D. Gotchev, P. Pitis, X-T. Chen, G. Liu and C. C. K. Yuan, Opioid receptor ligands and methods of using and making same. *United States Patent Application* 20120245181, 2012.
215. H. Mizoguchi, *et al.*, New vistas in opioid control of pain, *Curr. Opin. Pharmacol.*, 2012, **12**(1), 87–91.
216. J. B. Wang, *et al.*, mu opiate receptor: cDNA cloning and expression, *Proc. Natl. Acad. Sci. U. S. A.*, 1993, **90**(21), 10230–10234.
217. S. Kasai and K. Ikeda, Pharmacogenomics of the human μ-opioid receptor, *Pharmacogenomics*, 2011, **12**(9), 1305–1320.
218. R. M. van Rijn, J. L. Whistler and M. Waldhoer, Opioid-receptor-heteromer-specific trafficking and pharmacology, *Curr. Opin. Pharmacol.*, 2010, **10**(1), 73–79.

219. L. Diego, *et al.*, Novel opioid antagonists for opioid-induced bowel dysfunction, *Expert Opin. Invest. Drugs*, 2011, **20**(8), 1047–1056.
220. Nektar. NKTR-181, 2012, www. nektar. com/ product_pipeline/ cns_pain_nktr-181.html. Accessed 18 December 2012.
221. Nektar. NKTR-118, 2012, www. nektar. com/ product_pipeline/ cns_pain_oral_naloxegol_nktr118.html. Accessed 18 December 2012.
222. M. S. Wallace, Dwight Moulin, A. J. M. Clark, R. Wasserman, A. Neale, P. Morley-Forster, J-P. Castaigne and S. Teichman, A phase II, multicenter, randomized, double-blind, placebo-controlled crossover study of CJC-1008 a long acting, parenteral opioid analgesic in the treatment of postherpetic neuralgia. *J. Opioid Manage.*, 2006, **2**(3), 167–173.
223. K. A. Witt and Thomas P. Davis, CNS drug delivery: opioids and the blood-brain barrier, *AAPS J.*, 2006, **8**(1), E76–E88.
224. L. Arendt-Nielsen, *et al.*, Analgesic efficacy of peripheral κ-opioid receptor agonist CR665 compared to oxycodone in a multi-modal, multi-tissue experimental human pain model: selective effect on visceral pain, *Anesthesiology*, 2009, **111**(3), 616–624.
225. L. R. Roberts, *et al.*, Kappa agonist CovX-bodies, *Bioorg. Med. Chem. Lett.*, 2012, **22**(12), 4173–4178.
226. M. R. Hutchinson, *et al.*, Opioid activation of Toll-like receptor 4 contributes to drug reinforcement, *J. Neurosci.*, 2012, **32**(33), 11187–11200.

CHAPTER 3
Anti-Inflammatory Drugs

ARIAMALA GOPALSAMY* AND SUVIT THAISRIVONGS

Worldwide Medicinal Chemistry, Pfizer, Inc., 200 Cambridge Park Drive, Cambridge, MA 02140, USA
*Email: Ariamala.gopalsamy@pfizer.com

3.1 Introduction

Chronic pain disorder is a high unmet medical need. Pain differs in underlying causes, symptoms, and mechanisms. Inflammatory pain originates from all forms of inflammation.[1] Combating inflammation[2] slows the degenerative process (for example, of joint diseases) and the associated pain. Rheumatoid arthritis (RA), psoriasis, systemic lupus erythematosus (SLE), inflammatory bowel disease (IBD), Crohn's disease (CD), ankylosing spondylitis (AS), Multiple Sclerosis (MS) and chronic obstructive pulmonary disease (COPD) are some of the chronic inflammatory autoimmune diseases. Some of the current treatment strategies[3] for autoimmune diseases can be classified into various categories:

- corticosteroids
- non-steroidal anti-inflammatory drugs (NSAIDs)
- selective COX-2 inhibitors
- disease-modifying anti-rheumatic drugs (DMARDs)
- anti-cytokine therapies
- inhibition of intracellular signalling pathways
- T-cell inhibition
- B-cell depletion

RSC Drug Discovery Series No. 35
Pain Therapeutics: Current and Future Treatment Paradigms
Edited by Charlotte Allerton

Published by the Royal Society of Chemistry, www.rsc.org

In this chapter, we will exemplify our discussion on anti-inflammatory therapy[4] with RA since it represents the best model to summarise the rapid and constant progress in the field.

RA[5] is a debilitating, systemic inflammatory autoimmune disease associated with progressive joint destruction, significantly compromised quality of life and reduced survival. It commonly affects the 40–70 year old age group. RA is three times more common in females than males. It is characterised by chronic and generally symmetric joint inflammation, often involving the wrists and the small joints of the hands and feet. Persistent joint inflammation is associated with progressive joint destruction manifested by cartilage loss and joint space narrowing, erosive damage to juxta-articular bone, and resultant functional impairment. Tendon and bursal involvement are frequent and often clinically dominant in early disease. Synovitis involving the metacarpophalangeal, proximal interphalangeal and wrist joints causes a characteristic tender swelling on palpation with early severe motion impairment and no radiological evidence of bone damage. Advanced disease state leads to uncontrolled inflammation and destruction of bones and cartilages. In addition, substantial comorbidity is seen outside of the musculoskeletal system, most prominently cardiovascular and dyslipidaemic effects, and a propensity for infection is likely due in part to the immunological disturbance of the disease itself and the need for treatment with immunomodulatory agents. Infiltration of activated T- and B-lymphocytes, macrophages, mast cells and dendritic cells into the synovial tissue are central to the pathogenesis of RA. Overproduction of pro-inflammatory cytokines TNF-α, IL-1 and IL-6 has been identified in rheumatoid synovial tissues.

Treatment for RA[6] has both short-term and long-term goals. The main short-term goal is pain relief and reduction of joint swelling. The long-term goal is to stop or slow down the progression of joint damage and improve the joint function. In the long run, this goal can accomplish the short-term goal as well, since the control of inflammation can provide relief from associated pain. Listed in Table 3.1 are the currently approved small-molecule drugs for RA.

Apart from the traditional small-molecule DMARDs, a number of biologics have entered the market in the past decade.[7] These biologic agents are targeted therapies, thus more effective, and have become part of standard therapy.

Table 3.1 Approved RA drugs: small molecules.

Name	*RA drug class*
Sulfasalazine	DMARDs
Methotrexate	DMARDs
Hydroxychloroquine	DMARDs
Ciclosporin	DMARDs
Penicillamine	DMARDs
Cyclophosphamide	DMARDs
Leflunomide	DMARDs
Ibuprofen, naproxen	NSAIDs
Celecoxib	NSAIDs

Table 3.2 Approved RA drugs: biologics.

Name	*Mechanism of action*	*Company*	*Year*
Etanercept (Enbrel)	TNF Inhibitor	Amgen/Pfizer	1998
Infliximab (Remicade)	TNF Inhibitor	Centocor	1999
Adalimumab (Humira)	TNF Inhibitor	Abbott	2002
Anakinra (Kinaret)	IL-1 Inhibitor	Amgen	2001
Abatacept (Orencia)	T-cell co-stimulatory inhibitor	Bristol Myers Squibb	2005
Certolizumab (Cimzia)	TNF Inhibitor	UCB Inc.	2009
Golimumab (Symponi)	TNF Inhibitor	Centocor	2009
Rituximab (Rituxan)	Anti-CD20	Roche/Biogen	2006
Tocilizumab (Actemra)	IL-6 Inhibitor	Roche	2010

They target the underlying cause of the inflammation. Most of the biologics target the cytokine signalling pathways and T- and B-cell surface antigens, and are neutralising antibodies. For example, biologics targeting the pro-inflammatory cytokine TNF-α have revolutionised the treatment of diverse inflammatory disorders, such as RA, psoriasis and CD. These agents have made a significant impact on treatment responses in RA. Clinical responses to these agents as measured by the American College of Rheumatology 20% improvement criteria (ACR20), ACR50 and ACR70 response rates at 24 weeks were 50–70%, 20–40% and 10–20%, respectively, across clinical trials. The predominant risk associated with these agents is that of increased frequency of serious infections, not only bacterial but also viral, mycobacterial, and fungal. Increased incidences of lymphoma and cases of other malignancies also have been reported. It is estimated that ~30–40% of RA patients do not respond to anti-TNF-α therapies. This has prompted the search for alternate immuno-modulatory therapy such as antibody to antigen CD-20 and anti-IL-6 receptor antibody. In spite of these advances, biologics need to be injected frequently, are associated with high production costs and are not necessarily safer than small molecules. Listed in Table 3.2 are the currently approved biologics for treatment of RA.

3.2 Corticosteroids

Glucocorticoids (GCs) and mineralocorticoids, hormones produced by the adrenal cortex, are referred as corticosteroids. GCs that function by binding to glucocorticoid receptors (GR) are steroid hormones synthesised from cholesterol. GCs play an important role in the homeostatic function of the human immune system.[8] GCs inhibit many early-phase inflammatory responses including the expression of pro-inflammatory cytokines and chemokines that regulate recruitment and activation of leukocytes. They reduce the expression of many pro-inflammatory effectors in macrophages, natural killer (NK) cells, dendritic (DC) cells, neutrophils, B and T lymphocytes. These anti-inflammatory properties[9] are responsible for the therapeutic use of GCs which revolutionised the treatment of RA in the early 1900s. For over 50 years,

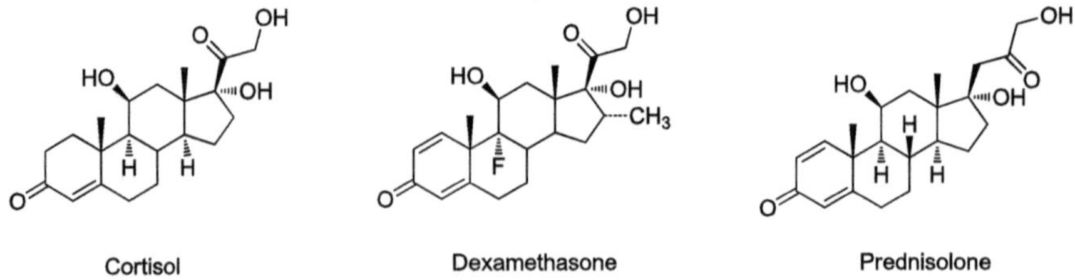

Figure 3.1 Glucocorticoids: cortisol and some synthetic analogues.

cortisol (hydroxycortisone), its oxidised product cortisone and their many synthetic analogues were the mainstay of second-line therapy for a number of inflammatory disorders. Inhaled GCs are the second-line treatment for asthma. Figure 3.1 shows the structure of cortisol and some of the synthetic analogues with improved pharmcokinetic/pharmacodynamic (PK/PD) profile.

3.2.1 Mechanism of Action

Corticosteroids regulate gene expression both positively and negatively. Cytosolic GR is activated upon binding to GCs. The resulting complex relocates to the nucleus and binds to the GC response element of the gene target. This result in regulation of gene expression and the process is called transactivation. The counter-mechanism is called transrepression. The activated receptor complex interacts with specific transcription factors such as NF-κB[10] and prevents the transcription of target genes. Hence GCs are capable of preventing the transcription of pro-inflammatory genes. Figure 3.2 describes the mechanisms of cellular actions of GCs.[11]

3.2.2 Resistance and Side Effects

A significant patient population can have severe resistance to glucocortiocids.[12] The resistance can be a result of the nature of the allergen (in the case of asthma), genetic predisposition or inflammatory pathways that bypass GCs, or variance in pharmacokinetics. Suppressing steroid biogenesis in adrenal cortex increases dependence upon exogenous steroids, reflecting the symptoms of cortisol excess seen in Cushing's syndrome. Systemic use of GCs for anti-inflammatory therapy is limited due to their side effects.[13] Hyperglycemia and insulin resistance due to increased gluconeogenesis, steroid-induced osteoporosis, accelerating atherosclerosis, hypertension, muscle wasting, skin thinning, glaucoma, and ophthalmic lesions are some of the side effects attributed to high doses or prolonged use of anti-inflammatory steroids.

3.2.3 Dissociated Ligand of GC Receptor

Regarding the mechanism of action of GCs, it is hypothesized that the anti-inflammatory effects of GCs are mostly due to the inhibition of the

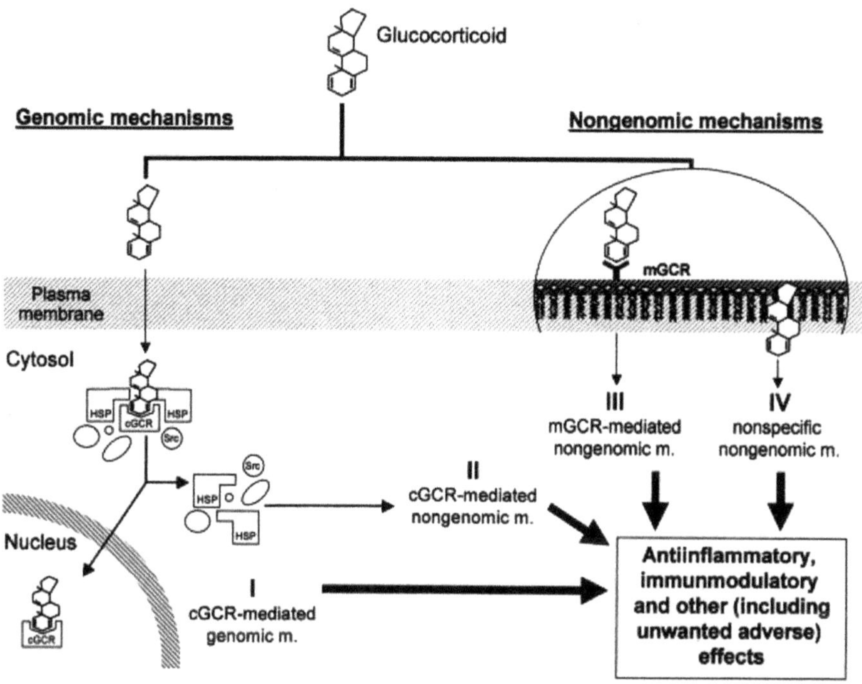

Figure 3.2 Mechanism of cellular actions of glucocorticoids (GCs). As lipophilic substances, GCs pass very easily through the cell membrane into the cell, where they bind to ubiquitously expressed cytosolic GC receptors (cGCR). This is followed by either the classic cGCR-mediated genomic effects (I) or by cGCR-mediated nongenomic effects (II). Moreover, the GC is very likely to interact with cell membranes either specifically, *via* membrane bound GC receptors (mGCR) (III), or *via* non-specific interactions with cell membranes (IV). HSP, heat-shock protein; m, mechanism.
(Reproduced from Ref. 11).

transcription of genes (transrepression) while the transactivation or activation of transcription is responsible for side effects.[14] This paves the way for uncoupling these two processes by identifying selective ligands for GRs that inhibit rather than activate gene transcription. This effort will provide a mean of uncoupling the anti-inflammatory effects from unwanted side effects. The resulting 'dissociated ligand of GR'[15] should be able to differentiate itself in the marketplace by its fewer side effects.

The first reported examples of such well-characterised dissociated ligands were steroid derivatives such as RU-24858. However, the translation of *in vitro* dissociation to *in vivo* was questionable, partially attributed to metabolism. A flurry of activity[16] by various pharmaceutical companies in early 2000 to test the transrepression/transactivation hypothesis resulted in a number of non-steroidal ligands,[17] as shown in Figure 3.3. Considering the large number of biological activities associated with GR, this strategy for drug development is

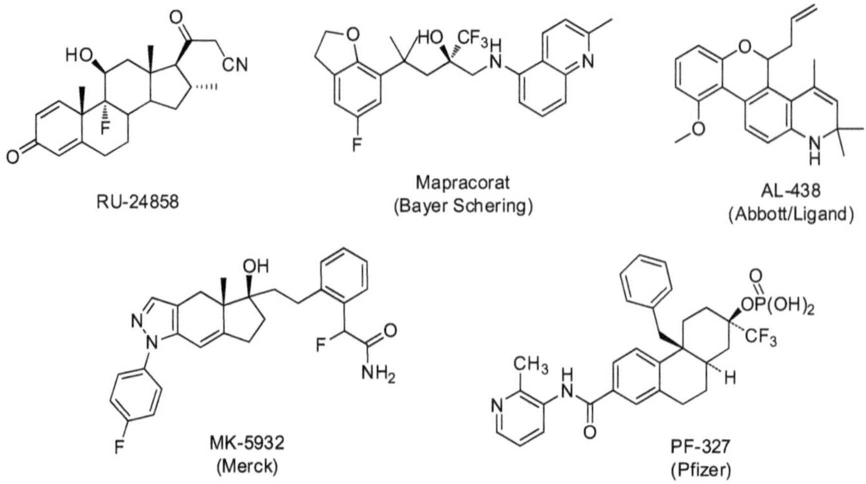

Figure 3.3 Examples of dissociated ligand of glucocorticoid receptors.

being experimented with cautiously in the industry. Nevertheless, over the years, many activities of GRs have been identified to be functionally distinct and any extent of dissociation should provide improved GR-based therapeutics.

To that effect, from single-dose studies of PF-327 in healthy volunteers, dose-dependent biomarker effects consistent with GR-mediated pharmacology were observed, including cortisol suppression, osteocalcin suppression, lymphopenia, eosinopenia, and neutophilia.[18] The data suggested that PF-327 modulated a biomarker of adverse bone effects differentially compared to biomarkers of anti-inflammatory activity when compared to prednisolone. The relative potency of PF-327 for osteocalcin suppression (biomarker for side effect) *versus* effects on other efficacy biomarkers (cortisol suppression, neutrophilia) was lower compared to prednisolone. This observed dissociation between adverse GC-mediated effects and the anti-inflammatory activity of PF-327 remains to be further characterised in more detail in additional long-term studies in patients.

3.3 Disease-Modifying Anti-Rheumatic Drugs

NSAIDs and COX-2 inhibitors possess analgesic effects apart from their anti-inflammatory effects. Hence they are primarily used for symptomatic pain relief. On the other hand, DMARDs—a broad and diverse set of therapeutic agents (Figure 3.4)—slow the progression of the disease and have an impact on the signs and symptoms of RA. Older DMARDs work by modifying the host immune response to inflammation. Use of DMARDs to treat RA can control synovitis and may slow, or even stop, radiographic progression, improve function and quality of life and normalise mortality rates. There are two classes of DMARDs: conventional small-molecule drugs and biologic agents. Because

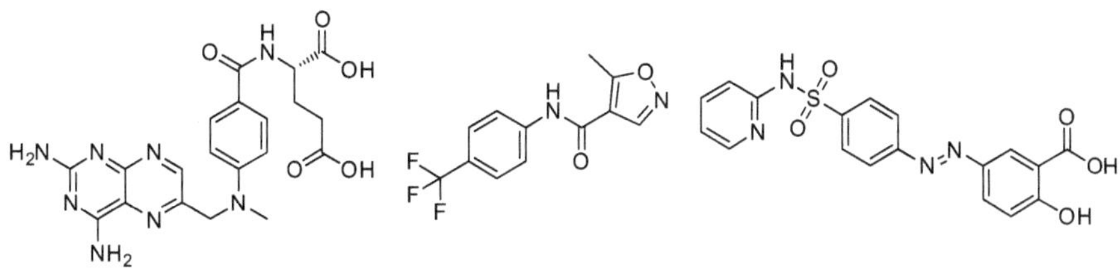

Figure 3.4 Representative examples of conventional DMARDs.

of their potency in modifying the course of disease, different DMARDs may be classified as mild, moderate or strong. Starting with symptomatic therapy and escalating to milder and thereafter stronger drugs (therapeutic pyramid) was favoured in the 1970s. The longer active disease persists, the less likely the patient is to respond to therapy. Most of the DMARDs are quite potent and the strategy with respect to start and change of the treatment paradigm may be even more important than the choice of the drug itself. Nowadays potent DMARD therapy with tight control, targeting remission, is the standard. De-escalation follows after achieving disease control.

3.3.1 Methotrexate

The most commonly used DMARD is methotrexate (MTX), because of its proven efficacy and long-term safety profile. Although MTX emerged at the same time as corticosteroids, initially it received little attention for treatment of rheumatic diseases. MTX is a potent anti-inflammatory and immunosuppressant agent[19] that acts by decreasing cell proliferation, increasing adenosine release.[20] It is a structural analogue of folic acid which competitively inhibits the binding of dihydrofolic acid to the enzyme dihydrofolate reductase (DHFR). It also modifies the expression of cellular adhesion molecules, alters production of cytokines, and has effects on humoral responses and on bone formation and deposition. Although the cellular action of MTX is understood, the mechanism by which it improves the signs and symptoms of RA is not well understood. MTX is the anchor DMARD for the treatment of RA and other types of inflammatory arthritis (psoriatic arthritis, juvenile idiopathic arthritis, *etc.*) because of its efficacy in decreasing articular inflammation and preventing joint damage. Since its use became widespread in the 1980s, MTX has dramatically improved RA outcomes. In the initial stage of the disease, MTX is prescribed as a standalone therapy. In the later stages of disease progress it is commonly combined with biologics to increase the clinical effectiveness. Due to the complexity of RA pathogenesis and the heterogeneity of disease manifestations and severity, there is substantial variability in how patients respond to each DMARD. Approximately 30–40% of patients do not have a good response to MTX despite optimal dosing regimens, and in some patients its use is associated with clinically significant side effects such as cytopenias, liver function test abnormalities, and rarely lymphoma, and other serious conditions.

3.3.2 Other DMARDs

Some of the other conventional small-molecule DMARDs that are commonly used in RA therapy are leflunomide (LEF), antimalarials, sulfasalazine (SSZ), cyclosporine, D-pencillamine, minocycline, doxycycline and azathioprine. LEF inhibits pyrimidine synthesis pathways. This oral drug is converted to its active form, a malononitrilamide known as teriflunomide, upon absorption. LEF is often used in patients intolerant to MTX or in combination with MTX in patients who have an inadequate response to MTX. It has a comparable side-effect profile. Hydroxychloroquine and chloroquine are used to treat rheumatic and connective tissue diseases, in particular in patients with early, mildly active RA. SSZ, although the mechanism is not fully defined, affects inflammation by increased release of adenosine and weakly inhibiting the arachidonic acid pathways. It is generally considered less potent than MTX or LEF and cannot be used in sulfa-allergic or aspirin-allergic patients.

3.3.3 Combination Therapies

MTX is still considered the anchor drug among the DMARDs. However, the efficacy of MTX to improve signs and symptoms of disease varies among individual patients. The ultimate therapeutic goal in treatment of RA is remission or at least low disease activity and this goal is not achieved in all patients with MTX monotherapy. Combination therapy of anti-inflammatory agents has become widely accepted in various therapeutic indications similar to those combination therapy practices in the treatment of viral/bacterial infections, cancer and hypertension. In the past two decades the use of drug combinations based on MTX for RA treatment has steadily increased. It maintains a significant role in the treatment of anti-rheumatic therapy in combination with other conventional DMARDs or biological DMARDs.

3.4 Non-Steroidal Anti-Inflammatory Drugs

NSAIDs possess analgesic effects apart from their anti-inflammatory effects. Hence they are primarily used for symptomatic pain relief.[21] They are some of the most widely used prescription and non-prescription analgesics worldwide. Early NSAIDs were identified by direct human testing without the benefit of pharmacology or toxicology. They are a chemically heterogeneous group of compounds which shares the same mechanism of action. Some of the commonly used NSAIDs and their structural classification are shown in Figure 3.5.

3.4.1 NSAIDs—Mechanism of Action

NSAIDs inhibit production of prostaglandins (PGs)[22] from polyunsaturated fatty acids by inhibiting the first enzyme in the biosynthetic pathway, cyclo-oxygenase (COX). COX catalyses double dioxygenation of arachidonic acid to

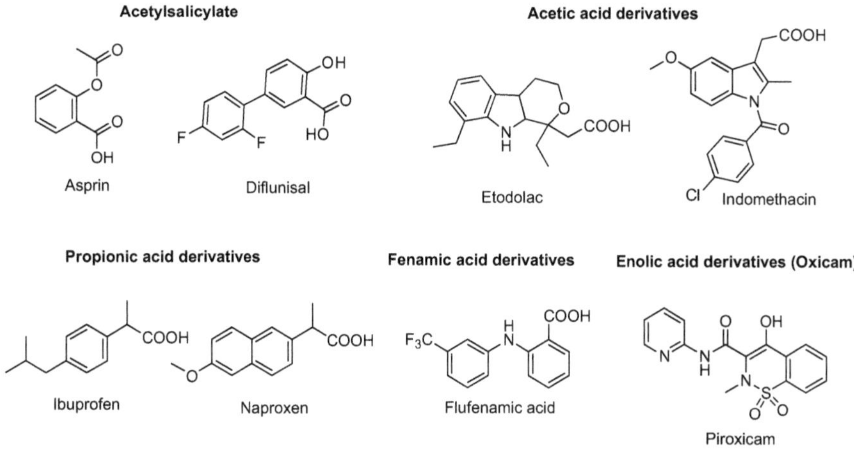

Figure 3.5 Examples of common NSAIDs and their structural classification.

Table 3.3 *In vitro* inhibition of COX-1 and COX-2 by NSAIDs.

NSAIDs	*COX-1 IC50 (μM)*	*COX-2 IC50 (μM)*
Indomethacin	0.1	1
Naproxen	32	93
Ibuprofen	38	17
Asprin	145	180
Diclofenac	0.03	0.01
Etodolac	>100	54
Nabumetone (6-MNA)	82	>1000
Piroxicam	679	662
Meloxicam	1300	1700

prostaglandin endoperoxides PGG2 and PGH2 and hence is also called prostaglandin G/H synthase. NSAIDs bind to the active site of cyclooxygenase and prevent arachidonic acid binding. Prostanoids prostaglandin E2 (PGE2), PGF2α, PGD2, thromboxane A2 (TXA2) and prostacyclin (PGI2) are biologically active derivatives of arachidonic acid and are lipid mediators that serve pivotal functions in inflammation and pain. Inhibition of prostanoid production is the major analgesic mechanism of NSAIDs, but also the source of their adverse side effects. Most NSAIDs are non-selective inhibitors of enzymes cyclooxygenase 1 (COX-1) and cyclooxygenase 2 (COX-2). Table 3.3 shows the non-selective inhibition of cyclooxygenase by NSAIDs *in vitro*.

3.4.2 NSAIDs—Adverse Effect

Long-term treatment with traditional NSAIDs is associated with significant risk of gastrointestinal complications such as bleeding and perforation, with a relative risk 5–6-fold higher than in individuals not treated with NSAIDs.

3.4.3 Selective COX-2 Inhibitors (Coxibs)

COX-1, which has the structural features of a housekeeping gene, is constitutively expressed in all tissues and is responsible for some of the physiological functions such as gastrointestinal cytoprotection, platelet aggregation and vascular smooth muscle tone modulation. In contrast, COX-2 is an inducible enzyme and is expressed in response to IL-1β and other inflammatory stimuli at macrophages, synoviocytes and endothelial cells. Selective inhibitors of COX-2 (Figure 3.6), also known as coxibs,[23] spare the gastrointestinal tract, while they exert anti-inflammatory and analgesic effects. NSAIDs inhibit COX by two enzymatic mechanisms: (1) a competitive and reversible mechanism and (2) a time-dependent (pseudo-reversible) mechanism. Selective COX-2 inhibitors are weak competitive inhibitors of COX-1, but time-dependent inhibitors of COX-2. Hence the selectivity is driven by differences in kinetics.

While the structures of COX-1 and COX-2 are highly conserved, the structural basis for selectivity can be attributed to the presence of a much larger inhibitor binding pocket in COX-2 compared to COX-1 (Figure 3.7). In the cyclooxygenase active site, COX-2 has a valine while COX-1 has a larger isoleucine residue. This deletion of a methylene at this position in COX-2 provides access to additional binding pocket that directly leads to solvent. A selective COX-2 inhibitor[24] like celecoxib exploits this additional binding pocket for enhanced binding through the phenyl sulfonamide moiety. Another

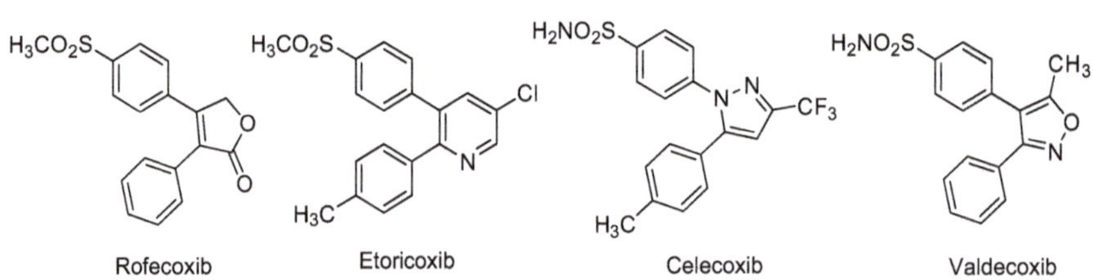

Figure 3.6 Examples of COX-2 inhibitors.

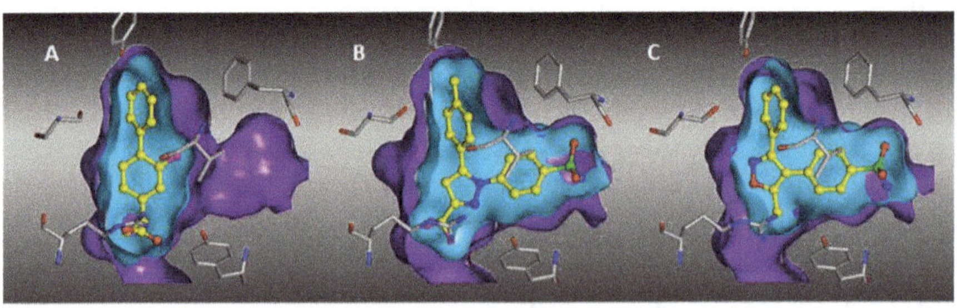

Figure 3.7 Structural differences in the active site of COX-1 and COX-2: (A) flurbiprofen in COX-1 active site; (B) celecoxib in COX-2 active site; (C) valdecoxib in COX-2 active site.

feature that distinguishes COX-2 from NSAIDs is the lack of carboxylic acid and the inability to form the salt bridge with the enzyme. This results in higher flexibility and conformational changes in the active site.

3.4.4 COX-2—Adverse Effects

Although coxibs appear to spare gastrointestinal tract,[25] there has been concern regarding adverse effects in renal and cardiovascular systems. The underlying mechanism that leads to increased risk of amyocardial ischaemic event is not fully elucidated. One of the simpler proposed theories is the 'imbalance' theory. Reduction of COX-2 dependent prostanoid PGI2 without affecting platelets and COX-1 dependent TXA2 will predispose to a prothrombotic state. COX-2 is the major source of systemic synthesis of PGI2 and PGI2 is the only prostanoid with a documented ability to restrain thrombotic events. Both rofecoxib and celecoxib suppress formation of PGI2 in healthy volunteers.[26]

3.4.5 Alternatives to NSAIDs and Coxibs

Nitric oxide (NO) and hydrogen sulfide (H_2S) are potent vasodilatory agents that maintain the mucosal integrity in the gastrointestinal tract. Aspirin and NSAIDs are reported to reduce H_2S synthesis, apparently through suppression of expression of cystathionine. This suppression of H_2S synthesis is suggested as an additional mechanism (apart from COX-1 inhibition) through which NSAIDs produce gastrointestinal damage. This reasoning is responsible for the ongoing development of several H_2S-releasing NSAIDs.[27] These compounds are hybrid molecules obtained by adding an 'H_2S-releasing moiety' such as 5-(4-hydroxyphenyl)-1,2-dithiol-3-thione (ADT-OH) to a conventional NSAID. ADT-OH releases H_2S enzymatically in biological fluids.

With the realisation that NO produced many of the same effects of PGs with respect to mucosal defence, NO-releasing derivatives of traditional NSAIDs (COX-inhibiting nitric oxide donor—CINOD) with the ability to spare the gastric mucosa while maintaining the COX-inhibiting ability of the parent NSAID, are being investigated. Naproxinod (nitroxybutyl ester of naproxen) is one of the most advanced CINODs.

3.5 Inflammatory Kinases

Signalling pathways in inflammatory disorders are regulated by kinases. Some of the kinases in the MAP, JAK and Tec families are specifically implicated in immune disorders. The kinases play a key role in signal transduction by catalysing the transfer of the terminal phosphate group of ATP to an appropriate substrate, leading to the activation of the substrate for its role in the next step of the signalling cascade. A large majority of kinase inhibitors that have been developed interact with the ATP-binding site. Structurally, the kinase domain, which includes the ATP-binding site, is highly conserved among typical protein kinases and that poses a great challenge in developing selective kinase inhibitors. Lack of selectivity is accountable for the undesired side effects. Considering that the anti-inflammatory drug treatment is chronic, developing a low-dose, highly kinome-selective inhibitor with exquisite safety profile is mandated from this approach. Successful development of an anti-inflammatory kinase inhibitor[28] potentially offers a unique advantage over biologics with respect to route of administration and cost. Some of the advanced kinase inhibitors targeting immune disorders are discussed below.

3.5.1 p38 Kinase Inhibitors

Ever since the disclovery of pyridinyl imidazoles as the first inhibitor of p38, almost all the major pharmaceutical companies have targeted p38 MAP kinase as a viable drug discovery target (Figure 3.8).[29] The anti-inflammatory action of p38 inhibitors is linked to their impact on the levels of pro-inflammatory mediators TNF-α, IL-1, IL-6 and IL-8. These cytokines regulate COX-2/PG-2, RANKL, intefrins, MMPs, some interleukins and leukotrienes which are involved in bone and cartilage destruction, joint swelling and pain.[30] p38 is a serine/threonine mitogen-activated kinase and there are four isoforms: p38α, p38β, p38δ, p38γ. Of these, the p38α isoform is implicated in the inflammatory process (Table 3.4). However, because of the highest homology between α and β isoforms, early p38 inhibitors are dual inhibitors.

ATP competitive p38 kinase inhibitors occupy the adenine-binding region of the ATP-binding site. There are two hydrophobic pockets adjacent to the ATP-binding site and one of the hydrophobic sites is open to the surrounding

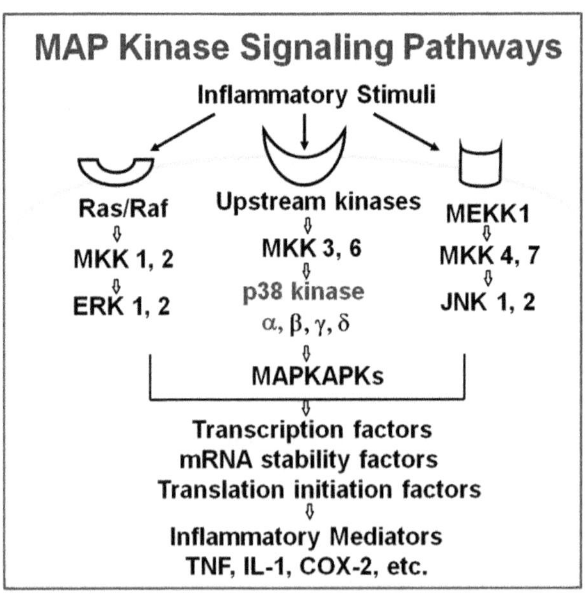

Figure 3.8 Role of p38 kinase in the MAP kinase signalling pathways.

Table 3.4 p38 isoforms and their cellular expression.

p38 isoform	*Expression profile*
p38α	All cell types
p38β	Endothelial cells, T cells
p38δ	Macrophages, neutrophils, T cells, monocytes
p38γ	Limited/no expression in immune system

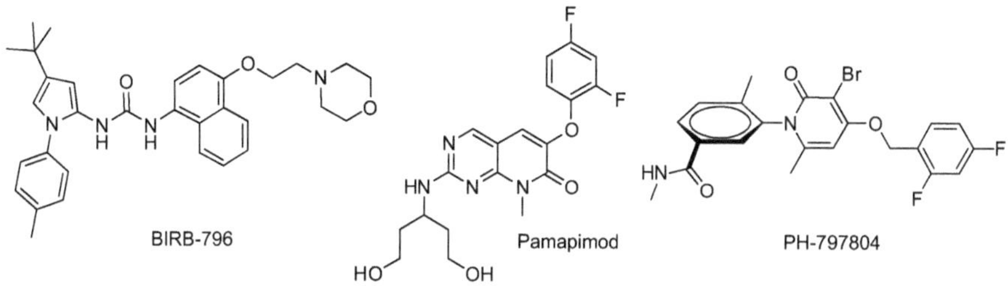

Figure 3.9 Examples of p38 inhibitors.

solvent. Many inhibitors have exploited these hydrophobic pockets, which are less conserved in kinases, to achieve selectivity over other kinases. Early inhibitors like pyridinyl imidazoles[31] feature a central core that can direct peripheral residues into the hydrophobic region. Another class of reported p38 kinase inhibitors is DFG-out binders or deep pocket binders. They bind to the inactive[32] enzyme wherein the DFG region of the enzyme is facing out into the solvent. Placing a suitable group in the DFG pocket pushes the DFG motif facing the adenosine-binding pocket. This change in conformation prevents the ATP from entering the ATP-binding site. One of the early p38 DFG-out binders is BIRB-796[33] (Figure 3.9). Although this binding mode can be described as partially allosteric, BIRB-796 has been recently shown to have considerable off-target activity. Some other compounds achieve selectivity by employing a glycine flip binding mode. After the initial binding of the compound to the hinge region, the enzyme rotates a peptide bond that is adjacent to Met 109. This provides a second hydrogen bonding potential. This rotation is possible only when a small glycine residue occupies the position adjacent to Met 109. Therefore, this strategy reduces the number of kinases that can accommodate this binding mode and enhances the kinome selectivity of the inhibitors. One such reported[34] inhibitor is PHA-797804. The inhibitor induces the glycine flip by establishing two hydrogen bonds with the kinase hinge region via the carbonyl oxygen. The difluorophenyl moiety is situated in a hydrophobic pocket (Figure 3.10). It showed no off-target activity when tested against a panel of 58 kinases. Selectivity was also examined at the cellular level. No inhibition of ERK or JNK was observed. Although PHA-797804 did not have a favourable outcome in a Phase 2 clinical trial for treating RA, it showed proof of concept for treating COPD.

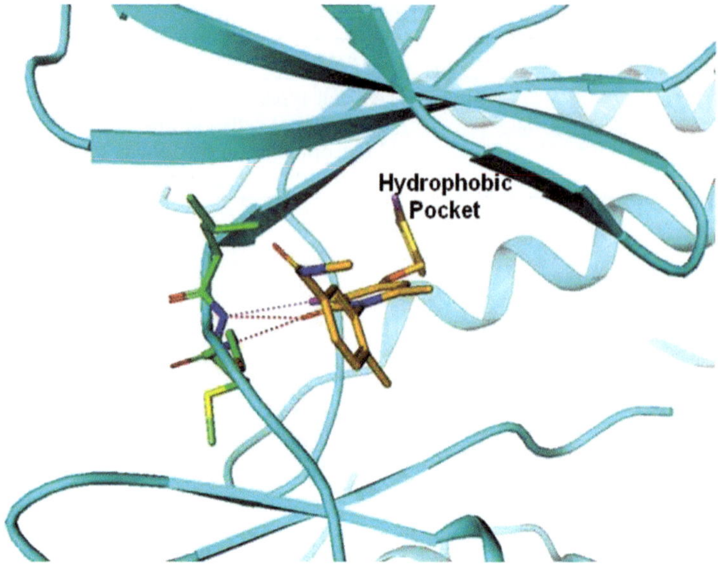

Figure 3.10 PHA-797804 binding site interactions with p38.

In spite of two decades of research and advancement of over 10 compounds into clinical evaluation, none of the p38 inhibitors progressed beyond Phase 2 studies due to adverse findings such as skin disorders, infections and lack of sustained efficacy in disease such as RA.[35] In many of these trials, a drop in the levels of - C-reactive protein (CRP), an inflammatory biomarker, was observed upon drug treatment. However, this was not sustained after 2 weeks of drug treatment, suggesting possible redundancy in the p38 pathway. This probably highlights the challenges of targeting mechanisms broadly utilised by multiple cell types for survival, proliferation and activation. Since RA is not the only disease in which p38 MAP kinase plays a pivotal role, several other indications like CD, psoriatic arthritis, ankolysing spondylitis and COPD[36] are potential clinical indications for p38 inhibitor use. Currently the studies that are testing old inhibitors for new indications outnumber the studies testing new inhibitors for RA.

3.5.2 Bruton's Tyrosine Kinase Inhibitors

In the recent past, the Tec family of non-receptor tyrosine kinases (BTK, TEC, BMX, ITK, and RLK) has been shown to play a key role in inflammation and bone destruction. Bruton's tyrosine kinase (BTK) is expressed in most haematopoietic cells such as B cells, mast cells, and macrophages and plays key roles in multiple cell signalling pathways (Figure 3.11).[37] Rituxan, a CD20 antibody, has impacted B cells in the pathogenesis of many autoimmune diseases, such as RA, SLE, and MS. This is responsible for the interest in small-molecule inhibitors of BTK,[38] being pursued by several pharmaceutical

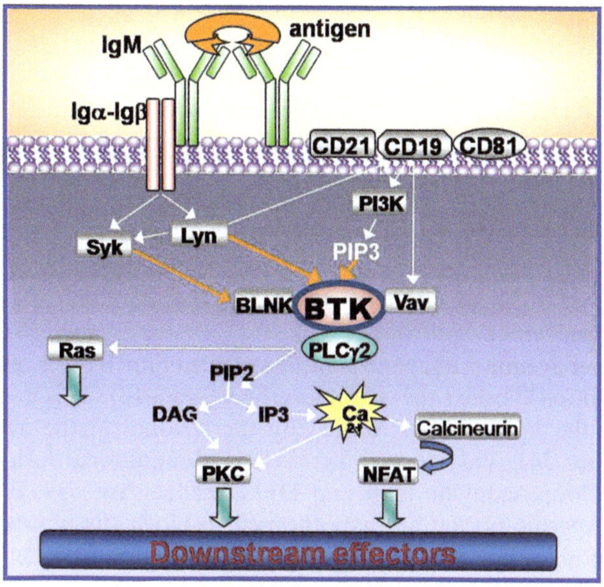

Figure 3.11 Central role of BTK signalling in B-cell development and effector function.

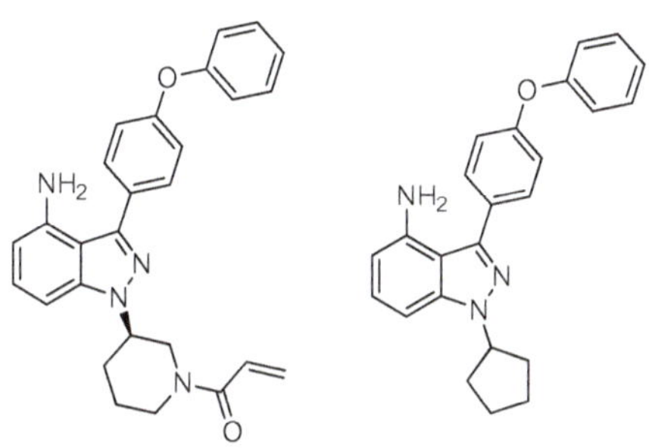

PCI-32765 PCI-29732

Figure 3.12 BTK inhibitors (reversible and irreversible) reported by Pharmacyclics.

companies. In the quest for kinome-selective BTK inhibitors, the development of covalent inhibitor and approaches that target the inactive conformation of the kinase[39] or the selectivity pocket are worth mentioning. Within the ATP-binding pocket, BTK has a unique amino acid residue Cys 481 that serves as a handle to design covalent inhibitors. Sequence analysis of human kinome indicates that only nine other kinases have cysteine in this position. Invoking irreversible interaction[40] with this unique cysteine residue could provide highly kinome-selective BTK inhibitors. A reported[41] irreversible inhibitor PCI-32765 (Figure 3.12) shows >100-fold *in vitro* selectivity over kinases that do not contain a cysteine or serine in this position.

BTK also has a smaller threonine residue, Thr 474, as the gatekeeper residue which provides access to the back pocket for the ATP competitive inhibitor design. However, rotation of the gatekeeper threonine side chain is not observed in BTK,[42] resulting in a smaller back pocket. While 20% of reported kinases have threonine as gatekeeper residue, only eight kinases have the unique combination of cysteine in the ATP-binding pocket along with threonine as gatekeeper. Thus, design of inhibitors that interact with both Cys 481 and Thr 474 could provide BTK specificity. The non-covalent analogue of PCI-32765, compound PCI-29732, is reported to show selectivity over a number of kinases, but for 13 kinases that contain threonine as gatekeeper residue.[43]

Yet another avenue pursued to achieve specificity for BTK is making use of a specificity pocket[44] called the 'H3 pocket' (Figure 3.13). This pocket is within reach from the hinge and is formed by the residues in the activation loop (Leu 542, Sser 543, Val 546 and Tyr 551), with additional residues from the glycine-rich loop, catalytic loop and DFG residue Asp 539. Inhibitors that occupy the specific pocket are hypothesised to block the phosphorylation of activation loop leading to high selectivity. However, this is pocket is not unique to BTK since a similar pocket created by the activation loop conformation is reported for SRC and HCK as well.[45]

At present none of the BTK inhibitors have entered the clinic for autoimmune disorders and the clinical experience has been limited to oncology. Nevertheless, the increased understanding of the structure biology of BTK offers a unique drug development opportunity to enter anti-inflammatory treatment paradigm.

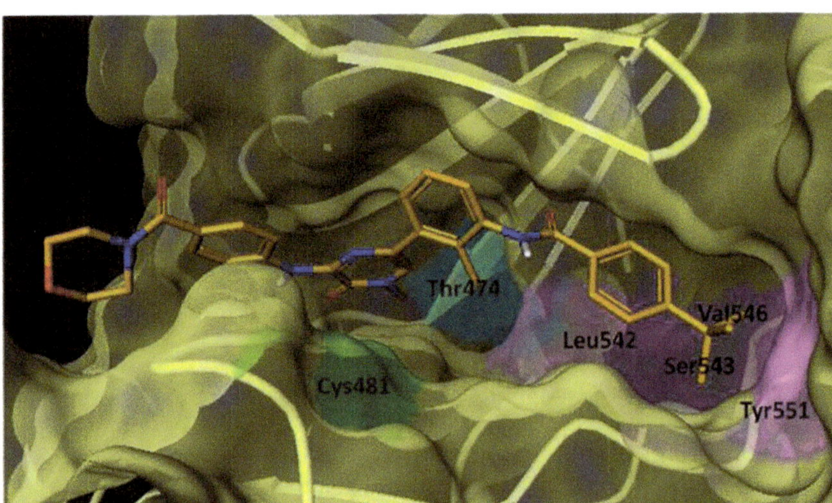

Figure 3.13 'H3 pocket' (specificity pocket) of BTK. Labelled residues are C481 (green), and the specificity pocket (pink).

3.5.3 Spleen Tyrosine Kinase (SYK) Inhibitors

Spleen tyrosine kinase (SYK) is a non-receptor tyrosine kinase in signal transduction having a key mediator role in immunoreceptor signalling in a host of inflammatory cells such as macrophages, neutrophils, mast cells and B cells.[46] Its aberrant regulation is associated with different allergic disorders and antibody-mediated autoimmune diseases such as RA,[47] asthma and allergic rhinitis.[48] SYK is associated with a number of receptors on the cell surfaces of immune system (Figure 3.14). These ligand-stimulated receptors activate SYK, which initiates different cellular processes, including cytokine production. SYK is ubiquitously expressed in haematopoietic cells, including macrophages, mast cells, leukocytes, platelets, and erythrocytes. SYK activation induces a cascade of events leading to the synthesis and release of inflammatory modulators responsible for acute allergic reactions. SYK is involved in activation of B cells and T cells, induced by antigen interaction with receptors, BCR and TCR. Modulation of SYK activity could represent a therapeutic intervention in a number of immune disorders.[49]

SYK contains dual SH2 domains in the N-terminus, separated by inter-domain A. The SH2 domain interacts with ITAM (immune tyrosine activation motif) sequences in the immunoglobulin domains of Fce and Fcg receptor complexes, leading to activation of several nodes of the inflammatory cascade. The interdomain A has a role in regulating SYK activity. The catalytic domain in the C-terminus is separated from the SH2 domain by interdomain B, which has auto- and trans-phosphorylation sites. SYK activation process is initiated when Src kinases phosphorylate conserve sequences within receptors known as ITAMs[50].

Rigel Pharmaceuticals has the most advanced SYK inhibitor, fostamatinib (R788), in clinical trials for the potential treatment of RA.[51] It is in the 2,4-pyridmidine-diamine class of kinase inhibitors (Figure 3.15). R788 is a

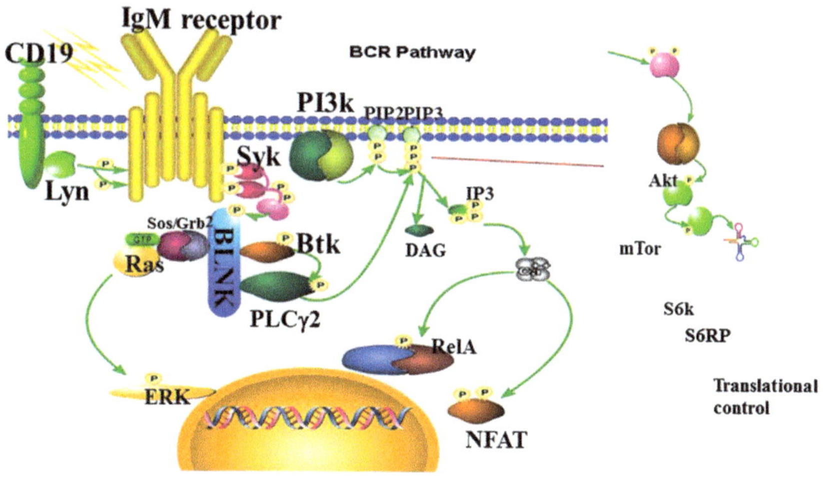

Figure 3.14 SYK signalling pathway.

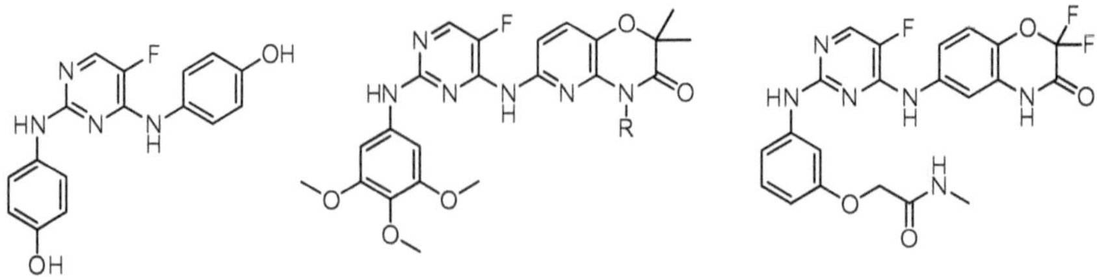

R112 R406, R = H
R788, R = -CH2OPO3H2 R343

Figure 3.15 SYK Inhibitor fostamatinib (R788) and related analogues.

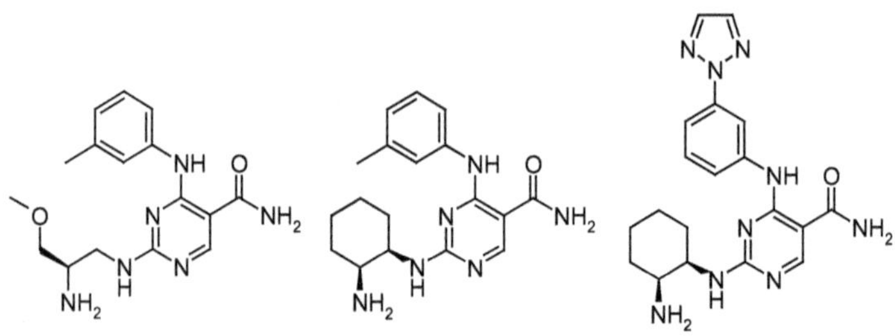

Figure 3.16 Diaminopyrimidines as SYK inhibitors (reported by Portola).

methylene phosphate-derived prodrug of R406, and is a potent inhibitor of SYK, with an IC_{50} value of 40 nM, and potently inhibits the BCR signalling pathway (IC_{50} value of 50 nM in inhibition of CD69 upregulation on primary human B cells). R406 is a potent inhibitor of IgE- and IgG-mediated activation of Fc receptor signalling with an IC_{50} value for degranulation of 60 nM. It inhibits phosphorylation of SYK substrate linker for activation of T cells in mast cells and B-cell linker protein. Preclinical studies of these inhibitors showed significant reduction in inflammatory mediators, leading to reduced inflammation and bone degradation of RA animal models. In the Phase 2 clinical trials of 229 patients with active RA, R788 showed ACR20 response rate of 65% and 72%, relative to 38% placebo response, at 100 mg and 150 mg twice daily, respectively.[52] The pseudo-symmetrical 2,4-diaminopyrimidine R112, with moderate SYK affinity (K_i 96 nM), was the first development compound that advanced to Phase 2 clinical trials for seasonal allergic rhinitis. The inhaled formulation of the later-generation compound R343 was reported to have entered clinical trials for allergic asthma.[53]

Astellas reported[54] a series of aniline-pyrimidines as SYK inhibitors that displayed selectivity over kinases such as ITK, BTK, and PKC isoforms. Kissei researchers designed[55] SYK inhibitors having chemical structures of 1,2,4-triazolo[4,3-c]pyrimidine and 1,2,4-triazolo[1,5-c]pyrimidine. A number of these compounds are dual inhibitors of SYK and ZAP-70, having IL-2 suppression activities *in vitro* and *in vivo*. Portola has published SYK inhibitors, including those with the pyrimidiine core (Figure 3.16). The indazole derivative inhibits SYK with an IC_{50} value of 10–100 nM. Portola is in clinical trial with the SYK inhibitor PRT062607, the chemical structure of which has not been published. The compound was described as being a highly selective SYK inhibitor in a broad panel of kinase and cellular assays.[46]

3.5.4 Janus Tyrosine Kinase (JAK) Inhibitors

Several aspects of inflammation and immunology rely on the critical roles played by cytokine signalling. Cytokines are a diverse group of secreted signalling molecules that regulate a wide range of cellular functions. Type I and type II cytokine receptors lack intrinsic enzyme activity and associate themselves with tyrosine kinases for mediating signal transduction. The hallmark of signalling by multiple cytokines is the activation of Janus kinase-signal transducer and activator transcription pathway (JAK-STAT pathway). Phosphorylation of STATs by JAK kinases triggers dimerisation, nuclear translocation and activation of gene transcription.[56] Many of the pro-inflammatory cytokines implicated in the pathogenesis of RA, including IL-6, IL-12, IL-2, IL-15, IL-23 and IFN-γ, utilise the JAK/STAT pathway[57] for signalling (Figure 3.17). On the basis of clinical validation of individual cytokines such as IL-6, IL-12 and IL-23 using neutralising antibodies against each cytokine in

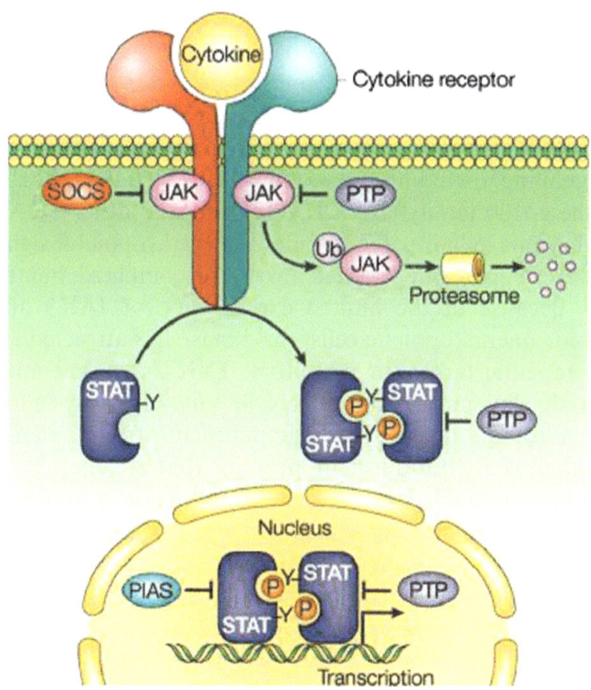

Figure 3.17 JAK-STAT signalling pathway. (Reproduced from Ref. 57).

Table 3.5 JAK family and cytokine associations.

JAK	*Cytokines*
JAK1	β and γ-C cytokines, type I interferons, interferon- γ and Gp130 cytokines
JAK2	EPO, TPO, GH, PRL, IL-12, and interferon- γ
JAK3	γ-C cytokines
TYK2	Type I interferons, Gp130 cytokines, IL-12 and IL-23

chronic autoimmune states, it was hypothesised that JAK inhibitors may offer the advantage of targeting multiple pathogenic cytokines[58] at once and hence may have a broad spectrum of activity compared with neutralising antibodies against single cytokines.

There are four members in the JAK family of kinases:[59] JAK-1, JAK-2, JAK-3 and TYK-2. A single JAK or combinations of JAK members selectively associate with different cytokine receptors for transmitting their signals[60] after the engagement of the corresponding receptor by a cytokine (Table 3.5). Cytokines that signal through heteromeric receptors containing the gp130 subunit primarily utilise JAK1 and JAK2. Type II cytokine receptors utilise JAK2 and TYK2 for signalling. Receptors for growth factors erythropoietin (EPO), thrombopoietin (TPO) or granulocyte-macrophage colony-stimulating factor (GM-CSF) use JAK2. Receptors for the IFN-γ receptor use JAK1 and JAK2, whereas cytokines IL-2, IL-4, IL-7, IL-9, IL-15 and IL-21 signal through gamma common (γc)-chain containing receptors JAK1 and JAK3. JAK1, JAK2 and Tyk2 exhibit broad patterns of expression, whereas JAK3 is limited to haematopoietic cells. A number of gene knockout studies have provided functional relevance for each member of the JAK family of kinases. Deletion of JAK-1 was lethal since a large number of cytokines are dependent on JAK-1 signalling, including the γc family (IL-2, IL-4, IL-7, IL-9, IL-15 and IL-21) and the gp130 family (IL-6, IL-11, IFNs and others). JAK2 deficiency results in embryonic lethality due to failure of erythropoietin signalling. JAK-3 knockouts were viable, but resulted in severe combined immune deficiency phenotype.[61] Because of the limited expression[62] of JAK3 and its lack of function outside haematopoietic cells, this kinase has attracted attention as an attractive therapeutic target for inhibitors. TYK-2-deficient mice were viable and did not display any major effects in any organ systems.[63] The only significant abnormality noted was the impairment of IL-12 signalling.

The JAK kinases are multidomain proteins that are characterised by seven homology domains (JH1–JH7).[64] The JH1 domain contains the catalytic protein tyrosine kinase, which conserves a high level of sequence identity among the four JAK isozymes. The catalytically active kinase domain is responsible for the physiological function of the JAK kinase, and typically small-molecule inhibitors have been designed to target the ATP-binding pocket of the JAK kinase. The Pfizer inhibitor CP-690550 (tofacitinib) is a potent, pan-JAK inhibitor[65] that exhibits significant selectivity against a panel of more than 300 other kinases. A series of structures of the JAK1 and JAK2 kinase domains in complex with several pan inhibitors, including

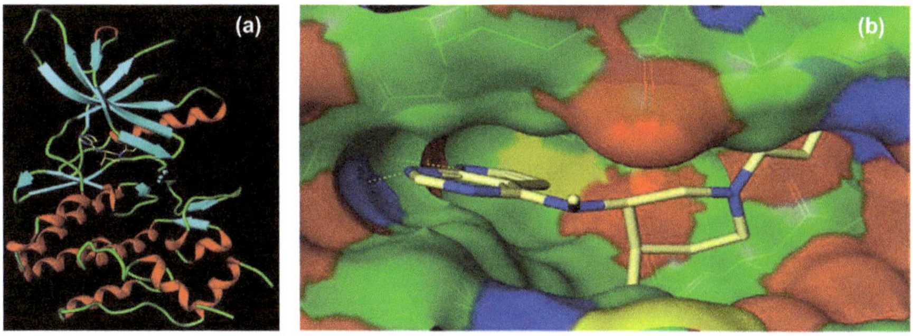

Figure 3.18 (a) X-ray structure of JAK-3 with CP-690550. (b) Binding site interaction of CP-690550 with the ATP-binding pocket of JAK-3.

CP-690550, have been reported, providing insights towards the development of potentially selective JAK inhibitors. CP-690550 displays a high level of shape complementarity with the JAK3 ATP-binding cleft (Figure 3.18). In crystal structures,[66] the pyrrolopyrimidine ring is situated against the hinge region, and forms two hydrogen bonds with the main chain atoms of Glu 903 and Leu 905. The piperidine ring is situated in the hydrophobic environment of the JAK3 binding cavity, forming hydrophobic interactions with Met 902, Val 836, Cys 909, Leu 956 and Ala 966. Additionally, the C7 methyl group of the piperidine ring is situated in an underlying pocket at the lower rim of the binding cavity of JAK3. Finally, the terminal cyano group packs against the glycine-rich loop.

It is challenging to design JAK isozyme-selective compounds, which is a result of the significant overall sequence homology among the four isozymes.[67] The JAK protein kinase domains share a high level of sequence homology: JAK3 shares 62% sequence identity with JAK2, 52% with JAK1 and 50% with TYK2. Further complicating the design of selective JAK3 inhibitors is the nearly identical amino acid composition in the ATP-binding cavity of JAK3 and JAK2, with only two major amino acid differences between the two JAK isoforms within 6 Å of the ATP-binding cavity: JAK2, JAK1 and TYK2 maintain serine at position 936, while JAK3 has cysteine at the equivalent position (909); and JAK2, JAK1 and TYK2 maintain glycine at position 993 while JAK3 has alanine at the equivalent position (966). Ser 936 resides in the C lobe of the JAK isozymes, and frames the bottom of the ligand-binding cavity, 3.9 Å away from CP-690550. Structurally, the subtle difference between the serine and cysteine residues would not favour the design of compounds that would have the capacity to selectively discriminate between JAK3 and JAK2. However, Cys 909 could be exploited through the design of irreversible covalent inhibitors that would not be predicted to interact with the serine at the equivalent position in JAK1, JAK2 or TYK2. However, Gly 993 of JAK2 is located immediately adjacent to the highly conserved DFG motif, and might provide additional opportunities to design in selectivity against JAK3, which has alanine at the equivalent position.

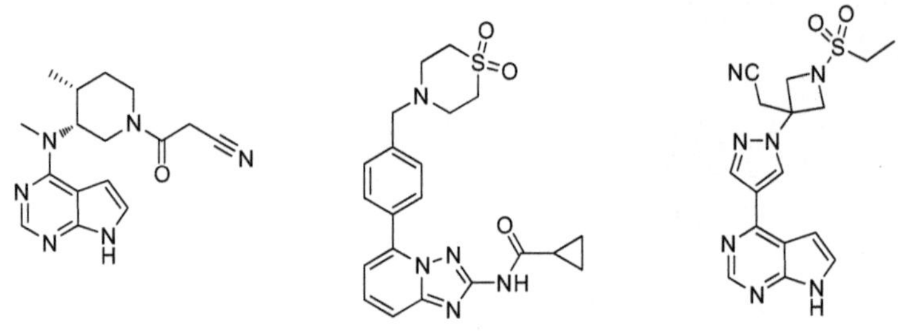

Tofacitinib GLPG0634 analog Baricitinib

Figure 3.19 Advanced JAK inhibitors with known structures.

Table 3.6 JAK platform clinical landscape for inflammation.

Compound	*Company*	*Indication*	*MoA*	*Status*
Tofacitinib	Pfizer	RA	Pan-JAK	Registration
Tofacitinib	Pfizer	Psoriasis	Pan-JAK	Phase 3
Baricitinib	Incyte/Lilly	RA/psoriasis	JAK1/2	Phase 3
Tofacitinib	Pfizer	Ulcerative colitis	Pan-JAK	Phase 2
Tofacitinib	Pfizer	Transplant	Pan-JAK	Phase 2
GLPG0634	Galapagos	RA	JAK1	Phase 2
VX509	Vertex	RA	JAK3	Phase 2
ASP015K	Astellas	RA/psoriasis	JAK3	Phase 2

There are several JAK inhibitors[68] (Figure 3.19) in clinical trials and under development, many of which inhibit multiple JAK family members (Table 3.6). Many of these have been found to be efficacious in preclinical models of autoimmune diseases. Xeljanz CP690550, tofacitinib is one of the first JAK inhibitors to enter the clinic[69] and was recently approved for treatment of RA. Tofacitinib inhibits JAK-1, JAK-2 and JAK-3 and to a lesser extent TYK-2. Consequently, tofacitinib potently inhibits cytokines[70] that bind to the γc chain, which associates with JAK-3. It has been shown[71] in a short-term (6-week) proof-of-concept trial to be highly efficacious *versus* placebo when used as monotherapy. Tofacitinib is also efficacious over periods of 12–24 weeks when used in combination with MTX in a Phase 2b trial. Potential exists for its use in other rheumatic diseases.

3.6 Conclusion

Despite advances in therapeutics, a 'cure' continues to be elusive; remission rates in RA remain difficult to achieve and the high cost and inconvenience of parenteral administration typical of biologic agents make them less attractive to some patients. In the light of the recent advances made in the field of personalized medicine, biomarker-driven individualized pharmacotherapy will provide the best treatment choice for RA patients with the goal of remission/low disease activity and optimized safety.

List of Abbreviations

BTK	Bruton's tyrosine kinase
CINOD	COX-inhibiting nitric oxide donor
COPD	chronic obstructive pulmonary disease
COX-1	cyclooxygenase-1
COX-2	cyclooxygenase-2
DMARD	disease-modifying anti-rheumatic drug
GC	glucocorticoid
GR	glucocorticoid receptor
IL-6	interleukin-6
JAK	Janus tyrosine kinase
MTX	methotrexate
NSAID	non-steroidal anti-inflammatory drug
PG	prostaglandin
RA	rheumatoid arthritis
SYK	spleen tyrosine kinase
TNF-α	tumour necrosis factor-α

Acknowledgement

We would like to thank Dr Li Xing (Pfizer), Dr Ravi Kurumbail (Pfizer) and Dr Jim Kiefer (currently at Genentech) for providing some of the figures in this chapter.

References

1. C. Luo, R.-G. Xie and S.-J. Hu, Anti-inflammatory anti-allergy agents, *Med. Chem.*, 2011, **10**, 43.
2. M. W. Whitehouse, *Curr. Med. Chem.*, 2005, **12**, 2931.
3. C. A. Dinarello, *Cell (Cambridge, MA)*, 2010, **140**, 935.
4. Y. M. Di, Z.-W. Zhou, C. G. Li and S.-F. Zhou, Anti-inflammatory anti-allergy agents, *Med. Chem.*, 2011, **10**, 92.
5. Available from http://emedicine.medscape.com/article/331715-overview-Rheumatoid Arthritis.
6. J. Liu and H. Mehmet, *Drug Dev. Res.*, 2011, **72**, 805.
7. C. A. Dinarello, *Cell*, 2010, **140**, 935.
8. S. G. Hillier, *J. Endocrinol.*, 2007, **195**, 1.
9. P. J. Barnes, *Br. J. Pharmacol.*, 2006, **148**, 245.
10. U. Muller-Ladner, R. E. Gay and S. Gay, *Curr. Rheumatol. Rep.*, 2002, **4**, 201.
11. F. Buttgereit, R. H. Straub, M. Wehling and G. R. Burmester, *Arthritis Rheum.*, 2004, **50**, 3408.

12. M. W. Whitehouse, *Inflammopharmacology*, 2011, **19**, 1.
13. H. Schacke, H. Rehwinkel and K. Asadullah, *Curr. Opin. Invest. Drugs*, 2005, **6**, 503.
14. J. Rosen and J. N. Miner, *Endocrine Rev.*, 2005, **26**, 452.
15. A. R. Clark and M. G. Belvisi, *Pharmacol. Therapeut.*, 2012, **134**, 54.
16. L. Buckbinder and R. P. Robinson, *Curr. Drug Targets Inflamm. Allergy*, 2002, **1**, 127.
17. C. M. Yates, P. J. Brown, E. L. Stewart, C. Patten, R. J. Austin, J. A. Holt, J. M. Maglich, D. C. Angell, R. Z. Sasse, S. J. Taylor, I. J. Uings and R. P. Trump, *J. Med. Chem.*, 2010, **53**, 4531.
18. X. Hu, S. Du, C. Tunca, T. Braden, K. R. Long, E. G. Lee, J. Webb, J. D. Dietz, S. Hummert, S. Rouw, S. G. Hegde, R. K. Webber and M. G. Obukowicz, *Endocrinology*, 2011, **152**, 3123.
19. B. N. Cronstein, *Pharmacol. Rev.*, 2005, **57**, 163.
20. M. C. Montesinos, A. Desai and B. N. Cronstein, *Arthritis Res. Ther.*, 2006, **8**, R53.
21. D. R. Lionberger and M. J. J. Brennan, *Pain Res.*, 2010, **3**, 223.
22. E. Ricciotti and G. A. FitzGerald, *Arterioscler. Thromb. Vasc. Biol.*, 2011, **31**, 986.
23. L. J. Marnett, *Annu. Rev. Pharmacol. Toxicol.*, 2009, **49**, 265.
24. R. G. Kurumbail, A. M. Stevens, J. K. Gierse, J. J. McDonald, R. A. Stegeman, J. Y. Pak, D. Gildehaus, J. M. Miyashiro, T. D. Penning, K. Seibert, P. C. Isakson and W. C. Stallings, *Nature (London)*, 1996, **384**, 644.
25. J. Meyer-Kirchrath and K. Schror, *Curr. Med. Chem.*, 2000, **7**, 1121.
26. M. C. Zweers, T. N. de Boer, J. van Roon, J. W. Bijlsma, F. P. Lafeber and S. C. Mastbergen, *Arthritis Res. Ther.*, 2011, **13**, 239.
27. S. Fiorucci and E. Distrutti, *Curr. Med. Chem.*, 2011, **18**, 3494.
28. S. Cohen and R. Fleischmann, *Curr. Opin. Rheumatol.*, 2010, **2**, 330.
29. A. Cuenda and S. Rousseau, *Biochim. Biophys. Acta*, 2007, **1773**, 1358.
30. S. Kumar, J. Boehm and J. C. Lee, *Nat. Rev. Drug Discov.*, 2003, **2**, 717.
31. J. C. Lee, J. T. Laydon, P. C. McDonnell, T. F. Gallagher, S. Kumar, D. Green, D. McNulty, M. J. Blumenthal, J. R. Heys, S. W. Landvatter, *et al.*, *Nature*, 1994, **372**, 739.
32. Y. Liu and N. S. Gray, *Nat. Chem. Biol.*, 2006, **2**, 358.
33. J. Regan, A. Capolino, P. F. Cirillo, T. Gilmore, A. G. Graham, E. Hickey, R. R. Kroe, J. Madwed, M. Moriak, R. Nelson, C. A. Pargellis, A. Swinamer, C. Torcellini, M. Tsang and N. Moss, *J. Med. Chem.*, 2003, **46**, 4676.
34. L. Xing, H. S. Shieh, S. R. Selness, R. V. Devraj, J. K. Walker, B. Devadas, H. R. Hope, R. P. Compton, J. F. Schindler, J. L. Hirsch, A. G. Benson, R. G. Kurumbail, R. A. Stegeman, J. M. Williams, R. M. Broadus, Z. Walden and J. B. Monahan, *Biochemistry*, 2009, **48**, 6402.
35. S. A. Doggrell and A. M. Christensen, *Exp. Opin. Pharmacother.*, 2010, **1**, 2437.
36. D. S. Millan, M. E. Bunnage, J. L. Burrows, K. J. Butcher, P. G. Dodd, T. J. Evans, D. A. Fairman, S. J. Hughes, I. C. Kilty, A. Lemaitre,

R. A. Lewthwaite, A. Mahnke, J. P. Mathias, J. Philip, R. T. Smith, M. H. Stefaniak, M. Yeadon and C. J. Phillips, *Med. Chem.*, 2011, **54**, 7797.
37. K. S. Currie, *Annu. Rep. Med. Chem.*, 2010, **45**, 175.
38. Y. Lou, T. D. Owens, A. Kuglstatter, R. K. Kondru and D. M. J. Goldstein, *Med. Chem.*, 2012, **5**, 4539.
39. D. J. Marcotte, Y. T. Liu, R. M. Arduini, C. A. Hession, K. Miatkowski, C. P. Wildes, P. F. Cullen, V. Hong, B. T. Hopkins, E. Mertsching, T. J. Jenkins, M. J. Romanowski, D. P. Baker and L. F. Silvian, *Protein Sci.*, 2010, **19**, 429.
40. J. Singh, R. C. Petter and A. F. Klugeq, *Curr. Opin. Chem. Biol.*, 2010, **14**, 475.
41. E. S. Winer, R. R. Ingham and J. Castillo, *J. Expert Opin. Invest. Drugs*, 2012, **21**, 355.
42. A. Kuglstatter, A. Wong, S. Tsing, S. W. Lee, Y. Lou, A. G. Villasenor, J. M. Bradshaw, D. Shaw, J. W. Barnett and M. F. Browner, *Protein Sci.*, 2011, **20**, 428.
43. Z. Pan, H. Scheerens, S. J. Li, B. E. Schultz, P. A. Sprengeler, L. C. Burrill, R. V. Mendonca, M. D. Sweeney, K. C. Scott, P. G. Grothaus, D. A. Jeffery, J. M. Spoerke, L. A. Honigberg, P. R. Young, S. A. Dalrymple and J. T. Palmer, *ChemMedChem*, 2007, **2**, 58.
44. J. A. Di Paolo, T. Huang, M. Balazs, J. Barbosa, K. H. Barck, B. J. Bravo, R. A. Carano, J. Darrow, D. R. Davies, L. E. DeForge, L. Diehl, R. Ferrando, S. L. Gallion, A. M. Giannetti, P. Gribling, V. Hurez, S. G. Hymowitz, R. Jones, J. E. Kropf, W. P. Lee, P. M. Maciejewski, S. A. Mitchell, H. Rong, B. L. Staker, J. A. Whitney, S. Yeh, W. B. Young, C. Yu, J. Zhang, K. Reif and K. S. Currie, *Nat. Chem. Biol.*, 2011, **7**, 41.
45. T. Schindler, F. Sicheri, A. Pico, A. Gazit, A. Levitzki and J. Kuriyan, *Mol. Cell*, 1999, **3**, 639.
46. M. Riccaboni, I. Bianchi and P. Petrillo, *Drug Discov. Today*, 2010, **15**, 517.
47. H. Okamoto and A. Kobayashi, *N. Engl. J. Med.*, 2011, **364**, 83.
48. J. Denyer and V. Patel, *Drug News Perspect.*, 2009, **2**, 146.
49. V. C. Kyttaris and G. C. Tsokos, *Clin. Immunol.*, 2007, **124**, 235.
50. J. M. Bradshaw, *Cell. Signal.*, 2010, **2**, 1175.
51. M. E. Weinblatt, A. Kavanaugh, M. C. Genovese, T. K. Musser, E. B. Grossbard and D. B. Magilavy, *N. Engl. J. Med.*, 2010, **363**, 1303.
52. M. C. Genovese, A. Kavanaugh, M. E. Weinblatt, C. Peterfy, J. DiCarlo, M. L. White, M. O'Brien, E. B. Grossbard and D. B. Magilavy, *Arthritis Rheum.*, 2011, **63**, 337.
53. W. J. Moore, D. Richard and A. Thorarensen, *Exp. Opin. Therapeut. Pat.*, 2010, **20**, 1703.
54. H. Hisamichi, R. Naito, A. Toyoshima, N. Kawano, A. Ichikawa, A. Orita, M. Orita, N. Hamada, M. Takeuchi, M. Ohta and S. Tsukamoto, *Bioorg. Med. Chem.*, 2005, **13**, 4936.

55. A. Hirabayashi, H. Mukaiyama, H. Kobayashi, H. Shiohara, S. Nakayama, M. Ozawa, E. Tsuji, K. Miyazawa, K. Misawa, H. Ohnota and M. Isaji, *Bioorg. Med. Chem.*, 2008, **16**, 9247.
56. Y. A. Ivanenkov, K. V. Balakin and Y. Lavrovsky, *MiniRev. Med. Chem.*, 2011, **1**, 55.
57. K. Shuai and B. Liu, *Nat. Rev. Immunol.*, 2003, **3**, 900.
58. Y. Bordon, *Nat. Rev. Immunol.*, 2012, **12**, 621.
59. K. Vaddi and M. Luchi, *Exp. Opin. Invest. Drugs*, 2012, **21**, 961.
60. J. J. O'Shea, M. Gadina and R. D. Schreiber, *Cell*, 2002, 109 Suppl, S121.
61. M. Pesu, F. Candotti, M. Husa, S. R. Hofmann, L. D. Notarangelo and J. J. O'Shea, *Immunol. Rev.*, 2005, **203**, 127.
62. M. Kawamura, D. W. McVicar, J. A. Johnston, T. B. Blake, Y. Q. Chen, B. K. Lal, A. R. Lloyd, D. J. Kelvin, J. E. Staples, J. R. Ortaldo, *et al.*, *Proc. Natl. Acad. Sci. U. S. A.*, 1994, **91**, 6374.
63. K. Shimoda, K. Kato, K. Aoki, T. Matsuda, A. Miyamoto, M. Shibamori, M. Yamashita, A. Numata, K. Takase, S. Kobayashi, S. Shibata, Y. Asano, H. Gondo, K. Sekiguchi, K. Nakayama, T. Nakayama, T. Okamura, S. Okamura, Y. Niho and K. Nakayama, *Immunity*, 2000, **13**, 561.
64. N. L. Alicea-Velazquez and T. J. Boggon, *Curr. Drug Targets*, 2011, **12**, 546.
65. M. E. Flanagan, T. A. Blumenkopf, W. H. Brissette, M. F. Brown, J. M. Casavant, S.-P. Chang, J. L. Doty, E. A. Elliott, M. B. Fisher, M. Hines, C. Kent, E. M. Kudlacz, B. M. Lillie, K. S. Magnuson, S. P. McCurdy, M. J. Munchhof, B. D. Perry, P. S. Sawyer, T. J. Strelevitz, C. Subramanyam, J. Sun, D. A. Whipple and P. S. Changelian, *J. Med. Chem.*, 2010, **53**, 8468.
66. J. E. Chrencik, A. Patny, I. K. Leung, B. Korniski, T. L. Emmons, T. Hall, R. A. Weinberg, J. A. Gormley, J. M. Williams, J. E. Day, J. L. Hirsch, J. R. Kiefer, J. W. Leone, H. D. Fischer, C. D. Sommers, H.-C. Huang, E. J. Jacobsen, R. E. Tenbrink, A. G. Tomasselli and T. E. Benson, *J. Mol. Biol.*, 2010, **400**, 413.
67. A. F. Wilks, *Semin. Cell Dev. Biol.*, 2008, **19**, 319.
68. J. S. Fridman, P. A. Scherle, R. Collins, T. C. Burn, Y. Li, J. Li, M. B. Covington, B. Thomas, P. Collier, M. F. Favata, X. Wen, J. Shi, R. McGee, P. J. Haley, S. Shepard, J. D. Rodgers, S. Yeleswaram, G. Hollis, R. C. Newton, B. Metcalf, S. M. Friedman and K. Vaddi, *J. Immunol.*, 2010, **184**, 5298.
69. C. A. Zerbini and A. B. Lomonte, *Expert Rev. Clin. Immunol.*, 2012, **8**, 319.
70. K. Ghoreschi, M. I. Jesson, X. Li, J. L. Lee, S. Ghosh, J. W. Alsup, J. D. Warner, M. Tanaka, S. M. Steward-Tharp, M. Gadina, C. J. Thomas, J. C. Minnerly, C. E. Storer, T. P. LaBranche, Z. A. Radi, M. E. Dowty, R. D. Head, D. M. Meyer, N. Kishore and J. J. O'Shea, *J. Immunol.*, 2011, **186**, 4234.
71. R. J. Riese, S. Krishnaswami and J. Kremer, *Best Pract. Res. Clin. Rheumatol.*, 2010, **24**, 513.

CHAPTER 4

Ion Channel Modulators

KARL R. GIBSON,*[a] PETER J. COX,[b] EDWARD STEVENS,[b] WENDY J. WINCHESTER,[b] AARON C. GERLACH,[c] BRIAN E. MARRON[c] AND RICHARD P. BUTT[b]

[a] Sandexis, Innovation House, Discovery Park, Sandwich, Kent, CT13 9ND, UK; [b] Pfizer Neusentis, Granta Park, Cambridge, CB21 6GS, UK; [c] Pfizer Neusentis, 4222 Emperor Blvd, Durham, NC 27703, USA
*Email: karl.r.gibson@sandexis.co.uk

4.1 Introduction—The Role of Ion Channels in Pain

Pain is defined by the International Association for the Study of Pain as 'an unpleasant sensory and emotional experience associated with actual or potential tissue damage or described in terms of such damage'. As such this definition describes the peripheral sensory input and central mechanisms that describe the patient's perception of pain. Many ion channels play a key role in the initiation and conduction of action potentials in nociceptive nerve fibres and in the synaptic transmission to second-order neurons in the spine (Figure 4.1). As such they have proven attractive targets for the discovery of novel therapeutic agents for the control of pain.

Without exception, current pain therapies offer limited efficacy to a limited number of subjects, and are all associated with side effects that limit their use. The search for novel analgesics therefore continues, with the priority targeting increased efficacy and fewer side effects through an increasing understanding of the molecular physiology of pain that is enabling targeting of key specific steps in the pain cascade.

RSC Drug Discovery Series No. 35
Pain Therapeutics: Current and Future Treatment Paradigms
Edited by Charlotte Allerton

Published by the Royal Society of Chemistry, www.rsc.org

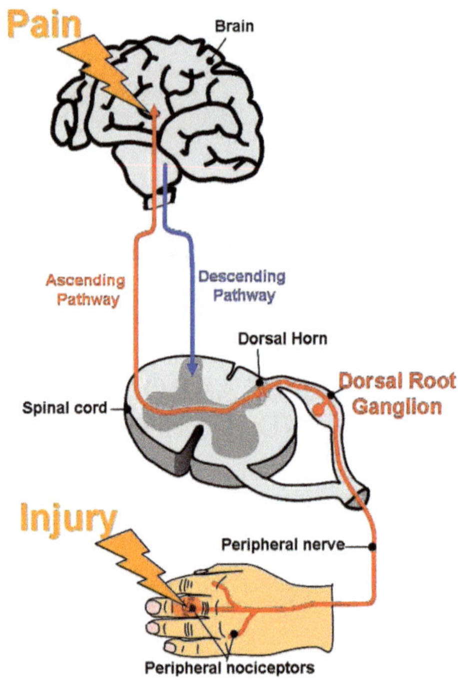

Figure 4.1 Pain pathway (courtesy of Neil Castle, Pfizer Neusentis, Raleigh, NC, USA).

Over the last decade an increased understanding of precise molecular mechanisms involved in generating pain signals following injury or sensitisation of the peripheral nociceptors, the transduction of these signals into action potentials (Figure 4.2), the conduction of these signals up sensory nociceptive neurons making up the peripheral nerve into the spinal cord, the transmission across the first synapse in the dorsal horn of the spinal cord, and the changes and key areas of the brain responsible for the perception of pain have been uncovered. With this growing knowledge the development of novel analgesics for the treatment of many subtypes of pain is a realistic long-term goal. The crucial role of ion channels in the generation, transduction and conduction of pain offer new targets for the treatment of acute and chronic pain.

The role of the peripheral pain-sensing fibres (the peripheral nociceptors) is to transmit external environmental stimuli or internal tissue injury into the spinal cord and then onwards for interpretation by the brain. Nociceptors are highly specialised, branched nerve fibres with defined receptive fields and represent a single-cell continuum from the periphery into the spinal cord that can extend to greater than 1 m in length. Primary nociceptors are found in many areas of the body including skin, cornea and in visceral tissues such as the gastrointestinal tract, muscles, joints and bladder. The role of the nociceptor is to generate an action potential, or a train of action potentials, when a noxious stimulus is detected. This process is entirely dependent on ion channels that

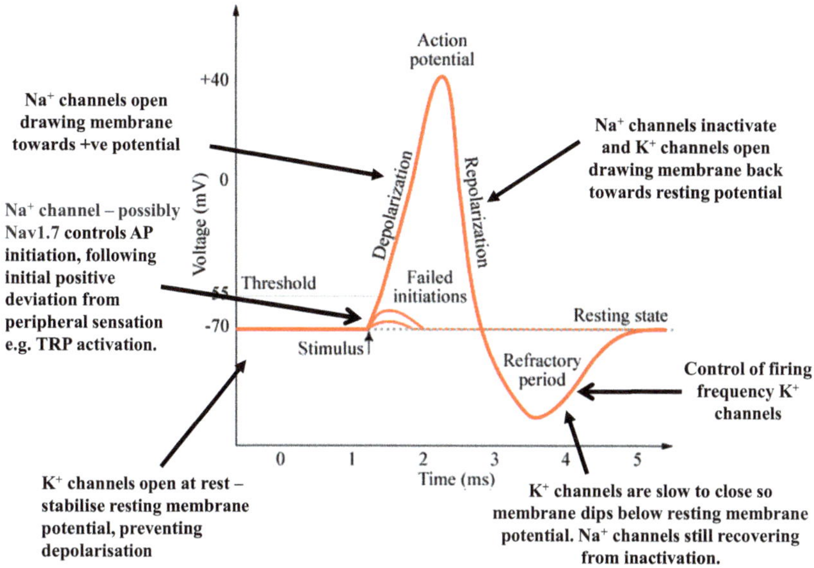

Figure 4.2 An action potential (Courtesy of Anne Phelan, Pfizer Neusentis, Cambridge, UK).

convert the stimulus into electrical signals that are conducted into the central nervous system (CNS).

A nociceptor at rest has a negative membrane potential, stabilised by open potassium channels. Through the involvement of a range of 'sensing' ion channels at the nociceptor terminals, a depolarisation from the resting membrane potential is generated in the presence of a stimulus. This small change in depolarisation is then amplified and conducted by sodium channels. Additionally, the sensitivity of the nociceptor can be changed through modulation as a consequence of the action of inflammatory mediators. As a threshold membrane potential is achieved sodium channels open, generating a large positive swing in membrane potential. Sodium channels then inactivate and potassium channels open, reducing the membrane potential, generating the rapid pulse of electrical activity that is the action potential. The slow closure of potassium channels leads to a dip below the resting potential forming a refractory period before the next action potential can fire. The size of action potentials remains constant; it is rather the rate of firing of action potentials that governs how painfully any stimulus is perceived to be.

Several distinct populations of nociceptors can be characterised by their electrical properties, but as yet have not been able to be characterised by the ion-channel components, even though these components must underlie their differential responsiveness to mechanical, chemical, thermal and polymodal stimuli. Two fundamentally different types of nociceptor have been identified with Aδ fibres being thinly myelinated, conducting at approximately 20 ms^{-1} and being responsible for sharp pain, and C fibres being unmyelinated, conducting slowly at $2\,ms^{-1}$ and being responsible for prolonged dull burning and aching pain.

As shown in Figure 4.1, these sensory action potentials from afferent nociceptive fibres are conducted to the spinal cord where they form synapses in the dorsal horn. Within the spinal cord at the first synapse neurotransmitter release is further regulated by additional different classes of ion channels such as the voltage-gated N-type calcium channel. Presynaptic release of neurotransmitters such as glutamate, substance P, glycine and GABA are then responsible for excitation and inhibition of ascending spinal second-order neurons that transmit pain signals into the brain.

In summary, while the complexity of nociceptor function is unravelling to the point we can now specifically target this type of neuron, at a molecular level the full complexity of pain sensing and the key role of ion channels is yet to be fully understood. This emerging knowledge is reflected in the comparison between current ion-channel-targeted pain medicines and those only now emerging from research laboratories. This chapter will deal with the drugs and drug discovery research targeting ion channels influencing the formation, conduction and synaptic transmission of pain-related action potentials.

4.2 TRP Channels as Nociceptor Terminal Excitators—Signal Generation

This section concentrates on TRP channels as candidates for the role of sensing ion channels as they have received extensive pharmaceutical research interest due to the weight of evidence as to their physiological importance. Other ion-channel families such as the P2X family or ASIC family may also play important roles, and these have been recently reviewed elsewhere.[1–3]

The TRP cation family consists of 28 mammalian members with 6 subfamilies, the vanilloid (TRPV), melastatin (TRPM), ankyrin (TRPA1), polycystin (TRPP), mucolipin (TRPML) and canonical (TRPC). The focus of this section is to concentrate on those TRP channels which have been implicated to play a role in pain processing which has resulted subsequently in pharmaceutical evaluation. Research into the role of the vast majority of this superfamily is in its infancy and a role for other family members may be revealed in the future.

It had long been established that sensitivity to capsaicin, the pungent chemical isolated from chilli, was a surrogate marker for somatic nociceptors. The discovery of the molecular structure of TRPV1[4] and subsequent demonstration of its sensitivity to candidate pain activators (heat, low pH)[5] resulted in an explosion in the pharmaceutical exploitation of the channel. The identification of other family members followed shortly afterwards: TRPV2,[6] TRPV3,[7–9] TRPV4,[10] TRPA1[11] and TRPM8,[12,13] and defined the TRP family as a set of environmental detectors to temperature across a broad temperature range from less than 10 °C to more than 45 °C. In addition to the thermal sensitivity of the channels, they each respond to additional environmental stimuli such as acidification and osmolarity as well as noxious and pungent chemicals and toxins.

TRP proteins are six-transmembrane domain subunits which exist as homotetramers or potentially heterotetramers to form a cationic channel. They are non-selective, although a number are Ca^{2+}-preferring ion channels, which upon activation leads to influx of cations into the cell resulting in subsequent cell activation. Each TRP channel has its own unique multiple modalities of activation. This section will only discuss the TRP channels with the strongest rationale for involvement in pain states: TRPV1, TRPA1, TRPV2, TRPV3, TRPV4 and TRPM8.

4.2.1 TRPV1

TRPV1 is localised on small-diameter C and Aδ afferent fibres within the peripheral nervous system and has been reported to have a functional pain role within the periphery as well as at the central afferent synapse.[14] It is activated by capsaicin, heat (>42 °C) and low pH.[15] In addition, it is modulated, through cellular activation and subsequent intracellular processing, by bradykinin, and nerve growth factor (NGF)[16] and ATP,[17] all of which are established pain mediators. Furthermore the by-products of the arachidonic acid cascade and lipoxygenase products cause phosphorylation of the channel.[18] This localisation and broad modulation by inflammatory mediators highlighted TRPV1 as a core target for the treatment of pain, particularly inflammatory pain.

The use of desensitising agonists, both capsaicin and the later identified resiniferatoxin (RTX) (Figure 4.3),[19] has been shown to reduce pain sensitivity preclinically[20] and clinically.[21] Capsaicin-mediated desensitisation, the mode of action for clinical topical formulations, is hampered by the initial excitation of the nociceptor. However, utilisation of the rapid desensitisation remains a therapeutically viable topical analgesic. Recent approval of the Astellas Qutenza® patch, with a high concentration of capsaicin (8%), suggests a remaining appetite for such therapies; in this therapy a topical local anaesthetic

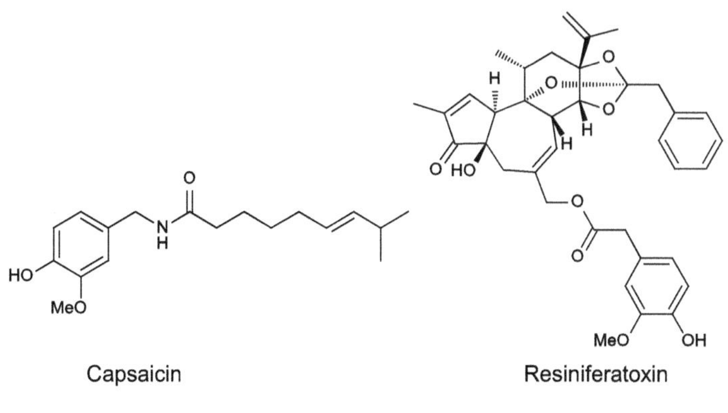

Figure 4.3 Capsaicin and resiniferatoxin.

is applied to the painful area prior to topical capsaicin to prevent the initial painful flare.

Given the preclinical and clinical experience of desensitising agonists, there was considerable pharmaceutical investment into the discovery of TRPV1-selective antagonists for the treatment of pain. Several companies discovered and developed TRPV1 antagonists which entered clinical development (Figure 4.4). These compounds had been demonstrated to cause inhibition of all modalities of activation (heat and pH as well as capsaicin activation). Further demonstration of *in vivo* efficacy in pain models[22–25] delivered promise that this mechanism may result in analgesia clinically. A relatively early observation of an antagonist-mediated hyperthermia had been observed preclinically. However, this was shown to be an acute effect, lost on repeated administration of compound,[26] and it was not clear whether this effect would be observed clinically, or whether it would be transient. Therefore, although caution was required during early clinical development, a number of companies continued to progress molecules into the clinic.

A significant clinical hyperthermia was demonstrated initially first with the Amgen compound AMG-517[27] and subsequently with the AstraZeneca compound AZD-1386[28] and ABT-102.[29] Only SB-705498 from GlaxoSmithKline (GSK) did not display marked hyperthermia compared to the other compounds,[30] but GSK certainly pursued a cautious clinical development path, and it remains to be demonstrated whether this approach led to the same level of pharmacological blockade achieved with the other molecules. With respect to clinical analgesia, AZD-1386 achieved a transient inhibition of dental pain[31] but this was not sustained. There was also a blockade of oesophageal pain in a human experimental model of gastroesophageal reflux disease.[32] Similarly, SB-705498 was effective against capsaicin-induced hypersensitivity[31] but further clinical data is not available.

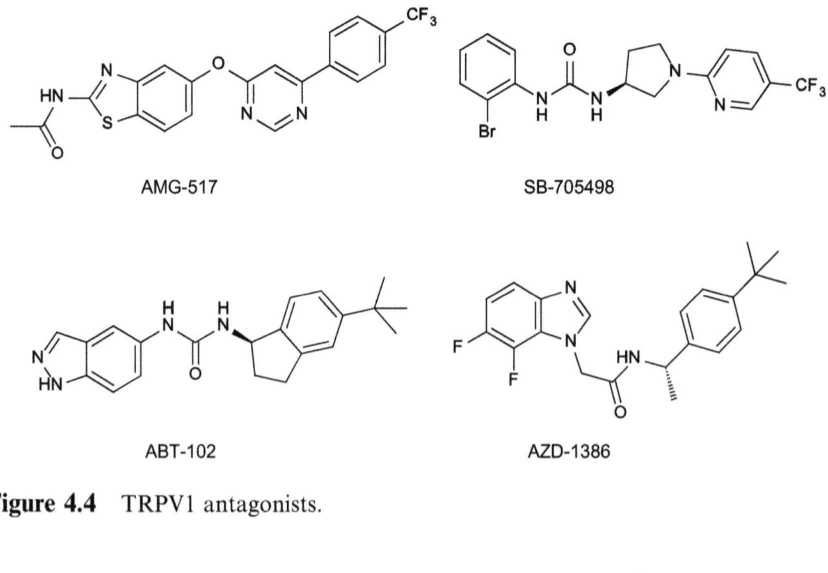

Figure 4.4 TRPV1 antagonists.

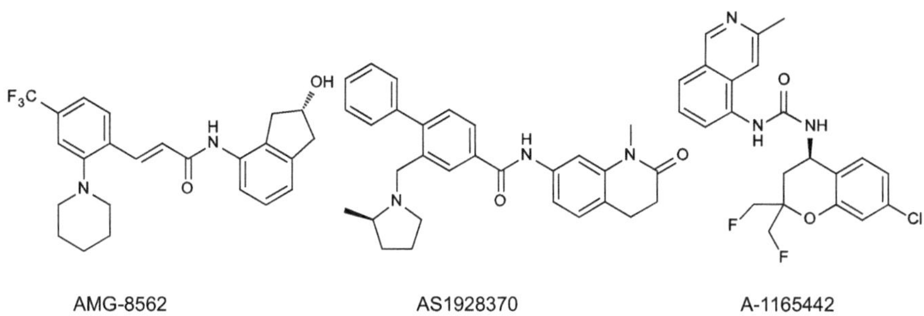

Figure 4.5 Modality-sparing TRPV1 antagonists.

Achieving significant, sustained inhibition of TRPV1 to evaluate analgesia for this mechanism has not been possible because of the hyperthermic activity of the first generation of blockers. Recent activity within the industry has refocused upon modality-selective or at least modality-sparing compounds such as AMG-8562,[33] AS1928370[34] and A-1165442[35] (Figure 4.5). The main focus of effort has been on developing compounds which block the activation of TRPV1 but not all mechanisms of activation, with proton or temperature activation being spared. This may afford analgesia without the concomitant hyperthermia demonstrated with the first generation of TRPV1 antagonists. Efficacy has been demonstrated in preclinical pain models without a concomitant increase in core body temperature, which supports this hypothesis.[36,37] However, whether this represents a viable option and whether efficacy translates from rodent to human still is untested. It will be very interesting to watch the further development of modality-selective blockers into the clinic.

4.2.2 TRPA1

TRPA1, initially named ANKTM1 due to the presence of N-terminal ankyrin repeats,[38] has a similar expression pattern to TRPV1 with localisation in small-diameter sensory neurons within dorsal root ganglia (DRG) as well as trigeminal ganglia,[11] and colocalised with CGRP and TRPV1. The channel is activated by cold (<15 °C), resulting in an influx of calcium. A number of natural noxious chemicals are known to activate TRPA1 through the covalent modification of cysteine residues, including mustard oil,[39] allyl isothiocyanate (AITC, a mustard oil constituent), cinnamaldehyde and formalin (Figure 4.6).[40] Formalin is also a known pain stimulus used to induce pain states in preclinical species. Like TRPV1, TRPA1 is also activated by bradykinin,[41] suggestive of a role in integrating inflammatory pain.

The TRPA1 channel has been directly linked to pain in humans by the discovery of an autosomal-dominant familial episodic pain syndrome, characterised by episodes of debilitating upper body pain. A single mutation of the amino acid residue N855S results in a gain-of-function mutation of TRPA1.[42] Preclinically, the evidence is mixed. TRPA1 antagonist HC-030031 shows beneficial effects in a spinal nerve ligation model of neuropathic pain,[43] but the

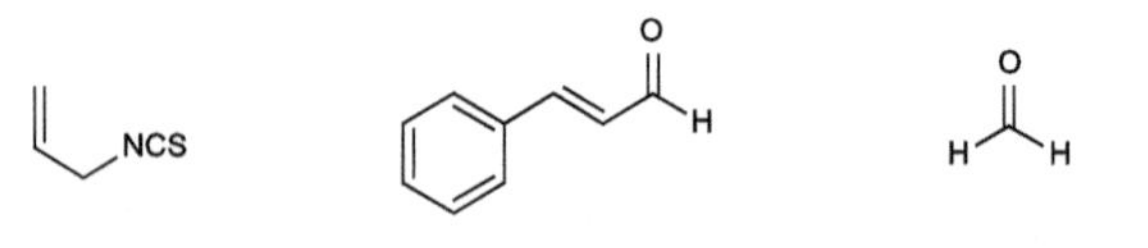

Mustard oil / allylisothiocyanate   Cinnamaldehyde   Formaldehyde

Figure 4.6 TRPA1 activators.

TRPA1 null mice still responded to nerve ligation. These mice did display an analgesic phenotype, however, as a decreased response to mechanical and bradykinin-induced pain was observed.[44] Unsurprisingly, given that formalin has been shown to be a direct activator of TRPA1, the formalin pain test has been demonstrated to be mediated *via* TRPA1 as sensitivity was ablated by either the gene deletion or HC-030031.[40]

To date, no results of pain trials with a TRPA1 antagonist have been reported. The most advanced clinical TRPA1 antagonist is Glemark's GRC17536.[45] The company reports that GRC has successfully completed Phase 1 trials. A Phase 2, placebo-controlled, 4-week trial in diabetic peripheral neuropathy is currently recruiting with an expected completion date of March 2013 and the primary endpoint of mean 24-hour average pain intensity (API) score. The outcome from this study is eagerly awaited.

Although the structure of GRC17536 has not been disclosed, a potent exemplar from one of Glenmark Pharmaceutical's patent applications (Figure 4.7),

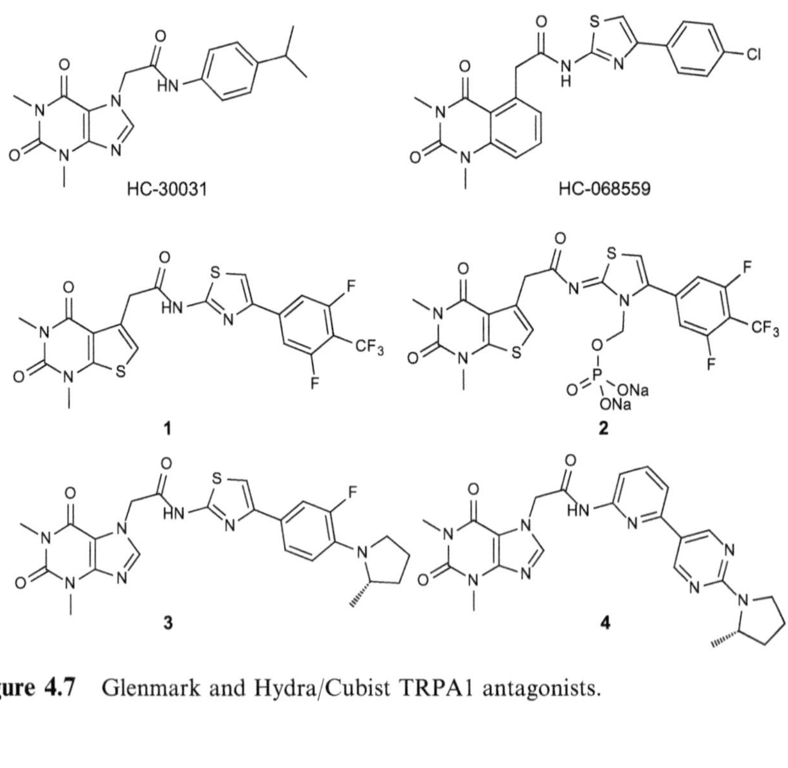

HC-30031   HC-068559

1   2

3   4

Figure 4.7 Glenmark and Hydra/Cubist TRPA1 antagonists.

compound **1**,[46] is reported to be a 2.5 nM inhibitor of TRPA1. Glenmark have filed patent applications where the 1,3-dimethylthienopyrimidine dione of **1** has been replaced with a number of other heterocycles.[47,48] A recent application[49] claims prodrugs whereby the prodrug moiety is attached on the acylthiazole nitrogen in its tautomeric form as shown with compound **2**. Compound **2** was reported to have a solubility of more than 1200 mg mL^{-1} compared to less than 0.2 mg mL^{-1} for the parent, and a substantial increase in plasma exposure was noted when analysing for the parent,**1**, after dosing the prodrug.

Hydra Biosciences has partnered their TRPA1 programme with Cubist Pharmaceuticals. HC-030031[50] and HC-068559[51] were early compounds from Hydra that have been widely used in pharmacology studies to elucidate the role of small-molecule TRPA1 inhibitors *in vivo*. Another patent application by Hydra Biosciences, describes compound **3**, which is a 4 nM TRPA1 blocker that demonstrated activity in the rat formalin and AITC-induced flinch models.[52] A recent joint application from Hydra Biosciences and Cubist Pharmaceuticals describe compound **4**, with an IC_{50} of 93 nM in patch clamp and equipotent activity against the rat and dog TRPA1 orthologues.[53] In the application they indicated that compound **3** displayed hepatotoxicity in preclinical studies as indicated by serum chemistry biomarkers and from 7-day rat toxicology studies whereas **4** did not. Compound **4** is reported to be more than 100-fold selective over other TRP channels, hERG and $Na_V1.5$ and active in several rodent pain models. Hydra/Cubist have reported completion of a single ascending-dose Phase 1 study with no safety signals attributable to the test article.

Several other groups have recently disclosed TRPA1 small-molecule blockers but it does not appear that any have entered the clinic at this time (Figure 4.8). Abbott has disclosed a hydroxyl amine series, AP-18,[54] with a structure that

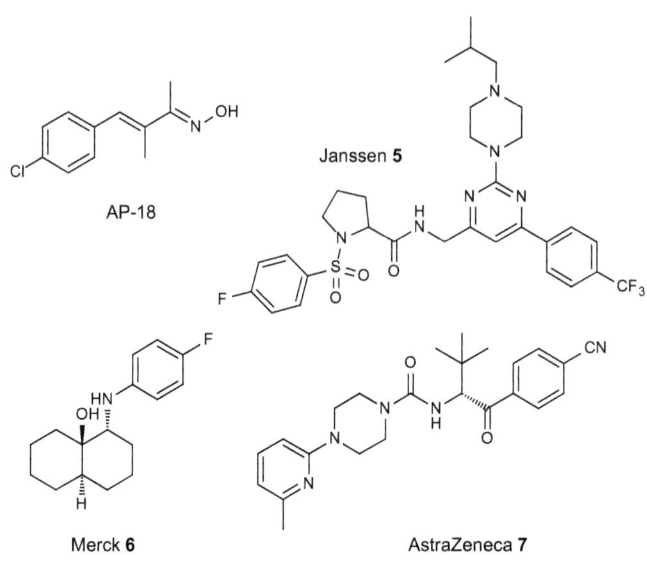

Figure 4.8 Non-clinical TRPA1 antagonists.

differs dramatically from the previously disclosed Hydra and Glenmark series. Janssen's disclosure, exemplified by **5**, covers a large number of analogues, many with less than 100 nM activity.[55] In another substantial departure from previous series Merck have disclosed decalin/hexahydro-chromene-ol compounds such as **6**.[56] Finally, Astra Zeneca has filed narrowly around **7**.[57] Compound **7** is reminiscent of TRPV1 compounds; however, no selectivity data is provided.

4.2.3 TRPV2

In spite of early identification of TRPV2[6] and broadly similar distribution of TRPV2 and TRPV1, no pharmaceutical investment has been made against identification of antagonists for this channel. Data suggesting that the channel is an osmosensor in blood vessels[58] and has a role in mast cell function[59] suggest that there will be limited future pharmaceutical exploitation of the channel for the treatment of pain.

4.2.4 TRPV3

TRPV3 is a calcium-permeable channel activated by warm temperatures (>31 °C) with localisation to keratinocytes and a subpopulation of small-diameter DRG neurons.[7,8] Following nerve injury, TRPV3 expression is significantly enhanced in human sensory neurons.[8,60] The keratinocyte localisation has also been shown to be enhanced in breast pain patients.[61] This raises the intriguing possibility that this target may represent an ideal candidate for a pain therapeutic with reduced side-effect liability in comparison to other TRP channels discussed here. TRPV3 has an additional property that it sensitises to repeated stimulation *via* a modulation of the N-terminal domain calmodulin binding site.[62] TRPV3 activation of keratinocytes results in the release of ATP which has been shown to subsequently activate sensory neurons and as such may act as a transducer of sensory inputs from the skin.[63,64] In addition to a potential role in pain modulation, a gain-of-function mutation in TRPV3 has been demonstrated in Olmsted syndrome (a dermal disorder).[65]

Identification of TRPV3 antagonists has enabled further exploration of a role of TRPV3 in pain processing preclinically. Thus, the Glemark compound GRC15300 (structure unknown) inhibited CFA-induced mechanical hyperalgesia.[66] Glenmark reported the completion of a Phase 1 study with GRC15300 in 2011. Following a deal with Sanofi-Aventis, a Phase 2 trial was opened in neuropathic pain in 2012 with predicted completion in 2013. Glenmark have filed multiple applications on a number of series (Figure 4.9). Examples of these are the pyrido-pyrimidine **8** and benzimidazole **9**[67] and the unrelated spiro-chroman chemotype **10**.[68]

Other patent activity for TRPV3 inhibitors has been primarily from Hydra Biosciences and Abbott (now AbbVie) (Figure 4.10). Hydra Biosciences have reported on a couple of series of substituted thio-dihydroisoquinolone ethones (**11** and **12**),[69,70] with selectivity claims in both applications over other TRP

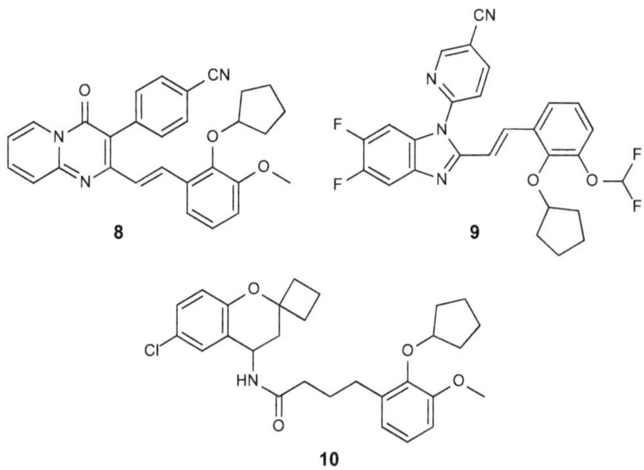

Figure 4.9 Glenmark TRPV3 antagonists.

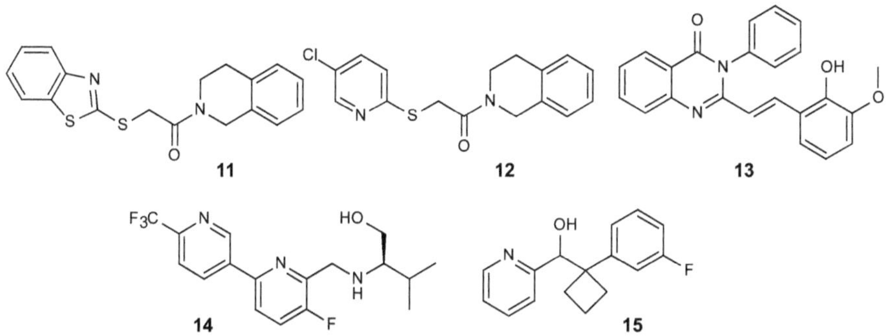

Figure 4.10 Hydra and Abbott TRPV3 antagonists.

channels. They have also reported a quinazolinone series **13** with similar structures to the Glenmark compounds;[71,72] however, it does not appear that they have advanced any compounds to the clinic at this time. Abbott have recently published applications based around a substituted pyridine core such as **14**[73] and **15**.[74] Compound **14** was reported to be active in the mouse chronic constriction injury (CCI) model of neuropathic pain when dosed orally at 100 mg kg^{-1}.

4.2.5 TRPV4

The TRPV4 channel[10] responds to temperatures in excess of 27 °C. In addition, it was identified as an osmosensor.[75] Localisation studies have identified broad distribution.[76] A role in visceral pain from the gastrointestinal (GI) tract,[77] pancreas[78] and bladder[79] has been demonstrated using the TRPV4 null mice.

There are two key factors which may have excluded significant investment by the major pharmaceutical companies. The first is the critical role of TRPV4 in cardiovascular control.[80,81] The second is its role in skeletal integrity, as evident from the human gain-of-function mutations.[82]

4.2.6 TRPM8

The final thermosensory channels to be considered as a pain therapeutic intervention is TRPM8. This channel was demonstrated to be the primary cold-sensing channel, with additional sensitivity to menthol and voltage.[12,13] Cold detection is significantly blunted in TRPM8 null mice.[83] TRPM8 is a potential candidate channel (in addition to TRPA1) for modulating cold allodynia, a feature of a number of pain syndromes. Recent publications have confirmed the upregulation of TRPM8 in chronic nerve injury,[84] with TRPM8 knockdown attenuating CCI-induced cold allodynia.[85] Similar effects to knockdown have been reported with the Janssen TRPM8 antagonists **16** and **17** (Figure 4.11).[86,87]

The identification, through a genome-wide association study (GWAS) of migraine, of a SNP close to the TRPM8 gene[88] subsequently replicated in an additional GWAS study,[89] raises the exciting possibility that TRPM8 may play a role in migraine progenesis.

However, in common with a number of the other TRP channels, already described above, TRPM8 seems to play additional, important, homeostatic role. Thus, pharmacological blockade results in a significant hypothermia.[90] This raises the question over whether inhibition of this channel can be achieved safely. Multiple companies have been active in this area (Amgen, Jannsen, Johnson & Johnson, Pfizer, Bayer, Mitubishi Tanabe, Senomyx, Glenmark and Raqualia) with no reported clinical activity to date.

In summary, considerable investment has been made into identifying compounds to target this superfamily of channels with a number of them progressing or poised for clinical evaluation. However, it remains to be demonstrated whether clinically relevant efficacy can be achieved without a concomitant detrimental effect upon homeostasis. The progression of compounds with selective modes of action in the case of TRPV1 and clinical trial outputs from the other classes will hopefully address these questions.

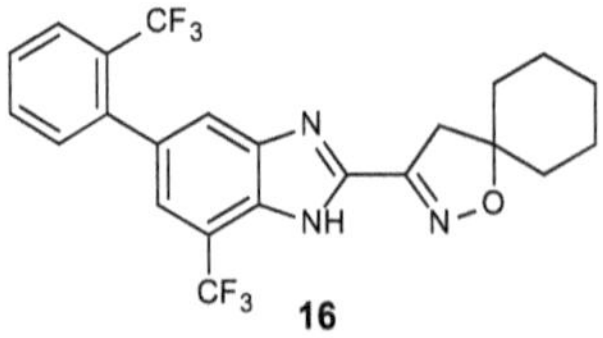

16

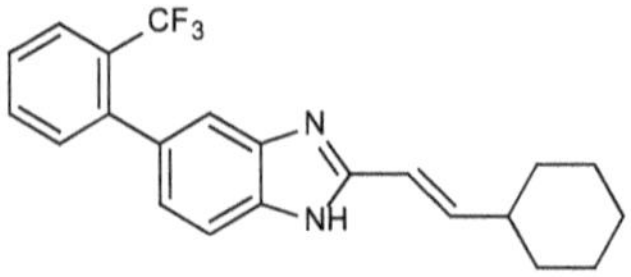

17

Figure 4.11 Janssen TRPM8 antagonists.

4.3 Voltage-Gated Sodium Channels—Signal Transduction and Conduction

Sodium channels are key regulators of cellular excitability, contributing to the generation and propagation of action potentials in neurons, muscle and heart tissues. Consequently, sodium channels are precedented drug targets as anti-dysrhythmics, anticonvulsants and anaesthetics.[91] Indeed, non-selective sodium channel blockers such as local anaesthetics (lidocaine, benzocaine and bupivacaine) have been widely used for many years to prevent pain in acute settings by silencing peripheral nerves as injectable or topical agents at the site of action. However, the use of non-selective sodium channel blockers in the treatment of pain is limited due to side effects arising from the blockade of sodium channels in other tissues in addition to the nociceptors.

In total there are nine voltage-gated sodium channels ($Na_V1.1$–$Na_V1.9$) of which several are proposed to play a role in nociceptive fibres. The limited and selective expression of sodium channel isoforms suggest that if the right combination of voltage-gated sodium channels can be blocked then nociceptor function and sensory pain input could be diminished, without the side effects of the current small-molecule non-selective sodium channel blockers or toxins acting on the cardiovascular ($Na_V1.5$), neuromuscular ($Na_V1.4$) and CNS ($Na_V1.1$, 1.2) sodium channels. In this section the sodium channels thought to be involved in nociceptor function (Na_Vs 1.3, 1.7, 1.8 and 1.9) are reviewed with respect to their biological function, structure–function relationships and the modulators known to be in drug development.

4.3.1 Na_V Channel Structure

Figure 4.12 shows an apo crystal structure of a voltage-gated sodium channel.[92] Although some drug binding sites have been deduced from site-directed mutagenesis experiments, there are currently no co-crystal structures of any ligand bound in a channel.

The sodium channel protein is comprised of four subunits that create a central pore with an ion selectivity filter, surrounded by four voltage-sensor domains. This voltage-sensing domain (VSD) responds to membrane depolarisation and through conformational change opens a gate at the base of the pore through movements of all four subunits. In this way, the ion channel moves from one state (*e.g.* closed) to another (open) through concerted movements and then through to an inactivated state as the channel ceases to conduct ions. As the voltage-gated sodium channel exists in distinct conformations for these different closed, open and inactivated states, it follows that ligands may exhibit different strengths of interaction (potencies) for these states. It follows from this that sodium channel modulators may exhibit state-dependent pharmacology where there may be selective interaction with one state, or conformation of the channel. If the ligand can selectively bind to and stabilise a particular channel state, then it also follows that such a modulator can perturb the channel/state population distribution and 'collect'

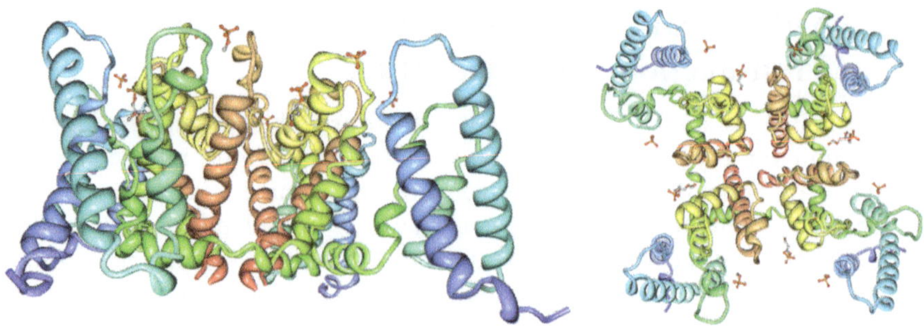

Figure 4.12 A bacterial Na_V channel, Na_VAb, PDB code 4EKW.

the channels into the state associated with this bound conformation. This will occur more readily for channels cycling through these states at a higher rate, for example in a nociceptor conducting a train of action potentials in response to a noxious stimulus. Thus it follows that this 'state-' and 'use-dependent' modulation of sodium channels may influence the channels in more rapidly firing pain neurons more than in other physiological systems, potentially allowing some therapeutic index to be observed even for ligands lacking subtype selectivity.

4.3.2 Non-selective Na_V Modulators

Tetrodotoxin (TTX) binds to the extracellular channel pore, and the sensitivity of the nine sodium channels to TTX blockade has been used to divide the family into two classes: (1) TTX-S (sensitive), IC_{50} <30 nM against Na_V1.1, 1.2, 1.3, 1.4, 1.6, 1.7 and (2) TTX-R (resistant), IC_{50} >30 nM against Na_V1.5, 1.8). This affinity difference is proposed to be due to a key cysteine residue in the TTX-binding site.[93] Local anaesthetics such as lidocaine, benzocaine and bupivacaine have been used for many years to silence peripheral nerves as injectable or topical agents at the site of action. From a physicochemical and structural point of view, all of these compounds are either weakly basic or neutral and bind to a site on the intracellular side of the channel pore.[94] This region of the channel shows a high degree of sequence conservation across subtypes, perhaps explaining the lack of selectivity observed across Na_V subtypes for these ligands. In addition, several more recent small-molecule modulators with general Na_V activity have entered clinical development (Figure 4.13). These include ralfinamide, a state- and use-dependent Na_V modulator from Newron which is reported to act through Na_V1.3, Na_V1.7, Na_V1.8 and the voltage-gated calcium channel Ca_V2.2, although this compound[95] failed to show superiority to placebo in a neuropathic lower back pain clinical trial.[96] Lacosamide from UCB enhances slow inactivation of Na_V channels. This compound also possesses subtype selectivity towards TTX-S

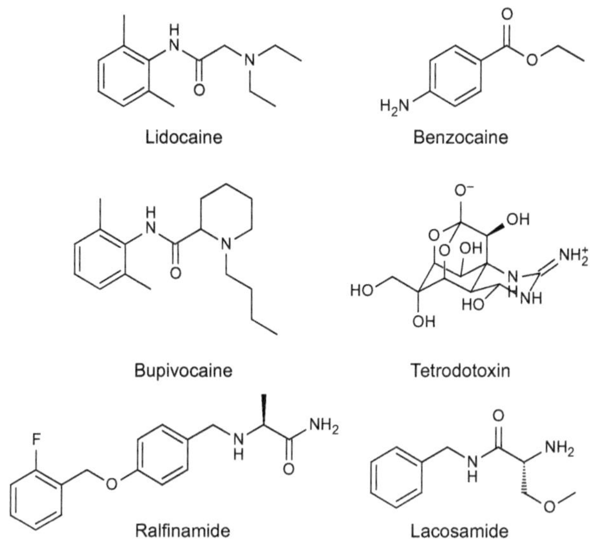

Figure 4.13 Non-selective Na_V channel modulators.

channels with $Na_V1.3$ IC_{50} 415 nM, $Na_V1.7$ IC_{50} 182 nM and $Na_V1.8$ IC_{50} 16 μM.[97] Lacosamide was approved in 2008 for the treatment of partial-onset seizures and diabetic neuropathic pain.

4.3.3 $Na_V1.3$

The biophysics of $Na_V1.3$ support a role in pain processing. $Na_V1.3$ is capable of generating a large ramp current, and capable of generating trains of action potentials required for pain sensing due to its rapid recovery from inactivation. These properties confer significant versatility and support a role in either nociceptor terminals in amplifying small currents/generating action potentials in the periphery and a potential function as a spinal amplifier in the CNS. Expression profiling and preclinical study of $Na_V1.3$ remains inconclusive with regards to its potential as a pain drug target. $Na_V1.3$ is predominantly expressed during development and not expressed under normal physiological states in adult animals or humans. The data supporting $Na_V1.3$ in pain states is equivocal with target upregulation observed following spinal cord injury in rats, but differing reports of the impact of target knockdown with antisense oligonuceotides.[98,99] Furthermore, neither the $Na_V1.3$ global knockout mouse, nor an inducible knockout of $Na_V1.3$ solely in nociceptive fibres expressing $Na_V1.8$, showed a neuropathic pain phenotype; and in these animals *in vitro* measurements of excitability were also unchanged.[100] More recently knockdown studies in peripheral DRG neurons using small hairpin RNA has suggested a role for peripherally expressed $Na_V1.3$ following spared nerve injury.[101] The only human data supporting a role for $Na_V1.3$ in neuropathic pain demonstrate target expression in human neuromas.[102]

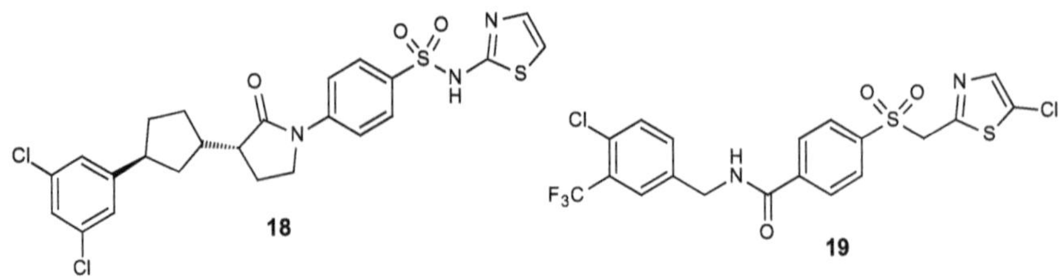

18 19

Figure 4.14 $Na_V1.3$ modulators.

Therefore, although the biophysical properties of the channel and the reported upregulation following injury in animals and humans is attractive, the uncertainty of the precise physiological role and function remain open questions and render $Na_V1.3$ an interesting but higher-risk drug target.

There are currently no selective modulators of $Na_V1.3$ in clinical development, but Vertex, Icagen and Pfizer have published several patents disclosing similar series containing acidic (estimated $pK_a \sim 7$), *N*-heteroaryl sulphonamides (Figure 4.14), with $Na_V1.3$ inhibitory activity (compound **18** reported to have $Na_V1.1$ and $Na_V1.3$ $IC_{50} < 2\,\mu M$)[103] (compound **19** reported to have $Na_V1.3$ IC_{50} 30 nM and >1000-fold selectivity over the cardiac channel $Na_V1.5$).[104,105]

4.3.4 $Na_V1.7$

In the nociceptive system $Na_V1.7$ is expressed solely in nociceptive and sympathetic neurons, making it an interesting target for all types of pain which rely on environmental sensory perception and noxious peripheral stimuli. The biophysics of $Na_V1.7$ also support this role, with a large ramp current and slow recovery from inactivation placing it ideally to amplify small generator currents in the peripheral nociceptor terminals, and to provide the 'ramp current' that depolarises the nociceptor membrane to enable other channels, such as $Na_V1.3$ or $Na_V1.8$, to generate trains of action potentials.[106] As such $Na_V1.7$ can be considered the 'gain' at the nociceptor terminal. The slow recovery from inactivation means $Na_V1.7$ cannot support repetitive or rapid firing of sensory neurons.

The major interest in $Na_V1.7$ as a drug target was triggered in 2006 with a publication that demonstrated that congenital insensitivity to pain (CIP) could be conferred solely by missense mutations in $Na_V1.7$.[107] These findings describe $Na_V1.7$ as being essential and non-redundant for the sensation of pain. Remarkably, all the other sensory functions in these subjects are intact. Sensation of warm, cool, vibration and touch remain normal with the only other consistent finding being a profound anosmia (lack of sense of smell). The mechanism of anosmia has recently been studied in the olfactory system where the role of $Na_V1.7$ is proposed to be as a presynaptic regulator of neuropeptide release at the first synapse in the olfactory system.[108]

Prior to the findings in CIP patients, gain-of-function mutations in $Na_V1.7$ had been described in the rare, extreme pain conditions of inherited primary erythromelagia (IEM)[109] and paroxysmal extreme pain disorder (PEPD).[110] In summary, loss of target functions causes loss of pain sensation, whereas gain-of-function mutations lead to the creation of spontaneous pain states. Preclinically extensive work has been published using *in vitro* electrophysiological techniques as well as *in vivo* models to demonstrate physiological function. *In vitro* studies have proposed a role for $Na_V1.7$ in amplifying small depolarising generator potentials caused by environmental factors and transduced through activation of, for example, TRP channels or GPCRs. This amplification is then proposed to cause further depolarisation that activates other sodium channels, such as $Na_V1.8$ and $Na_V1.3$, to generate action potentials. The described biophysics of Na_V 1.7 support this hypothesis.[106] Preclinical *in vivo* data in support of $Na_V1.7$ as a pain target is less impressive. The global knockout mouse is not viable after birth, reportedly due to anosmia resulting in a lack of feeding instinct. The nociceptor specific knockout animal in which $Na_V1.7$ is ablated in neurons expressing $Na_V1.8$ shows a pain phenotype to inflammatory and nociceptive stimuli, but no neuropathic pain phenotype.[111] Additionally the $Na_V1.7$ and $Na_V1.8$ combined knockout animal also has no neuropathic pain phenotype.[112] This data is consistent with similar findings from the $Na_V1.3$ and $Na_V1.8$ knockout animals in which very limited pain phenotypes are observed, and raises questions about the translation from these preclinical models to the clinical setting, and also with respect to cross-species differences in the fine details of pain processing and perception.

The clinical genetics findings described above have triggered a huge pharmaceutical investment in the discovery of selective $Na_V1.7$ blockers for the treatment of all types of pain (Figure 4.15). The challenge in this field will be to derive true pharmacological selectivity for $Na_V1.7$ over the other Na_V subtypes expressed in the heart and the CNS, and to demonstrate clinical analgesic efficacy for selective channel blockers in patient populations without causing unwanted anaesthesia. A number of companies have moved compounds with $Na_V1.7$ activity into clinical trials including the proline derivative **20** from Convergence[113,114] and a neutral spiro-oxindole template from Xenon, exemplified by XEN402 which has effectively treated pain in a patient with congenital erythromelalgia.[115] Pfizer have also advanced a $Na_V1.7$ compound into clinical trials for the treatment of pain,[116] from a joint discovery collaboration with Icagen, which has reported aminoheterocycle sulfonamides (*e.g.* **21**) with $Na_V1.7$ potency in the single digit nanomolar range.[117,118] Still more $Na_V1.7$ modulators have been described preclinically, with a broad array of structural types reported. Benzazepinone **22** has been reported by Merck to be a state-dependent inhibitor of $Na_V1.7$ (IC_{50} 90 nM) and $Na_V1.8$ (IC_{50} 680 nM) displaying ~10-fold use-dependent selectivity over $Na_V1.5$.[119] Finally, AstraZeneca have published several sodium channel inhibitor patents covering chromane-based inhibitors such as **23** with $Na_V1.7$ IC_{50} 66 nM, $Na_V1.5$ IC_{50} 13 μM, $Na_V1.2$ IC_{50} >33 μM.[120,121] With the exception of Xenon's treatment of patients carrying $Na_V1.7$ gain-of-function mutations there have yet to be

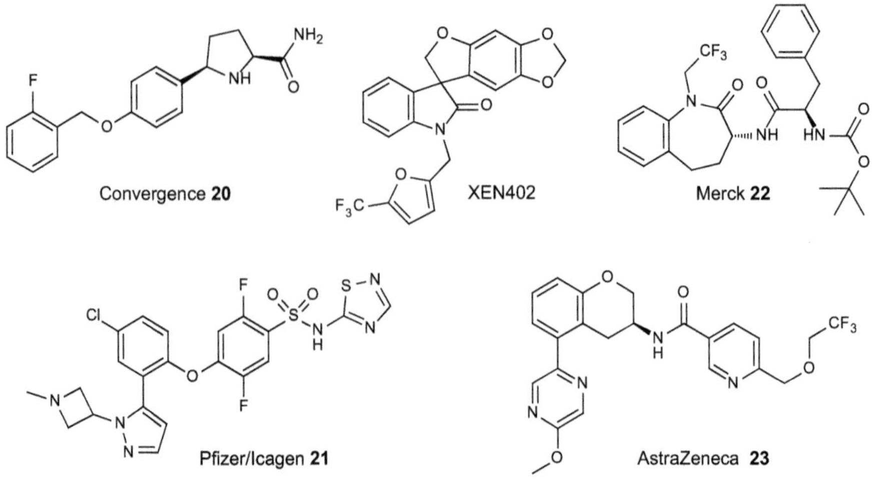

Figure 4.15 Na$_V$1.7 modulators.

peer-reviewed reports of efficacy of selective Na$_V$1.7 modulators in human pain. Convergence have recently reported positive interim data from the open-label phase of a trigeminal neuralgia study. With a number of clinical trials ongoing across the industry, further outcomes are eagerly awaited in the field.

4.3.5 Na$_V$1.8

Na$_V$1.8 is preferentially expressed in nociceptive neurons and is considered responsible for generating the current represented by the upstroke of the action potential.[122] The biophysics of the channel support this role, evidenced by activation and inactivation at depolarised membrane potentials with rapid recovery from inactivation. The combination of expression profile and biophysical properties underlie the proposed role for Na$_V$1.8 in driving the repetitive firing in nociceptors that is critical for the sensation of pain.

There is a wealth of preclinical data supporting a key role for Na$_V$1.8 in both inflammatory and neuropathic pain responses. This is most convincingly demonstrated following the discovery of selective Na$_V$1.8 blockers exemplified by A-803467 (Figure 4.16).[123] However, a role for Na$_V$1.8 in neuropathic pain was once again not observed in the Na$_V$1.8 knockout mouse.[124] The difference in these observations with pharmacological *versus* knockout studies may reflect compensatory changes in other ion channels and these results should be treated with caution in translating to the clinical situation.

Interestingly, there is a significant volume of data demonstrating that Na$_V$1.8 is modulated in expression and function through the action of intracellular signalling cascades. Visceral inflammation, NGF, IL-1β and TNF-α and pathways downstream of these mediators such as ceramide and p38 increase Na$_V$1.8 expression and/or current density. Furthermore PKA and PKC produce hyperpolarising shifts in voltage-dependent activation. Finally, DRG

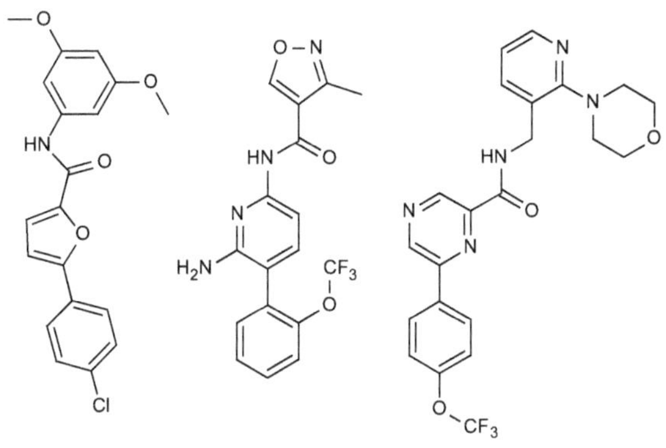

A-803467 **24** **25**

Figure 4.16 $Na_V1.8$ modulators.

neurons from diabetic rats show increased current amplitude and hyperpolarising shifts of activation and steady-state inactivation, and more recently data from diabetic neuropathy patients with increased circulating methyl glyoxal has been reported to be associated with pain in this patient groups through posttranslational modification of $Na_V1.8$.[125] This body of data clearly demonstrates a regulation of $Na_V1.8$ that may play a role in modifying the threshold or intensity of pain sensation mediated through $Na_V1.8$.

Although much of the focus in the development of sodium channel blockers was previously in the development of selective $Na_V1.8$ blockers, much of the focus has shifted to $Na_V1.7$ based on the human genetic rationale for that target. However, $Na_V1.8$ remains a key ion channel in the nociceptive pain cascade and the human pharmacological consequence in pain states of selective $Na_V1.8$ block has not yet been investigated. Some progress has been reported in the preparation of $Na_V1.8$-selective sodium channel modulators, although no clinical trial results have been reported to date. Abbott in collaboration with Icagen reported the first $Na_V1.8$ selective modulator A-803467 which has become the standard preclinical pharmacological tool in the field.[126] Pfizer has reported entering clinical trials with a $Na_V1.8$-selective compound[91] and have patented a 6,6-biaryl series exemplified by **24** reported to selectively modulate $Na_V1.8$ with a whole-cell $Na_V1.8$ IC_{50} of 260 nM and selectivity of more than 20-fold over $Na_V1.1$, $Na_V1.5$ and $Na_V1.7$.[127,128] Interestingly, many of the reported modulators possess a relatively rigid biaryl motif, although the aryl rings differ in size and substitution. A collaboration between Abbott and Icagen has resulted in alternative 6,6- and 6,5-biaryls with pyrazine **25** showing $Na_V1.8$ IC_{50} of 30 nM and good selectivity over $Na_V1.2$ and $Na_V1.5$.

4.3.6 $Na_V1.9$

$Na_V1.9$ is also preferentially expressed in small-diameter, nociceptive DRG neurons. The $Na_V1.9$ current has a hyperpolarised voltage dependency of activation compared to $Na_V1.7$ and $Na_V1.8$ with widely overlapping activation and inactivation curves enabling a persistent current. The observed ultra-slow inactivation suggests no contribution to the action potential, but rather a role in setting the resting membrane potential and the response to subthreshold stimuli. Significant discrepancies in the literature exist with the $Na_V1.9$ knockout animals from different groups with some findings suggesting a role in inflammatory and visceral pain[129] while other studies show diminished or no effect of $Na_V1.9$ knockout or knockdown.[130]

Together with the uncertain physiological role, this preclinical literature remains confused, and it is difficult to reach a full interpretation on the validity of $Na_V1.9$ as a therapeutic drug target. As far as the authors are aware there is no ongoing drug development research into $Na_V1.9$. Although it is an interesting potential target, research in this field is compromised by the inability to stably express $Na_V1.9$ recombinantly in cells to enable a successful screening campaign.

In general, while the effectiveness of local anaesthetics and the human genetics surrounding $Na_V1.7$ make selective sodium channel modulators exciting targets for the treatment of pain states, to date subtype-selective sodium channel blockers have not demonstrated clinical efficacy. Surprisingly, considering the human genetics of sodium channels, sodium channel knockout animals have been remarkable for their lack of robust pain phenotypes, not only for $Na_V1.3$, but also for $Na_V1.7$ and $Na_V1.8$, and combination knockouts. This suggests that in addition to the inherent challenges in discovering subtype-selective sodium channel modulators the translational research from preclinical stage to clinical stage drug development will be complex.

4.4 Potassium Channels as Pain Targets

4.4.1 Introduction to Voltage-Gated Potassium Channels

Potassium channels (K^+ channels) are a class of ion channels that modulate excitability of a variety of cell types including nociceptive neurons. They are defined by their selectivity for conducting potassium ions across cell membranes. K^+ flux across a neuronal cell membrane drives the membrane to more negative potentials, resulting in suppression of excitability. Hence, in neurons K^+ channels serve to dampen neuronal excitability by either stabilising the membrane potential at rest or by repolarising/hyperpolarising the membrane potential following action potential firing (see Figure 4.2). For this reason, K^+ channel openers are attractive therapeutic targets for the treatment of conditions that involve neuronal hyperexcitability, such as pain.

There are a number of plausible K^+ channels in nociceptors, but K_V7 channels are the most understood K^+ channels as relates to pain processing. K_V7 channels are gated by membrane voltage and K_V7 openers have demonstrated clinical effects in the treatment of pain although significant unanswered questions about the role of K^+ channels, and K_V7, in pain remain. Although other K^+ channel targets are gaining momentum as pain targets, this work is in its infancy and additional studies are required.

4.4.2 K_V7 channels

K_V7 channels encompass a gene family consisting of five members denoted $K_V7.1$–7.5. $K_V7.2$–7.3 and $K_V7.5$ channels are widely expressed in neurons, including cortical, hippocampal, spinal and dorsal root ganglion neurons. Each gene encodes a six-transmembrane subunit of a voltage-gated K^+ channel. The individual subunits contain a charged S4 region and single P-loop that self-assemble as tetramers to form a voltage-dependent, K^+-selective ion channel. The assembled tetramers include homomeric and heteromeric combinations that increase K_V7 channel diversity. These channel subtypes are important signalling proteins and potential drug targets for new therapies to treat pain and other disorders of neuronal excitability.[131] By contrast, $K_V7.4$ channel expression is more restricted to neurons involved in auditory and dopaminergic signalling, while $K_V7.1$ is expressed primarily in non-neuronal tissues, including cardiac tissue where it contributes to repolarisation of the cardiac action potential.[132] Thus for any K_V7 opener drug discovery programme, obtaining selectivity over $K_V7.1$ is important.

There is a genetic association of K_V7 channel mutations and neuronal hyperexcitability observed in benign familial neonatal convulsions (BFNC), a rare epilepsy. Mutations in $K_V7.2$ and $K_V7.3$ are observed in patients with this disorder.[133] To date this seizure disorder has been linked to over 35 missense, frame-shift and splice-site mutations in $K_V7.2$ or $K_V7.3$ that compromise but do not eliminate channel function. This linkage with neuronal hyperexcitability suggests that K_V7 channels may also modulate the excitability of pain-processing neurons as $K_V7.2$, 7.3 and 7.5 channels are expressed in pain pathway neurons such as DRG, spinal dorsal horn and thalamic neurons.[134–137] Thus, in theory K_V7 channel openers could reduce the hyper-excitablity of pain pathway neurons by driving the membrane potential more negative in relation to action potential threshold.

Figure 4.17 illustrates the structures of several agents which activate or block K_V7 channels. These modulators have been useful experimental tools in dissecting the role of these channels in different systems. A few of these modulators are currently approved for use in humans.

Flupirtine is a K_V7 channel opener that was first approved in Germany as an analgesic in the 1980s. Subsequently, it has been approved in a number of other European countries (Italy, Portugal, Russia, Slovakia, Estonia. Latvia, Lithuania) and also in Brazil to treat various pain states. The approved pain indications differ between countries and include musculoskeletal pain, neuralgias and neuritis, headache, and postoperative, cancer and trauma pain. In addition to use in pain management, flupirtine displays muscle relaxant and anti-convulsant activity. However, only recently has this drug been shown to be an activator of K_V7 channels.[138] It represents the first drug in the class of analgesics denoted as 'selective neuronal potassium channel openers' (SNEPCO).

Retigabine (known as ezogabine in the United States), a close flupirtine analogue, was approved in 2011 for the treatment of partial-onset seizures in epilepsy and has been the most widely used tool to explore both the *in vitro* and

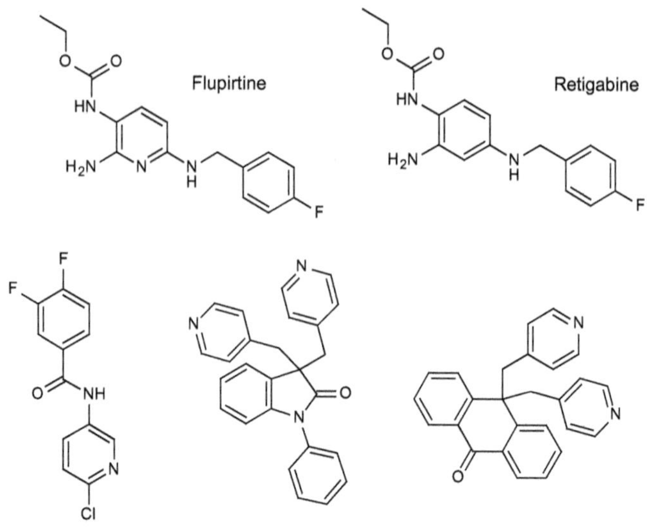

Figure 4.17 K_V7 channel modulators.

in vivo effects of K_V7 activation. It was initially shown to increase K^+ channel activity in a neuroblastoma cell line[139,140] and subsequently was demonstrated to have direct effects on K_V7 channels.[141–143] The compound, like most known K_V7 openers, exerts its effect by shifting the voltage-dependence of K_V7 activation to more hyperpolarised potentials. Thus, at physiological resting membrane potentials, retigabine increases K^+ conductance resulting in hyperpolarisation of the membrane potential away from action potential threshold. Retigabine interacts with a putative gating-hinge region in the channel and mutation of a single amino acid, Trp 236, completely eliminates the agonist activity on $K_V7.2$.[144]

ICA-27243 is a structurally distinct K_V7 opener that is subtype selective for $K_V7.2/7.3$ channels and has been reported to alleviate pain in preclinical models.[145–147] Two widely used K_V7 blockers are linopirdine[148] and its more potent analogue XE-991.[149] These compounds have been instrumental in delineating the role of K_V7 channels in native systems and confirming the selectivity of K_V7 openers.

Carlsson and Jurna demonstrated that intravenous (IV) administration of flupirtine inhibited neuronal activity following nerve stimulation.[150] Responses were measured from both C and Aδ fibres with reductions in both spontaneous activity in both populations and also a reduction of evoked C-fibre activity. This study is one of the first reports suggesting pharmacological modulation of K_V7 channels can directly affect C-fibre activity. Passmore *et al.*[137] were able to demonstrate that pharmacological activation of K_V7 channels by direct application of retigabine to the spinal cord inhibited both C-fibre and Aδ-fibre-mediated signalling. Consistent with this observation and in C-type fibres of human sural nerve, Lang *et al.* found retigabine produced a reduction of excitability.[151]

Rivera-Arconada *et al.* further demonstrated effects on nociceptive neuronal reflex activity using retigabine.[152] The activity on nociceptive reflex loops was long lasting and reversed by the K_V7 blocker XE-991. The same authors went on to show that retigabine hyperpolarises the membrane potential of primary afferents and increases the stimulation threshold.[153] Therefore K_V7 openers could potentially be useful analgesics.

Finally, Wladyka and Kunze demonstrated expression of K_V7 channels in visceral sensory neurons from nodose ganglion and showed depolarisation upon block of K_V7 channels with XE-991 and hyperpolarisation following administration of retigabine.[154] These results are consistent with other reports demonstrating functional expression of K_V7 currents in sensory neurons. Interestingly, block of K_V7 channels often has little effect on resting membrane potential in many DRG neurons suggesting little constitutive activation. In summary, there is significant functional expression of K_V7 channels in the peripheral nervous system and that modulation of these channels affects neuronal firing patterns involved in pain signalling.

Despite these aggregate data from genetics, expression profiles, biophysical studies and *in vitro* pharmacology there have been relatively few studies reported in peer-reviewed journals testing the effects of K_V7 openers in animal pain models and the majority of these studies have used retigabine.

In the spinal nerve ligation (SNL or Chung) model of neuropathic pain, retigabine was effective in reversing tactile allodynia with a minimum effective dose (MED) of 10 mg kg^{-1} (oral) and was also effective on tactile hyperalgesia in the SNL model.[155] This effect was completely reversed by the K_V7 blocker linopirdine consistent with an on-target effect at K_V7 channels. Roeloffs *et al.* reported similar efficacy on tactile allodynia in the SNL model using ICA-27243.[156] In other surgically induced neuropathic pain models, however, the results are less clear. Blackburn-Munro and Jensen found no effect on tactile allodynia (assessed with von Frey filaments) in the CCI model, but did observe an effect on tactile hyperalgesia induced by pinprick following administration of retigabine.[157] Cold allodynia was also reduced. Similar effects were seen in the spared nerve injury (SNI) model, where retigabine was not effective on tactile allodynia, but did show efficacy against tactile hyperalgesia. In the SNI model, however, there was no effect on cold allodynia.

A variety of other preclinical pain models have also been used to examine the effects of activation of K_V7 channels. Both retigabine and ICA-27243 are efficacious in the formalin model, particularly in reducing Phase 2 responses. The efficacy of retigabine in the carrageenan model of inflammatory pain using weightbearing as an endpoint (5 mg kg^{-1} oral dose) has been reported. The effect was completely reversed by XE-991.[137] Retigabine reduced the visceral pain induced by intracolonic injection of capsaicin and acid-induced muscle pain in rodents.[158] Despite the small number of reports, the results clearly support the role for K_V7 channels in pain signalling and by extension a potential role for K_V7 openers as pain therapeutics.

A number of studies have been published on the clinical effects of flupirtine. However, these data are limited in being generated from relatively small controlled trials or open-label studies. These trials have been reviewed in detail elsewhere.[159] Despite these limitations, across trials and open-label studies there is consistency with flupirtine being efficacious for a variety of pain states and improved tolerability compared to opiates. Flupirtine is not selective enough to exclusively ascribe the activity in humans to K_V7 activation. However, at plasma levels achieved in clinical studies, K_V7 channels should be activated and based on the preclinical body of data are likely to contribute, to the clinical activity observed. The primary reported side effects of flupirtine are dizziness, nausea, drowsiness, headache and fatigue.

As noted above, retigabine has been the most extensively studied K_V7 opener in preclinical pain models. In 2011, it was approved for the adjunctive treatment of partial-onset seizures in adult patients.[160–162] The most common reported side effects were dizziness, somnolence, fatigue and headache. Subsequently, a Phase 2 proof-of-concept clinical trial was conducted to evaluate the efficacy of retigabine in pain from post-herpetic neuralgia (PHN). Retigabine was ineffective at reducing pain scores, in that the change from baseline pain scores were similar for retigabine and placebo. Secondary endpoints were also not significantly different between groups. The lack of efficacy of retigabine in PHN is surprising given the data supporting a key role for this K^+ channel in regulating pain, as well as the clinical efficacy in pain seen with the less potent modulator, flupirtine (although there were clinical trial limitations with these studies). Retigabine was efficacious in epilepsy trials and was maintained at the maximum tolerated dose (MTD) in the PHN trial, which suggests that retigabine engaged and opened K_V7 channels in the pain trial. Although the clinical trials for flupirtine are less rigorous, it does appear to be better tolerated than retigabine. In preclinical animal models the efficacious dose for K_V7 openers is typically lower for epilepsy compared to pain. Considering that retigabine was efficacious in the epilepsy trial at doses close to MTD, it is possible the CNS side effects were dose limiting for efficacy in PHN. Other possibilities include that PHN is not a viable pain indication for K_V7 openers (flupirtine has not been studied in this indication), or that flupirtine possesses unidentified off-target analgesic actions that are absent in retigabine. Despite the retigabine PHN results, the body of literature supports the development K_V7 openers as therapeutic strategy for pain.

4.5 Calcium Channel Modulators as Pain Treatments

4.5.1 Introduction to Voltage-Gated Calcium Channels

Voltage-gated calcium channels (VGCCs) are macromolecular protein assemblies consisting of a pore-forming α subunit (that is permeable to calcium) and auxiliary $\alpha_2\delta$, β and (in some instances) γ subunits. The auxiliary subunits generally play a modulatory role by altering the biophysical properties and/or trafficking of the channel. There are 10 human α subunits, 4 $\alpha_2\delta$ subunits, 4 β

subunits and 8 γ subunits. VGCCs are broadly classified into five different types, L, N, P/Q, R and T types, based on a combination of biophysical and pharmacological properties. VGCCs open in response to changes in membrane potential and therefore function in controlling and responding to changes in cellular excitability in the nervous system, cardiac tissue and muscle. A variety of mutations in various VGCC subunits have been described in humans, as have naturally occurring and engineered genetic defects in mice that result in phenotypes affecting these physiological systems.[163] Interestingly, none of the described mutations in humans suggest a role for VGCCs in painful conditions other than migraine.

However, basic research has clearly demonstrated that VGCCs play a key role in transmitting painful signals into the CNS, particularly at the first nociceptor synapse in the dorsal horn of the spinal cord where the activity of the N-type channel $Ca_V2.2$ modulates neurotransmitter release.[164] The discovery of VGCC-based treatments for pain has been somewhat serendipitous with the identification of various natural toxins (conotoxins) and the highly successful $\alpha_2\delta$ targeted drugs pregabalin (Lyrica®) and gabapentin (Neurontin®) for the treatment of chronic or intractable pain conditions. However, with advances in ion-channel medicinal chemistry and screening, more small-molecule candidate drugs are emerging that target the pore-forming α subunits of N- and T- type channels, the latter of which have also been shown to play a potential role in pain.

4.5.2 N-type Calcium Channels

Nociceptive neurons consisting of either C or Aδ nerve fibres relay pain signals from the peripheral tissue to the superficial laminae of dorsal horn of the spinal cord where the first nociceptor synapse is located. $Ca_V2.2$ is co-expressed with key peptide neurotransmitters CGRP and substance P in sensory neurons[165] and plays a key role in regulating neurotransmitter release at the central terminal.[166] A number of studies have demonstrated modulation of $Ca_V2.2$ protein expression, distribution and function in sensory neurons from rodent models of inflammatory and neuropathic pain.[167–169] These data have led to the hypothesis that increased trafficking of these channels from nociceptor cell bodies to central terminals is a key plastic change mediating pain in specific rodent models. Specific knockdown of the predominant nociceptor variant of $Ca_V2.2$ ameliorates pain in chemically induced and neuropathic pain models.[170]

Pharmacology studies on the role of these channels in pain were aided by the discovery of cone snail toxins, specifically ω-conotoxins, which are cysteine knot proteins that selectively interact with N-type calcium channels. In neuropathic pain models, spinal administration of ω-conotoxin reduced pain behaviours and the activity of dorsal horn neurons in response to evoked noxious and non-noxious stimuli.[171,172] The ω-conotoxin MVIIA can affect allodynia mediated by pre- but not postsynaptic mechanisms, providing further evidence of presynaptic function for $Ca_V2.2$ in pain.[173]

Synthetic versions of naturally occurring ω-conotoxins have been developed into clinical agents for the treatment of chronic pain; however, their use is limited by significant nervous system, sympathetic and/or cardiac side effects.[174] Ziconotide (Prialt®; Figure 4.18), is a synthetic form of ω-conotoxin MVIIA with demonstrated efficacy in a number of chronic pain conditions.[175] Circumvention of side effects is, to some extent, achieved by the intrathecal delivery of the treatment. Although this is the only approved route of administration, delivery of ziconotide *via* this route is still associated with significant side effects and has to be performed and monitored in a specialised setting.

Efforts have been made to develop systemic small-molecule N-type inhibitors to overcome these limitations.[176] These molecules are hypothesised to show less of the side effects associated with ω-conotoxins as they display both use- and state-dependent blockade of the channel and should preferentially inhibit rapid firing rates associated with injured or sensitised sensory neurons rather than blocking all channel firing.

Two small-molecule inhibitors of $Ca_V2.2$ (Figure 4.19), CNV2197944 (likely to be from the same family as **26**[177] and Z-160/NMED-160)[178] have been progressed to Phase 1 studies. So far both compounds have demonstrated a relatively clean safety profile in preclinical species and in Phase 1 studies, and

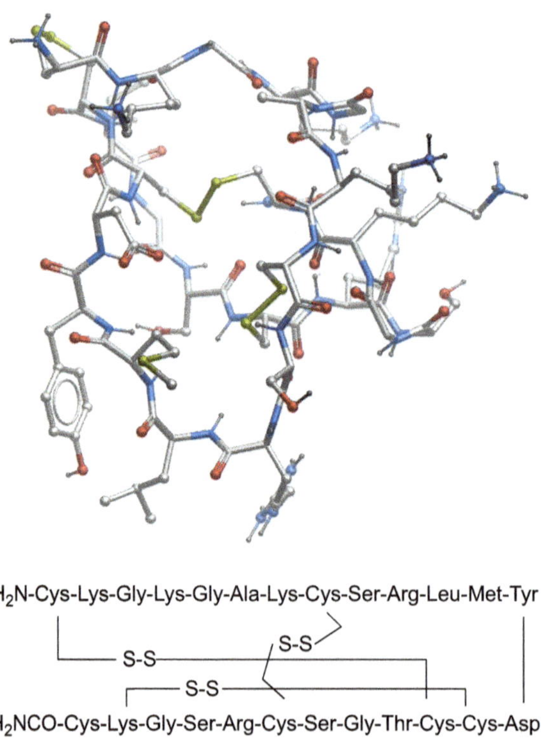

Figure 4.18 Structure of the ω-conotoxin ziconotide.

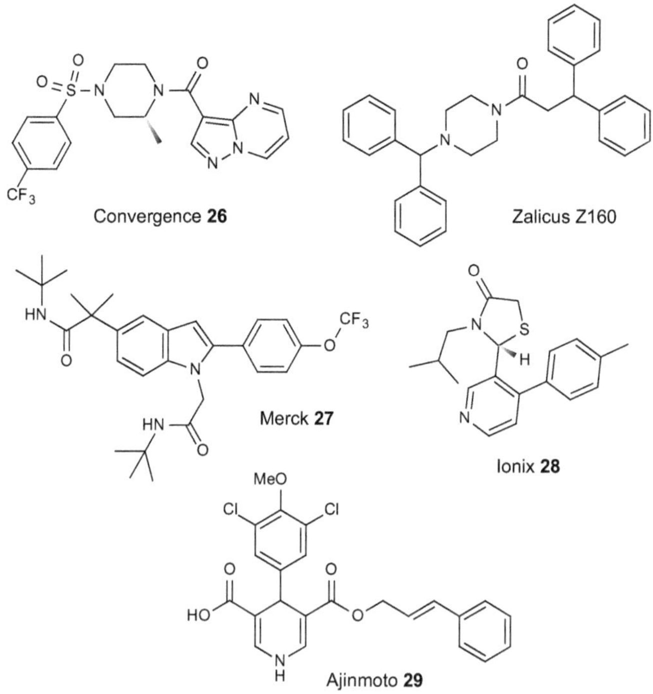

Figure 4.19 $Ca_V2.2$ blockers.

offer the prospect of effective and safe N-type calcium channel based treatments, if efficacy and safety is proven in a clinical setting.

In preclinical discovery, several groups have disclosed a variety of series including Merck (**27**) which has $Ca_V2.2$ IC_{50} 150 nM and $Ca_V1.2$ IC_{50} 910 nM,[179] Ionix (**28**) which has $Ca_V2.2$ IC_{50} 730 nM and Ca_V1 IC_{50} 23 μM,[180] and Ajinomoto Pharma (**29**) which has $Ca_V2.2$ IC_{50} 720 nM and Ca_V1 IC_{50} 44 μM.[181]

4.5.3 Other VGCCs as Pain Targets

Although T-type calcium channels have been linked specifically to pain,[182] no clinical drug candidates for the treatment of pain exist for these targets and there has been limited drug discovery activity.

4.5.4 $\alpha_2\delta$ Ligands as VGCC Modulators

The gabapentinoid drugs gabapentin and pregabalin are the most successful VGCC-based drug therapies for chronic pain conditions. Pregabalin represents the gold standard treatment option for chronic pain associated with diabetic peripheral neuropathy. The $\alpha_2\delta$ VGCC subunit proteins were identified as the high affinity binding sites for gabapentin in 1996,[183] but the exact mechanism of action of these drugs has remained elusive.

The $\alpha_2\delta$ subunits are expressed as a single contiguous protein which is cleaved into separate α_2 and δ proteins that become re-linked by disulfide bonds to form mature $\alpha_2\delta$ subunits.[184] There are four subtypes: $\alpha_2\delta$-1, $\alpha_2\delta$-2, $\alpha_2\delta$-3 and $\alpha_2\delta$-4. The δ portion is believed to remain membrane bound while the α_2 portion of the protein is believed to be extracellular. When co-expressed with pore-forming α subunits, $\alpha_2\delta$ subunits affect the expression and/or biophysical properties of the VGCCs.[185] Efficacy of gabapentinoids in chronic pain conditions is supported by preclinical studies demonstrating effects of these drugs in various models of neuropathic pain.[186,187] However, high concentrations of drug, much greater than the affinity of gabapentinoids for $\alpha_2\delta$, were required to affect all the aforementioned endpoints *in vivo*. Furthermore, direct effects of these drugs on the biophysical characteristics of VGCCs were not demonstrable. More recently it has been shown that gabapentin inhibits the cell surface trafficking of $\alpha_2\delta$ subunits, thereby affecting Ca_V channel plasma membrane levels. This mechanism requires that gabapentin has intracellular access to $\alpha_2\delta$ subunits and high intracellular gabapentin concentrations were required to affect trafficking.[188] *In vivo* studies provide evidence for a similar mechanism. Chronic application of pregabalin reduced $\alpha_2\delta$-1 trafficking to presynaptic sites in the superficial and deeper layers of the spinal cord.[189] However, none of these observations explain the relatively acute effects of these drugs in preclinical pain models.

Improvements in gabapentinoid-based therapies have been made with the advent of pregabalin, and although this drug affords better dose titration than gabapentin, somnolence and sedation remain significant side effects, and approximately 40–50% of patients do not achieve adequate pain relief. The lack of clarity over the mode of action of gabapentinoids has undoubtedly hindered efforts to discover follow-on drugs in this class.

Existing gabapentinoids bind with similar affinity to both $\alpha_2\delta$-1 and $\alpha_2\delta$-2 subtypes of the $\alpha_2\delta$ subunit, leading to the hypothesis that subtype-selective drugs might have better efficacy and/or side-effect profiles. Although subtype-selective compounds have been identified,[190] results from mice engineered to express mutant forms of $\alpha_2\delta$-1 or $\alpha_2\delta$-2 clearly demonstrate that both efficacy and sedation are mediated by $\alpha_2\delta$-1, at least in a preclinical setting.[191,192]

In the absence of a clear mechanism of action upon which to base drug discovery, follow-up to the early gabapentinoids has largely been confined to the modulation of pharmacokinetic properties through the use of novel formulations or prodrugs (Figure 4.20). DepoMed is currently developing a once-daily, extended-release tablet formulation of gabapentin using its gastric retention technology.[193] Gabapentin enacarbil[194] was approved in 2011 by the U.S. Food and Drug Authority (FDA) for the once-daily oral treatment of moderate-to-severe restless legs syndrome[195] and more recently has been approved for the treatment of PHN. The prodrug was designed to improve the limited and variable absorption rates of gabapentin.[196]

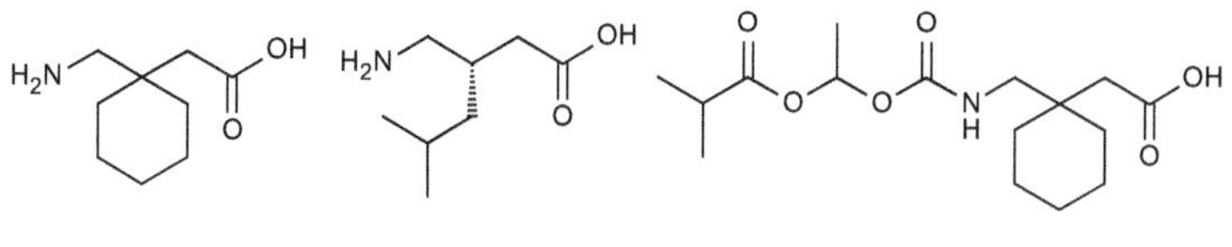

Figure 4.20 Gabapentin and analogues.

This section has focused entirely on VGCC-based targets and treatments that have advanced beyond preclinical testing; however, a recent proteomic analysis of the VGCC interactome[197] and specific studies on other auxiliary subunits, most notably $\alpha_2\delta$-3[198] and β-3,[199] illustrates that there may be many additional points of intervention for the development of VGCC-based pain treatments. Efforts to improve the number of patients that experience adequate pain relief and develop treatments with reduced side effects are needed and may one day be delivered through a clearer understanding of the mechanism of action of these highly successful drugs.

4.6 Challenges and Future Perspectives

The current range of analgesic treatments available to pain physicians all suffer from a poor response rate and efficacy that is often limited by adverse effects driven through off-target interactions. The significant developments in the ion channel field over the past few years have enabled the design and discovery of the next generation of ion-channel blockers for the treatment of pain. These advances include the recent publication of the first crystal structure of a sodium channel, advances in high-throughput and high-integrity electrophysiology screening platforms, and a dramatic increase in our knowledge of the role of specific ion-channel families and subtypes in the pain pathway through basic biology and genetics. These advances are enabling drug discovery teams to design and discover subtype-selective blockers of many of the target classes described above that offer the potential to provide both increased efficacy and a decreased side-effect burden.

With these advances the ion channel and pain fields are now experiencing a genuine increase in resource, interest and optimism that next-generation analgesics will provide real differentiation from their less selective predecessors. The challenge will be to rapidly move each of these target classes into clinical proof-of-concept studies to define which mechanisms provide genuinely differentiable efficacy in which patient populations.

References

1. T. Nakatsuka, T. Fujita and E. Kumamoto, P2X receptors and pain sensations, in E. Kumamoto (ed.) *Cellular and Molecular Mechanisms for the Modulation of Nociceptive Transmission in the Peripheral and Central Nervous Systems*, Research Signpost, Trivandrum, India, 2007, pp. 69–86.

2. R. J. Gum, B. Wakefield and M. F. Jarvis, *Purinergic Signal.*, 2012, **8**, 41–56.
3. X-P. Chu, C. J. Papasian, J. Q. Wang and Z-G. Xiong, *Int. J. Physiol. Pathophysiol. Pharmacol.*, 2011, **3**, 288–309.
4. M. J. Caterina, M. A. Schumacher, M. Tominaga, T. A. Rosen, J. D. Levine and D. Julius, *Nature*, 1997, **389**(6653), 816–824.
5. M. Tominaga, M. J. Caterina, A. B. Malmberg, T. A. Rosen, H. Gilbert, K. Skinner, B. E. Raumann, A. I. Basbaum and D. Julius, *Neuron*, 1998, **21**(3), 531–543.
6. M. J. Caterina, T. A. Rosen, M. Tominaga, A. J. Brake and D. Julius D, *Nature*, 1999, **298**, 436–441.
7. A. M. Peier, A. J. Reeve, D. A. Andersson, A. Moqrich, T. J. Earley, A. C. Hergarden, G. M. Story, S. Colley, J. B. Hogenesch, P. McIntyre, S. Bevan and A. Patapoutian, *Science*, 2002, **296**(5575), 2046–2049.
8. G. D. Smith, M. J. Gunthorpe, R. E. Kelsell, P. D. Hayes, P. Reilly, P. Facer, J. E. Wright, J. C. Jerman, J. P. Walhin, L. Ooi, J. Egerton, K. J. Charles, D. Smart, A. D. Randall, P. Anand and J. B. Davis, *Nature*, 2002, **418**(6894), 186–190.
9. H. Xu, I. S. Ramsey, S. A. Kotecha, M. M. Moran, J. A. Chong, D. Lawson, P. Ge, J. Lilly, I. Silos-Santiago, Y. Xie, P. S. DiStefano, R. Curtis and D. E. Clapham, *Nature*, 2002, **418**(6894), 181–186.
10. A. D. Güler, H. Lee, T. Iida, I. Shimizu, M. Tominaga and M. Caterina, *J. Neurosci.*, 2002, **22**(15), 6408–6414.
11. G. M. Story, A. M. Peier, A. J. Reeve, S. R. Eid, J. Mosbacher, T. R. Hricik, T. J. Earley, A. C. Hergarden, D. A. Andersson, S. W. Hwang, P. McIntyre, T. Jegla, S. Bevan and A. Patapoutian, *Cell*, 2003, **112**(6), 819–829.
12. D. D. McKemy, W. M. Neuhausser and D. Julius, *Nature*, 2002, **416**, 52–58.
13. A. M. Peier, A. Moqrich, A. C. Hegarden, A. J. Reeve, D. A. Andersson, G. M. Story, T. J. Earley, I. Dragoni, P. McIntyre, S. Bevan and A. Patapoutian, *Cell*, 2002, **108**(5), 705–715.
14. Y. H. Jin, T. W. Bailey, B. Y. Li, J. H. Schild and M. C. Andresen, *J. Neurosci.*, 2004, **24**(20), 4709–4717.
15. M. Tominaga, M. J. Caterina, A. B. Malmberg, T. A. Rosen, H. Gilbert, K. Skinner, B. E. Raumann, A. I. Basbaum and D. Julius, *Neuron*, 1998, **21**(3), 531–543.
16. H. Chuang, E. Prescott, H. Kong, S. Sheilds, S. Jordt, A. Basbaum, M. Chao and D. Julius, *Nature*, 2001, **411**, 957–962.
17. M. Tominaga, M. Wada and M. Masu, *Proc. Natl. Acad. Sci. U. S. A.*, 2001, **98**, 6951–6956.
18. G. Bhave, W. Zhu, H. Wang, D. J. Brasier, G. S. Oxford and R. W. Gereau, *Neuron*, 2002, **35**(4), 721–731.
19. A. Szallasi and P. M. Blumberg, *Brain Res.*, 1990, **524**, 106–111.
20. A. Baamonde, A. Lastra, L. Juarez, A. Hidalgo and L. Menéndez, *Brain Res. Bull.*, 2005, **67**(6), 476–481.

21. C. T. Hartrick, C. Pestano and N. Carlson, *Clin. Drug Investig.*, 2011, **31**(12), 877–882.
22. K. M. Walker, L. Urban, S. J. Medhurst, S. Patel, M. Panesar, A. J. Fox and P. McIntyre, *J. Pharmacol. Exp. Ther.*, 2003, **304**(1), 56–62.
23. N. R. Gavva, R. Tamir, Y. Qu, L. Klionsky, T. J. Zhang, D. Immke, J. Wang, D. Zhu, T. W. Vanderah, F. Porreca, E. M. Doherty, M. H. Norman, K. D. Wild, A. W. Bannon, J. C. Louis and J. J. Treanor, *J. Pharmacol. Exp. Ther.*, 2005, **313**(1), 474–484.
24. D. M. Swanson, A. E. Dubin, C. Shah, N. Nasser, L. Chang, S. L. Dax, M. Jetter, J. G. Breitenbucher, C. Liu, C. Mazur, B. Lord, L. Gonzales, K. Hoey, M. Rizzolio, M. Bogenstaetter, E. E. Codd, D. H. Lee, S. P. Zhang, S. R. Chaplan and N. I. Carruthers, *J. Med. Chem.*, 2005, **48**(6), 1857–1872.
25. A. Varga, J. Németh, A. Szabó, J. J. McDougall, C. Zhang, K. Elekes, E. Pintér, J. Szolcsányi and Z. Helyes, *Neurosci. Lett.*, 2005, **385**(2), 137–42.
26. P. Honore, P. Chandran, G. Hernandez, D. M. Gauvin, J. P. Mikusa, C. Zhong, S. K. Joshi, J. R. Ghilardi, M. A. Sevcik, R. M. Fryer, J. A. Segreti, P. N. Banfor, K. Marsh, T. Neelands, E. Bayburt, J. F. Daanen, A. Gomtsyan, C. H. Lee, M. E. Kort, R. M. Reilly, C. S. Surowy, P. R. Kym, P. W. Mantyh, J. P. Sullivan, M. F. Jarvis and C. R. Faltynek, *Pain*, 2009, **142**(1–2), 27–35.
27. N. R. Gavva, J. J. Treanor, A. Garami, L. Fang, S. Surapaneni, A. Akrami, F. Alvarez, A. Bak, M. Darling, A. Gore, G. R. Jang, J. P. Kesslak, L. Ni, N. H. Norman, G. Palluconi, M. J. Rose, M. Salfi, E. Tan, A. A. Romanovsky, C. Banfield and G. Davar, *Pain*, 2008, **136**(1–2), 202–210.
28. R. Karlsten, B. Jonzon, H. Quiding, M. Carlsson, M. Segerdahl, R. Malamut, M. Bjornsson, F. Miller, presented in part at the *13th World Congress on Pain*, Montreal, September, 2010.
29. A. A. Othman, W. Nothaft, W. M. Awni and S. Dutta, *Br. J. Clin. Pharmacol.*, 2013, **75**(4), 1029–1040.
30. B. A. Chizh, M. B. O'Donnell, A. Napolitano, J. Wang, A. C. Brooke, M. C. Aylott, J. N. Bullman, E. J. Gray, R. Y. Lai, P. M. Williams and J. M. Appleby, *Pain*, 2007, **132**, 132–141.
31. H. Quiding, L. Webster, B. Jonzon, A. Karin, R. Karlsten, A. Reimfelt and M. Segerdahl, presented in part at the *13th World Congress on Pain*, Montreal, September, 2010.
32. A. L. Krarup, L. Ny, M. Astrand, A. Bajor, F. Hvid-Jensen, M. B. Hansen, M. Simrén, P. Funch-Jensen and A. M. Drewes, *Aliment. Pharmacol. Ther.*, 2011, **33**(10), 1113–1122.
33. S. G. Lehto, R. Tamir, H. Deng, L. Klionsky, R. Kuang, A. Le, D. Lee, J.-C. Louis, E. Magal, B. H. Manning, J. Rubino, S. Surapaneni, N. Tamayo, T. Wang, J. Wang, W. Wang, B. Youngblood, M. Zhang, D. Zhu, M. H. Norman and N. R. Gavva, *J. Pharmacol. Exp. Ther.*, 2008, **326**, 218–229.

34. T. Watabiki, T. Kiso, T. Kuramochi, K. Yonezawa, N. Tsuji, A. Kohara, S. Kakimoto, T. Aoki and N. Matsuoka, *J. Pharmacol. Exp. Ther.*, 2011, **336**(3), 743–750.
35. R. M. Reilly, H. A. McDonald, P. S. Puttfarcken, S. K. Joshi, L. Lewis, M. Pai, P. H. Franklin, J. A. Segreti, T. R. Neelands, P. Han, J. Chen, P. W. Mantyh, J. R. Ghilardi, T. M. Turner, E. A. Voight, J. F. Daanen, R. G. Schmidt, A. Gomtsyan, M. E. Kort, C. R. Faltynek and P. R. Kym, *J. Pharmacol. Exp. Ther.*, 2012, **342**(2), 416–428.
36. S. G. Lehto, R. Tamir, H. Deng, L. Klionsky, R. Kuang, A. Le, D. Lee, J.-C. Louis, E. Magal, B. H. Manning, J. Rubino, S. Surapaneni, N. Tamayo, T. Wang, J. Wang, W. Wang, B. Youngblood, M. Zhang, D. Zhu, M. H. Norman and N. R. Gavva, *J. Pharmacol. Exp. Ther.*, 2008, **326**, 218–229.
37. T. Watabiki, T. Kiso, T. Kuramochi, K. Yonezawa, N. Tsuji, A. Kohara, S. Kakimoto, T. Aoki and N. Matsuoka, *J. Pharmacol. Exp. Ther.*, 2011, **336**(3), 743–750.
38. D. Jaquemar, T. Schenker and B. Trueb, *J. Biol. Chem.*, 1999, **274**, 7325–7333.
39. L. J. Macpherson, A. E. Dubin, M. J. Evans, F. Marr, P. G. Schultz, B. F. Cravatt and A. Patapoutian, *Nature*, 2007, **445**, 541–545.
40. C. R. McNamara, J. Mandel-Brehm, D. M. Bautista, J. Siemens, K. L. Deranian, M. Zhao, N. J. Hayward, J. A. Chong, D. Julius, M. M. Moran and C. M. Fanger, *Proc. Natl. Acad. Sci. U. S. A.*, 2007, **104**, 13525–13530.
41. M. Bandell, G. M. Story, S. W. Hwang, V. Viswanath, S. R. Eid, M. J. Petrus, T. J. Earley and A. Patapoutian, *Neuron*, 2004, **41**, 849–857.
42. B. Kremeyer, F. Lopera, J. J. Cox, A. Momin, F. Rugiero, S. Marsh, C. G. Woods, N. G. Jones, K. J. Paterson, F. R. Fricker, A. Villegas, N. Acosta, N. G. Pineda-Trujillo, J. D. Ramírez, J. Zea, M. W. Burley, G. Bedoya, D. L. Bennett, J. N. Wood and A. Ruiz-Linares, *Neuron*, 2010, **66**(5), 671–680.
43. S. R. Eid, E. D. Crown, E. L. Moore, H A. Liang, K. C. Choong, S. Dima, D. A. Henze, S. A. Kane and M. O. Urban, *Mol. Pain*, 2008, **4**, 48.
44. K. Y. Kwan, A. J. Allchorne, M. A. Vollrath, A. P. Christensen, D. S. Zhang, C. J. Woolf and D. P. Corey, *Neuron*, 2006, **50**(2), 277–289.
45. I. Mukhopadhyay, P. Gomes, S. Aranake, M. Shetty, P. Karnik, M. Damle, S. Kuruganti, S. Thorat and N. Khairatkar-Joshi, *J. Recept. Signal. Transduct. Res.*, 2011, **31**(5), 350–358.
46. *Int. Pat. Appl.* WO2010/109334, 2010.
47. *Int. Pat. Appl.* WO2010/125469, 2010.
48. *Int. Pat. Appl.* WO2011/132017, 2011.
49. *Int. Pat. Appl.* WO2012/085662, 2012.
50. *Int. Pat. Appl.* WO2007/073505, 2007.
51. *Int. Pat. Appl.* WO2010/004390, 2010.
52. *Int. Pat. Appl.* WO2010/075353, 2010.

53. *Int. Pat. Appl.* WO2013/023102, 2013.
54. *Int. Pat. Appl.* WO2009/089083, 2009.
55. *Int. Pat. Appl.* WO2010/141805, 2010.
56. *Int. Pat. Appl.* WO2011/043954, 2011.
57. *Int. Pat. Appl.* WO2010/141805, 2010.
58. K. Muraki, Y. Iwata, Y. Katanosaka, T. Ito, S. Ohya, M. Shigekawa and Y. Imaizumi, *Circulation Res.*, 2003, **93**, 829–838.
59. A. J. Stokes, L. M. N. Shimoda, M. Koblan-Huberson, C. N. Adra and H. Turner, *J. Exp. Med.*, 2004, **200**(2), 137–147.
60. P. Facer, M. A. Casula, G. D. Smith, C. D. Benham, I. P. Chessell, C. Bountra, M. Sinisi, R. Birch and P. Anand, *BMC Neurol.*, 2007, **7**, 11.
61. P. Gopinath, E. Wan, A. Holdcroft, P. Facer, J. B. Davis, G. D. Smith, C. Bountra and P. Anand, *BMC Womens Health*, 2005, **5**(1), 2.
62. R. Xiao, J. Tang, C. Wang, C. K. Colton, J. Tian and M. X Zhu, *J. Biol. Chem.*, 2008, **283**, 6162–6174.
63. S. Mandadi, T. Sokabe, K. Shibasaki, K. Katanosaka, A. Mizuno, A. Moqrich, A. Patapoutian, T. Fukumi-Tominaga, K. Mizumura and M. Tominaga, *Pflugers Arch.*, 2009, **458**(6), 1093–1102.
64. Z. Lin, Q. Chen, M. Lee, X. Cao, J. Zhang, D. Ma, L. Chen, X. Hu, H. Wang, X. Wang, P. Zhang, X. Liu, L. Guan, Y. Tang, H. Yang, P. Tu, D. Bu, X. Zhu, K. Wang, R. Li and Y. Yang, *Am. J. Hum. Genet.*, 2012, **90**(3), 558–564.
65. J. E. Lai-Cheong, G. Sethuraman, M. Ramam, K. Stone, M. A. Simpson and J. A. McGrath, *Br. J. Dermatol.*, 2012, **167**(2), 440–442.
66. S. Gullapi, A. Thomas, S. S. Chaudhari, V. S. P. R. Lingham, V. Kattige, G. S. Gudi, Z. Godsafe, N. Maharaj, P. Ward, N. Khairatkar-Joshi, presented in part at the *World Pharmaceutical Congress*, Philadelphia, 2010.
67. *Int. Pat. Appl.* WO2009/084034, 2009.
68. *Int. Pat. Appl.* WO2010/004379, 2010.
69. *Int. Pat. Appl.* WO2006/122156, 2006.
70. *Int. Pat. Appl.* WO2008/033564, 2008.
71. *Int. Pat. Appl.* WO2007/056124, 2007.
72. *Int. Pat. Appl.* WO2008/140750, 2008.
73. *Int. Pat. Appl.* WO2012/129491, 2012.
74. *Int. Pat. Appl.* WO2012/019315, 2012.
75. W. Liedtke, Y. Choe, M. A. Martí-Renom, A. M. Bell, C. S. Denis, A. Sali, A. J. Hudspeth, J. M. Friedman and S. Heller, *Cell*, 2000, **103**(3), 525–535.
76. W. Everaerts, B. Nilius and G. Owsianik, *Prog. Biophys. Mol. Biol.*, 2010, **103**(1), 2–17.
77. N. Cenac, C. Altier, K. Chapman, W. Liedtke, G. N. Zamponi and N. Vergnolle, *Gastroenterology*, 2008, **135**(3), 937–946.
78. E. Ceppa, F. Cattaruzza, V. Lyo, S. Amadesi, J. C. Pelayo, D. P. Poole, N. Vaksman, W. Liedtke, D. M. Cohen, E. F. Grady, N. W. Bunnett and

K. S. Kirkwood, *Am. J. Physiol. Gastrointest. Liver Physiol.*, 2010, **299**(3), G556–571.

79. W. Everaerts, X. Zhen, D. Ghosh, J. Vriens, T. Gevaert, J. P. Gilbert, N. J. Hayward, C. R. McNamara, F. Xue, M. M. Moran, T. Strassmaier, E. Uykal, G. Owsianik, R. Vennekens, D. De Ridder, B. Nilius, C. M. Fanger and T. Voets, *Proc. Natl. Acad. Sci. U. S. A.*, 2010, **107**(44), 19084–19089.
80. W. Liedtke and J. M. Friedman, *Proc. Natl. Acad. Sci. U. S. A.*, 2003, **100**(23), 13698–13703; J. McHugh, N. R. Keller, M. Appalsamy, S. A. Thomas, S. R. Raj, A. Diedrich, I. Biaggioni, J. Jordan and D. Robertson, *Hypertension*, 2010, **55**(6), 1438–1443.
81. S. A. Mendoza, J. Fang, D. D. Gutterman, D. A. Wilcox, A. H. Bubolz, R. Li, M. Suzuki and D. X. Zhang, *Am. J. Physiol. Heart. Circ. Physiol.*, 2010, **298**(2), H466–476.
82. J. Dai, T. J. Cho, S. Unger, E. Lausch, G. Nishimura, O. H. Kim, A. Superti-Furga and S. Ikegawa, *J. Hum. Genet.*, 2010, **55**(7), 400–402.
83. D. M. Bautista, J. Siemens, J. M. Glazer, P. R. Tsuruda, A. I. Basbaum, C. L. Stucky, S. E. Jordt and D. Julius, *Nature*, 2007, **448**(7150), 204–208.
84. H. Xing, M. Chen, J. Ling, W. Tan and J. G. Gu, *J. Neurosci.*, 2007, **27**(50), 13680–13690.
85. L. Su, C. Wang, Y-H. Yu, Y-Y. Ren, K-L. Xie and G. Wang, *BMC Neuroscience*, 2011, **12**, 120.
86. D. J. Parks, W. H. Parsons, R. W. Colburn, S. K. Meegalla, S. K. Ballentine, C. R. Illig, N. Qin, Y. Liu, T. L. Hutchinson, M. L. Lubin, D. J. Stone, J. F. Baker, C. R. Schneider, J. Ma, B. P. Damiano, C. M. Flores and M. R. Player, *J. Med. Chem.*, 2011, **54**(1), 233–247.
87. R. R. Calvo, S. K. Meegalla, D. J. Parks, W. H. Parsons, S. K. Ballentine, M. L. Lubin, C. Schneider, R. W. Colburn, C. M. Flores and M. R. Player, *Bioorg. Med. Chem. Lett.*, 2012, **22**(5), 1903–1907.
88. D. I. Chasman, M. Schürks, V. Anttila, B. de Vries, U. Schminke, L. J. Launer, G. M. Terwindt, A. M. van den Maagdenberg, K. Fendrich, H. Völzke, F. Ernst, L. R. Griffiths, J. E. Buring, M. Kallela, T. Freilinger, C. Kubisch, P. M. Ridker, A. Palotie, M. D. Ferrari, W. Hoffmann, R. Y. Zee and T. Kurth, *Nat. Genet.*, 2011, **43**(7), 695–698.
89. T. Freilinger, V. Anttila, B. de Vries, R. Malik, M. Kallela, G. M. Terwindt, P. Pozo-Rosich, B. Winsvold, D. R. Nyholt, W. P. J. van Oosterhout, V. Artto, U. Todt, E. Hämäläinen, J. Fernández-Morales, M. A. Louter, M. A. Kaunisto, J. Schoenen, O. Raitakari, T. Lehtimäki, M. Vila-Pueyo, H. Göbel, E. Wichmann, C. Sintas, A. G. Uitterlinden and A. Hofman, *Nat. Genet.*, 2012, **44**(777–782).
90. N. R. Gavva, C. Davis, S. G. Lehto, S. Rao, W. Wang and D. X. Zhu, *Molecular Pain*, 2012, **8**(1), 36–452012; M. C. Almeida, T. Hew-Butler, R. N. Soriano, S. Rao, W. Wang, J. Wang, N. Tamayo, D. L. Oliveira, T. B. Nucci, P. Aryal, A. Garami, D. Bautista, N. R. Gavva and A. A. Romanovsky, *J. Neurosci.*, 2012, **32**(6), 2086–2099.

91. M. I. Kemp, *Prog. Med. Chem.*, 2010, **49**, 81–111.
92. J. Payandeh, T. M. G. El-Din, T. Scheuer, N. Zheng and W. A. Catterall, *Nature*, 2012, **486**, 135–139.
93. D. S. Ragsdale, J. C. McPhee, T. Scheuer and W. A. Catterall, *Proc. Natl. Acad. Sci. U. S. A.*, 1996, **93**(17), 9270–9275.
94. S. England and M. J. de Groot, *Br. J. Pharmacol.*, 2009, **158**(6), 1413–1425.
95. F. Cattabeni, *IDrugs*, 2004, **7**(10), 935–939.
96. A. C. Errington, T. Stohr, C. Heers and G. Lees, *Mol. Pharmacol.*, 2008, **73**(1), 157–169.
97. P. L. Sheets, C. Heers, T. Stoehr and T. R. Cummins, *J. Pharmacol. Exp. Ther.*, 2008, **326**(1), 89–99.
98. B. C. Hains, C. Y. Saab, J. P. Klein, M. J. Craner and S. G. Waxman, *J. Neurosci.*, 2004, **24**(20), 4832–4839.
99. J. A. Lindia, M. G. Kohler, W. J. Martin and C. Abbadie, *Pain*, 2005, **117**(1–2), 145–153.
100. M. A. Nassar, M. D. Baker, A. Levato, G. Malluci, S. B. McMahon and J. N. Wood, *Mol. Pain*, 2006, **19**(2), 33.
101. O. A. Samad, A. M. Tan, C. Xiiaoyang, E. Foster, S. Dib-Hajj and S. G. Waxman, *Mol. Ther.*, 2013, **21**(1), 49–56.
102. J. A. Black, L. Nikolajsen, K. Kroner, T. S. Jensen and S. G. Waxman, *Ann. Neurol.*, 2008, **64**(6), 644–653.
103. *Int. Pat. Appl.* WO2010/002956, 2010.
104. *Int. Pat. Appl.* WO2009/012241, 2009.
105. *Int. Pat. Appl.* WO2008/118758, 2008.
106. A. M. Rush, T. R. Cummins and S. G. Waxman, *J. Physiol.*, 2007 15, **579**(Pt 1), 1–14.
107. J. J. Cox, F. Reimann, A. K. Nicholas, G. Thornton, E. Roberts, K. Springell, G. Karbani, H. Jafri, J. Mannan, Y. Raashid, L. Al-Gazali, H. Hamamy, E. M. Valente, S. Gorman, R. Williams, D. P. McHale, J. N. Wood, F. M. Gribble and C. G. Woods, *Nature*, 2006 14, **444**, 894–898.
108. M. Weiss, E. Pyrski, B. Jacobi, V. Bufe, B. Willnecker, P. Schick, S. J. Zizzari, C. A. Gossage, T. Greer, C. G. Leinders-Zufall, J. N. Woods, Wood and F. Zufall, *Nature*, 2011, **472**, 186–190.
109. Y. Yang, Y. Wang, S. Li, Z. Xu, H. Li, L. Ma, J. Fan, D. Bu, B. Liu, Z. Fan, G. Wu, J. Jin, B. Ding, X. Zhu and Y. Shen, *J. Med. Genet.*, 2004, **41**(3), 171–174.
110. R. Fertleman, M. D. Baker, K. A. Parker, S. Moffatt, F. V. Elmslie, B. Abrahamsen, J. Ostman, N. Klugbauer, J. N. Wood, R. M. Gardiner and M. Rees, *Neuron*, 2006, **7**, 767–774.
111. M. A. Nassar, L. C. Stirling, G. Forlani, M. D. Baker, E. A. Matthews, A. H. Dickenson and J. N. Wood, *Proc. Natl. Acad. Sci. U. S. A.*, 2004, **101**(34), 12706–12711.
112. M. A. Nassar, A. Levato, L. C. Stirling and J. N. Wood, *Mol. Pain.*, 2005, **22**, 1–24.

113. *Int. Pat. Appl.* WO2007/042239, 2007.
114. *Int. Pat. Appl.*WO2007/042250, 2007.
115. Y. P. Goldberg, N. Price, R. Namdari, C. J. Cohen, M. H. Lamers, C. Winters, J. Price, C. E. Young, H. Verschoof, R. Sherrington, S. N. Pimstone and M. R. Hayden, *Pain*, 2012, **153**(1), 80–85.
116. P. G. Nantermet and D. A. Henze, *Annu. Rep. Med. Chem.*, 2011, **46**, 19–32.
117. *Int. Pat. Appl.* WO2010/079443, 2010.
118. *Int. Pat. Appl.* WO2010/0106, 2010.
119. B. S. Williams, P. J. Felix, B. T. Priest, R. M. Brochu, K. Dai, S. B. Hoyt, C. London, Y. S. Tang, J. L. Duffy, W. H. Parsons, G. J. Kaczorowski and M. L. Garcia, *Biochemistry*, 2007, **46**(50), 14693–14703.
120. *Int. Pat. Appl.* WO2012/039657, 2012.
121. *Int. Pat. Appl.* WO2012/100922, 2012.
122. N. T. Blair and B. P. Bean, *J. Neurosci.*, 2002, **22**(23), 10277–10290.
123. M. F. Jarvis, P. Honore, C. C. Shieh, M. Chapman, S. Joshi, X. F. Zhang, M. Kort, W. Carroll, B. Marron, R. Atkinson, J. Thomas, D. Liu, M. Krambis, Y. Liu, S. McGaraughty, K. Chu, R. Roeloffs, C. Zhong, J. P. Mikusa, G. Hernandez, D. Gauvin, C. Wade, C. Zhu, M. Pai, M. Scanio, L. Shi, I. Drizin, R. Gregg, M. Matulenko, A. Hakeem, M. Gross, M. Johnson, K. Marsh, P. K. Wagoner, J. P. Sullivan, C. R. Faltynek and D. S. Krafte, *Proc. Natl. Acad. Sci. U. S. A*, 2007, **104**(20), 8520–8525.
124. A. N. Akopian, V. Souslova, S. England, K. Okuse, N. Ogata, J. Ure, A. Smith, B. J. Kerr, S. B. McMahon, S. Boyce, R. Hill, L. C. Stanfa, A. H. Dickenson and J. N. Wood, *Nat. Neurosci.*, 1999, **2**(6), 541–548.
125. A. Bierhaus, T. Fleming, S. Stoyanov, A. Leffler, A. Babes, C. Neacsu, S. K. Sauer, M. Eberhardt, M. Schnölzer, F. Lasitschka, W. L. Neuhuber, T. I. Kichko, I. Konrade, R. Elvert, W. Mier, V. Pirags, I. K. Lukic, M. Morcos, T. Dehmer, N. Rabbani, P. J. Thornalley, D. Edelstein, C. Nau, J. Forbes, P. M. Humpert, M. Schwaninger, D. Ziegler, D. M. Stern, M. E. Cooper, U. Haberkorn, M. Brownlee, P. W. Reeh and P. P. Nawroth, *Nat. Med.*, 2012, **18**(6), 926–33.
126. M. E. Cort, I. Drizin, R. J. Gregg, M. J. C. Scanio, L. Shi, M. F. Gross, R. N. Atkinson, M. S. Johnson, G. J. Pacofsky, J. B. Thomas, W. A. Carroll, M. J. Krambis, D. Liu, C-C. Shieh, X-F. Zhang, G. Hernandez, J. P. Mikusa, C. Zhong, S. Joshi, P. Honore, R. Roeloffs, K. C. Marsh, B. P. Murray, J. Liu, S. Werness, C. R. Faltynek, D. S. Krafte, M. F. Jarvis, M. L. Chapman and B. E. Marron, *J. Med. Chem.*, 2008, **51**(3), 407–416.
127. *Int. Pat. Appl.* WO 2008/135830, 2008.
128. *Int. Pat. Appl.* WO2008/0421, 2008.
129. S. Lolignier, M. Amsalem, F. Maingret, F. Padilla, M. Gabriac, E. Chapuy, A. Eschalier, P. Delmas and J. Busserolles, *PLoS One*, 2011, **6**(8), e23083.

130. Y. Q. Yu, F. Zhao, S. M. Guan and J. Chen, *J, PLoS One*, 2011, **6**(5), e19865.
131. A. D. Wickenden and G. McNaughton-Smith, *Curr. Pharm. Des.*, 2009, **15**, 1773–1798.
132. Q. Wang, M. E. Curran, I. Splawski, T. C. Burn, J. M. Millholland, T. J. VanRaay, J. Shen, K. W. Timothy, G. M. Vincent, T. de Jager, P. J. Schwartz, J. A. Toubin, A. J. Moss, D. L. Atkinson, G. M. Landes, T. D. Connors and M. T. Keating, *Nat. Genet.*, 1996, **12**, 17.
133. H. S. Wang, Z. Pan, W. Shi, B. S. Brown, R. S. Wymore, I. S. Cohen, J. E. Dixon and D. McKinnon, *Science*, 1998, **282**, 1890–1893.
134. E. C. Cooper, K. D. Aldape, A. Abosch, N. M. Barbaro, M. S. Berger, W. S. Peacock, Y. N. Jan and L. Y. Jan, *Proc. Natl. Acad. Sci. U. S. A.*, 2000, **97**, 4914–4919.
135. W. P. Yang, P. C. Levesque, W. A. Little, M. L. Conder, P. Ramakrishnan, M. G. Neubauer and M. A. Blanar, *J. Biol. Chem.*, 1998, **273**, 19419–19423.
136. C. Rundfeldt and R. Netzer, *Neurosci. Lett.*, 2000, **282**, 73–76.
137. G. M. Passmore, A. A. Selyanko, M. Mistry, M. Al-Qatari, S. J. Marsh, E. A. Matthews, A. H. Dickenson, T. A. Brown, S. A. Burbidge, M. Main and D. A. Brown, *J. Neurosci.*, 2003, **23**, 7227–7236.
138. G. Munro and W. Dalby-Brown, *J. Med. Chem.*, 2007, **50**, 2576–2582.
139. C. Rundfeldt, *Eur. J. Pharmacol.*, 1997, **336**, 243–249.
140. C. Rundfeldt, *Epilepsy Res.*, 1999, **35**, 99–107.
141. C. Rundfeldt and R. Netzer, *Neurosci. Lett.*, 2000, **282**, 73–76.
142. M. J. Main, J. E. Cryan, J. R. Dupere, B. Cox, J. J. Clare and S. A. Burbidge, *Mol. Pharmacol.*, 2000, **58**, 253–262.
143. A. D. Wickenden, W. Yu, A. Zou, T. Jegla and P. K. Wagoner, *Mol. Pharmacol.*, 2000, **58**, 591–600.
144. T. V. Wuttke, G. Seebohm, S. Bail, S. Maljevic and H. Lerche, *Mol. Pharmacol.*, 2005, **67**, 1009–1017.
145. A. D. Wickenden, J. L. Krajewski, B. London, P. K. Wagoner, W. A. Wilson, S. Clark, R. Roeloffs, G. McNaughton-Smith and G. C. Rigdon, *Mol. Pharmacol.*, 2008, **73**, 977–986.
146. R. Roeloffs, A. D. Wickenden, C. Crean, S. Werness, G. McNaughton-Smith, J. Stables, J. O. McNamara, N. Ghodadra and G. C. Rigdon, *J. Pharmacol. Exp. Ther.*, 2008, **326**, 818–828.
147. K. Padilla, A. D. Wickenden, A. C. Gerlach and K. McCormack, *Neurosci. Lett.*, 2009, **465**, 138–142.
148. S. P. Aiken, B. J. Lampe, P. A. Murphy and B. S. Brown, *Br. J. Pharmacol.*, 1995, **115**, 1163–1168.
149. M. E. Schnee and B. S. Brown, *J. Pharmacol. Exp. Ther.*, 1998, **286**, 709–717.
150. K. H. Carlsson and I. Jurna, *Eur. J. Pharmacol.*, 1987, **143**, 89–99.
151. P. M. Lang, J. Fleckenstein, G. M. Passmore, D. A. Brown and P. Grafe, *Neuropharmacology*, 2008, **54**, 1271–1278.

152. I. Rivera-Arconada, J. Martinez-Gomez and J. A. Lopez-Garcia, *Neuropharmacology*, 2004, **46**, 598–606.
153. I. Rivera-Arconada and J. A. Lopez-Garcia, *Neuropharmacology*, 2006, **51**, 756–763.
154. C. L. Wladyka and D. L. Kunze, *J. Physiol.*, 2006, **575**, 175–189.
155. R. Dost, A. Rostock and C. Rundfeldt, *Naunyn-Schmiedeberg's Arch. Pharmacol.*, 2004, **369**, 382–390.
156. R. Roeloffs, A. D. Wickenden, C. Crean, S. Werness, G. McNaughton-Smith, J. Stables, J. O. McNamara, N. Ghodadra and G. C. Rigdon, *J. Pharmacol. Exp. Ther.*, 2008, **326**, 818–828.
157. G. Blackburn-Munro and B. S. Jensen, *Eur. J. Pharmacol.*, 2003, **460**, 109–116.
158. K. Hirano, K. Kuratani, M. Fujiyoshi, N. Tashiro, E. Hayashi and M. Kinoshita, *Neurosci. Lett.*, 2007, **413**, 159–162.
159. J. Devulder, *CNS Drugs*, 2010, **24**, 867–881.
160. C. E. Stafstrom, S. Grippon and P. Kirkpatrick, *Nat. Rev. Drug Discov.*, 2011, **10**, 729–730.
161. G. Orhan, T. V. Wuttke, A. T. Nies, M. Schwab and H. Lerche, *Expt. Opin. Pharmacother.*, 2012, **13**, 1807–1816.
162. M. Y. Splinter, *Clin. Ther.*, 2012, **34**, 1845–1856 e1841.
163. S. M. Cain and T. P. Snutch, *Biofactors*, 2011, **37**, 197–205.
164. A. Meir, S. Ginsburg, A. Butkevich, S. G. Kachalsky, I. Kaiserman, R. Ahdut, S Demirgoren and R. Rahamimoff, *Physiol. Rev.*, 1999, **79**, 1019–1088.
165. R. E. Westenbroek, L. Hoskins and W. A. Catterall, *J. Neurosci.*, 1998, **18**, 6319–6330.
166. A. Meir, S. Ginsburg, A. Butkevich, S. G. Kachalsky, I. Kaiserman, R. Ahdut, S Demirgoren and R. Rahamimoff, *Physiol. Rev.*, 1999, **79**, 1019–1088.
167. S. G. Lu, X. L. Zhang, Z. D. Luo and M. S. Gold, *Pain*, 2010, **151**, 633–643.
168. K. Yokoyama, T. Kurihara, K. Makita and T. Tanabe, *Anesthesiology*, 2003, **99**, 1364–1370.
169. D. Cizkova, J. Marsala, N. Lukacova, M. Marsala, S. Jergova, J. Orendacova and T. L. Yaksh, *Exp. Brain Res.*, 2002, **147**, 456–463.
170. C. Altier, C. S. Dale, A. E. Kisilevsky, K. Chapman, A. J. Castiglioni, E. A. Matthews, R. M. Evans, A. H. Dickenson, D. Lipscombe and N. Vergnolle, *J. Neurosci.*, 2007, **27**, 6363–6373.
171. S. R. Chaplan, J. W. Pogrel and T. L. Yaksh, *J. Pharmacol. Exp. Therap.*, 1994, **269**, 1117–1123.
172. E. A. Matthews and A. H. Dickenson, *Pain*, 2001, **92**, 235–246.
173. M. Tsukamoto, T. Kiso, Y. Shimoshige, T. Aoki and N. Matsuoka, *Eur. J. Pharmacol.*, 2010, **634**, 40–45.
174. A. Schmidtko, J. Lotsch, R. Freynhagen and G. Geisslinger, *Lancet*, 2010, **375**, 1569–1577.
175. J. G. McGivern, *Neuropsychiatr. Dis. Treat.*, 2007, **3**, 69–85.

176. T. Yamamoto and A. Takahara, *Curr. Top. Med. Chem.*, 2009, **9**(4), 377–395.
177. V. Morisset, presented in part at the *IASP 14th World Congress of Pain*, Milan, Italy, 2012.
178. T. Snutch, presented in part at *Division of Medicinal Chemistry Abstracts, 226th National Meeting of the American Chemical Society*, New York, 2003.
179. S. Tyagarajan, P. K. Chakravarty, M. Park, B. Zhou, J. B. Herrington, K. Ratliff, R. M. Bugianesi, B. Williams, R. J. Haedo, A. M. Swensen, V. A. Warren, M. Smith, M. Garcia, G. J. Kaczorowski, O. B. McManus, K. A. Lyons, X. Li, M. Madeira, B. Karanam, M. Green, M. J. Forrest, C. Abbadie, E. McGowan, S. Mistry, N. Jochnowitz and J. L. Duffy, *Bioorg. Med. Chem. Lett.*, 2011, **21**(2), 869–873.
180. L. J. S. Knutsen, C. J. Hobbs, C. G. Earnshaw, A. Fiumana, J. Gilbert, S. L. Mellor, F. Radford, N. J. Smith, P. J. Birch, J. Russell Burley, S. D. C Ward and I. F. James, *Bioorg. Med. Chem. Lett.*, 2007, **17**(3), 662–667.
181. T. Yamamoto, S. Niwa, M. Tokumasu, T. Onishi, S. Ohno, M. Hagihara, H. Koganei, S. Fujita, T. Takeda, Y. Saitou, S. Iwayama, A. Takahara, S. Iwata and M. Shoji, *Bioorg. Med. Chem. Lett.*, 2012, **22**(11), 3639–3642.
182. M. O. Jacus, V. N. Uebele, J. J. Renger and S. M. Todorovic, *J. Neurosci.*, 2012, **32**, 9374–9382.
183. N. S. Gee, J. P. Brown, V. U. Dissanayake, J. Offord, R. Thurlow and G. N. Woodruff, *J. Biol. Chem.*, 1996, **271**, 5768–5776.
184. A. Davies, L. Douglas, J. Hendrich, J. Wratten, A. Tran Van Minh, I. Foucault, D. Koch, W. S. Pratt, H. R. Saibil and A. C. Dolphin, *J. Neurosci.*, 2006, **26**, 8748–8757.
185. A. C. Dolphin, *Curr. Opin. Neurobiol.*, 2009, **19**, 237–244.
186. M. J. Field, E. F. Holloman, S. McCleary, J. Hughes and L. Singh, *J. Pharmacol. Expt. Ther.*, 1997, **282**, 1242–1246.
187. M. J. Field, R. J. Oles, A. S. Lewis, S. McCleary, J. Hughes and L. Singh, *Br. J. Pharmacol.*, 1997, **121**, 1513–1522.
188. J. Hendrich, A. T. Van Minh, F. Heblich, M. Nieto-Rostro, K. Watschinger, J. Striessnig, J. Wratten, A. Davies and A. C. Dolphin, *Proc. Natl. Acad. Sci. U. S. A.*, 2008, **105**, 3628–3633.
189. C. S. Bauer, M. Nieto-Rostro, W. Rahman, A. Tran-Van-Minh, L. Ferron, L. Douglas, I. Kadurin, Y. Sri Ranjan, L. Fernandez-Alacid and N. S. Millar, *J. Neurosci.*, 2009, **29**, 4076–4088.
190. J. W. Myatt, M. P. Healy, G. S. Bravi, A. Billinton, C. N. Johnson, K. L. Matthews, K. S. Jandu, W. Meng, A. Hersey and D. G. Livermore, *Bioorg. Med. Chem. Lett.*, 2010, **20**, 4683–4688.
191. M. J. Field, P. J. Cox, E. Stott, H. Melrose, J. Offord, T. Z. Su, S. Bramwell, L. Corradini, S. England and J. Winks, *Proc. Natl. Acad. Sci. U. S. A.*, 2006, **103**, 17537–17542.
192. S. M. Lotarski, S. Donevan, A. El-Kattan, S. Osgood, J. Poe, C. P. Taylor and J. Offord, *J. Pharmacol. Exp. Ther.*, 2011, **338**, 615–621.

193. M. Abramovicz, *Med. Lett.*, 2011, **53**(1378), 94–95.
194. K. C. Cundy, R. Branch, T. Chernov-Rogan, T. Dias, T. Estrada, K. Hold, K. Koller, X. Liu, A. Mann, M. Panuwat, S. P. Raillard, S. Upadhyay, Q. Q. Wu, J. N. Xiang, H. Yan, N. Zerangue, C. X. Zhou, R. W. Barrett and M. A. Gallop, *J. Pharmacol. Exp. Ther.*, 2004, **311**(1), 315–23.
195. K. C. Cundy, S. Sastry, W. Luo, J. Zou, T. L. Moors and D. M. Canafax, *J. Clin. Pharmacol.*, 2008, **48**(12), 1378–1388.
196. H. N. Bockbrader, D. Wesche, R. Miller, S. Chapel, N. Janiczek and P. A. Burger, *Clin. Pharmacokinet.*, 2010, **49**(10), 661–669.
197. C. S. Muller, A. Haupt, W. Bildl, J. Schindler, H. G. Knaus, M. Meissner, B. Rammner, J. Striessnig, V. Flockerzi and B. Fakler, *Proc. Natl. Acad. Sci. U. S. A.*, 2010, **107**, 14950–14957.
198. G. G. Neely, A. Hess, M. Costigan, A. C. Keene, S. Goulas, M. Langeslag, R. S. Griffin, I. Belfer, F. Dai and S. B. Smith, *Cell*, 2010, **143**, 628–638.
199. L. Li, X. H. Cao, S. R. Chen, H. D. Han, G. Lopez-Berestein, A. K. Sood and H. L. Pan, *J. Biol. Chem.*, 2012, **287**, 6002–6013.

CHAPTER 5

Neurotransmitters and Other Non-Opiate Central Mechanisms

SIMON E. WARD

Professor of Medicinal Chemistry & Director of Translational Drug Discovery Group, School of Life Scisences, University of Sussex, Brighton BN1 9QJ, UK
Email: simon.ward@sussex.ac.uk

5.1 Introduction

Pain comes in many forms and originates from a large number of factors, pathologies and processes, which collectively represent a health problem with unmet medical need causing significant economic cost to society and seriously impaired quality of life for sufferers. Across the spectrum of pain conditions—inflammatory, neuropathic, dysfunctional and cancer—many patients are poorly managed by current therapies which carry downsides including low levels of efficacy, abuse liabilities or therapy-limiting side effects. Diagnosis and treatment selection by the physician often follows an empirical approach and the most effective pain medications are only able to reduce global pain scores by around one-third, for those that respond at all.[1]

Importantly, three main groups of pain can be considered—nociceptive, a high threshold protective pain-sensing mechanism that analgesic medication (other than for anaesthetic purposes) should not impair; inflammatory, a protective mechanism for repair of damaged tissue but which needs moderating in conditions with prolonged inflammation such as arthritis; pathological, including neuropathic pain involving damage to nerve tissue and dysfunctional

RSC Drug Discovery Series No. 35
Pain Therapeutics: Current and Future Treatment Paradigms
Edited by Charlotte Allerton

Published by the Royal Society of Chemistry, www.rsc.org

pain with no obvious nerve damage (includes a range of conditions such as fibromyalgia).[1b]

It has long been recognised that the sensing of pain is not a straightforward and uniform detection, but is under active regulation in response to diverse endogenous and exogenous factors, such as mood, memory and stress levels. Thus, the control of responses to painful stimuli is regulated by a number of processes and mechanisms that are found both in the spinal cord and in supraspinal brain centres. Importantly, the communication between sites in the periphery that detect sensory inputs and the higher brain centres is bidirectional, and the transmission of this message from the periphery, through the spinal cord to the higher brain centres, can involve mechanisms that both facilitate or inhibit the signalling.

A number of empirical sources of evidence confirm the role of central control in the perception of pain. First, many individuals, in particular soldiers or athletes, who have sustained major trauma or injury report no or minimal pain despite the severity of their injuries, leading to the assumption that the stress of the particular situation had blocked the usual perception of pain.[2] Secondly, and with also considerable negative impact on the ability to run effective and reproducible clinical trials, is the major positive placebo or negative placebo (or nocebo) effect observed in individuals.[3] Many studies have shown significant effect of the power of suggestion for increasing or decreasing the expected pain stimulus, and more interestingly, later studies have now been able to associate placebo response to activation of specific brain regions, which overlap with brain sites that are activated by opioid analgesics.[4]

This chapter reviews the role of central mechanisms in pain, excluding opiate analgesics and voltage-gated ion channels comprising both central and peripheral components, which are covered in Chapters 2 and 4 respectively.

5.2 Role of Central Mechanisms in Pain

5.2.1 Ascending Pain Pathways

Sensory inputs from nociceptive primary afferent fibres are transmitted from the periphery along fibres which are made up of the axons of T-shaped sensory neurons located in a spinal ganglion. and enter the spinal cord and form synaptic connections in the spinal dorsal horn onto transmission neurons. (On entering the spinal cord, the fibres initially divide, travelling both up and down the spinal cord, creating the structure known as Lissauer's tract.) The different fibres (Aα, Aβ, Aδ and C) all form synapses at spatially distinct regions in the dorsal horn; in particular the Aβ fibres connect both directly to the ipsilateral dorsal column of the spinal cord and also to the deep layers of the ventral horn (thus connecting with motor neurons and enabling the withdrawal reflex) and the Aδ and C fibres synapse onto specific and non-specific nociceptive neurons located in layers I/II and V respectively.[5] The details of these projections are reviewed extensively in standard neuroscience anatomy textbooks, and their onward projections are described in terms of either phylogenetic or anatomic classifications. The transmission at these synapses is mediated by various neurotransmitters, such as glutamate, substance P and gamm-aminobutyric acid (GABA), allowing the dorsal horn to modulate inputs to either enhance or attenuate the signal before communication with higher brain centres. These transmission neurons then project up to the thalamus, with additional projections targeting a range of other systems within the spinal cord, namely, the mesencephalic nuclei (including the dorsal reticular nucleus and rostral ventral medulla) and the midbrain periaqueductal grey region.

The signal reaching the thalamus is sent both to cortical brain regions (and subsequently on to the amygdala) as well as directly to the amygdala.

5.2.2 Descending Pain Pathways

Initial attempts to rationalise and explain the ability of individuals to respond differently to equivalent pain stimuli resulted in the model of gate theory described by Melazack and Wall.[6] This theory sought to explain the subjective and contextual nature of pain by invoking control mechanisms present in the brain and spinal cord. This work developed into the concepts of diffuse noxious inhibitory controls (DNICs) induced by pain stimuli,[7] which identified that nociceptive responses could be inhibited by applying painful stimuli elsewhere on the body. This reduction in nociceptive signalling relies on the fact that the ascending pain transmission makes connections to the periaqueductal grey matter and the nucleus raphe magnus of the midbrain and brainstem which connect back through efferent inhibitory interneurons.

Since then, considerable effort has defined an anatomical network which enables the regulation of nociceptive processing, principally within the dorsal horn.[8] The amygdala inputs detailed above are received into the lateral and basolateral amygdala, and subsequently signals are passed out from the amygdala, starting these descending pathways, from the central amygdala (CeA). These circuits that comprise the periaqueductal grey region in the upper brainstem, the nucleus reticularis gigantocellularis, nucleus raphe magnus and the locus coeruleus are components of the descending inhibitory pain pathways which serve to block incoming pain information at the level of the spinal cord inputs. For importance for the following specific therapeutic approaches, it is clear that the serotonergic pathways play a major role in the regulation of the descending facilitatory pathways. Considerable effort has been expended in attempting to map out and identify all the sites in the ascending and descending pathways that make up the pain neuronal circuitry. Of note are the studies which have either used direct injection of opiates into specific areas of the brain, or which have subjected cell subpopulations to electrical stimulation. These have led to detailed understanding of the various pathways, and also to preclinical and even clinical approaches, such as the electrical stimulation of the periaqueductal grey region, which has now largely been abandoned due to side effects. Crucially, the pain reversal that could be induced by electrical stimulation of PAG could be reversed by naloxone.[9] This connection between the periaqueductal grey region, the cortex and the amygdala has been also confirmed by imaging studies.[10]

Given the evidence above, the periventricular and periaqueductal grey matters are sites of investigation for potential therapeutic intervention to modulate descending pain pathways. Building on this, the grey matter regulation is mediated primarily through the serotonergic neurons of the raphe neurons, with an important contribution also made through the noradrenergic neurons of the locus coeruleus and the dopaminergic neurons of the ventral tegmental area. All of these neurons make up a group of descending fibres termed the dorsolateral funiculus, which synapse at inhibitory interneurons in the dorsal horn (Figures 5.1 and 5.2).

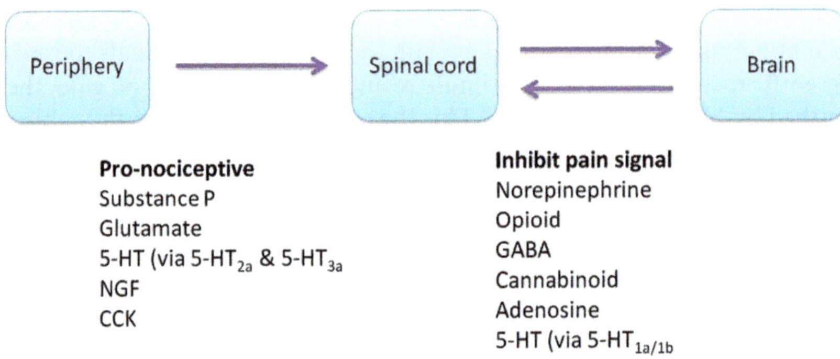

Figure 5.1 Neurotransmitters/signalling molecules known to increase or decrease pain signalling pathways.

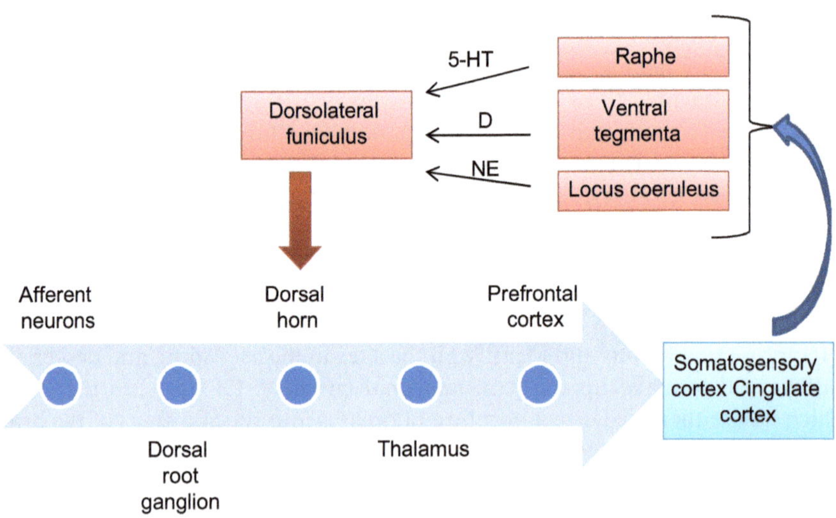

Figure 5.2 Schematic of neuronal circuitry involved in processing pain signals. 5-HT, 5-hydroxytryptamine/serotonin; NE, norepinephrine/noradrenaline; D, dopamine. Arrow left to right indicates ascending pain pathway; all other arrows represent descending pain pathways.

5.3 Reuptake Inhibitors

As mentioned earlier, one of the main contributors to persistent pain states is the reinforcement of descending inhibitory pathways, in particular by elevated concentrations of noradrenaline (norepinephrine, NE) and/or 5-HT in the synaptic cleft at spinal and supraspinal levels. Early investigation on the opioid analgesics confirmed both that although many agents did block 5-HT and NE reuptake (*e.g.* levomethorphan, levorphanol) others do not (morphine, naloxone).[11] Furthermore, studies in mice confirmed that mice with impaired 5-HT synthesis responded less effectively to opiate analgesia, which promoted interest in the ability of molecules increasing 5-HT levels to treat pain.[12] However, although the descending spinal pathway is inhibitory on post-synaptic 5-$HT_{1B/D}$ receptors, it is facilitatory *via* dorsal horn excitatory 5-HT_3 receptors, which potentially contributes to the overall poor efficacy of the selective serotonin reuptake inhibitors (SSRIs) as discussed below.[13]

Much work has been carried out using various 5-HT and NE tools in different animal models to understand their relative contributions, both to rationalise empirical clinical observation and practice, and also to enable the rational, prospective design of new agents. Specifically, it has been shown that 5-HT and NE agonists administered intrathecally can alleviate pain-driven responses[14] presumably *via* inhibition of descending pathways; and conversely, intrathecal administration of the 5-HT_3 antagonist ondanestron also blocked pain responses *via* block of the excitatory facilitation mentioned.[15]

A considerable body of preclinical and clinical data exists for molecules which produce elevated concentrations of these neurotransmitters by blocking their respective reuptake proteins. Clearly, the largest volume of data available is for the major class of antidepressant drugs, the so-called tricyclic antidepressants (TCAs), which have proven efficacy in a number of pain state clinical trials. This cross-over of medication is not surprising given the clinical convergence of mood disorders with various persistent pain states,[16] and the high concordance of pain and depression in the general population.[17] In particular, data from older studies confirmed that up to 80% of patients diagnosed with a depressive disorder experience persistent pain of head, neck, lower back or abdomen[18] and conversely up to 40% of fibromyalgia sufferers experience depression.[17a,19]

However, although the TCAs give some insight into mechanisms of pain relief, these molecules, while undoubtedly having significant activity at blocking neurotransmitter reuptake, also have wide-ranging pharmacology including being antagonists of various receptors (such as adrenergic α_1), blockers of various ion channels (NMDA, sodium, and calcium), activators of various ion channels and GPCRs (such as potassium, $GABA_B$) as well as modulating adenosine release and activating mu- and delta-opioid receptors. As such, it is more helpful to restrict consideration to the newer generation of selective single- or dual-reuptake inhibitors below.

An additional avenue of exploration is also the use of the TCAs as adjuvant therapy with morphine. Two main clinical studies have been reported, firstly combining nortriptyline and controlled-release morphine[20] in chronic lumbar root pain *versus* the individual treatments and placebo, and secondly adding either amitriptyline or placebo to patients on morphine therapy for neuropathic cancer pain. Neither study gave any significant difference in primary outcome measure.[21]

Later studies have looked at combinations of gabapentin with nortriptyline in mixed neuropathic pain populations and found, for a study of 56 individuals, that the mean daily pain at the maximum tolerated dose was significantly lower for the combination *versus* individual treatments.[22]

Lastly, fluphenazine (a typical antipsychotic) has been studied with amitryptyline[23] or nortriptyline[24] in postherpetic neuralgia (PHN) and painful diabetic neuropathy respectively. Neither showed any meaningful additional impact on pain measures.

5.3.1 Selective Serotonin Reuptake Inhibitors

Many SSRIs have been evaluated in chronic pain states, and although the SSRIs have replaced the TCAs as front line treatment for several depression indications, the comparison is less favourable for pain, with SSRIs being significantly less efficacious in the studies that have been reported. A search of clinical trials reveals at least 20 such studies, with a far greater number of patients being treated off-label, despite little evidence of their effectiveness (Table 5.1 lists common clinical practice for off-label prescription of reuptake inhibitors[25]) (Figures 5.3 and 5.4).

Table 5.1 Serotonin and noradrenaline (norepinephrine) reuptake in inhibitors and their commonly prescribed off-label use in treating pain states.

Drug	*Off-label typical use*	*Doses prescribed*
SSRIs		
Fluoxetine	Fibromyalgia	20 mg day^{-1}
Citalopram	Neuropathic pain Fibromyalgia	20–40 mg day^{-1}
Escitalopram	Neuropathic pain Fibromyalgia	20–40 mg day^{-1}
Paroxetine	Diabetic neuropathy	Initially 10–20 mg day^{-1}
Sertraline	Neuropathic pain Fibromyalgia	Initially 50 mg day^{-1}
Fluvoxamine	Neuropathic pain Fibromyalgia	Initially 50 mg day^{-1}
NRIs		
Bupropion	Neuropathic pain	100–400 mg day^{-1}
Maprotiline	Chronic pain Neuropathic pain Fibromyalgia	75 mg day^{-1}
SNRIs		
Venlafaxine	Neuropathic pain	75 mg day^{-1}
Milnacipran	Neuropathic pain *(FDA approved for fibromyalgia)*	100–200 mg day^{-1}
Duloxetine	Non-diabetic neuropathic pain *(FDA approved for diabetic neuropathy and fibromyalgia)*	60 mg one or twice daily

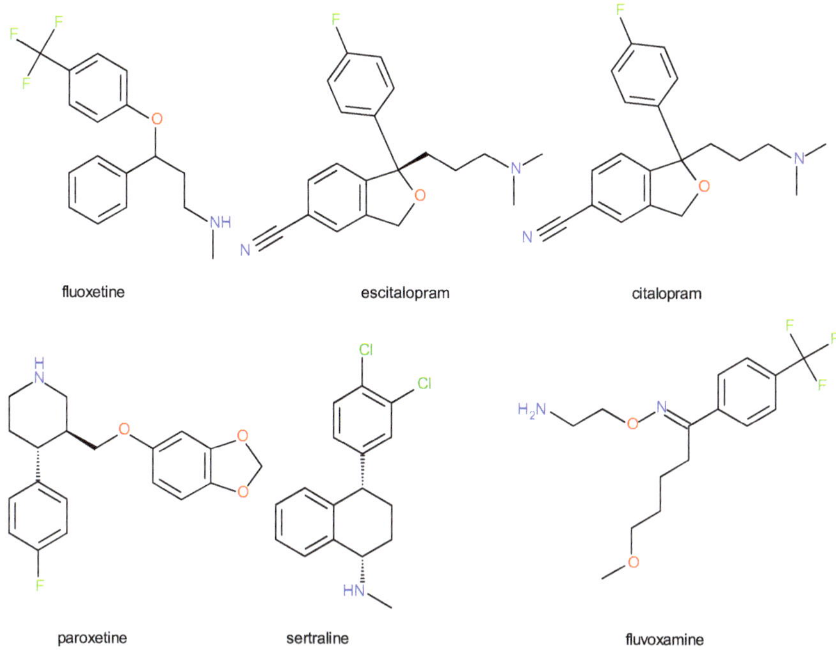

Figure 5.3 Structures of SSRIs reported for use in pain medication.

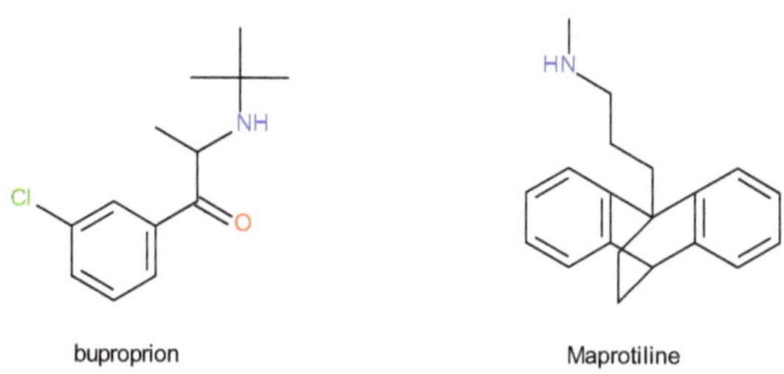

Figure 5.4 Structures of explored NRIs.

A comprehensive review published in 2007 which evaluated and summarised data for various SSRIs, NRIs, and SNRIs across rodent models of pain concluded that NE has a more significant role to play that 5-HT in mediating pain responses.[26] Associated reports of direct injection of SSRIs into specific brain regions have not produced effects that have been widely replicated or supported.

An earlier clinical review published in 1997 evaluated the clinical data reported across 19 studies on various pain states (headache, diabetic neuropathy, fibromyalgia, mixed-chronic pain) and found clear evidence for efficacy in just the mixed-chronic pain subjects.[27] These data have been supported by more recent studies, from which meta-analyses have demonstrated relative effect sizes across the class. For this analgesic effect size (ES) it is defined as the number obtained by subtracting the placebo analgesic effect from the drug analgesic effect and dividing the difference by the pooled estimated standard deviation. The ES is ranked 0.2–0.3 as small, 0.5 as medium and 0.8 or more as large. For the pooled, placebo-controlled studies in neuropathic pain, fluoxetine was judged to be the most effective (ES = 1.08), ahead of citalopram (ES = 0.45) then paroxetine (ES = 0.19).[25b] Furthermore, no effect was confirmed for paroxetine in lower back pain, and more generally for SSRIs in osteoarthritis, rheumatoid arthritis or headache.[28]

5.3.2 Mixed Serotonin and Norepinephrine Reuptake Inhibitors

Preclinical studies have compared the relative effectiveness of blocking both reuptake of serotonin or NE. A comparison of the NRIs desipramine and reboxetine with SSRIs fluoxetine and paroxetine found that inhibiting reuptake of NE was more effective than inhibiting reuptake of serotonin in consistently reversing tactile allodynia in rodents, with equivalent efficacy to gabapentin.[26] Furthermore, the SNRIs duloxetine, venlafaxine and milnacipram were effective in several neuropathic pain models to a greater extent than either selective blockade mechanism.[29]

Clinically, the three approved SNRIs duloxetine, venlafaxine and milnacipram have been assessed in multiple acute and chronic pain trials. For venlafaxine, seven out of the eight randomised controlled trials were positive—in acute postoperative pain (equivalent to gabapentin),[30] human experimental pain,[31] painful diabetic neuropathy (NNT 4.5 for >50% pain relief),[32] painful neuropathy (NNT 5.2 for >50% pain relief),[33] migraine[34] and tension-type headache (Figure 5.5).[35]

Duloxetine achieved the first antidepressant approval for fibromyalgia (and the only other agent at the time other than pregabalin to be approved for fibromyalgia in the United States) after demonstrating efficacy in two pivotal trials. A meta-analysis indicates improvement against both pain and quality of life measures.[36] A further three studies in painful diabetic neuropathy were also positive (NNT 6 for >50% pain relief *versus* NNH 15).[37](NNT is the number needed to treat for one patient to benefit compared to control; NNH is the

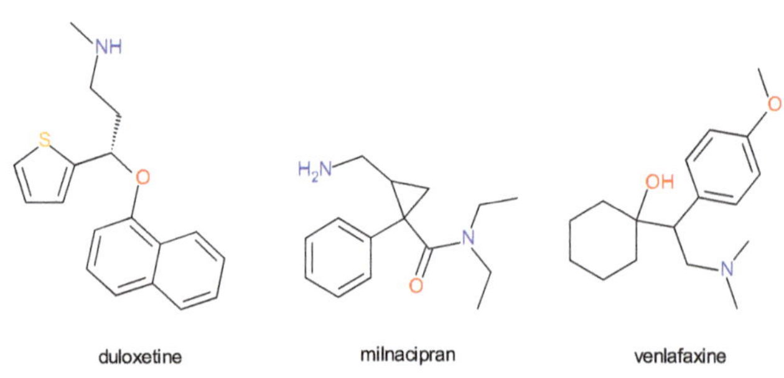

Figure 5.5 Structures of SNRIs.

number needed to be exposed to the drug to cause harm in one patient compared to control.)

Finally, milnacipram was also approved by the U.S. Food and Drug Authority (FDA) for fibromyalgia in 2009 following data from Phase 3 trials involving over 2000 patients, with positive outcomes on both pain and functional measures (no comparative NNT data available).[38]

Overall, robust clinical trial data support the use of SNRIs in neuropathic pain and fibromyalgia, and specifically duloxetine in lower back pain.[39] However, despite improved tolerability over the pre-existing TCAs, the NNT values tend to be larger. For further clinical guidance, more directly comparative pain trials are required between molecules with different pharmacology

5.3.3 Next-Generation Reuptake Inhibitors

Over the last 5 years, many papers have been reported in the medicinal chemistry literature seeking to build on the profiles and successful drug registrations of the SNRIs. These papers either cover the next generation of SNRIs with improved properties (selectivity, safety or pharmacokinetics), such as reported by Pfizer (see reference for example work and lead-in to all cited and citing papers from same group[40]) or combinations of SNRIs with other pharmacology. Of major interest has been the work which has sought to dial in additional monoamine reuptake inhibition by creating the so-called triple reuptake inhibitors blocking serotonin, NE and dopamine reuptake. In contrast to the studies reported above for 5-HT and NE, the level of understanding of the role of dopamine in regulating ascending and descending pain signals, and in particular of its action in the spinal cord, is less well developed. Recent publications discuss signals from the substantia nigra and hypothalamus providing dopaminergic innervations to the spinal cord as well as dopamine-synthesising cells being present in the dorsal root.[41]

In the last couple of years, many publications have described distinct chemical classes of triple reuptake inhibitors with potential therapeutic application in major depression and pain disorders. Figure 5.6 shows a representative sample of structure from each of these classes. Many of these molecules are potent blockers of all three monoamine reuptake systems, with good PK and evidence for receptor occupancy *in vivo*, although interestingly, the occupancy levels assessed *ex vivo* in rat cortical synaptosomes was lower than

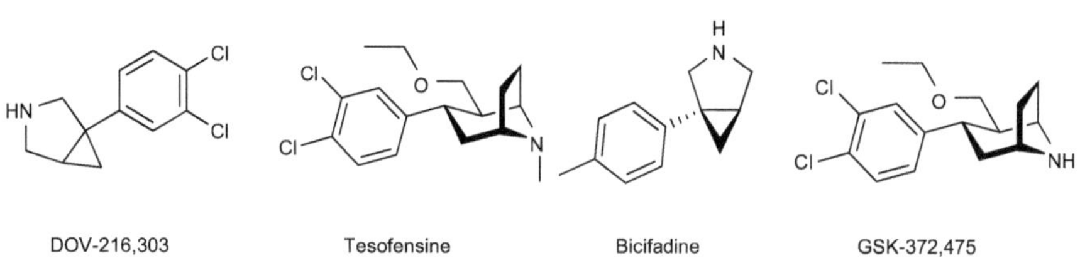

Figure 5.6 Structures of recent triple reuptake inhibitors.

that achieved with SSRIs or NRIs for the same behavioural outcome. Most of the behavioural data reported focused on the development of agents for major depressive disorders, although extensive data have been reported for the triple reuptake inhibitor bicifadine across a range of acute inflammatory, persistent inflammatory and chronic neuropathic pain models.[42] These behavioural data were correlated with *in vitro* binding affinities and *in vivo* microdialysis monitoring. A second molecule, NS7051, was also as effective as tramadol in a range of pain models.[43] Some clinical investigation has been initiated for this class, and tolerability and/or early efficacy data have been reported for DOV216303,[44] tesofensine,[45] bicifadine[46] and GSK372475[47], although particularly for the latter compound, there was a clear lack of antidepressant efficacy coupled with poor tolerability when compared to placebo, which raises concerns for the potential clinical utility of this triple reuptake class.

5.4 Alpha$_{2A}$ Agonists

Three α_2-adrenergic receptor subtypes (α_{2A}, α_{2B} and α_{2C}) have been cloned and are widely distributed in the peripheral and central nervous systems, although the highest density of expression is found in the locus coeruleus. α_{2A}-Adrenergic receptors have been investigated as potential targets to achieve antinociceptive activity, in particular with agents such as dexmedetomidine (approved for sedation of critically ill or injured patients in an intensive care unit setting). Rodent studies using various α_2-adrenergic receptor subtype knockouts supported the hypothesis that the α_{2A}-adrenergic receptor is the predominant subtype involved in mediating not only the antinociceptive properties, but also the sedative and hypothermic actions of dexmedetomidine (Figure 5.7).[48]

Due to the challenge of separating the desired and undesired properties of α_{2A}-adrenergic receptors for pain, effort has also focused on exploring the synergistic interactions of α_{2A}-adrenergic receptor agonists with opioid and cannabinoid receptor agonists. In particular, dexmedetomidine was studied in combination with CP55,940, a full, non-selective CB1/2 agonist, in the rodent and found that in the mouse tail flick assay, effects were additive (*i.e.* potentially mediate antinociception *via* independent pathways) whereas in the hot plate assay, effects were synergistic.[49] Clinical studies of α_{2A}-adrenergic receptor agonists have explored the potential for reduction in morphine consumption, and a report from 2012 conducted a meta-analysis of nearly 1800 patients across 30 studies and concluded that there was evidence for postoperative

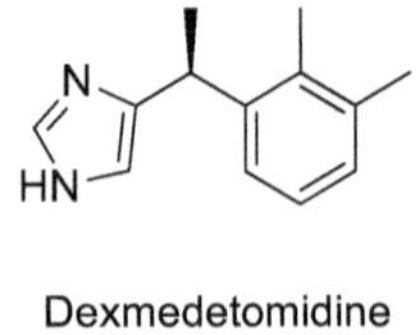

Dexmedetomidine

Figure 5.7 Alpha$_{2A}$ agonist studied clinically.

sparing of morphine at 24 hours following administration of either dexmedetomidine or clonidine as well as reductions in pain intensity, although increases in both bradycardia and arterial hypotension were observed.[50]

5.5 NMDA Receptor Antagonists

The signalling of acute pain is mediated by glutamate release from nociceptor terminals which act on receptors on the postsynaptic terminals to generated excitatory postsynaptic potentials (EPSPs) in the second-order dorsal horns by activation of fast excitatory ionotropic glutamate receptors, AMPA and kainate subtypes. Following sufficient depolarisation of the postsynaptic cell, the magnesium block of the NMDA receptor subtype is relieved which leads to calcium influx (which can also be reinforced by mGluR and NK1 receptor activation) and hence activation of multiple downstream pathways *via* protein kinase mediators such as MAPK, PKA, PKC, PI3K and Src. This onward activation serves to strengthen the synaptic connections between the nociceptors and the dorsal horn, and has been described as long-term potentiation (LTP) in pain[51] and operates in an analogous fashion to the activation of LTP in learning and memory.[52] This is reinforced by observations that molecules that are able to inhibit LTP in the spinal pathways are able to reduce hyperalgesia following tissue injury.

Data have been reported in various preclinical models for both the NMDA receptor channel blockers ketamine and dextromethorphan and also the non-competitive NMDA receptor antagonists neramexane and memantine, with the latter group of agents showing improved tolerability. Neramexane and memantine are both use-dependent blockers of the NMDA channel, and by only blocking the channel with moderate affinity in its open state, with strong voltage dependence, they are thought to be effective at reversing pain states caused by persistent channel activation or opening (with improved tolerability). Neramexane has been shown to be effective in a rat model of diabetic neuropathic pain, achieving levels of antinociceptive effect comparable to that of gabapentin and memantine but with improved tolerability.[53]

A recent review surveyed the literature of clinical trials measuring spontaneous pain in acute and chronic pain states (see cited references for original trial literature).[54] For inclusion in the review, the trials were required to be blinded, randomised and placebo-controlled and to report using numerical scales. The authors then also subjectively assessed the trial quality for inclusion, which resulted in a consideration of 28 trials of NMDA antagonists for the treatment of neuropathic pain. Only four groups of trials contained sufficient data to allow an assessment of the summary ESs, from which the only positive was ketamine IV in post-amputation pain. The other analyses, ketamine IV in complex regional pain syndrome and memantine orally in PHN and post-amputation pain, showed no significant effect. Overall, although both ketamine and dextromethorphan have shown efficacy in clinical pain symptoms,[53,55] both drugs are also poorly tolerated at therapeutic doses with side effects including hallucinations, dysphoria, cognitive impairment and motor function impairment.[56]

Ketamine has however also been explored in adjunctive therapy settings—in particular with morphine. A recent review has studied the various clinical trials that have been run, following contradictory reported results.[57] From the 11 studies analysed, the findings were that 6 showed a significant improvement in postoperative analgesia compared to morphine alone (and also allowed for reduced morphine consumption), and 5 showed no effect.

For the use-dependent blockers, neramexane has been found to be effective in a human experimental pain model of induction of acute pain by an intradermal injection of capsaicin.[58]

Newer approaches have sought to capitalise on the analgesia demonstrated preclinically and clinically by the various NMDA receptor antagonists but seeking to prepare and characterise subtype-selective ligands in the hope of improving their therapeutic utility through reduced side effects. Functional NMDA receptors are composed of an ion channel-forming NR1 unit and at least one NR2 unit (subtypes NR2A/B/C/D exist). In particular, many groups have pursued agents which target the NR2B subunit specifically due to its localisation in the forebrain and superficial lamina of the spinal cord, as compared to the wider distribution of NR2A in the brain and spinal cord.[59] It is hoped that this restricted distribution will lead to agents with improved tolerability (Figure 5.8).

To this end, a series of molecules were developed from the less selective parent, ifenprodil, including the most widely characterised molecules, traxoprodil (CP-101,606) and Ro25-6981, which were both able to reduce the pain induced in both acute and chronic inflammatory[60] and neuropathic injury in rodents.[59a] Furthermore, in two different models of spinal cord injury, both ifenprodil and Ro25-6981 injected intrathecally were able to reduce mechanical nociception in a dose-dependent manner, although the former agent also impaired motor responses at higher doses. These studies were coupled with receptor expression studies which indicated that GluN2B/NR2B was expressed to a significantly higher level following spinal cord injury.[59b] Ifenprodil[61] and CP-101 606[62] are also effective in models of neuropathic pain with diminished CNS side effects. Furthermore, CP-101 606 has also been reported to have analgesic efficacy when administered to chronic neuropathic pain subjects in both spinal cord injury and radiculopathy settings *via* 24-hour continuous IV infusion. These data, however, are only reported through a conference abstract with no peer-reviewed publication data available.[63] Conversely, radiprodil

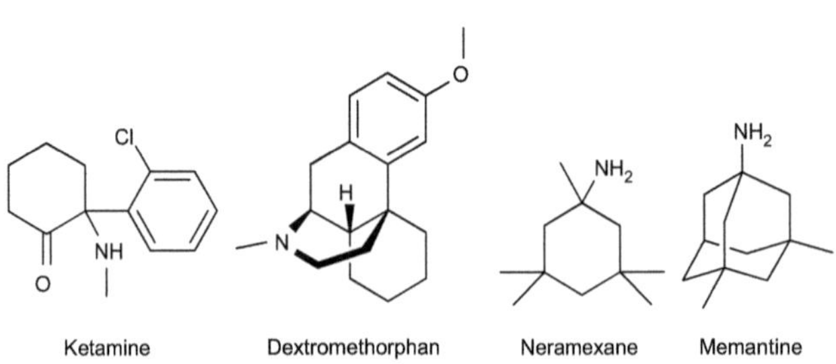

Figure 5.8 Structures of competitive NMDA antagonists and dextromethorphan.

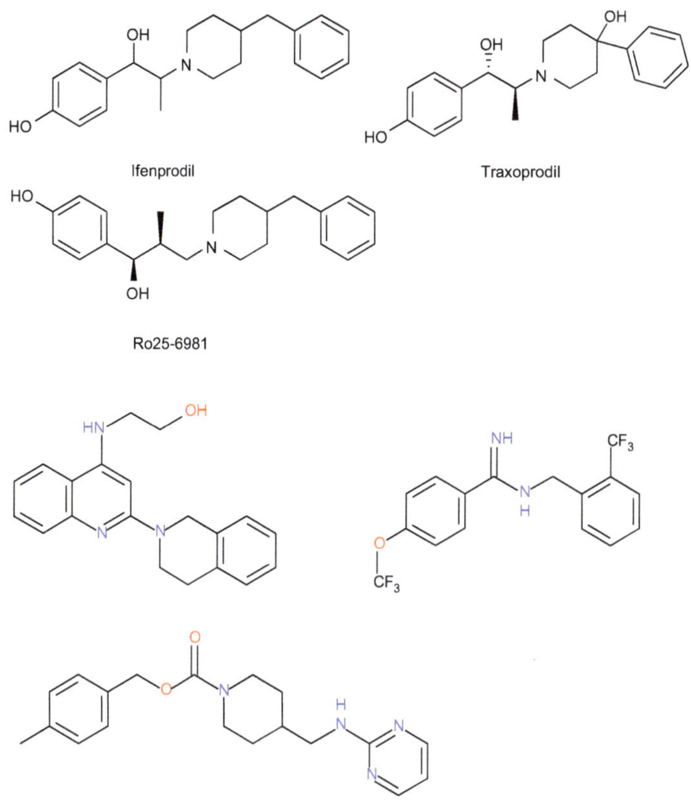

Figure 5.9 Structures of GluN2B selective NMDA antagonists and recently disclosed novel chemotypes.

(RGH-896, Gedeon-Richter & Forest) failed to achieve its positive endpoints in a Phase 2b study in neuropathic pain (see company press releases). A number of molecules from this class have, however, progressed into clinical development for other indications including treatment-resistant depression for which CP101,606 achieved proof of concept.[64] Additionally, AZD6765 (mixed GluN2A/B antagonist) is being studied in a Phase 2b treatment-resistant depression trial, MK-0657 (GluN2B antagonist) was studied in Phase 1 Parkinson's disease and depression (although development has been discontinued, potentially due to observed cardiovascular effects) and EVT-101 and EVT-103 (GluN2B antagonists), were terminated in Phase 2 and have completed Phase 1 respectively.[65]

Many groups are still actively seeking improved NR2B-selective agents and Figure 5.9 shows a range of recent chemotypes.

5.6 GABA Receptor Modulators

Given the counterbalancing role GABA plays to glutamate by mediating the majority of fast inhibitory neurotransmission, it is unsurprising that the GABA receptor subtypes have been studied extensively for their potential to deliver therapeutic strategies in pain. GABA-ergic inhibitory interneurons are present at high levels in the superficial dorsal horn, and as such, a long-standing theory indicates that loss of these inhibitory cells would lead to increased pain (the original gate theory of pain mentioned earlier[6]). In support of this, spinal administration of bicuculline ($GABA_A$ competitive receptor antagonist) evokes hypersensitivity to touch in normal rats.[66] Furthermore, spinal administration of muscimol or isoguvacine (non-selective $GABA_A$ receptor agonists) can reduce pain-driven behaviours in a range of rodent neuropathic pain models.[67] Various mechanisms have been postulated for these observations, including peripheral nerve-injury induced cell death of GABA-ergic interneurons[68] or altered storage and release of GABA.[69] Supporting data is found in PKCγ-positive neurons which are only found in the innermost part of the spinal cord lamina and are essential for the expression of nerve injury-evoked persistent pain (deletion of encoding gene in mouse reduces nerve-injury evoked mechanical hypersensitivity).[52] Following block of Gly receptor-mediated inhibition with strychnine, these PKCγ-positive neurons are activated by hindpaw brushing.

More recent studies have concentrated on postsynaptic $GABA_A$ receptors which are localised on the intrinsic dorsal horn neurons. Peripheral nerve injury causes downregulation of KCC2 (potassium–chloride cotransporter protein) in lamina I projection neurons such that $GABA_A$ receptor activation results in depolarisation rather than hyperpolarisation, hence, paradoxically, enhancing excitability. This interpretation was supported by blocking or downregulating KCC2 in the rat and observing mechanical allodynia.[70]

The $GABA_A$ receptor is a multimeric protein with five subunits which cover a wide group of proteins (α_{1-6}, β_{1-3}, γ_{1-3}, δ, ε, θ, π), and many groups have striven to produce subtype-selective GABA receptor modulators for both pain and psychiatric indications. In support of this, transgenic mice have been used to try to tease apart the desired pharmacology to maximise efficacy and minimise side effects, although the interpretation of these data do not allow direct assessment of the preferred target pharmacological profile.[71] From behavioural studies it has been shown that NS11394 (α_3, α_5-subtype-preferring $GABA_A$ receptor positive allosteric modulator, PAM) is effective in reducing pain-induced behaviours in models of inflammatory and neuropathic pain at levels equivalent to the non-selective $GABA_A$ receptor full potentiator diazepam, but without the profound motor impairment observed with the latter.[72] This raises the potential for appropriately subtype-selective ligands to be better-tolerated agents than the currently clinical available molecules. Further data for a $GABA_A$ receptor partial PAM, bretazenil (partial PAM at all of α_{1-6} subunits), was unable to reverse hypersensitivity in the rat at anxiolytic doses which raises the possibility that a threshold level of PAM activity is required at the α_2- and α_3-containing $GABA_A$ receptors for analgesia.[71]

Clinically, a number of the benzodiazepines have been studied in clinical trials or used *ad hoc* and off-label to treat various chronic pain states; however, little

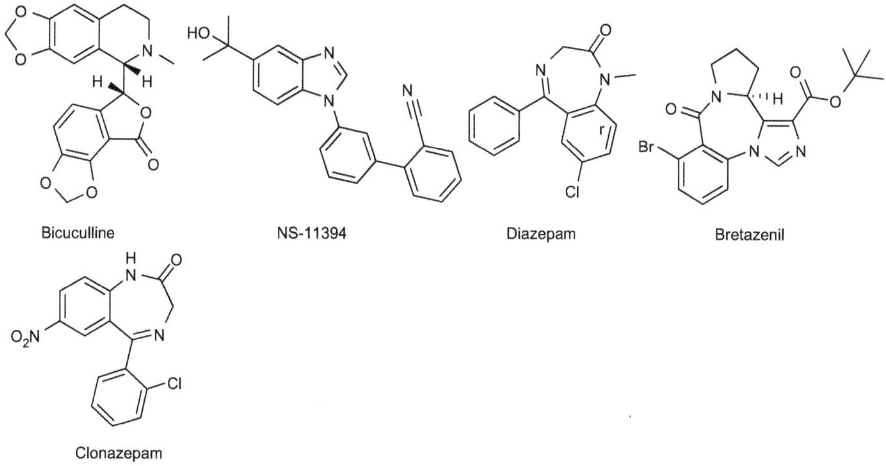

Figure 5.10 Structures of non-selective GABA modulators.

evidence for their efficacy has been recorded, in contrast to the wealth of data for other anticonvulsant agents such as gabapentin. Clonazepam is a benzodiazepine with a relatively long half-life (approximately 35 hours) and marked anticonvulsant effects with less pronounced sedation compared to other benzodiazepines. It has been studied in three reported clinical studies in pain, including topical use to treat stomatodynia,[73] but from a recent review, none of these were judged to be sufficiently well-designed or informative to support the use of clonazepam in chronic neuropathic pain or fibromyalgia (Figure 5.10).[74]

5.7 Cannabinoid Receptor Agonists/Modulators

Cannabis or marijuana (dried blossom tips) and hashish (dried blossom resin) from *Cannabis sativa* have been used both for relief of pain and recreationally throughout human recorded history, with many ancient civilisations describing medicinal uses of different marijuana preparations.[75] However, despite the evident analgesic properties of cannabis, its effects on cognitive, affective and motor functions limit its therapeutic potential. Nonetheless, despite legal restrictions in many countries and states, cannabis is used widely to control pain in many disease states.[76] In addition to anecdotal and empirical assessments, there have been many attempts to evaluate the effectiveness of various preparations of cannabis in patient populations. One recent review evaluated 18 trials of cannabis in chronic pain and concluded that cannabis treatment is moderately efficacious for treatment of chronic pain;[77] however, it also indicated that this positive effect is counterbalanced by observations of adverse events and poor tolerance.[78] In particular, smoked cannabis has been demonstrated to be efficacious in HIV-induced neuropathic pain,[79] and in treating pain (and spasticity) caused by multiple sclerosis (such as through the approved agent Sativex®) (Figure 5.11).[80]

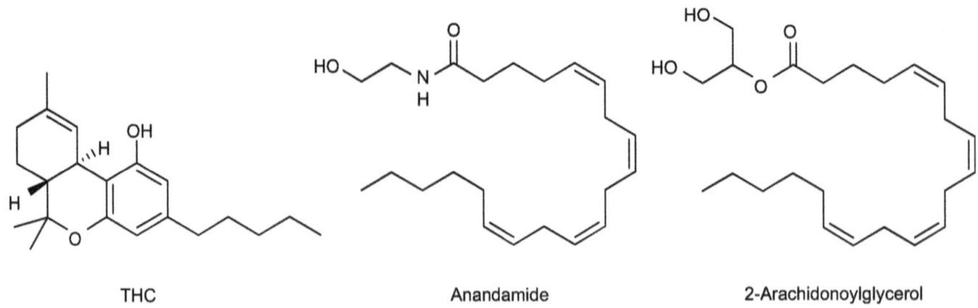

Figure 5.11 Structures of endocannabinoids.

Unsurprisingly, given these observations, the discoveries relating to both the constituent components of the plant responsible for analgesic and centrally mediated effects, the cloning of their target receptors and more recently the identification of their endogenous substrates has prompted an enormous effort within the research community to obtain well-tolerated cannabinoid derivatives. The principal neuroactive cannabis constituent, Δ^9-tetrahydrocannabinol (THC), acts *via* cannabinoid type-1 and type-2 (CB1 and CB2) receptors, which are GPCRs which are widely expressed in neurons (CB1) and immune cells and microglia (CB2). Research efforts have both targeted agents at the CB1 and CB2 receptors, and also investigated approaches to modulating the biosynthesis and inactivation of their natural ligands, the endocannabinoids such as *N*-arachidonoyl-ethanolamine (anandamide) and 2-arachidonylglycerol (2-AG). From these understandings, dronabinol (Marinol), a synthetic THC approved by the FDA in the treatment of anorexia in AIDS patients, as well as for refractory nausea and vomiting of patients undergoing chemotherapy, is currently under evaluation in Phase 3 neuropathic pain trials. Additionally, nabiximols (Sativex), a combination of THC and cannabidiol, has been approved for multiple sclerosis and neuropathic pain associated with cancer.

CB1 receptors are highly expressed in neurons that are involved in transmission of pain signals, such as the periaqueductal grey and dorsal horn of the spinal cord, and to a lesser extent in the dorsal root ganglia. It has been demonstrated that CB1 receptors are present in approximately 75% of nociceptive neurons in the dorsal root ganglion,[81] and a range of studies have demonstrated that CB1 receptor agonists inhibit both GABA-ergic and glycinergic neurotransmission presynaptically in lamina II neurons. Although the demonstration of reductions of pain-induced behaviours with THC and CB1-selective receptor agonists is complicated by the concomitant psychomotor activity, there is nonetheless clear evidence that the cannabinoids produce antinociception at both spinal and supraspinal sites.[82] Furthermore, although the acute activity of cannabinoids is mediated *via* central CB1 receptors in naive animals, the evidence suggests that both central and peripheral CB1 receptors are involved in mediating the effects in animal models of chronic pain. Numerous studies then attempted to dissect the actions at the central and peripheral CB1 receptors in the hope of dissecting the effects on pain behaviours from motor and cognitive effects. Preclinical proof of principle was first demonstrated by molecules with potent CB1 receptor agonist activity which had low penetration into the CNS, and a range of molecules was reported by various pharmaceutical companies, several of which have now entered clinical development (see Table 5.2).[75] In addition to these clinical and preclinical discovery projects, other approaches have been investigated, such as the targeted upregulation of CB1 receptors *via* partial agonists or the specific delivery of a CB1 agonist to a localised area such as skin or spinal cord, for which preliminary evidence supports a separation of the analgesic from classical side effects.[83]

Complementary to these activities focused on CB1 receptor activation and upregulation has been investigation into CB2 receptor agonists, in particular because their mainly peripheral distribution and lack of association with CNS side effects makes them an attractive alternative. Increased expression of CB2 receptors are observed in preclinical pain models and following tissue injury and many groups have worked to identify selective CB2 agonists which have been efficacious in various preclinical models, although concerns are still raised

Table 5.2 CB1 receptor agonists in clinical evaluation.

Compound	*Sponsor/company*	*Status*
Dronabinol (THC)	GW Pharmaceuticals	Launched
Nabiximols (THC + cannabidiol)	Unimed Pharmaceuticals	Launched
KN-38–7271 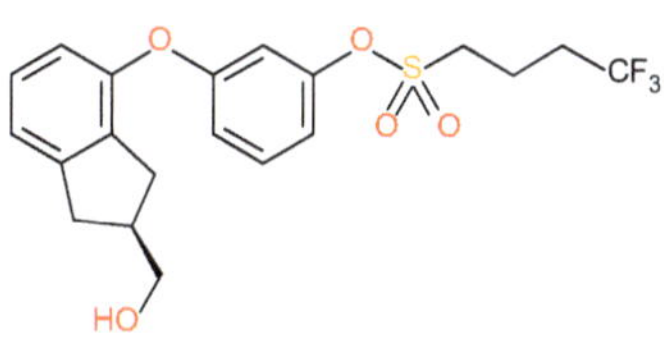	Bayer	Phase 2
Ajulemic acid 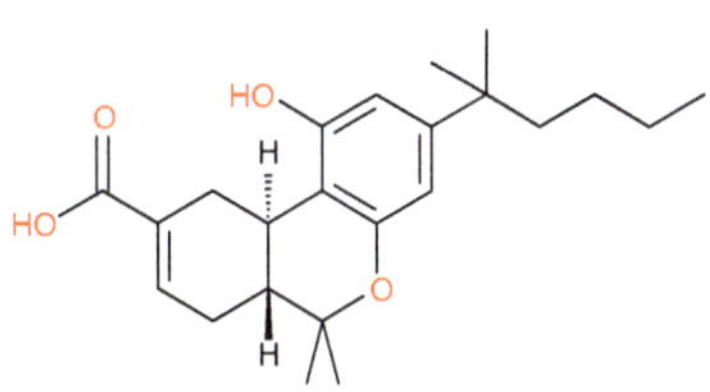	Cervelo	Phase 2
AZD-599 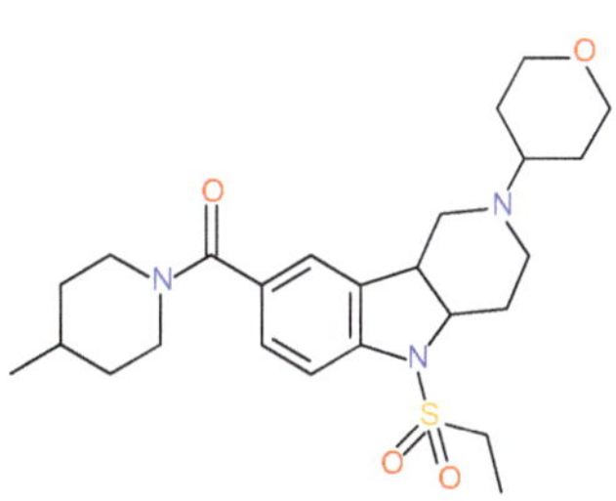	AstraZeneca	

about the potential for immunosuppressive action on chronic administration.[84] However, despite this body of data showing efficacy in rodent models, no molecule has yet been able to demonstrate a positive proof of concept in a Phase 2 clinical trial, of at least five that have entered clinical development.

An alternative approach to both these receptors lies in their endogenous ligands the endocannabinoids, which are locally produced and released to activate CB1 and CB2 receptors and also the extent of this production and release is altered during specific tissues with localised chronic or inflammatory pain.[85] This suggests that there may be therapeutic potential in modulating the levels of these endocannabinoids so that they only have effects in active neural pathways, thereby reducing the potential for side effects that result from widespread systemic cannabinoid receptor activation.[86]

A widely explored approach to be able to modulate the levels and duration of action of these endocannabinoids is to target the enzymes responsible for their hydrolysis, in particular fatty acid amide hydrolase (FAAH) and monoacylglyceride lipase (MGL). FAAH is an integral membrane serine hydrolase protein for which the crystal structure has been solved and detailed studies have been reported on its mechanism of action.[87] Unlike the majority of serine hydrolases, FAAH functions *via* a Ser-Ser-Lys catalytic triad and both covalent and non-covalent inhibitors have been identified.[88] These inhibitors were developed from initial substrate-mimicking skeletons into molecules with improved selectivity across the other lipid hydrolases. All of these next-generation inhibitors possessed carbamate groups which formed covalent bonds to the nucleophilic serine residue in the catalytic triad, whereas subsequent generations explored a wider range of structures which included non-covalent inhibitors, with extensive discussion in the literature as to their respective merits (Figure 5.12).

Similarly, with a view to specifically regulating levels of 2-AG, MGL has been investigated due to its major role in inactivating 2-AG in whole brain homogenates and whole cells.[89] MGL possesses a catalytic triad of Ser-Asp-His and a range of initially non-selective, and more recently selective, covalent inhibitors have been described.[90]

Inhibitors of both enzymes, and in particular of FAAH, have carried great therapeutic promise due to the hope that they would modulate pain pathways but be devoid of the central CB1 and potentially immunosuppressive CB2 activation-mediated side effects. In support of this, both FAAH and MGL inhibitors have shown robust and marked efficacy in many rodent models of inflammatory and chronic pain (as well as activity in a range of other disease models[88]). These molecules have generated clear data linking target modulation (direct assessment of both FAAH inhibition and AEA concentration increases) in the brain with the behavioural effects, facilitated by excellent pharmacokinetic profiles. From these encouraging preclinical data, molecules have been progressed into clinical development from Pfizer, Ironwood and Infinity Pharmaceuticals,[91] and more recently a PET ligand, [^{18}F]-PF-9811 has also been reported.[92] The most advanced of the molecules in development, PF-04457845, reported data in 2012 from a Phase 2 study in patients with pain

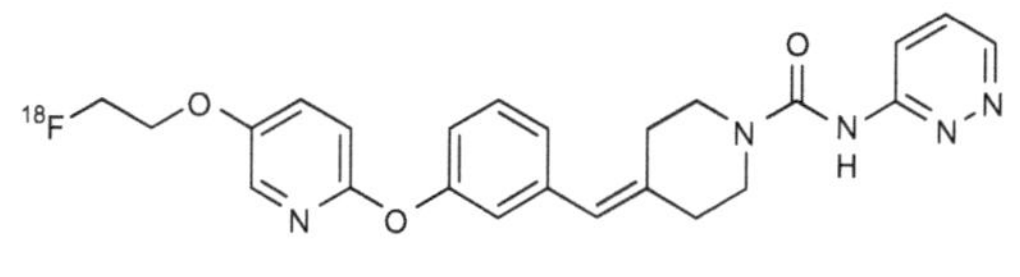

^{18}F-PF9811

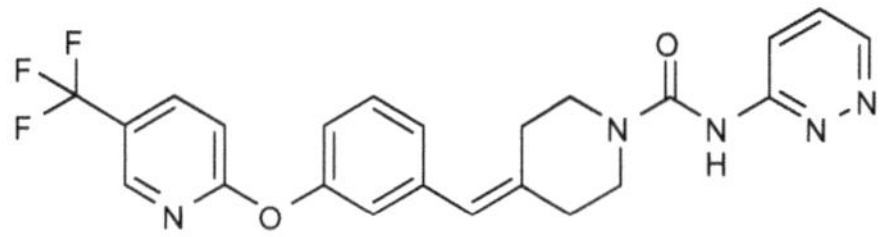

PF-04457845

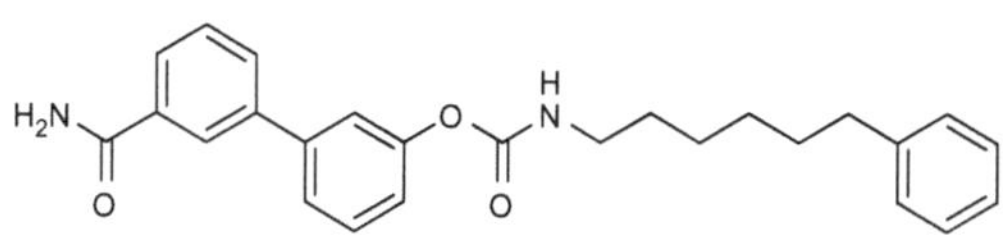

JP-83

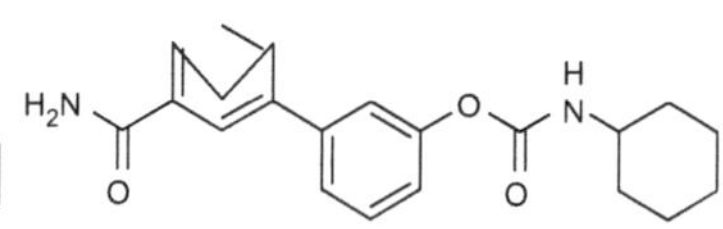

URB-597

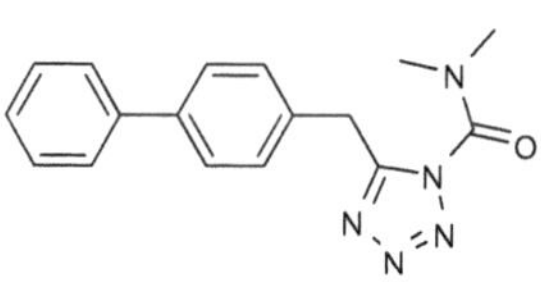

LY-2183240

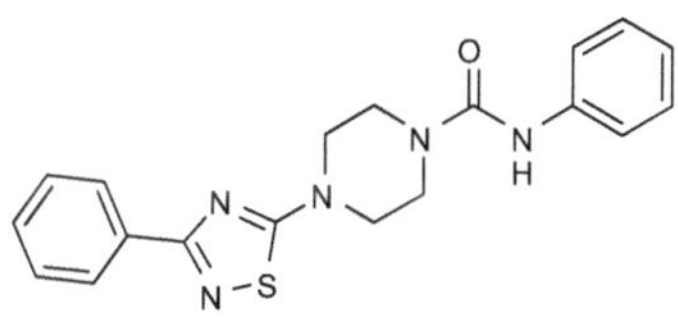

JNJ-1661010

Figure 5.12 Structures of inhibitors of FAAH.

due to osteoarthritis of the knee which failed to demonstrate any analgesic effect.[93] PF-04457845 is a potent, selective FAAH inhibitor which has demonstrated wide-spectrum analgesic effects in rodents and had reported clear Phase 1 data indicating both good tolerability and ability to modulate endocannabinoid levels.[94] FAAH1 activity was almost completely inhibited following doses of at least 0.3 mg (single dose) and 0.5 mg once daily (multiple dose) of PF-04457845 giving rise to elevations in fatty acid amide concentrations (3.5- to 10-fold). Furthermore, there was no evidence of effects of PF-04457845 on cognitive function. The trial was randomised and placebo-controlled and compared 4 mg PF-04457845 one daily with 500 mg naproxen twice daily using the WOMAC pain score as primary endpoint. Despite a clear signal for naproxen, no effect was observed with the FAAH inhibitor, despite reduction in FAAH activity by more than 96% and marked reductions in the levels of the fatty acid amides.

These clinical data clearly represent a significant setback for proponents of endocannabinoid therapeutic potential, and despite discussions that inhibition of FAAH may have had greater potential in a non-inflammatory, neuropathic or fibromyalgia patient group, this clear species difference coupled with the redundancies that exist in the biochemistry and pharmacology of the endocannabinoid pathways and concerns are going to make it challenging to gain sufficient confidence to invest in future clinical studies in this area.[95]

5.8 Adenosine Agonists

Adenosine has been studied in animals and humans and has been found to have antinociceptive properties. Adenosine can act *via* four purinergic P1 receptor subtypes, A_1, A_{2a}, A_{2b} and A_3 receptors, of which the A_1 receptor has been widely studied due to its expression in nociceptive and spinal cord neurons.[96] Although adenosine administered intrathecally has demonstrated positive benefits in patients with chronic pain,[97] it is not currently used in clinical practice because of its questionable efficacy against spontaneous pain and concern over side effects (back pain, headache) when dosed intrathecally. Furthermore, actions on peripheral adenosine receptors can trigger pronociceptive and vasodilatory effects as well as reducing heart rate.[96a]

In order to understand the differential effects on allodynia and hyperalgesia *versus* spontaneous pain, rodent studies studied the differential effects of clonidine, ω-conotoxin and adenosine and found that although all reduced allodynia, adenosine alone was unable to inhibit spontaneous pain-driven behaviours, despite its ability to inhibit N-type calcium channels.[98] This was confirmed in humans in a study in which intrathecal adenosine reduced hyperalgesia and allodynia but had no effect on spontaneous pain.[99] Conflicting data have been generated for the study of adenosine as a pre-emptive analgesic administered before major surgery with observations in some studies of reduced opioid consumption[97,100] contradicted by later negative studies.[101]

There are a number of clinical studies in progress (Table 5.3), as well as more recent discoveries seeking creative solutions to the challenge of mechanism-driven side effects of A_1 receptor activation. In particular, groups are focusing

Table 5.3 Adenosine A1 receptor agonists in clinical evaluation.

Trial	*Sponsor*	*Status*
Clonidine intrathecally *versus* adenosine to treat neuropathic pain 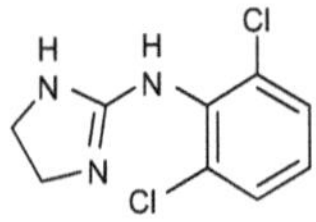 Clonidine	Wake Forest University	Ongoing
Adenosine IV in perioperative pain	Xsira	No significant effect
GW493838 (A1 agonist) in peripheral neuropathic pain	GSK	Completed 2006
BVT 11959 orally (A2a agonist) in diabetic neuropathic pain	Biovitrum	No significant effect

(Table 5.3 adapted from Zylka[96b])

on partial agonists, or attempts to increase local concentrations of adenosine, such as *via* ectonucleotidases.[96b]

5.9 Future Perspectives

It is clear that a detailed understanding of the molecular mechanisms underlying various pain states has the potential to change the future of pain management, although this needs to be coupled with well-designed, informative clinical studies which allow questions to be answered around individual mechanisms and drug combinations. At present, the traditional model of discovery places an over-reliance on validation through animal models with repeated failures of translational to the clinical setting. Furthermore, the absence of clear pharmacokinetic/pharmacodynamic information and patient heterogeneity frustrates interpretation of the clinical 'negatives' and reduces the opportunity to refine potential approaches.

Nonetheless, our considerably increased understanding of the central mechanisms underlying transmission, facilitation and inhibition of pain signals remains a fruitful area for continued exploration, both into the underlying biology, and into experimental agents to study in humans.

References

1. (a) T. S. Jensen, C. S. Madsen and N. B. Finnerup, *Curr. Opin. Neurol.*, 2009, **22**, 467–474; (b) C. J. Woolf, *Nat. Med.*, 2010, **16**, 1241–1247.
2. (a) M. H. Ossipov, G. O. Dussor and F. Porreca, *J. Clin. Invest.*, 2010, **120**, 3779–3787; (b) H. K. Beecher, *Bull. U. S. Army. Med. Dep.*, 1946, **5**, 445–454; (c) U. Bingel and I. Tracey, *Physiology. (Bethesda)*, 2008, **23**, 371–380.
3. J. D. Levine, N. C. Gordon and H. L. Fields, *Lancet.*, 1978, **2**, 654–657.
4. (a) P. Petrovic, E. Kalso, K. M. Petersson and M. Ingvar, *Science.*, 2002, **295**, 1737–1740; (b) J. K. Zubieta, J. A. Bueller, L. R. Jackson, D. J. Scott, Y. Xu, R. A. Koeppe, T. E. Nichols and C. S. Stohler, *J. Neurosci.*, 2005, **25**, 7754–7762.
5. (a) P. Rainville, *Curr. Opin. Neurobiol.*, 2002, **12**, 195–204; (b) A. D. Craig, *Curr. Opin. Neurobiol.*, 2003, **13**, 500–505; (c) I. Tracey, *Pain.*, 2005, **116**, 173–174.
6. R. Melzack and P. D. Wall, *Science*, 1965, **150**, 971–979.
7. (a) D. Le Bars, A. H. Dickenson and J. M. Besson, *Pain*, 1979, **6**, 283–304; (b) D. Le Bars, A. H. Dickenson and J. M. Besson, *Pain*, 1979, **6**, 305–327.
8. H. L. Fields, A. I. Basbaum and M. M. Heinricher, *Central Nervous System Mechanisms of Pain Modulation*, Elsevier Health Sciences, Burlington, MA, 2005.
9. (a) Y. Hosobuchi, J. E. Adams and R. Linchitz, *Science*, 1977, **197**, 183–186; (b) D. E. Richardson and H. Akil, *Neurosurgery*, 1977, **1**, 199–202; (c) D. E. Richardson and H. Akil, *J. Neurosurg.*, 1977, **47**,

178–183; (d) D. E. Richardson and H. Akil, *J. Neurosurg.*, 1977, **47**, 184–194.

10. (a) I. Tracey, *Curr. Opin. Neurobiol.*, 2005, **15**, 478–487; (b) J. Brooks and I. Tracey, *J. Anat.*, 2005, **207**, 19–33.
11. R. Vallejo, R. L. Barkin and V. C. Wang, *Pain Physician*, 2011, **14**, E343–360.
12. J. S. Mogil and J. E. Grisel, *Pain*, 1998, **77**, 107–128.
13. M. P. Davis, F1000. *Med. Rep.*, 2010, **2**, 63.
14. D. A. Fishbain, M. J. Detke, J. Wernicke, A. S. Chappell and D. K. Kajdasz, *Curr. Med. Res. Opin.*, 2008, **24**, 3105–3115.
15. A. M. Gray, D. M. Pache and R. D. Sewell, *Eur. J. Pharmacol.*, 1999, **378**, 161–168.
16. A. Tsang, M. Von Korff, S. Lee, J. Alonso, E. Karam, M. C. Angermeyer, G. L. Borges, E. J. Bromet, K. Demytteneare, G. de Girolamo, R. de Graaf, O. Gureje, J. P. Lepine, J. M. Haro, D. Levinson, M. A. Oakley Browne, J. Posada-Villa, S. Seedat and M. Watanabe, *J. Pain*, 2008, **9**, 883–891.
17. (a) M. Von Korff, P. Crane, M. Lane, D. L. Miglioretti, G. Simon, K. Saunders, P. Stang, N. Brandenburg and R. Kessler, *Pain*, 2005, **113**, 331–339; (b) L. A. McWilliams, R. D. Goodwin and B. J. Cox, *Pain*, 2004, **111**, 77–83; (c) L. A. McWilliams, B. J. Cox and M. W. Enns, *Pain*, 2003, **106**, 127–133.
18. L. J. Kirmayer, J. M. Robbins, M. Dworkind and M. J. Yaffe, *Am. J. Psychiatr.*, 1993, **150**, 734–741.
19. J. I. Hudson, M. S. Hudson, L. F. Pliner, D. L. Goldenberg and H. G. Pope, Jr., *Am. J. Psychiatr.*, 1985, **142**, 441–446.
20. S. Khoromi, L. Cui, L. Nackers and M. B. Max, *Pain*, 2007, **130**, 66–75.
21. S. Mercadante, E. Arcuri, W. Tirelli, P. Villari and A. Casuccio, *Tumori*, 2002, **88**, 239–242.
22. I. Gilron, J. M. Bailey, D. Tu, R. R. Holden, A. C. Jackson and R. L. Houlden, *Lancet*, 2009, **374**, 1252–1261.
23. S. B. Graff-Radford, L. R. Shaw and B. N. Naliboff, *Clin. J. Pain*, 2000, **16**, 188–192.
24. (a) F. J. Gomez-Perez, J. A. Rull, H. Dies, J. G. Rodriquez-Rivera, J. Gonzalez-Barranco and O. Lozano-Castaneda, *Pain*, 1985, **23**, 395–400; (b) F. J. Gomez-Perez, R. Choza, J. M. Rios, A. Reza, E. Huerta, C. A. Aguilar and J. A. Rull, *Arch. Med. Res.*, 1996, **27**, 525–529.
25. (a) G. T. Clark and R. Dionne, *Orofacial Pain: A Guide to Medications and Management*, Wiley-Blackwell, Oxford, 2012; (b) P. Dharmshaktu, V. Tayal and B. S. Kalra, *J. Clin. Pharmacol.*, 2012, **52**, 6–17.
26. L. Leventhal, V. Smith, G. Hornby, T. H. Andree, M. R. Brandt and K. E. Rogers, *J. Pharmacol. Exp. Ther.*, 2007, **320**, 1178–1185.
27. A. C. Jung, T. Staiger and M. Sullivan, *J. Gen. Intern. Med.*, 1997, **12**, 384–389.
28. (a) J. F. Wernicke, Y. L. Pritchett, D. N. D'Souza, A. Waninger, P. Tran, S. Iyengar and J. Raskin, *Neurology*, 2006, **67**, 1411–1420; (b) B. Verdu,

I. Decosterd, T. Buclin, F. Stiefel and A. Berney, *Drugs*, 2008, **68**, 2611–2632; (c) S. Perrot, R. M. Javier, M. Marty, C. Le Jeunne and F. Laroche, *Rheum. Dis. Clin. North. Am.*, 2008, **34**, 433–453; (d) P. L. Moja, C. Cusi, R. R. Sterzi and C. Canepari, *Cochrane Database Syst. Rev.*, 2005, CD002919.

29. S. Iyengar, A. A. Webster, S. K. Hemrick-Luecke, J. Y. Xu and R. M. Simmons, *J. Pharmacol. Exp. Ther.*, 2004, **311**, 576–584.
30. Y. M. Amr and A. A. Yousef, *Clin. J. Pain*, 2010, **26**, 381–385.
31. A. Yucel, S. Ozyalcin, G. Koknel Talu, E. Kiziltan, B. Yucel, O. K. Andersen, L. Arendt-Nielsen and R. Disci, *Eur. J. Pain*, 2005, **9**, 407–416.
32. M. C. Rowbotham, V. Goli, N. R. Kunz and D. Lei, *Pain*, 2004, **110**, 697–706.
33. S. H. Sindrup and T. S. Jensen, *Neurology*, 2000, **55**, 915–920.
34. S. N. Ozyalcin, G. K. Talu, E. Kiziltan, B. Yucel, M. Ertas and R. Disci, *Headache*, 2005, **45**, 144–152.
35. N. P. Zissis, S. Harmoussi, N. Vlaikidis, D. Mitsikostas, T. Thomaidis, G. Georgiadis and K. Karageorgiou, *Cephalalgia*, 2007, **27**, 315–324.
36. (a) L. M. Arnold, Y. Lu, L. J. Crofford, M. Wohlreich, M. J. Detke, S. Iyengar and D. J. Goldstein, *Arthritis. Rheum.*, 2004, **50**, 2974–2984; (b) L. M. Arnold, Y. L. Pritchett, D. N. D'Souza, D. K. Kajdasz, S. Iyengar and J. F. Wernicke, *J. Womens Health (Larchmt)*, 2007, **16**, 1145–1156; (c) L. M. Arnold, A. Rosen, Y. L. Pritchett, D. N. D'Souza, D. J. Goldstein, S. Iyengar and J. F. Wernicke, *Pain*, 2005, **119**, 5–15.
37. M. P. Lunn, R. A. Hughes and P. J. Wiffen, *Cochrane Database Syst. Rev.*, 2009, CD007115.
38. C. U. Pae, D. C. Marks, C. Han, A. A. Patkar and P. S. Masand, *Biomed. Pharmacother.*, 2009, **63**, 69–71.
39. C. P. Watson, I. Gilron, J. Sawynok and M. E. Lynch, *Pain*, 2011, **152**, 2206–2210.
40. G. A. Whitlock, P. V. Fish, M. J. Fray, A. Stobie and F. Wakenhut, *Bioorg. Med. Chem. Lett.*, 2008, **18**, 2896–2899.
41. (a) M. J. Millan, *Prog. Neurobiol.*, 2002, **66**, 355–474; (b) G. Hache, F. Coudore, A. M. Gardier and B. P. Guiard, *Pharmaceuticals*, 2011, **4**, 285–342.
42. A. S. Basile, A. Janowsky, K. Golembiowska, M. Kowalska, E. Tam, M. Benveniste, P. Popik, A. Nikiforuk, M. Krawczyk, G. Nowak, P. A. Krieter, A. S. Lippa, P. Skolnick and E. Koustova, *J. Pharmacol. Exp. Ther.*, 2007, **321**, 1208–1225.
43. G. Munro, C. A. Baek, H. K. Erichsen, A. N. Nielsen, E. O. Nielsen, J. Scheel-Kruger, P. Weikop and D. Peters, *Neuropharmacology*, 2008, **54**, 331–343.
44. B. Beer, J. Stark, P. Krieter, P. Czobor, G. Beer, A. Lippa and P. Skolnick, *J. Clin. Pharmacol.*, 2004, **44**, 1360–1367.
45. A. Astrup, D. H. Meier, B. O. Mikkelsen, J. S. Villumsen and T. M. Larsen, *Obesity (Silver Spring)*, 2008, **16**, 1363–1369.

46. P. A. Krieter, M. Gohdes, T. J. Musick, F. P. Duncanson and W. E. Bridson, *Drug. Metab. Dispos.*, 2008, **36**, 252–259.
47. S. Learned, O. Graff, S. Roychowdhury, R. Moate, K. R. Krishnan, G. Archer, J. G. Modell, R. Alexander, S. Zamuner, A. Lavergne, G. Evoniuk and E. Ratti, *J. Psychopharmacol.*, 2012, **26**, 653–662.
48. J. C. Hunter, D. J. Fontana, L. R. Hedley, J. R. Jasper, R. Lewis, R. E. Link, R. Secchi, J. Sutton and R. M. Eglen, *Br. J. Pharmacol.*, 1997, **122**, 1339–1344.
49. S. M. Tham, J. A. Angus, E. M. Tudor and C. E. Wright, *Br. J. Pharmacol.*, 2005, **144**, 875–884.
50. G. Blaudszun, C. Lysakowski, N. Elia and M. R. Tramer, *Anesthesiology*, 2012, **116**, 1312–1322.
51. R. Drdla and J. Sandkuhler, *Mol. Pain*, 2008, **4**, 18.
52. A. I. Basbaum, D. M. Bautista, G. Scherrer and D. Julius, *Cell*, 2009, **139**, 267–284.
53. S. R. Chen, G. Samoriski and H. L. Pan, *Neuropharmacology*, 2009, **57**, 121–126.
54. S. Collins, M. J. Sigtermans, A. Dahan, W. W. Zuurmond and R. S. Perez, *Pain Med.*, 2010, **11**, 1726–1742.
55. C. G. Parsons, *Eur. J. Pharmacol.*, 2001, **429**, 71–78.
56. (a) P. Cvrcek, *Pain Med.*, 2008, **9**, 253–257; (b) C. N. Sang, S. Booher, I. Gilron, S. Parada and M. B. Max, *Anesthesiology*, 2002, **96**, 1053–1061.
57. M. Carstensen and A. M. Moller, *Br. J. Anaesth.*, 2010, **104**, 401–406.
58. T. Klein, W. Magerl, A. Hanschmann, M. Althaus and R. D. Treede, *Eur. J. Pain*, 2008, **12**, 17–29.
59. (a) S. Boyce, A. Wyatt, J. K. Webb, R. O'Donnell, G. Mason, M. Rigby, D. Sirinathsinghji, R. G. Hill and N. M. Rupniak, *Neuropharmacology*, 1999, **38**, 611–623; (b) Y. Kim, H. Y. Cho, Y. J. Ahn, J. Kim and Y. W. Yoon, *Pain*, 2012, **153**, 1022–1029.
60. A. W. Wilson, S. J. Medhurst, C. I. Dixon, N. C. Bontoft, L. A. Winyard, K. T. Brackenborough, J. De Alba, C. J. Clarke, M. J. Gunthorpe, G. A. Hicks, C. Bountra, D. S. McQueen and I. P. Chessell, *Eur. J. Pain*, 2006, **10**, 537–549.
61. K. Williams, *Curr. Drug. Targets*, 2001, **2**, 285–298.
62. (a) B. L. Chenard, J. Bordner, T. W. Butler, L. K. Chambers, M. A. Collins, D. L. De Costa, M. F. Ducat, M. L. Dumont, C. B. Fox and E. E. Mena, *et al.*, *J. Med. Chem.*, 1995, **38**, 3138–3145; (b) K. Taniguchi, K. Shinjo, M. Mizutani, K. Shimada, T. Ishikawa, F. S. Menniti and A. Nagahisa, *Br. J. Pharmacol.*, 1997, **122**, 809–812.
63. C. N. Sang, J. J. Weaver, L. Jinga, J. Wouden and M. D. Saltarelli, The NR2B subunit-selective NMDA receptor antagonist, CP-101,606, reduces spontaneous pain intensity in patients with central and peripheral neuropathic pain, in *Society for Neuroscience Annual Meeting*, New Orleans, LA, 2003.
64. S. H. Preskorn, B. Baker, S. Kolluri, F. S. Menniti, M. Krams and J. W. Landen, *J. Clin. Psychopharmacol.*, 2008, **28**, 631–637.

65. B. Szewczyk, A. Palucha-Poniewiera, E. Poleszak, A. Pilc and G. Nowak, *Expert Opin. Investig. Drugs*, 2012, **21**, 91–102.
66. E. Polgar, D. I. Hughes, A. Z. Arham and A. J. Todd, *J. Neurosci.*, 2005, **25**, 6658–6666.
67. (a) T. P. Malan, H. P. Mata and F. Porreca, *Anesthesiology*, 2002, **96**, 1161–1167; (b) M. J. Eaton, M. A. Martinez and S. Karmally, *Brain Res.*, 1999, **835**, 334–339.
68. J. Scholz, D. C. Broom, D. H. Youn, C. D. Mills, T. Kohno, M. R. Suter, K. A. Moore, I. Decosterd, R. E. Coggeshall and C. J. Woolf, *J. Neurosci.*, 2005, **25**, 7317–7323.
69. (a) K. A. Moore, T. Kohno, L. A. Karchewski, J. Scholz, H. Baba and C. J. Woolf, *J. Neurosci.*, 2002, **22**, 6724–6731; (b) H. Baba, R. R. Ji, T. Kohno, K. A. Moore, T. Ataka, A. Wakai, M. Okamoto and C. J. Woolf, *Mol. Cell. Neurosci.*, 2003, **24**, 818–830.
70. T. J. Price, F. Cervero and Y. de Koninck, *Curr. Top. Med. Chem.*, 2005, **5**, 547–555.
71. N. R. Mirza and G. Munro, *Drug News Perspect.*, 2010, **23**, 351–360.
72. G. Munro, J. A. Lopez-Garcia, I. Rivera-Arconada, H. K. Erichsen, E. O. Nielsen, J. S. Larsen, P. K. Ahring and N. R. Mirza, *J. Pharmacol. Exp. Ther.*, 2008, **327**, 969–981.
73. C. Gremeau-Richard, A. Woda, M. L. Navez, N. Attal, D. Bouhassira, M. C. Gagnieu, J. F. Laluque, P. Picard, P. Pionchon and S. Tubert, *Pain*, 2004, **108**, 51–57.
74. R. Corrigan, S. Derry, P. J. Wiffen and R. A. Moore, *Cochrane Database Syst. Rev.*, 2012, **5**, CD009486.
75. R. Talwar and V. K. Potluri, *CNS Neurol. Disord. Drug Targets*, 2011, **10**, 536–544.
76. L. Leung, *J. Am. Board Fam. Med.*, 2011, **24**, 452–462.
77. M. A. Ware, T. Wang, S. Shapiro, A. Robinson, T. Ducruet, T. Huynh, A. Gamsa, G. J. Bennett and J. P. Collet, *CMAJ*, 2010, **182**, E694–701.
78. E. Martin-Sanchez, T. A. Furukawa, J. Taylor and J. L. Martin, *Pain Med.*, 2009, **10**, 1353–1368.
79. R. J. Ellis, W. Toperoff, F. Vaida, G. van den Brande, J. Gonzales, B. Gouaux, H. Bentley and J. H. Atkinson, *Neuropsychopharmacology*, 2009, **34**, 672–680.
80. V. I. Leussink, L. Husseini, C. Warnke, E. Broussalis, H. P. Hartung and B. C. Kieseier, *Ther. Adv. Neurol. Disord.*, 2012, **5**, 255–266.
81. J. Ahluwalia, L. Urban, M. Capogna, S. Bevan and I. Nagy, *Neuroscience*, 2000, **100**, 685–688.
82. W. J. Martin, N. K. Lai, S. L. Patrick, K. Tsou and J. M. Walker, *Brain Res.*, 1993, **629**, 300–304.
83. M. Karst and S. Wippermann, *Expert Opin. Investig. Drugs*, 2009, **18**, 125–133.
84. K. D. Patel, J. S. Davison, Q. J. Pittman and K. A. Sharkey, *Curr. Med. Chem.*, 2010, **17**, 1393–1410.

85. (a) V. Di Marzo, *Nat. Rev. Drug Discov.*, 2008, **7**, 438–455; (b) S. Petrosino, E. Palazzo, V. de Novellis, T. Bisogno, F. Rossi, S. Maione and V. Di Marzo, *Neuropharmacology*, 2007, **52**, 415–422.
86. L. J. Alvarez-Jaimes and J. A. Palmer, *Curr. Pharm. Biotechnol.*, 2011, **12**, 1644–1659.
87. (a) M. H. Bracey, M. A. Hanson, K. R. Masuda, R. C. Stevens and B. F. Cravatt, *Science.*, 2002, **298**, 1793–1796; (b) M. K. McKinney and B. F. Cravatt, *Annu. Rev. Biochem.*, 2005, **74**, 411–432.
88. S. Petrosino and V. Di Marzo, *Curr. Opin. Investig. Drugs*, 2010, **11**, 51–62.
89. J. L. Blankman, G. M. Simon and B. F. Cravatt, *Chem. Biol.*, 2007, **14**, 1347–1356.
90. (a) T. Bisogno, G. Ortar, S. Petrosino, E. Morera, E. Palazzo, M. Nalli, S. Maione and V. Di Marzo, *Biochim. Biophys. Acta.*, 2009, **1791**, 53–60; (b) J. Z. Long, W. Li, L. Booker, J. J. Burston, S. G. Kinsey, J. E. Schlosburg, F. J. Pavon, A. M. Serrano, D. E. Selley, L. H. Parsons, A. H. Lichtman and B. F. Cravatt, *Nat. Chem. Biol.*, 2009, **5**, 37–44.
91. I. K. Khanna and C. W. Alexander, *CNS Neurol. Disord. Drug Targets*, 2011, **10**, 545–558.
92. M. B. Skaddan, L. Zhang, D. S. Johnson, A. Zhu, K. R. Zasadny, R. V. Coelho, K. Kuszpit, G. Currier, K. H. Fan, E. M. Beck, L. Chen, S. E. Drozda, G. Balan, M. Niphakis, B. F. Cravatt, K. Ahn, T. Bocan and A. Villalobos, *Nucl. Med. Biol.*, 2012, **39**, 1058–1067.
93. J. P. Huggins, T. S. Smart, S. Langman, L. Taylor and T. Young, *Pain*, 2012, **153**, 1837–1846.
94. (a) K. Ahn, S. E. Smith, M. B. Liimatta, D. Beidler, N. Sadagopan, D. T. Dudley, T. Young, P. Wren, Y. Zhang, S. Swaney, K. Van Becelaere, J. L. Blankman, D. K. Nomura, S. N. Bhattachar, C. Stiff, T. K. Nomanbhoy, E. Weerapana, D. S. Johnson and B. F. Cravatt, *J. Pharmacol. Exp. Ther.*, 2011, **338**, 114–124; (b) G. L. Li, H. Winter, R. Arends, G. W. Jay, V. Le, T. Young and J. P. Huggins, *Br. J. Clin. Pharmacol.*, 2012, **73**, 706–716.
95. (a) J. L. Blankman and B. F. Cravatt, *ACS Chem. Neurosci.*, 2012, **3**, 337–338; (b) F. Piscitelli and V. Di Marzo, *ACS Chem. Neurosci.*, 2012, **3**, 356–363.
96. (a) K. A. Jacobson and Z. G. Gao, *Nat. Rev. Drug Discov.*, 2006, **5**, 247–264; (b) M. J. Zylka, *Trends. Mol. Med.*, 2011, **17**, 188–196.
97. M. Hayashida, K. Fukuda and A. Fukunaga, *J. Anesth.*, 2005, **19**, 225–235.
98. T. King, L. Vera-Portocarrero, T. Gutierrez, T. W. Vanderah, G. Dussor, J. Lai, H. L. Fields and F. Porreca, *Nat. Neurosci.*, 2009, **12**, 1364–1366.
99. J. C. Eisenach, R. L. Rauck and R. Curry, *Pain*, 2003, **105**, 65–70.
100. T. J. Gan and A. S. Habib, *Anesth. Analg.*, 2007, **105**, 487–494.
101. A. S. Habib, H. Minkowitz, T. Osborn, B. Ogunnaike, K. Candiotti, E. Viscusi, J. Gu, M. R. Creed and T. J. Gan, *Anesthesiology*, 2008, **109**, 1085–1091.

CHAPTER 6

Biotherapeutics for the Treatment of Pain

GREGORY J. CARVEN[a] AND LEE R. ROBERTS*[b]

[a] Pfizer Global Biotherapeutics Technologies, Cambridgepark Drive, Cambridge, MA 02140, USA; [b] Pfizer Worldwide Medicinal Chemistry, Cambridgepark Drive, Cambridge, MA 02140, USA
*Email: lee.roberts@pfizer.com

6.1 Introduction

Despite numerous breakthroughs in the treatment of pain over the past 50 years, chronic pain remains a huge area of unmet medical need. It is estimated that current therapies are poorly effective, with only one in four patients achieving adequate relief.[1] Such therapies include narcotic (opioids) and non-narcotic analgesics, non-steroidal anti-inflammatory drugs (NSAIDs) and heat therapy. Complicating the matter is the fact that current treatments are also associated with adverse tolerability profiles, unfavourable side effects, concerns with long-term use and the potential for misuse or abuse. Thus, the challenge remains to develop new pain medicines which are more effective with fewer unwanted side effects.

Biological therapies represent an emerging class of novel analgesics which has the potential to overcome shortfalls of the current modalities used for treatment of pain. For small-molecule drugs, efficacy is often limited by off-target activity. Biologics, such as monoclonal antibodies, however, offer exquisite target specificity and thus may be better suited to provide adequate target engagement with limited off-target toxicity. While antibodies form the mainstay of biological approaches for many disease indications, several

RSC Drug Discovery Series No. 35
Pain Therapeutics: Current and Future Treatment Paradigms
Edited by Charlotte Allerton

Published by the Royal Society of Chemistry, www.rsc.org

promising new approaches are currently being explored which may offer distinct advantages over antibody therapeutics. These include recombinant proteins, peptides, small interfering RNA (siRNA) and gene therapy, each of which has recently been used to target specific pain pathways in preclinical and clinical studies. The current status and future opportunities for biological therapies as approaches for treating pain are reviewed.

6.2 Antibodies and Antibody-Related Molecules

Monoclonal antibodies (mAbs) have revolutionized the treatment of cancer and autoimmune diseases and represent the fastest-growing product segment in the pharmaceutical industry.[2] Application of antibody therapeutics to neurological disorders such as pain has been somewhat limited to date; however, several new approaches may provide greater analgesic efficacy without the limitations associated with current pharmacological therapies.

There are many considerations necessary for targeting pain with antibodies or other large-molecule therapeutics. The large size (150 kDa) and the ability to be recycled *via* interaction with neonatal Fc receptors on endothelial cells give antibodies a pharmacokinetic half-life (on the order of 7–14 days)[3] which is significantly greater than that of small-molecule drugs. Antibodies provide high affinity and excellent specificity for target recognition, rarely interacting with proteins other than the intended therapeutic target.

By virtue of their endogenous function in the immune system, antibodies are well suited to the binding and modulation of extracellular signalling systems which are typically characterized by receptor–ligand interactions. This is in stark contrast to small-molecule drugs, which often bind to small hydrophobic pockets most often found in enzymes, nuclear receptors, ion channels and G-protein coupled receptors (GPCRs). Thus the targets which are accessible or druggable by antibodies are quite different than those by small-molecule drugs.[4,5] Importantly, when considering application to pain or other central nervous system (CNS) disorders, one must consider the level (or lack) of penetration of the blood–brain barrier (BBB). Generally, access to the CNS is restricted to molecules with a molecular mass less than 500 Da with high lipid solubility. Thus, large molecules such as antibodies do not cross the BBB readily. Typical antibody concentrations within the CNS compartment after chronic dosing have been measured at less than 0.5% of the dose given.[6] Therefore, optimal targets for mAbs may be soluble mediators, cytokines and growth factors and may be localised in readily-accessible compartments such as blood.

Over the past decade, smaller antibody fragments and engineered variants have become viable alternatives offering some advantages over full-length antibodies (see Figure 6.1). As single entities, these fragments allow broader bioavailability and faster pharmacokinetic clearance. Antibody fragments retain the target specificity of full-length antibodies, are active as monomers and, owing to their small size and inherent stability, can be formatted into larger molecules to create drugs with prolonged serum half-lives or other pharmacological activities as needed.

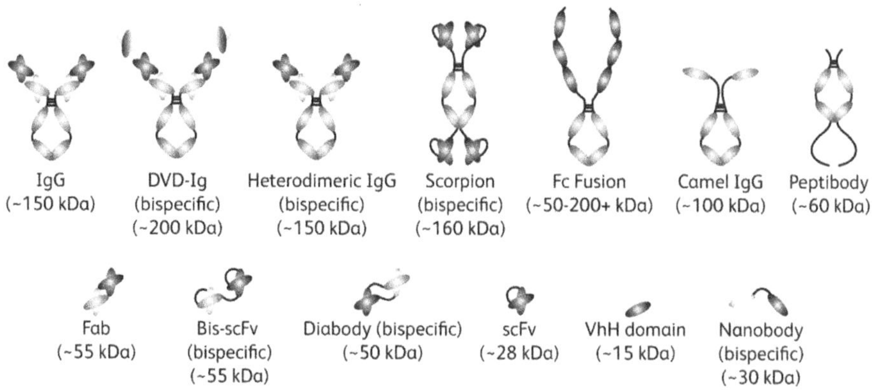

Figure 6.1 Representation of different biologic formats. Antibodies (IgG) form the mainstay of biologic modalities. Antibody fragments comprising the variable regions of IgG, such as Fabs and scFv retain antigen binding and can be used to create engineered molecules such as bispecifics. The Fc portion of IgG can be used to extend the half-life of non-antibody therapeutics including Fc fusion proteins and peptibodies.

Many advances in biotherapeutic delivery across the BBB have recently been described (reviewed by Rajayhyaksha[7]). However, to date, there is no single delivery technology which can overcome all of the obstacles with respect to safely and effectively delivering large-molecule drugs to the CNS. In addition to utilizing CNS-focused delivery systems, a number of protein–antibody fusions have been designed to trigger transport across the BBB *via* a receptor-mediated transport system.[8] Such BBB-penetrating fusion proteins take advantage of endogenous transporters of circulating peptides such as insulin, transferrin or insulin-like growth factors (IGFs). Coupling biologic modalities to antibody binding domains specific for insulin receptor, transferrin receptor, or IGF receptor allows transcytosis across the BBB with release into brain interstitial space. To date, none of these novel fusions has been validated clinically, but continued investment in this area may yield valuable therapeutic options for the treatment of pain and other CNS diseases.

6.3 Targeting Neurotrophins and Neurotrophin Receptors

Neurotrophins are an important family of regulatory proteins which are essential for the development and survival of sympathetic and sensory neurons (reviewed by Skaper).[9] The effects of neurotrophin proteins are mediated *via* activation of tropomyosin-related kinase (Trk) family of receptor tyrosine kinases. In addition, all neurotrophins activate the structurally distinct p75 neurotrophin receptor, a member of the tumour necrosis factor (TNF) receptor superfamily. Considerable evidence suggests that neurotrophic factors, particularly nerve growth factor (NGF), act as peripheral and central pain mediators. Blockade of neurotrophin signalling is an effective analgesic mechanism in many models of persistent pain.

6.3.1 Antibodies Against Nerve Growth Factor

NGF is one of the most studied neurotrophins. It is a key regulator of neuron excitability and is an important mediator of nociceptive and neuropathic pain.[10] NGF participates in a complex inflammatory cytokine network (reviewed by Seidel).[11] Proinflammatory cytokines such as TNFα and IL-1β stimulate the release of NGF from both structural and inflammatory cell types. NGF binds TrkA, inducing dimerisation and phosphorylation of tyrosine residues in its kinase domain, initiating a signalling cascade ultimately leading to the upregulation and release of neurotransmitters, including substance P, calcitonin gene-related peptide (CGRP) and BDNF, as well as the sensitisation of ion channels present on nerve cells (such as TRPV1 and ASIC3).

NGF is overexpressed in many inflammatory and degenerative rheumatic diseases. In humans, increased levels of NGF have been reported in patients suffering from arthritis,[12–16] pancreatitis[17] and cancer pain.[18,19] Additionally, a rare genetic disorder, CIPA (congenital insensitivity to pain with anhidrosis), is caused by polymorphisms resulting in loss-of-function mutations in the gene encoding the NGF receptor TrkA.[20]

NGF sequestration is an area of intensive research, with many large industrial players seeking to develop antibody candidates. As of 2012, five anti-NGF antibodies have progressed into clinical trials for the management of moderate to severe nociceptive pain, three of which have progressed to late stage development (www.clinicaltrials.gov). Furthest advanced is the humanised antibody tanezumab, developed by Pfizer's Rinat Neuroscience division, which has been or is currently being studied in as many as 26 human trials. If approved, tanezumab would represent the first biologic drug to be approved for the management of pain. Other anti-NGF antibodies include fulranumab (Johnson & Johnson), RGN475 (Regeneron Pharmaceuticals and SanofiAventis), PG110 (Pangenetics, licensed to Abbott) and MEDI578 (Astra Zeneca). These other agents are earlier in development and should report data in the near future.

In Phase 2 proof-of-concept studies, tanezumab demonstrated significant radiographic and pain reductions in patients with moderate to severe knee osteoarthritis (OA).[21] Phase 3 studies were performed to evaluate tanezumab in combination with NSAIDs and suggest greater reduction in joint pain with the combination of the NGF inhibitor and an NSAID, although the combination was associated with higher incidence of serious adverse effects including suspected osteonecrosis and worsening OA.[22]

Due to reports of patients experiencing worsening OA symptoms leading to joint replacement, the U.S. Food and Drug Authority (FDA) placed development of tanezumab and other anti-NGF agents on clinical hold in mid-2010. In 2012, after extensive review citing data from over 30 anti-NGF trials including over 11 000 patients, an FDA arthritis advisory committee voted unanimously that there is a role for the ongoing development of NGF inhibitors, and to continue development for anti-NGF treatments.

Despite well-demonstrated efficacy, the clinical and commercial success of anti-NGF therapy will depend on the long-term side-effect profile. To date, the studies with anti-NGF agents suggest that the treatment avoids typical side effects associated with opioids, such as constipation, respiratory depression and psychological dependence. The most frequently observed side effects of NGF therapies included peripheral neuropathy (tingling and numbness), dysaesthesia (abnormal sense of touch) and paraesthesia (tingling, pricking or numbness) which are likely due to the mode of action in targeting nerve cells.

6.3.2 Nerve Growth Factor Blockade by other Modalities

In addition to antibody blockade, NGF sequestration can be achieved using a competitive binding molecule containing the NGF-binding region of the natural receptor TrkA. Preclinical work with a recombinant Fc fusion of two TrkA extracellular domains (TrkA-IgG) produced a sustained chemical and thermal hypoalgesic effect in rats which were chronically administered TrkA-IgG.[23] Structural studies have revealed that only a single extracellular domain of TrkA (TrkA domain 5 or TrkAd5) is required for binding and sequestration of NGF.[24,25] TrkAd5 can be produced as a single recombinant entity which is expressed in *E. coli* and refolded to generate a stable therapeutic protein.[26] In animal studies, TrkAd5 has been shown to reduce bladder overactivity associated with cyclophosphamide-induced inflammation.[27] Additionally, TrkAd5 administration reduced hyperalgesia and mechanical allodynia observed in UVB-treated rats. This effect was durable and lasted for up to 4 days.[28]

Amgen has also reported the generation of AMG819, a 'peptibody' comprised of a phage display-derived NGF-binding peptide fused to the C-terminus of an antibody Fc (see Figure 6.1). Subcutaneous administration of the molecule reduced thermal nociception in rats and reversed tactile allodynia in rat models of neuropathic pain.[29] AMG819 was tested in a Phase 1 dose escalation study in healthy volunteers to assess pharmacokinetics of escalating doses. Mean terminal half-life of the peptibody was 85 hours,[30] which is significantly less than that of a typical antibody. Development of AMG819 has not progressed.

Non-antibody approaches to blockade or sequestration of NGF represent an interesting novel class of pain biotherapeutics. Given the advanced state of the anti-NGF antibodies currently in clinical testing, coupled with the potential safety risks identified during their study, it will likely be difficult to develop an alternative biologic in this space. To differentiate, novel NGF-directed therapies should target different patients, indications, or severities than those targeted by tanezumab and the other monoclonals.

6.3.3 Modulation of TrkA

Targeting the NGF/TrkA interaction site with an antibody against the TrkA receptor may also be a therapeutically viable option for modulation of pain.

Early studies with a neutralising anti-TrkA antibody, MNAC13, durably reduced pain behaviours in a murine model of formalin-induced inflammatory pain.[31] The same authors show that injection of MNAC13 induces significant functional recovery in mice subjected to sciatic nerve ligation, suggesting that treatment with anti-TrkA also produces a significant antiallodynic effect on neuropathic pain.

Glenmark Pharma, under licence from Lay Line Genomics, is currently investigating a humanised version of MNAC13, called GBR-900, for the treatment of various forms of chronic pain as well as other pathological conditions associated with inflammation and persistent pain. Despite its discovery over a decade ago, clinical development of this antibody has been slow to progress. The antibody is currently listed as preclinical according to Glenmark.[32]

Despite the recent successes of antibody approaches to modulation of NGF-mediated signalling, some drawbacks may exist such as extracellular targeting of the receptor, as described above. Numerous small-molecule approaches have been undertaken to identify compounds which can prevent signalling *via* the TrkA kinase domain, but most are not specific for TrkA relative to other kinases.[33] Thus the development of small peptide antagonists may be beneficial. Tabata *et al.* describe the generation of a cell-penetrating peptide inhibitor (IPTRK3) based on the amino acid sequence of the TrkA activation loop combined with the cell-penetrating peptide sequence from HIV-1 transactivator of transcription protein (TAT).[34] It was found that this synthetic peptide inhibits TrkA activity and suppresses both noxious pain in a rat model of inflammation[35] and neuropathic pain in a mouse model of partial sciatic nerve ligation.[36] Additionally, this peptide is also capable of suppression of melanoma-induced pain in a mouse skin cancer pain model.[37]

Pain often accompanies most forms of cancer and current therapies for treating cancer pain typically have significant unwanted side effects. Developing targeted therapies that relieve cancer pain without the side effects of current analgesics has significant potential to improve the quality of life for cancer patients. In addition to their roles in pain sensation, Trk kinases play important roles in tumour cell growth and survival signalling. In pancreatic cancer in particular, it is thought that paracrine and/or autocrine NGF-Trk interactions may both influence the invasive behaviour of the cancer, and contribute to significant cancer related pain. Thus, biologic inhibitors of Trk receptor kinases might provide targeted cancer treatment while adding the beneficial effect of reducing cancer-associated pain.

6.4 Targeting Proinflammatory Cytokines with Biotherapeutics

Over the past decades, it has become clear that inflammatory and immune mechanisms play an important role in neuropathic pain.[38] Dysregulation of the interplay between the immune and nervous systems results in the development and maintenance of neuropathic pain. Infiltration of inflammatory cells and activation of resident immune cells in response to tissue injury or nervous system damage lead to subsequent production and secretion of characteristic proinflammatory cytokines, such as IL-1β, IL-6 and TNF. The hyperalgesic effects of these cytokines are typically indirect, as they do not act directly on pain receptors but induce nociceptive sensitisers such as PGE_2. Several anti-cytokine biotherapeutics, such as anti-cytokine antibodies, signalling inhibitors and neuron stabilisers have shown promising efficacy in clinical and preclinical pain studies.

6.4.1 Tumour Necrosis Factor

A large body of evidence links TNF and pain. It is well documented that patients with painful disease conditions such as rheumatoid arthritis (RA), ankylosing spondylitis and fibromyalgia have increased circulating levels of TNFα compared to healthy individuals.[39–41] TNFα is also implicated in the pathogenesis of multiple sclerosis and other peripheral demyelinating neuropathies.[42] Biologic therapies targeting TNF, including enantercept, infliximab and adalimumab, are powerful disease-modifying agents in RA and more than 16 other indications. These agents can retard or prevent disease progression and, thus, joint destruction and subsequent loss of function. However, despite significant preclinical evidence for the role of TNF in pain, very few clinical studies have directly evaluated the effect of TNF blockade on pain.

Enantercept is a recombinant fusion protein consisting of the soluble portion of human TNF receptor-p75 fused to the Fc portion of an antibody. It acts as a competitive inhibitor of TNFα as well as TNFβ (also called lymphotoxin). Enantercept was approved by the FDA in 1998 and has been shown to produce significant improvements in RA-associated pain measured using a standardised questionnaire in multiple double-blind, placebo-controlled clinical studies.[43,44] Similarly, infliximab, a chimeric anti-TNF antibody, reduced pain scores in RA patients at multiple doses.[45] A large retrospective study including 12 000 patients with RA and treated with either enantercept or infliximab showed significant improvement in self-reported pain scores.[46]

Given that improvement in pain symptoms and positive disease modification by anti-TNF therapies are not mutually exclusive, the use of anti-TNF therapy for the treatment of inflammatory pain has warranted further study. Recently, it was shown that neutralisation of TNFα acts through the CNS earlier than it affects the underlying joint inflammation.[47] Using functional MRI (fMRI) in human patients infused with infliximab, Hess *et al.* demonstrate blockade of nociceptive CNS activity in the thalamus and somatosensoric cortex within 24 hours of TNF blockade. Pain-associated neuronal activity was also significantly reduced as early as 24 hours after infusion. In contrast, clinical and laboratory markers of inflammation, such as joint swelling, were not affected by anti-TNF at these early time points.

The treatment of inflammatory conditions with TNF blockading agents is a very significant market. In 2012, adalimumab is expected to become the top-selling drug worldwide, with forecast annual sales of over $9 billion which will likely continue to grow. Clinical and commercial success of TNF blockers have spawned the development of dozens of follow-on and next generation anti-TNF biologic therapies. Despite similar modes of action among the available anti-TNF drugs, patient responses to these therapeutics remain variable. The observed lack of response, loss of response over time and tolerability issues suggest the need for additional treatment options.

Certolizumab is a humanised antigen-binding fragment (Fab) of a mAb that has been conjugated to polyethylene glycol. Certolizumab is currently approved for Crohn's disease and moderate to severe RA. In Phase 3 trials, treatment of RA patients with certolizumab resulted in significant improvement in patient-reported outcomes for function, fatigue and pain as well as measurable improvements in productivity both at work and at home.[48]

Engineered TNF inhibitors are increasingly being tested in the clinic. Placulumab is a domain antibody (dAb) against TNFα in Phase 2 trials for the treatment of sciatica. Domain antibodies or nanobodies are derived from antibody-like molecules found in camelidae, which contain single heavy-chain variable region (VHH) domains that do not require light chains for functionality (see Figure 6.1). Unlike a traditional monoclonal antibody, the nanobody construct is quite small, only 11–15 kDa, which might allow deeper penetration into tissues, or across the BBB. Recruitment has completed and results from this study are expected in mid-2013.

Another engineered molecule, ozoralizumab, an anti-TNF nanobody, recently demonstrated clinical proof of concept in a Phase 2 trial.[49] Monthly administration of ozoralizumab resulted in statistically significant improvement of ACR20 responses. Although pain scores were not directly reported, improvement in tenderness and swelling as measured by disease activity scores (DAS28) was observed.

6.4.2 Interleukin-6

IL-6 is another cytokine which plays an important role in the regulation of inflammatory immune responses. IL-6 levels are increased in patients suffering from RA[50–53] as well as other painful conditions including pancreatitis,[54] appendicitis,[55] severe abdominal pain,[54] migraine,[56,57] diabetic peripheral neuropathic pain[58] and chronic cancer pain.[59] Preclinically, IL-6 knockout animals show reduced pain in models of nerve injury[60,61] and neutralisation of IL-6 in wild-type animals has been shown to mimic the knockout phenotype.[62,63]

Currently, the only biologic therapy approved for IL-6 blockade is tocilizumab, a monoclonal antibody against the IL-6 receptor; however, there are at least nine additional monoclonal antibodies targeting IL-6 or IL6R in clinical development.[64] Like anti-TNF, few studies have focused directly on pain as a direct readout. In trials of tocilizumab in RA, administration of the antibody significantly reduced pain scores compared to placebo,[65,66] but whether this is due to modulation of the overall disease, or is a direct analgesic effect of the antibody is not known.

In addition to monoclonal antibodies, other biologic approaches targeting the IL-6 pathway are being considered. Ablynx is developing ALX-0061, a nanobody-based therapeutic targeting the IL-6 receptor (currently in Phase 1/2 studies).[64] ALX-0061 is a fusion of two VHH domains, one specific for the IL6-R, and a second which binds to human serum albumin to extend the half-life of the molecule.

Other modalities which are being studied include a soluble fusion protein comprising two extracellular IL-6 receptor glycoprotein130 (gp130) domains dimerised using an IgG-Fc[67] as well as an aptamer approach.[68] Finally, Cytos Biotechnology is pursuing a therapeutic vaccine using IL-6 as an immunogen for treatment of RA.[64] While not specifically aimed at analgesia, it will be interesting to follow the results of these studies to determine if vaccination against proinflammatory cytokines may be a valid route of therapeutic intervention for RA or other painful conditions.

6.4.3 Interleukin-1β

Like TNF and IL-6, interleukin-1β (IL-1β) is a pleiotropic proinflammatory cytokine which has been implicated in pain conditions. IL-1β has been shown to act both indirectly through upregulation of pro-nociceptive mediators such as nerve growth factor (NGF)[69,70] as well as directly *via* IL-1R1 expressed on sensory neurons.[71] Blockade of IL-1β has been shown to attenuate pain in numerous preclinical models (reviewed by Ren and Torres).[72]

Novartis has developed and launched canakinumab, a human anti-IL-1β monoclonal antibody, for the treatment of cryopyrin-associated periodic syndromes and Muckle–Wells syndrome. Canakinumab has been shown to be effective for the treatment of refractory gout, reducing pain associated with acute gout attacks, and reducing the frequency of attacks by 60–70%.[73,74] Despite the effectiveness of the drug, the FDA declined to approve canakinumab for treating gouty arthritis, requesting additional clinical data citing the need to evaluate the risk–benefit profile in refractory patients (news release, Novartis).

In addition to canakinumab, there are four monoclonal antibodies to IL-1β reported in the clinic as well as numerous preclinical alternative biologic modalities against this target. Rilonacept, a soluble receptor Fc fusion which inhibits both IL-1α and IL-1β, significantly reduced acute gout flares that occur when initiating uric acid-lowering therapy.[75] Abbott is studying two multimeric antibody molecules based on their dual-variable domain immunoglobulin (DVD-Ig) platform. One incorporates binding domains to both IL-1α and IL-1β. The other binds both IL-1β and IL-17, another proinflammatory cytokine which contributes to the regulation of immune cell infiltration and glial activation after peripheral nerve injury.[76]

6.5 Targeting Ion Channels with Antibodies

Ion channels play critical roles by modulating cellular function and are targets for a number of pain pathways. Modulators of ion channel function typically include small-molecule and peptide toxins derived from plants and venoms as well as those derived from screening efforts; however, selectivity for any given isoform within a channel class is very difficult to achieve. Antibodies have the potential advantage of offering high specificity for their targets though most existing antibodies against ion channels do not modulate function. Advances in structural elucidation of various ion channels, specifically the extracellular domains, will help guide future efforts in antibody design. Antibody targeting of ion channels, however, is considerably further behind mAb intervention of soluble mediators such as cytokines, growth factors and inflammatory proteins.

Voltage-gated ion channels which include voltage-gated potassium, sodium and calcium channels as well as the TRP channels comprise of six-membrane-spanning domains. Potassium and most cation channels consist of four separate six-transmembrane subunits which can form homo- or heterotetramers whereas sodium and calcium channels consist of four six-transmembrane-domain subunits in a single peptide chain. These six-transmembrane domains are interconnected by three extracellular loops, termed E1–3, which can be highly variable in their amino acid sequence and could help provide antibody specificity. These extracellular loop regions, however, are small and often glycosylated. In many cases, voltage-gated channels have no obvious extracellular surface, compounding the difficulty of raising functional antibodies.

Empirically, it was discovered that by targeting the E3 (turret) region between transmembrane helix 5 and the pore loop, functional polyclonal antibodies could be generated to TRPC5 and $Na_V1.5$, and since this pioneering work, further polyclonal antibodies against other voltage-gated ion channels have been generated (see Figure 6.2).[77] Importantly, functional monoclonal antibodies have been generated against the E3 extracellular loop of $K_V10.1$ (Eag1) and more recently $K_V1.3$.[78,79] A recent patent also describes functional

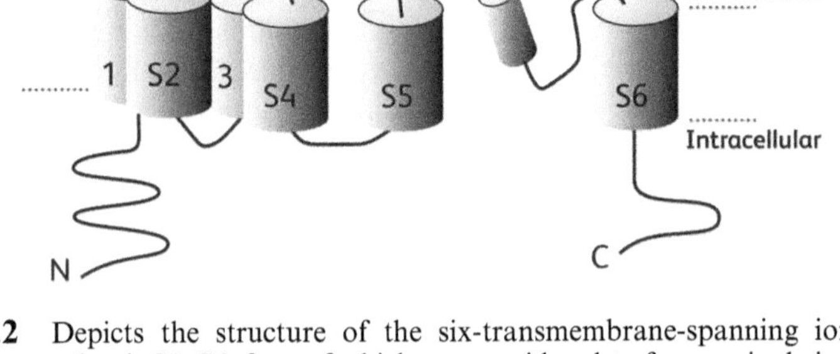

Figure 6.2 Depicts the structure of the six-transmembrane-spanning ion channel subunit S1–S6, four of which are considered to form a single ion channel. The extracellular loop 3 (E3) is highlighted, which has been targeted by a number of antibody approaches.

monoclonals against $Na_V1.7$ generated using synthetic peptide antigens comprised of the extracellular loop region sequences.[80]

An interesting approach for state-dependent blockade has been described by Dolphin *et al.* in which an antibody against the α1A subunit of a rat brain calcium channel was raised against an exofacial epitope at the mouth of the pore, which is not exposed unless cells are depolarized. Thus, antibodies would only bind when cells were polarized sufficiently to open the channel and reveal the antigenic sequences.[81,82]

Extracellular epitopes of ligand-gated ion channels (LGICs) have also been targeted. These channels are opened or closed in response to the binding of a chemical messenger. Often, the ligand-binding site is located on a different portion of the protein (an allosteric binding site) distinct from the ion conduction pore. As the ligand interaction sites typically contain more extracellular epitope space than do voltage-gated channels, targeting LGICs may less challenging.

Functional mAbs against the purinergic P2X7 were raised through immunisation using cells expressing the P2X7 receptor in Balb/c mice.[83]

Ion channels may be an ideal target for single-domain antibodies, such as those derived from camels, llamas or alpacas. Camelid antibody sequences are capable of adopting diverse structures which are not classifiable by established canonical rules.[84] Furthermore, camelid sequences contain longer CDRH3 loops which may be capable of penetrating channel pores. The small nature of nanobodies may also allow for deeper penetration into the peripheral nerve ending, the site of action of many pain-associated ion channels. Ablynx recently described the discovery of inhibitors of P2X7 using their camelid-derived VHH nanobody platform.[85] Many of the nanobodies described were capable of blocking the ATP- and NAD-induced CD62L shedding. This approach remains very early in the preclinical space; however, it will be interesting to follow the development of VHH domain antibody fragments for targeting proteins which have been traditionally refractory to antibody modulation.

6.6 Peptide Conjugates

For intracellular delivery of peptides, two main approaches have evolved: either peptides which contain a defined tertiary structure and facilitate transfer through a cell membrane or peptides that are capable of entering the cell as a linear sequence *via* active uptake mechanisms (for a review see Johnson *et al.*[86] and Tao *et al.*[87]). The latter approach enables the use of the linear peptide sequences to be used as carriers for cargoes which are not themselves cell-permeable. A number of membrane-permeable carriers have been developed of which the most frequently used is the polyarginine-containing protein transduction domain (PTD) of the human HIV-1 Tat protein (residues 47–57 of HIV-1 Tat).[88] Another reported peptide sequence is Transportan which has no arginine residues but is efficiently transported into cells *via* perturbation of the cell membrane.[89] Cell-penetrating peptide conjugates have been investigated for the ability to alleviate neuropathic and inflammatory pain.

Of the five PKC isoforms present in sensory neurons, εPKC is the predominant form that translocates to the cell membrane in response to inflammatory mediators, such as bradykinin. PKC activation triggers translocation to specific intracellular sites where binding to anchoring proteins termed RACKs (receptor for activated C kinases) are integral to the downstream signal transduction. This translocation of εPKC in dorsal root ganglion (DRG) neurons is associated with nociceptor sensitisation and can be suppressed by the εPKC inhibitor peptide εV1-2 (EAVSLKPT), which mimics the RACK binding domain of εPKC and thus competitively blocks the binding of εPKC to RACK away from the ATP binding site giving the compound isozyme selectivity.[90] The effects of multiple activators of sensory neurons that also induce hyperalgesia in whole animals, *e.g.* adrenaline (epinephrine), acetic acid, carrageenan, NGF and substance P can be attenuated in εPKC KO mice and/or normal animals using PKC inhibitors, including the selective εPKC inhibitor peptide εV1-2.[91] A cell-penetrating peptide version of this compound (KAI-1678) was developed by Kai Pharmaceuticals.[92] The compound was reported to be effective at reducing acute pain produced *via* inflammation, incision or capsaicin and has entered Phase 2 clinical trials for the treatment of moderate to severe postoperative pain and neuropathic pain. In early 2012, Amgen acquired Kai Pharmaceuticals, and since that time, no additional development has been reported.

Another example of targeting intracellular protein–protein interactions includes a peptide inhibitor which blocks the interaction between the receptor tyrosine kinase Src and the adapter protein ND2, both of which form a complex with the *N*-methyl-*D*-aspartate receptor (NMDAR), an ion channel that binds the excitatory neurotransmitter glutamate.[93] This peptide is currently being developed by NoNO Inc. Direct blockade of the NMDA receptor is undesirable due to its essential role in the nervous system and blockade has been linked to respiratory and locomotor side effects. A peptide derived from a region of the Src kinase which interacts with ND2 was conjugated to $TAT_{47–57}$ and tested in rodent models of neuropathic and inflammatory pain with the intention of eliciting NMDA receptor-mediated analgesia. Intravenous and intrathecal administration of $Src_{40–49}$Tat dose-dependently decreased evoked pain behaviour in the formalin model as well as significantly decreased mechanical and cold hyperalgesia caused by peripheral nerve injury. This work showed that the compound was able to suppress chronic pain through targeting of Src-mediated activity of NMDARs without targeting NMDAR directly and affecting Src catalytic activity.

Similarly, postsynaptic density protein-95 (PSD-95) also interacts with NMDAR signalling through the PDZ2 domain on PSD-95 and the NR2 subunit of NMDAR.[94] A peptide consisting of this PDZ2 domain was conjugated to TAT and injection of Tat-PSD-95 PDZ2 fusion peptide significantly diminished CFA-induced chronic inflammatory pain and reduced halothane MAC in mice.[95]

For extracellular targets, transport across the cell membrane is not an issue, but peptide stability and resistance to degradation are major hurdles for peptide therapeutics. Free peptides are often rapidly broken down by peptidases or cleared through the kidney *via* glomerular filtration. One general approach has been to conjugate a peptide pharmacophore to a carrier protein which can potentially improve pharmacokinetic properties of the peptides, protecting them from degradation and extending half-life.

Conjuchem has conjugated dynorphin A, a mixed mu and kappa agonist, to human serum albumin using their DAC™ and PC-DAC™ technologies.[96] Centrally mediated side effects, which are a major hurdle for opioid therapies, are not expected of the dynorphin–albumin conjugate as this moiety would be expected to have little CNS penetration. Additionally, the conjugate shows enhanced peripheral analgesic effect for an extended period of time due to its extended pharmacokinetic half-life. A Phase 2 study using CJC-1008 provided evidence of a greater analgesic effect compared to placebo in patients with postherpetic neuralgia. Unfortunately the effect lasted for only 8 hours and was diminished by 24 hours.[97] No further development has been reported with this conjugate.

Other approaches to prolong the half-life of peptide therapeutics include the use of poly(ethylene)glycol (PEG) or antibody conjugates such as a kappa peptide agonist which was recently conjugated to a modified IgG antibody.[98]

6.7 Gene Therapy

Gene therapy enables the transfer of defined genetic material to a target cell to either prevent or alter a particular disease state. The therapeutic actually has three components: healthy copies of the mutated gene present in the diseased individual, a carrier or delivery vehicle to deliver the gene to the cell and DNA elements that turn on the healthy gene in the desired cells at the correct levels. Advances in gene transfer technology have provided methods to target affected areas of the CNS involved in neuropathic pain. Such gene products recently described include proteins that have direct analgesic effects but can also be proteins that interfere with pain-associated biochemical changes in nerve tissues underlying the disease process.[99,100]

Gene therapeutic targeting of traditional opioid pathway targets have the potential to avoid the side-effect profile often observed with traditional therapies. For example, this pathway has been targeted by Diamyd Medical for the treatment of chronic pain and cancer pain and is currently in Phase 2 studies after a Phase 1 study showed dose-responsive pain relief.[101] The drug candidate NP-2 (HCMV-IEp-hPPE), given by intradermal injection, is a replication-defective herpes simplex virus expressing human proenkephalin-A under the control of human cytomegalovirus immediate early promoter.[102–104] This gene transfer vector expresses enkephalin, one of the body's natural opioid peptides, directly in sensory neurons to block pain signals before they are

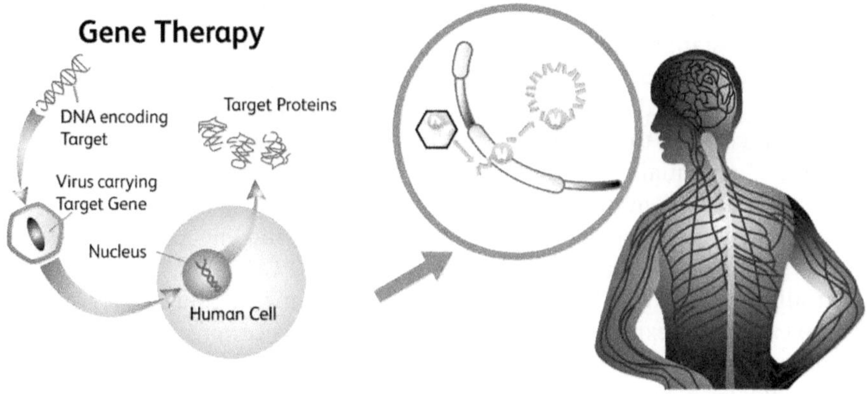

Figure 6.3 The gene transfer vector is injected into the skin in the area of pain from where the vector is carried into sensory nerves and releases the inhibitory neurotransmitter locally in the spinal cord.

transmitted through the spinal cord to the brain (see Figure 6.3). NP-2 is part of Diamyd Medical's nerve targeting drug delivery system (NTDDS) gene therapy platform, which works by delivering drugs directly to nerve cells, providing a localised effect in targeted cells by taking advantage of HSV's tropism for nerve cells. Since the drug is delivered to the nervous system and does not enter the bloodstream, fewer side effects are expected compared with current systemic pain treatments. A number of other gene therapy approaches have been reported for various pain indications but are further behind in development.[105]

RNA interference has been used for pain therapy and involves the use of double-stranded RNA (dsRNA) to invoke gene silencing. Once inside a cell, the dsRNA is processed into short 21–23 nucleotide small interfering RNAs (siRNA) that are used in a sequence-specific manner to recognise complementary RNAs. There are several classes of small RNA species including siRNAs, short hairpin RNAs (shRNAs) and microRNAs (miRNAs). P2X3 purinergic receptor is highly expressed in nociceptive sensory neurons and plays a crucial role in ATP-induced inflammatory and neuropathic pain. A siRNA complementary to P2X3, when given *via* intrathecal infusion into rats, showed diminished pain responses compared with untreated controls in a neuropathic pain model.[106] Delivery to lumbar dorsal root ganglia by injection of siRNA targeting the voltage-gated ion channel $Na_V1.8$ was evaluated *in vivo* and shown to reverse tactile allodynia in the rat chronic constriction nerve injury (CCI) model.[107] In a separate study, shRNA targeting $Na_V1.8$ delivered by lentivirus was shown to control neuronal hyperexciteability and pathological pain.[108] The same ion channels have also been targeted using antisense oligonucleotides. A number of other ion channels have been targeted by antisense and RNA therapies but have yet to move beyond preclinical studies.

6.8 Non-Antibody Protein Therapeutics

One of the most interesting protein therapeutics for pain therapy is botulinum toxin (BTX), a large protein complex that is decomposed in the physiological state to give two 150 kDa neurotoxin chains. Local intradermal injection of BTX-A, one of the serotypes of the toxin, is helpful in chronic focal painful neuropathies. BTX is a strong suppressant of acetylcholine and provides pain relief and anti-inflammatory effects by suppressing secretion of other neuropeptides and neurotransmitters.[109] The analgesic effects are not dependent on changes in muscle tone and benefits persist for at least 3 months from the time of administration. In 2010, Allergan commercialised a product for prophylactic therapy in adult patients with chronic migraine. At present, the neurotoxin is under development in several clinical trials, including Phase 3 trials for the treatment of shoulder pain, palmar hyperhidrosis and chronic prostatitis/chronic pelvic pain syndrome. Further trials are under way for the treatment of pain due to arthritis, stroke, prophylaxis for chronic daily headache, benign prostatic hyperplasia (BPH), idiopathic overactive bladder (OAB) and knee prosthesis pain. Results from these trials are expected to read out over the next 2–3 years. The compound had previously been under development for the treatment of lower back pain associated with muscle spasm. In those trials, paravertebral administration of BTX relieved pain and improved function in patients with chronic low back pain for up to 4 months.[110,111]

6.9 Future Directions for Pain Biotherapeutics

Current treatments for inflammatory and neuropathic pain remain unsatisfactory despite recent progress. Biotherapeutics offer a new approach for targeting pain pathways which have proved intractable with small-molecule approaches. Many of the modalities described above offer target specificity and selectivity that is well above traditional analgesics. As a result, several approaches have achieved clinical proof-of-concept (anti-NGF antibodies and BTX) and others are progressing in clinical and preclinical space.

Large-molecule approaches to targeting traditional pain targets such as ion channels and GPCRs have so far been a technical challenge, but recent advances suggest that modulation of these classes of protein is possible with biotherapeutic approaches. Technologies which facilitate transport of proteins across the BBB are enabling therapeutic intervention by new alternative modalities, which are not restricted to Lipinski's rules of druggability. In the coming years, biotherapeutics have the potential to transform pain therapy as they have transformed the treatment of cancer and autoimmune disease over the past decade.

References

1. I. Melnikova, *Nat. Rev. Drug. Discov.*, 2010, **9**, 589–590.
2. *Monoclonal Antibodies 2011*, Datamonitor, 2011.

3. R. Jefferis, *Arch. Biochem. Biophys.*, 2012, **526**, 159–166.
4. J. Drews, *Nat. Biotechnol.*, 1996, **14**, 1516–1518.
5. C. A. Lipinski, F. Lombardo, B. W. Dominy and P. J. Feeney, *Adv. Drug Delivery Rev.*, 1997, **23**, 3–25.
6. M. Bacher, C. Depboylu, Y. Du, C. Noelker, W. H. Oertel, T. Behr, G. Henriksen, M. Behe and R. Dodel, *Neurosci. Lett.*, 2009, **449**, 240–245.
7. M. Rajadhyaksha, T. Boyden, J. Liras, A. El-Kattan and J. Brodfuehrer, *Curr. Drug Discov. Technol.*, 2011, **8**, 87–101.
8. W. M. Pardridge and R. J. Boado, *Methods Enzymol.*, 2012, **503**, 269–292.
9. S. D. Skaper, *Methods Mol. Biol.*, 2012, **846**, 1–12.
10. M. Theodosiou, R. A. Rush, X. F. Zhou, D. Hu, J. S. Walker and D. J. Tracey, *Pain*, 1999, **81**, 245–255.
11. M. F. Seidel, M. Herguijuela, R. Forkert and U. Otten, *Semin. Arthritis Rheum.*, 2010, **40**, 109–126.
12. L. Aloe, M. A. Tuveri, U. Carcassi and R. Levi-Montalcini, *Arthritis Rheum.*, 1992, **35**, 351–355.
13. E. Dicou, C. Masson, W. Jabbour and V. Nerriere, *Neuroreport*, 1993, **5**, 321–324.
14. D. A. Halliday, C. Zettler, R. A. Rush, R. Scicchitano and J. D. McNeil, *Neurochem. Res.*, 1998, **23**, 919–922.
15. F. Falcini, M. Matucci Cerinic, A. Lombardi, S. Generini, A. Pignone, P. Tirassa, M. Ermini, L. Lepore, G. Partsch and L. Aloe, *Ann. Rheum. Dis.*, 1996, **55**, 745–748.
16. L. Aloe, L. Manni, G. Sebastiani and M. A. Tuveri, *Clin. Exp. Rheumatol.*, 1999, **17**, 632–633.
17. H. Friess, Z. W. Zhu, F. F. di Mola, C. Kulli, H. U. Graber, A. Andren-Sandberg, A. Zimmermann, M. Korc, M. Reinshagen and M. W. Buchler, *Ann. Surg.*, 1999, **230**, 615–624.
18. Z. Zhu, H. Friess, F. F. diMola, A. Zimmermann, H. U. Graber, M. Korc and M. W. Buchler, *J. Clin. Oncol.*, 1999, **17**, 2419–2428.
19. Y. Ye, D. Dang, J. Zhang, C. T. Viet, D. K. Lam, J. C. Dolan, J. L. Gibbs and B. L. Schmidt, *Mol. Cancer. Ther.*, 2011, **10**, 1667–1676.
20. Y. Indo, M. Tsuruta, Y. Hayashida, M. A. Karim, K. Ohta, T. Kawano, H. Mitsubuchi, H. Tonoki, Y. Awaya and I. Matsuda, *Nat. Genet.*, 1996, **13**, 485–488.
21. N. E. Lane, T. J. Schnitzer, C. A. Birbara, M. Mokhtarani, D. L. Shelton, M. D. Smith and M. T. Brown, *N. Engl. J. Med.*, 2010, **363**, 1521–1531.
22. Y. Yazici, E. F. Ekman, H. S. Greenberg, M. D. Smith, M. T. Brown, C. R. West and K. M. Verburg, *Arthritis. Rheum.*, 2011, **63**(Suppl), 828.
23. S. B. McMahon, D. L. Bennett, J. V. Priestley and D. L. Shelton, *Nat. Med.*, 1995, **1**, 774–780.
24. P. H. Holden, V. Asopa, A. G. Robertson, A. R. Clarke, S. Tyler, G. S. Bennett, S. D. Brain, G. K. Wilcock, S. J. Allen, S. K. Smith and D. Dawbarn, *Nat. Biotechnol.*, 1997, **15**, 668–672.
25. T. Wehrman, X. He, B. Raab, A. Dukipatti, H. Blau and K. C. Garcia, *Neuron*, 2007, **53**, 25–38.

26. J. J. Watson, M. S. Fahey, E. van den Worm, F. Engels, F. P. Nijkamp, P. Stroemer, S. McMahon, S. J. Allen and D. Dawbarn, *J. Pharmacol. Exp. Ther.*, 2006, **316**, 1122–1129.
27. V. Y. Hu, P. Zvara, A. Dattilio, T. L. Redman, S. J. Allen, D. Dawbarn, R. P. Stroemer and M. A. Vizzard, *J. Urol.*, 2005, **173**, 1016–1021.
28. T. Bishop, D. W. Hewson, P. K. Yip, M. S. Fahey, D. Dawbarn, A. R. Young and S. B. McMahon, *Pain*, 2007, **131**, 70–82.
29. C. Wu, M. A. Erickson, J. Xu, K. D. Wild and T. J. Brennan, *Anesthesiology*, 2009, **110**, 140–149.
30. L. Nguyen, M. Brumm, E. Carithers, B. DeSilva, M. Gonzalez, K. Beck and K. D. Wild, *AAPS. J.*, 2005, **7**(Suppl), 002866.
31. G. Ugolini, S. Marinelli, S. Covaceuszach, A. Cattaneo and F. Pavone, *Proc. Natl. Acad. Sci. U. S. A.*, 2007, **104**, 2985–2990.
32. *Glenmark Pharmaceuticals company worldwide website and pipeline*, Sep 2012.
33. T. Wang, D. Yu and M. L. Lamb, *Expert Opin. Ther. Pat.*, 2009, **19**, 305–319.
34. M. Hirose, M. Takatori, Y. Kuroda, M. Abe, E. Murata, T. Isada, K. Ueda, K. Shigemi, M. Shibazaki, F. Shimizu, M. Hirata, K. Fukazawa, M. Sakaguchi, K. Kageyama and Y. Tanaka, *J. Pharmacol. Sci.*, 2008, **106**, 107–113.
35. K. Ueda, M. Hirose, E. Murata, M. Takatori, M. Ueda, H. Ikeda and K. Shigemi, *J. Pharmacol. Sci.*, 2010, **112**, 438–443.
36. W. Y. Ma, E. Murata, K. Ueda, Y. Kuroda, M. H. Cao, M. Abe, K. Shigemi and M. Hirose, *J. Pharmacol. Sci.*, 2010, **114**, 79–84.
37. M. Tabata, E. Murata, K. Ueda, N. Kato-Kogoe, Y. Kuroda and M. Hirose, *J. Anesth.*, 2012, **26**, 545–551.
38. G. Moalem and D. J. Tracey, *Brain. Res. Rev.*, 2006, **51**, 240–264.
39. A. Bal, E. Unlu, G. Bahar, E. Aydog, E. Eksioglu and R. Yorgancioglu, *Clin. Rheumatol.*, 2007, **26**, 211–215.
40. C. Tetta, G. Camussi, V. Modena, C. Di Vittorio and C. Baglioni, *Ann. Rheum. Dis.*, 1990, **49**, 665–667.
41. H. Wang, M. Moser, M. Schiltenwolf and M. Buchner, *J. Rheumatol.*, 2008, **35**, 1366–1370.
42. H. P. Hartung, S. Jung, G. Stoll, J. Zielasek, B. Schmidt, J. J. Archelos and K. V. Toyka, *J. Neuroimmunol.*, 1992, **40**, 197–210.
43. H. Y. Kim, P. N. Hsu, M. Barba, W. Sulaiman, D. Robertson, B. Vlahos, R. Khandker, H. Nab, B. Freundlich and A. Koenig, *Int. J. Rheum. Dis.*, 2012, **15**, 188–196.
44. L. W. Moreland, S. W. Baumgartner, M. H. Schiff, E. A. Tindall, R. M. Fleischmann, A. L. Weaver, R. E. Ettlinger, S. Cohen, W. J. Koopman, K. Mohler, M. B. Widmer and C. M. Blosch, *N. Engl. J. Med.*, 1997, **337**, 141–147.
45. R. N. Maini, F. C. Breedveld, J. R. Kalden, J. S. Smolen, D. Davis, J. D. Macfarlane, C. Antoni, B. Leeb, M. J. Elliott, J. N. Woody, T. F. Schaible and M. Feldmann, *Arthritis. Rheum.*, 1998, **41**, 1552–1563.

46. F. Wolfe and K. Michaud, *J. Rheumatol.*, 2007, **34**, 1674–1683.
47. A. Hess, R. Axmann, J. Rech, S. Finzel, C. Heindl, S. Kreitz, M. Sergeeva, M. Saake, M. Garcia, G. Kollias, R. H. Straub, O. Sporns, A. Doerfler, K. Brune and G. Schett, *Proc. Natl. Acad. Sci. U. S. A.*, 2011, **108**, 3731–3736.
48. J. M. Hazes, P. Taylor, V. Strand, O. Purcaru, G. Coteur and P. Mease, *Rheumatology (Oxford)*, 2010, **49**, 1900–1910.
49. R. Fleischmann, S. Nayiager, I. Louw, B. Rojkovich, C. Fu, C. Udata and P. Fardipour, *Arthritis Rheum.*, 2011, **63**(Suppl), 2630.
50. M. Brozik, I. Rosztoczy, K. Meretey, G. Balint, M. Gaal, Z. Balogh, M. Bart, M. Mituszova, V. Velics and A. Falus, *J. Rheumatol.*, 1992, **19**, 63–68.
51. V. A. Danis, G. M. Franic, D. A. Rathjen, R. M. Laurent and P. M. Brooks, *Ann. Rheum. Dis.*, 1992, **51**, 946–950.
52. J. K. Lacki, K. Klama, S. H. Mackiewicz, U. Mackiewicz and W. Muller, *Inflamm. Res.*, 1995, **44**, 24–26.
53. H. Okamoto, M. Yamamura, Y. Morita, S. Harada, H. Makino and Z. Ota, *Arthritis. Rheum.*, 1997, **40**, 1096–1105.
54. U. Wereszczynska-Siemiatkowska, A. Dabrowski, A. Siemiatkowski, B. Mroczko, W. Laszewicz and A. Gabryelewicz, *Pancreas*, 2003, **26**, 144–152.
55. G. Gurleyik, E. Gurleyik, F. Cetinkaya and S. Unalmiser, *A. N. Z. J. Surg.*, 2002, **72**, 665–667.
56. A. Kocer, E. Kocer, R. Memisogullari, F. M. Domac and H. Yuksel, *Clin. J. Pain*, 2010, **26**, 690–693.
57. A. Kocer, R. Memisogullari, F. M. Domac, A. Ilhan, E. Kocer, S. Okuyucu, B. Ozdemir and H. Yuksel, *Pain Pract*, 2009, **9**, 375–379.
58. L. N. Yu, X. S. Yang, Z. Hua and W. Xie, *Zhonghua Yi Xue Za Zhi*, 2009, **89**, 469–471.
59. A. Koch, K. Zacharowski, O. Boehm, M. Stevens, P. Lipfert, H. J. von Giesen, A. Wolf and R. Freynhagen, *Inflamm. Res.*, 2007, **56**, 32–37.
60. P. G. Murphy, M. S. Ramer, L. Borthwick, J. Gauldie, P. M. Richardson and M. A. Bisby, *Eur. J. Neurosci.*, 1999, **11**, 2243–2253.
61. M. S. Ramer, P. G. Murphy, P. M. Richardson and M. A. Bisby, *Pain*, 1998, **78**, 115–121.
62. N. Kiguchi, T. Maeda, Y. Kobayashi, T. Kondo, M. Ozaki and S. Kishioka, *Eur. J. Pharmacol.*, 2008, **592**, 87–92.
63. W. Ma and R. Quirion, *J. Neurochem.*, 2006, **98**, 180–192.
64. *IDdb*, Thomson Pharma Partnering [internet database], 2012.
65. J. S. Smolen, A. Beaulieu, A. Rubbert-Roth, C. Ramos-Remus, J. Rovensky, E. Alecock, T. Woodworth and R. Alten, *Lancet*, 2008, **371**, 987–997.
66. Y. Yazici, J. R. Curtis, A. Ince, H. Baraf, R. L. Malamet, L. L. Teng and A. Kavanaugh, *Ann. Rheum. Dis.*, 2012, **71**, 198–205.
67. S. Smale, S. Carty, R. Goodfellow, E. Choy, S. Rose-John, S. Jones, S. Williams and A. S., *Arthritis Rheum.*, 2011, **63**(Suppl.), 1146.

68. C. Meyer, K. Eydeler, E. Magbanua, T. Zivkovic, N. Piganeau, I. Lorenzen, J. Grotzinger, G. Mayer, S. Rose-John and U. Hahn, *RNA Biol.*, 2012, **9**, 67–80.
69. D. Lindholm, R. Heumann, B. Hengerer and H. Thoenen, *J. Biol. Chem.*, 1988, **263**, 16348–16351.
70. D. Lindholm, R. Heumann, M. Meyer and H. Thoenen, *Nature*, 1987, **330**, 658–659.
71. J. C. Copray, I. Mantingh, N. Brouwer, K. Biber, B. M. Kust, R. S. Liem, I. Huitinga, F. J. Tilders, A. M. Van Dam and H. W. Boddeke, *J. Neuroimmunol.*, 2001, **118**, 203–211.
72. K. Ren and R. Torres, *Brain Res. Rev.*, 2009, **60**, 57–64.
73. N. Schlesinger, E. Mysler, H. Y. Lin, M. De Meulemeester, J. Rovensky, U. Arulmani, A. Balfour, G. Krammer, P. Sallstig and A. So, *Ann. Rheum. Dis.*, 2011, **70**, 1264–1271.
74. N. Schlesinger, R. E. Alten, T. Bardin, H. R. Schumacher, M. Bloch, A. Gimona, G. Krammer, V. Murphy, D. Richard and A. K. So, *Ann. Rheum. Dis.*, 2012.
75. H. R. Schumacher, Jr., J. S. Sundy, R. Terkeltaub, H. R. Knapp, S. J. Mellis, N. Stahl, G. D. Yancopoulos, Y. Soo, S. King-Davis, S. P. Weinstein and A. R. Radin, *Arthritis Rheum.*, 2012, **64**, 876–884.
76. C. F. Kim and G. Moalem-Taylor, *J. Pain*, 2011, **12**, 370–383.
77. J. Naylor and D. J. Beech, *Open Drug Discov. J.*, 2009, **1**, 36–42.
78. D. Gomez-Varela, E. Zwick-Wallasch, H. Knotgen, A. Sanchez, T. Hettmann, D. Ossipov, R. Weseloh, C. Contreras-Jurado, M. Rothe, W. Stuhmer and L. A. Pardo, *Cancer Res.*, 2007, **67**, 7343–7349.
79. X. F. Yang, Y. Yang, Y. T. Lian, Z. H. Wang, X. W. Li, L. X. Cheng, J. P. Liu, Y. F. Wang, X. Gao, Y. H. Liao, M. Wang, Q. T. Zeng and K. Liu, *PLoS One*, **7**, e36379.
80. *Patent application*: WO2011051350, 2011.
81. C. N. Wyatt, V. Campbell, J. Brodbeck, N. L. Brice, K. M. Page, N. S. Berrow, K. Brickley, C. M. Terracciano, R. U. Naqvi, K. T. MacLeod and A. C. Dolphin, *J. Physiol.*, 1997, **502**(Pt 2), 307–319.
82. N. L. Brice, N. S. Berrow, V. Campbell, K. M. Page, K. Brickley, I. Tedder and A. C. Dolphin, *Eur. J. Neurosci.*, 1997, **9**, 749–759.
83. G. Buell, I. P. Chessell, A. D. Michel, G. Collo, M. Salazzo, S. Herren, D. Gretener, C. Grahames, R. Kaur, M. H. Kosco-Vilbois and P. P. Humphrey, *Blood*, 1998, **92**, 3521–3528.
84. A. Sircar, K. A. Sanni, J. Shi and J. J. Gray, *J. Immunol.*, 2011, **186**, 6357–6367.
85. *Patent application*: WO2010070145, 2010.
86. R. M. Johnson, S. D. Harrison and D. Maclean, *Methods Mol. Biol.*, **683**, 535–551.
87. F. Tao and R. A. Johns, *Drug Dev. Res.*, **71**, 99–105.
88. P. A. Wender, D. J. Mitchell, K. Pattabiraman, E. T. Pelkey, L. Steinman and J. B. Rothbard, *Proc. Natl. Acad. Sci. U. S. A.*, 2000, **97**, 13003–13008.

89. S. W. Jones, R. Christison, K. Bundell, C. J. Voyce, S. M. Brockbank, P. Newham and M. A. Lindsay, *Br. J. Pharmacol.*, 2005, **145**, 1093–1102.
90. J. A. Johnson, M. O. Gray, C. H. Chen and D. Mochly-Rosen, *J. Biol. Chem.*, 1996, **271**, 24962–24966.
91. S. G. Khasar, Y. H. Lin, A. Martin, J. Dadgar, T. McMahon, D. Wang, B. Hundle, K. O. Aley, W. Isenberg, G. McCarter, P. G. Green, C. W. Hodge, J. D. Levine and R. O. Messing, *Neuron*, 1999, **24**, 253–260.
92. S. Harrison, K. Das, F. Karim, D. Maclean and D. Mendel, *Expert Opin. Drug Discov.*, 2008, **3**, 761–774.
93. X. J. Liu, J. R. Gingrich, M. Vargas-Caballero, Y. N. Dong, A. Sengar, S. Beggs, S. H. Wang, H. K. Ding, P. W. Frankland and M. W. Salter, *Nat. Med.*, 2008, **14**, 1325–1332.
94. F. Tao and R. A. Johns, *Anesthesiology*, 2008, **108**, 882–887.
95. H. Wu and F. Tao, *J. Biomater. Nanobiotechnol.*, **2**, 596–600.
96. D. L. Holmes, K. Thibaudeau, B. L'Archeveque, P. G. Milner, A. M. Ezrin and D. P. Bridon, *Bioconjug. Chem.*, 2000, **11**, 439–444.
97. M. S. Wallace, D. Moulin, A. J. Clark, R. Wasserman, A. Neale, P. Morley-Forster, J. P. Castaigne and S. Teichman, *J. Opioid. Manag.*, 2006, **2**, 167–173.
98. L. R. Roberts, K. Brady, A. Brown, D. Davey, L. Feng, H. Huang, D. Liu, L. Malet, G. McMurray, A. Phelan, K. Saunders and A. Bhat, *Bioorg. Med. Chem. Lett.*, **22**, 4173–4178.
99. K. K. Jain, *Expert Opin. Biol. Ther.*, 2008, **8**, 1855–1866.
100. W. F. Goins, J. B. Cohen and J. C. Glorioso, *Neurobiol. Dis.*, **48**, 255–270.
101. D. J. Fink, J. Wechuck, M. Mata, J. C. Glorioso, J. Goss, D. Krisky and D. Wolfe, *Ann. Neurol.*, **70**, 207–212.
102. J. R. Goss, M. Mata, W. F. Goins, H. H. Wu, J. C. Glorioso and D. J. Fink, *Gene. Ther.*, 2001, **8**, 551–556.
103. J. R. Goss, C. F. Harley, M. Mata, M. E. O'Malley, W. F. Goins, X. Hu, J. C. Glorioso and D. J. Fink, *Ann. Neurol.*, 2002, **52**, 662–665.
104. S. Hao, M. Mata, W. Goins, J. C. Glorioso and D. J. Fink, *Pain*, 2003, **102**, 135–142.
105. Y. Huang, X. Liu, L. Dong, Z. Liu, X. He and W. Liu, *Pain Res. Treat.*, **2011**, 968218.
106. G. Dorn, S. Patel, G. Wotherspoon, M. Hemmings-Mieszczak, J. Barclay, F. J. Natt, P. Martin, S. Bevan, A. Fox, P. Ganju, W. Wishart and J. Hall, *Nucleic Acids Res.*, 2004, **32**, e49.
107. X. W. Dong, S. Goregoaker, H. Engler, X. Zhou, L. Mark, J. Crona, R. Terry, J. Hunter and T. Priestley, *Neuroscience*, 2007, **146**, 812–821.
108. M. Mikami and J. Yang, *Anesthesiology*, 2005, **103**, 828–836.
109. W. S. Sim, *Korean J. Pain*, **24**, 1–6.
110. L. Foster, L. Clapp, M. Erickson and B. Jabbari, *Neurology*, 2001, **56**, 1290–1293.
111. B. Jabbari, J. Ney, A. Sichani, W. Monacci, L. Foster and M. Difazio, *Pain Med.*, 2006, **7**, 260–264.

CHAPTER 7

Peptide Therapeutics from Venomous Creatures

VINCENT LAVERGNE, GLENN F. KING, RICHARD J. LEWIS AND PAUL F. ALEWOOD*

Institute for Molecular Bioscience, The University of Queensland, St Lucia, Qld 4072, Australia
*Email: p.alewood@imb.uq.edu.au

7.1 Introduction

7.1.1 Peptide Toxins

Toxins are engineered molecules of choice for numerous uni- and multicellular organisms, from bacteria to plants and animals. They are used for defence, predation and competitor deterrence by acting as endogenous drugs that can disturb homeostasis and induce a wide variety of pathophysiological conditions by contact, absorption or injection into the soft tissue of animals.[1]

Non-peptide toxins are mainly produced as secondary metabolites and notably include phycotoxins produced by prokaryotic or eukaryotic microalgae like cyanobacteria, dinoflagellates, diatoms, haptophytes and raphydophytes.[2–4] Among the most studied molecules of this group are the neurotoxic azaspiracids;[5] domoic acid, a glutamate receptor agonist;[6,7] okadaic acid which inhibits serine/threonine protein phosphatases 1 and 2A;[8] ciguatoxins and brevetoxins, which activate voltage-gated sodium (Na_V) channels;[9–11] and saxitoxins, which block Na_V channels.[12] Tetrodotoxin, another potent and selective Na_V channel blocker, is produced by symbiotic bacteria.[13] The accumulation of guanidinium toxins in pufferfish and filter-feeding bivalves can

RSC Drug Discovery Series No. 35
Pain Therapeutics: Current and Future Treatment Paradigms
Edited by Charlotte Allerton

Published by the Royal Society of Chemistry, www.rsc.org

lead to severe food poisoning after consumption by humans. Plants also produce a broad range of non-peptide phytotoxins that are a proven source of pharmaceutical drugs including alkaloids (vinblastine, vincristine,[14] cocaine,[15] morphine,[16] ephedrine,[17] codeine[18]), terpenes such as taxol (diterpene)[19] or salicin (glucoside)[20] for instance, and phenolic compounds including furanocoumarin, lignins and tannins.[21,22] Finally, non-peptide toxins can also be found to a lesser extent in animals such as poison dart frogs that accumulate lipophilic alkaloids such as batrachotoxin and epibatine. These molecules are of particular interest because of their antinociceptive properties resulting from their interaction with Na_V channels and nicotinic acetylcholine receptors (nAChRs).[23–27]

Conversely, peptide toxins generally constitute the main components of most animal venoms, with non-elapid snakes being the exception. These highly efficient secretions, sometimes considered as an evolved saliva,[28] can only be administered *via* specialised envenomation organs with sharp body features that facilitate a traumatic subcutaneous, intramuscular or intravenous mode of delivery into prey or predators. The composition of venom is generally highly complex, comprising an array of salts, small molecules, bioactive peptide toxins, and various enzymes, leading to a broad range of molecular interactions in envenomed prey and predators across the phylogenetic spectrum.[29]

In this chapter, particular attention will be devoted to peptide toxins that can be elaborated by venomous eumetazoa from diverse taxa across marine invertebrates (marine cone snails, sea anemones, stingrays), centipedes, insects (ants, bees, wasps), arachnids (spiders, scorpions), reptiles (snakes, lizards) and mammals (platypus), with an emphasis relating to their structure and function. We also briefly describe the technologies and methods employed to date for the characterisation and elaboration of peptide toxins. Finally, we examine the pharmacological diversity of peptide toxins with potential antinociceptive activity.

7.1.2 Bioactive Peptide Toxins: Therapeutic and Diagnostic Potential

Despite their variable size (10–90 amino acid residues), most venom peptides have a high cysteine content which enables them to form specific disulfide bonds that provide them with a high level of thermodynamic stability as well as resistance to proteases. It appears that conserved cysteine patterns and disulfide-bond connectivities with only a limited number of folds are found among the venom peptides of probably several thousands of different venomous animal species. However, the remarkable hypervariability of their primary structures and accompanying post-translational modifications (PTMs) make venom peptides molecular ligands of unparalleled richness. In addition, their high selectivity and potency towards key receptors, channels, transporters or enzymes of crucial biochemical signalling pathways involved in organism homeostasis make them an unrivalled and unexplored source of leads for the development of molecules with therapeutic and diagnostic potential.[30–32]

The modes of action of venom peptides are broad. They may act on different body systems (circulatory, lymphatic, muscular, nervous, endocrine, immune, respiratory) and target specific tissues. At the molecular level, venom peptides are directed towards a large repertoire of target receptors. Indeed, they have been shown to selectively modulate the activity of precise subtypes of enzymes,[33–35] voltage- and ligand-gated sodium, potassium, calcium or chloride channels,[36–38] transporters,[39–41] and G-protein coupled receptors (GPCRs).[42,43] Some venom peptides form pores in cell membranes, enabling them to exhibit cytolytic or antimicrobial activities.[44–48]

7.1.3 Discovery and Characterisation of Peptide Toxins

Isolation of peptide toxins is usually carried out from crude venom obtained by either dissection or milking. The first method consists of dissecting the venom glands from the animal and subsequently extracting its contents *via* organic aqueous mixed solvents. This method is commonly used, but the crude material extracted contains many components and further purification steps are generally required to isolate the individual toxins. The milking procedure yields anywhere from a few microlitres of venom from venomous arthropods to millilitre quantities from larger snakes, and is commonly restricted to the larger animals.

Subsequent isolation of pure peptide toxins is then generally achieved using reversed-phase HPLC (RP-HPLC) followed by sequence determination using Edman chemistry in conjunction with tandem mass spectrometry.[49,50] Modern strategies are now taking a faster and complementary approach by integrating venom-gland transcriptomes with Edman sequencing and tandem mass spectrometry.[51,52] Once a partial peptide sequence has been obtained using either of the latter two techniques, a search of the transcriptomic data usually reveals the sequence of the entire transcript encoding the peptide of interest, and in many cases the sequence of paralogues as well.

Only rarely is sufficient native peptide available for complete structural, functional, and *in vivo* characterisation. However, once the amino acid sequence of the peptide has been determined, it can be produced using synthetic or recombinant techniques.[53] Production of venom peptides may be technically challenging because of the large number of specific disulfide bonds; thus, a toxin with three or four disulfide bonds is theoretically capable of forming 15 or 105 different disulfide-bond isomers, respectively. Longer polypeptide toxins can be accessed by coupling solid-phase peptide synthesis (SPPS) with native chemical ligation (NCL)[54] where the high cysteine content actually increases the efficiency of chemical synthesis.

Venom peptides can also be produced by overexpression in bacteria, yeast or insect cells.[53] Recombinant protein production is generally less time- and cost-effective than SPPS but when optimised enables facile site-directed mutagenesis for structure–activity relationship (SAR) studies and isotopic labelling for multidimensional NMR studies. This approach generally precludes the introduction of PTMs found in the native toxin and is limited to genetically-encoded L-amino acids.

In order to assess the potential activities within the venom it is usual to test the crude material directly on isolated tissues, cells, cell lines, or oocytes expressing specific targets at their surface. Activity-guided fractionation may then be performed to precisely identify the bioactive peptidic components. Each fraction separated by RP-HPLC or *via* some other chromatographic technique is then individually tested for its activity on particular receptors, transporters, or channels. Classical activity assays comprise binding and electrophysiology experiments on various targets such as voltage- and ligand-gated channels, nicotinic receptors, and GPCRs.

7.1.4 Optimisation of Peptide Toxins: from Candidates to Drug Leads

Determination of the three-dimensional structure of an active peptide toxin using NMR spectroscopy or X-ray crystallography often gives precise information about its potential pharmacophore and mode of interaction with its target, which may be further examined by *in silico* docking studies. As a consequence, the precise location of the residues involved in target binding can be used to guide the elucidation of SAR and the design of analogues with enhanced properties for system-specific human therapeutics.

We mentioned earlier that venom peptides are perfectly designed to accomplish their natural task. The formation of disulfide bonds and other PTMs not only facilitates the formation of a highly stable tertiary structure necessary for optimal potency and activity, but in many cases also provides resistance against extremes of pH and temperature as well as proteases that are present in the circulatory system and tissues of target organisms. However, these natural structural advantages are not always favourable for therapeutic purposes. Venom peptides may also have reduced bioavailability and can be subject to limitations due to their large size and hydrophilic nature, which prevent passive diffusion through hydrophobic epithelial layers including the blood–brain barrier. Nevertheless, if required, vectors such as homing peptides (HMs) or cell-penetrating peptides (CPPs) are able to deliver macromolecular cargoes across cell membranes, thus improving their selectivity, efficiency and tolerance *in vivo*.[55,56]

7.2 Pharmacology of Peptide Toxins in the Modulation of Nociception

Because of their ample structural diversity, venom peptides are directed towards a broad range of heterologous molecular targets, including soluble and membrane proteins. The proof of concept of the therapeutic utilisation of peptides isolated from animal venoms was demonstrated some time ago with the commercialisation of drugs for the treatment of various human pathophysiological conditions (see Table 7.1).

In the remaining part of this chapter we review selected well-characterised peptide toxins that are already available on the market for pain management, as well as molecules considered as future potential antinociceptive drugs

Table 7.1 Venom peptides used as medicines or with therapeutic potential. The source organism, peptide name, drug name (with the name of the distributing company or clinical phase), molecular function, indication, and reference of approved venom peptides (A), toxin analogues (B), peptides under clinical (C) or preclinical (D) trials, and with therapeutic potential (E).

Source organism		*Peptide names*	*Commercial or trial names (Company)*	*Molecular function*	*Indication*	*Reference*
A Approved toxin drugs						
Cone snail	*Conus magus*	ω-Conotoxin MVIIA (ziconotide)	Prialt® (Azur Pharma)	$Ca_V2.2$ channel inhibitor	Severe chronic pain	Wermeling (2005)
Lizard	*Heloderma suspectum*	Extendin-4 (exenatide)	Byetta® (Amylin/Eli Lilly)	GLP1R activator	Type 2 diabetes	Murphy (2012)
B Toxin analogue drugs						
Leech	*Hirudo medicinalis*	Hirudin	Bivalirudin/Angiomax® or Angiox® (The Medicines Company)	Direct thrombin inhibitor	Anticoagulant	Van de Car *et al.* (2010)
Snake	*Bothrops jararaca*	Bradykinin-potentiating peptide 9a (teprotide)	Captopril (*non-pepzide mimetic*)/Capoten® (Bristol-Myers Squibb)	ACE inhibitor	Hypertension; congestive heart failure	Camargo *et al.* (2012)
Snake	*Echis carinatus sochureki*	Disintegrin echistatin-α	Tirofiban (*non–peptide mimetic*)/Aggrastat® (Iroko Cardio; Medicure Pharma)	Integrin αIIβ3 inhibitor	Acute coronary syndromes	Menozzi *et al.* (2005)
Snake	*Sistrusus miliarius barbouri*	Disintegrin barbourin	Eptifibatide (*cyclic heptapeptide*)/ Integrilin® (Millenium Pharmaceuticals)	Integrin αIIβ3 inhibitor	Acute coronary syndromes	Koh *et al.* (2012)
C Venom peptides in clinical trials						
Cone snail	*Conus catus*	ω-conotoxin CVID (leconotide)	Phase 2 (AMRAD)/ NMED-160 (CNSBio)	$Ca_V2.2$ channel inhibitor	Neuropathic pain	Kolosov *et al.* (2010)
Cone snail	*Conus geographus*	Contulakin-G	Phase 2a/CGX-1160 (Cognetix—University of Utah)	NTR1 agonist	Chronic pain	Allen *et al.* (2007)

Table 7.1 (*Continued*)

Source organism		Peptide names	Commercial or trial names (Company)	Molecular function	Indication	Reference
Cone snail	*Conus geographus*	Conantokin-G	Withdrawn Phase 2/ CGX-1007 (Cognetix—University of Utah)	NR2B inhibitor	Convulsion; epilepsy	Teichert *et al.* (2007)
Cone snail	*Conus marmoreus*	χ-conotoxin MrIA	Phase 2b/Xen2174 (Xenome Ltd.)	NET inhibitor	PO pain	Brust *et al.* (2009)
Scorpion	*Leiurus quinquestriatus*	Chlorotoxin	Phase 2/TM-601 (Sinai Medical Center and TransMolecular)	Annexini II ligand; Cl channel blocker	Anticancer	Stroud *et al.* (2011)
Sea anemone	*Stoichactis helianthus*	Potassium channel toxin ShK	Phase 1/ShK-186 (Kineta)	$K_V1.3$ channel inhibitor	Autoimmune disease	Chi *et al.* (2012)
Snake	*Dendroaspis angusticeps*	Natriuretic peptide (DNP)	Phase 2/Cendreritide (*chymeric human CNP—snake DNP*) (Nile Therapeutics)	NPR-B activator	Congestive heart failure	Lee *et al.* (2009)
Snake	*Naja kaouthia*	Long neurotoxin 1 (α-cobratoxin)	Phase 1/RPI-78M (Nutra Pharma)	nAchR antagonist	Multiple sclerosis	Reid (2007)
D Venom peptides in preclinical trials						
Cone snail	*Conus marmoreus*	μ-conotoxin MrVIB	CGX-1002 (Cognetix—University of Utah)	$Na_V1.4$ & $Na_V1.8$ channel blocker	Anaesthesia; analgesia	Dawson *et al.* (2012)
Cone snail	*Conus purpurascens*	κ-conotoxin FVIIA	CGX-1051 (Cognetix—University of Utah)	Shaker K_V channel inhibitor	Myocardial ischaemia/ reperfusion	Lubbers *et al.* (2005)
Cone snail	*Conus victoriae*	α-conotoxin VclA	Cyclic Vc1.1 (University of Queensland)	α9/α10 nAchR inhibitor	Neuropathic pain	Carstens *et al.* (2011)
Sea anemone	*Anthopleura elegantissima*	Toxin APETx2	THA 902 (Theralpha)	ASIC3 inhibitor	Inflammatory, ischaemic, PO pain	Jensen *et al.* (2012)

Snake	*Dendroaspis polylepis*	Mambalgin-1	THA 904 (Theralpha)	ASICla blocker	Potential analgesic	Diochot *et al.* (2012)
Snake	*Ophiophagus hannah*	Long neurotoxin OH-55 (hannalgesin)	THA 903 (Theralpha)	NOS activator (and AchR inhibitor)	Chronic, acute pain	Pu *et al.* (1995)
Spider	*Psalmopoeus cambridgei*	π-theraphotoxin-Pc1a (psalmotoxin 1)	THA 901 (Theralpha)	ASIC1a blocker	Neuropathic, cancer, PO pain	Dwyer *et al.* (2009)
E Potential therapeutic leads						
Cone snail	*Conus purpurascens*	μ-conotoxin PIIIA	–	$Na_V1.4$ channel inhibitor	Potential analgesic	Holford *et al.* (2009)
Cone snail	*Conus regius*	α-conotoxin RglA	–	α9/α10 nAchR inhibitor	Muscle relaxation	Ellison *et al.* (2006)
Scorpion	*Centruoides margaritatus*	Potassium channel toxin α-KTx 2.2 (margatoxin)	–	$K_V1.3$ channel inhibitor	Autoimmune disease	Nikouee *et al.* (2012)
Snake	*Oxyuranus microlepidotus*	Taipan natriuretic peptide (TNP-c)	–	NPR-C; GC-A	Smooth muscle relaxation	Fry *et al.* (2005)
Spider	*Haplopelma schmidti*	μ-theraphotoxin-Hh2a(huwentoxin-IV)	–	$Na_V1.7$ channel inhibitor	Neurotransmission blockage	Xiao *et al.* (2008)
Spider	*Phoneutria nigriventer*	δ-ctenitoxin-Pn2a (PnTx2–6)	–	Neuronal Na_V channel inhibitor	Erectile dysfunction	Nunes *et al.* (2012)
Spider	*Thrixopelma pruriens*	β/ω-theraphotoxin-Tp2a (ProTx-II)	–	$Na_V1.7$ channel inhibitor (lethal at 'therapeutic' levels to rats by IV or IT administration)	Neurotransmission blockage	Schmalhofer (2008)

ACE, angiotensin converting enzyme; ASIC, acid-sensing ion channel; Ca_V, voltage-gated calcium channel; GC-A, type-A guanylate cyclase; GLP1R, glucagon-like peptide receptor; K_V, voltage-gated potassium channel; nAchR, nicotinic acetylcholine receptor; Na_V, voltage-gated sodium channel; NET, noradrenaline transporter; NOS, nitric oxide synthase; NPR-B, natriuretic peptide receptor B; NPR-C, natriuretic peptide receptor C; NR2B, *N*-methyl-D-aspartate receptor; NTR, neurotensin receptor; PO, postoperative.

undergoing advanced and early-stage clinical trials. Finally we will describe some structural and functional aspects of molecules still at the preclinical research phase.

7.2.1 Marketed Toxin Analogue Drug: Ziconotide

To our knowledge, the only venom peptide approved for clinical use in the treatment of acute pain is ziconotide, the non-opioid synthetic equivalent of ω-conotoxin MVIIA (hereafter called MVIIA), sold under the name Prialt by Azur Pharma.[57] Isolation of native ω-conotoxin MVIIA from the venom of the piscivorus cone snail *Conus magus* was first reported in 1982 by McIntosh *et al.*[58] The mature peptide contains 25 residues with an amidated C-terminus.[59,60] MVIIA contains six cysteine residues arranged in a C(6)C(6)CC(3)C(4)C pattern, with the number between parentheses representing the number of non-cysteine amino acids within each of the four intercystine loops. The cysteine residues are connected in I-IV, II-V, III-VI pattern that forms an inhibitor cystine knot (ICK) motif (Figure 7.1).[61] The ICK or 'knottin' fold, which is abundant in the venom of some animals, especially spiders,[62] is of particular interest for drug design because it confers thermodynamic and biological stability, most importantly resistance to proteases.[63] The three-dimensional structure of MVIIA is composed of a triple-stranded β-sheet segment involving residues Ala^{6}-Lys^{7}-Cys^{8} (β strand 1), Cys^{20}-Arg^{21} (β strand 2), and Lys^{24}-Cys^{25} (β strand 1).[64,65]

The pharmacologically active ω-conotoxins show potential analgesic properties because of their ability to inhibit, with high potency and selectivity, different subtypes of mammalian voltage-gated calcium (Ca_V) channels that are

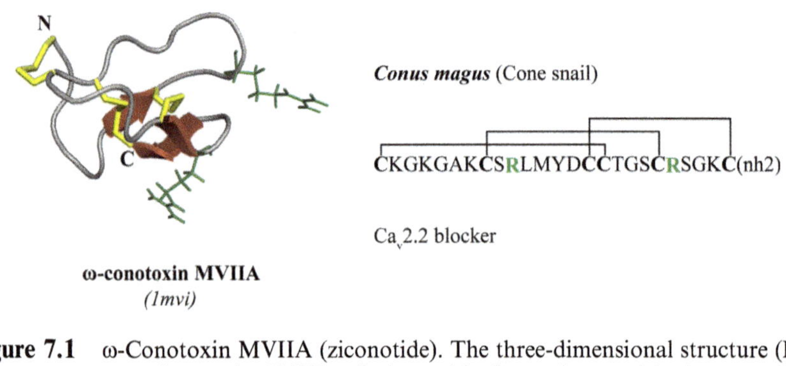

Figure 7.1 ω-Conotoxin MVIIA (ziconotide). The three-dimensional structure (PDB accession code 1MVI) of ziconotide from the magician's cone *Conus magus* is shown in the left panel. β-Strands and disulfide bonds are shown in red and yellow, respectively, and pharmacophore residues are highlighted in green. The N- and C-termini are labelled. In the amino acid sequence of the toxin shown on the right, cysteine residues are bold, pharmacophore residues are highlighted in bold green font, and the disulfide connectivity is shown as black lines above the sequence. Conotoxin MVIIA is a selective blocker of Ca_V2.2 channels. 'nh2' indicates C-terminal amidation.

expressed in sensory neurons of ascending pain pathways, particularly the neuronal P/Q-type ($Ca_V2.1$) and N-type ($Ca_V2.2$) channels.[66,67] It has been reported that ω-conotoxins bind preferentially the S5–S6 linker of domain III in the external vestibule of the channels.[68] Moreover, the kinetics and affinity of ω-conotoxin/$Ca_V2.2$ interactions were suggested to be driven by intracellular domains of the human channel as well.[69,70] Mutagenesis experiments have determined the critical role played by the conserved Tyr^{13} and Lys^{2} basic residues in ω-conotoxin that facilitate high-affinity binding to $Ca_V2.2$.[71–74]

In the case of ziconotide, the toxin is ~10 000-fold more selective for N-type over P/Q-type Ca_V channels and it appears to bind irreversibly to the former.[75] The channel affinity of MVIIA seems to be driven also by Arg^{21} in the fourth loop of the toxin, whereas Arg^{10} in the second loop may be responsible for channel subtype selectivity.[76] Though ziconotide is a potent and selective Ca_V channel antagonist that avoids the opiate side effects of addiction and tolerance, its intrathecal route of administration and dose-limiting neurological side effects on the central and peripheral nervous system influence its therapeutic index and limit its clinical use. Indeed, intrathecal ziconotide may cause dizziness, nystagmus, confusion, abnormal gait, somnolence, speech difficulties, amblyopia, ataxia, amnesia or abnormal thought processes.[77]

7.2.2 Venom Peptides in Clinical Trials

7.2.2.1 ω-conotoxin CVID

Another ω-conotoxin called CVID or leconotide was isolated from the crude venom of the piscivorous *Conus catus* by assay-guided fractionation and gene cloning.[75] The peptide completed a Phase 2 clinical trial with AMRAD and is being commercialised by CNSBio as NMED-160. CVID contains 27 residues with an amidated C-terminus and six cysteine residues arranged in a C(6)C(6)CC(3)C(6)C pattern. This relatively hydrophilic knottin peptide has three disulfide bonds (I–IV, II–V, III–VI) linked in the same way as MVIIA.[78] However, a unique characteristic of the fold comes from the presence of two hydrogen bonds (from the backbone amide protons of Lys^{10} and Leu^{11}) between the second and fourth loops, which leads to a unique orientation of loop 4 and which may provide an entropic advantage for binding to the target channel. In contrast to ziconotide, CVID is able to inhibit neurotransmitter release from the preganglionic cholinergic neurons innervating the rat submandibular ganglia, which is usually resistant to selective antagonists of Ca_V1, $Ca_V2.1$, $Ca_V2.2$, and Ca_V3 channels.[79] To date, CVID (together with its CVIE and CVIF isoforms) is the most selective ω-conotoxin for $Ca_V2.2$ over $Ca_V2.1$ channels (100-fold more selective than MVIIA).[75] In addition, compared to MVIIA, CVID has better *in vitro* reversibility, a faster onset/offset of action, and sustains antinociception with a better therapeutic index (≈ 5-fold) with no significant cardiovascular side effects.[80–82] These differences may be due to the fact that although CVID possesses the conserved Tyr^{13} residue essential for binding (inhibition of the rat brain $Ca_V2.1$ channel is reduced by 400-fold with a [Y13F]CVID analogue),[79] it lacks the Arg^{21} residue located in loop 4 of MVIIA and utilises instead an exposed Val^{24} for interaction with its target,[75] as well as Lys^{10} which is considered to be the preferred residue for inhibition of resistant-type calcium channel and neurotransmitter release in presynaptic neuronal ends.[79] These structural and functional differences compared to ziconotide suggest that CVID might make potentially a more effective and safer analgesic agent following systemic administration.

7.2.2.2 Contulakin-G

An interesting venom peptide directed to an unusual target for cone snail toxins is contulakin-G. Contulakin-G underwent a small Phase 2a clinical trial with Cognetix as CGX-1160. This 16-residue glycopeptide is devoid of cysteine residues, is N-terminally blocked by a pyroglutamic acid, and is *O*-glyclosylated on Thr^{10} (disaccharide β-D-Gal*p*-(1→3)-α-D-Gal*p*NAc-(1→).[83] Interestingly, the unmodified native toxin can be chemically synthesised and mammalian uridine diphospho-*N*-acetyl-D-galactosamine:polypeptide *N*-acetyl-galactos-aminyltransferase T1 (EC 2.4.1.41) can be used to efficiently and specifically transfer GalNAc to Thr^{10} of the peptide.[84,85]

The C-terminal part of contulakin-G closely resembles some members of the neurotensin neurotransmitter family such as bovine hypothalamus neurotensin, porcine spinal cord neuromedin N, *Xenopus laevis* xenopsin and human gastric mucosa xenin with identities of 66%, 83%, 66% and 66% respectively.[83] Contulakin-G has agonistic properties against cloned human neurotensin receptor (NTR)1, rat NTR1 and NTR2, as well as mouse NTR3, with slightly lower binding affinities than the endogenous neurotensins. The presence of the glycan at position 10 strongly enhances binding to the neurotensin GPCRs. Moreover, *in vivo* injection of contulakin-G induces the same sluggishness in mice as neurotensin. Recent preclinical trials revealed that, although its mechanism of action is still unknown, contulakin-G is a potent antinociceptive when delivered intrathecally with no observable motor or cardiovascular side effects in acute pain models in rats (formalin test) and dogs (thermal skin twitch model).[86] As a consequence, contulakin-G optimisation could lead to a promising therapeutic candidate and may provide an alternative to opioid spinal analgesics.

7.2.2.3 χ-conotoxin MrIA

χ-Conotoxin MrIA was isolated from the venom of the molluscivorous marble cone *Conus marmoreus*.[87] The stable pyroglutamate analogue is currently undergoing a Phase 2b double-blind clinical trial under the name Xen2174 with Xenome Ltd.[41] This short 13-residue peptide displays a compact cysteine pattern, (3)CC(4)C(2)C, with disulfide bonds defined by a 'ribbon' connectivity (I–IV, II–III) (Figure 7.2).[88] The disulfide-bond connectivity combined with the length of the intercystine loops and the presence of a hydroxyproline residue located prior to the last cysteine make χ-conotoxins structurally unique. The two-loop MrIA toxin adopts a β-hairpin structure with close N- and C-terminal

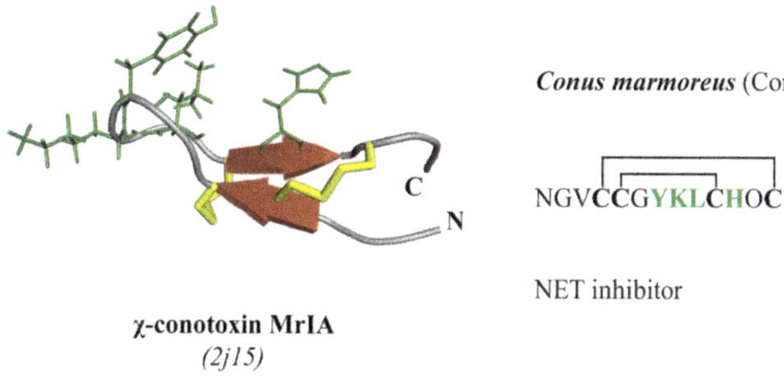

Figure 7.2 χ-Conotoxin MrIA. Left panel shows the three-dimensional structure of the cone snail venom peptide χ-conotoxin MrIA (PDB accession code 2JL5). β-strands and disulfide bonds are coloured red and yellow, respectively, and the side chains of pharmacophore residues are highlighted in green. The right panel shows the primary structure with disulfide connectivity shown as black lines and residues responsible for interaction with the norepinephrine transporter (NET) highlighted in bold green font. O, hydroxyproline.

ends stabilised by the hydrophobic interaction of Val^3 and hydroxyproline at position 12. The three-dimensional structure also shows a flexible γ-turn made by residues Gly^6-Tyr^7-Lys^8.[88]

MrIA is a reversible and selective inhibitor of the noradrenaline (norepinephrine) transporter (NET), where it acts in a non-competitive or allosteric manner.[89] The Na^+- and Cl^--dependent NET, one of the main mechanisms of elimination of noradrenaline from chemical synapses, is an important therapeutic target for various neurological disorders. Also, given the role of noradrenaline and multiple antidepressants targeting NET in antinociception, this transporter has significant potential as a target for pain management.[90–92] MrIA shows considerable selectivity for NET over other members of the monoamine neurotransmitter transporter family such as dopamine and serotonin transporters.[93] SAR experiments identified the region starting from the γ-turn up to the C-terminus as the active site of MrIA responsible for its interaction with NET, highlighting the crucial role of the Tyr^7-Lys^8-Leu^9 and His^{11} residues located in the second intercystine loop.[93,94] Interestingly, several analogues containing non-natural amino acid residues, while enhancing affinity for NET, also produced undesirable side effects or had reduced efficacy in a rat neuropathic pain model.[94]

7.2.3 Venom Peptides in Preclinical Trials

7.2.3.1 μ-conotoxin MrVIB

Voltage-gated sodium (Na_V) channels are expressed at high levels in nociceptive neurons.[95,96] Numerous peptide toxins isolated from the venom of spiders (theraphotoxins, thomitoxins, hexatoxins, ctenitoxin), scorpions (α and β toxins), or cone snails (μ- and δ-conotoxins) have been demonstrated to be efficient inhibitors of Na_V channels involved in pain pathways.[67,97,98] Their low specificity for Na_V channels has made most of these ligands unsuitable leads for drug design. However, recent studies have indicated that $Na_V1.7$ and $Na_V1.8$ subtype-selective inhibitors could have therapeutic potential.[99,100]

μ-Conotoxin MrVIB (hereafter called MrVIB) is a 31-residue venom peptide from the marine cone snail *Conus marmoreus* that has analgesic potential due to its ability to inhibit subtypes of Na_V channels. MrVIB was licenced to Cognetix under the name CGX-1002. It contains a large number of non-polar residues as well as six cysteine residues arranged in an ICK motif (1)C(6)C(9)CC(4)C(4)C(1) (Figure 7.3).[101,102] Recent advances in regioselective synthesis using selenocysteine residues has allowed a more facile production of synthetic MrVIB and determination of its three-dimensional structure.[103,104] The toxin displays a hydrophobic surface with two short β-strands comprising Ile^{24}-Cys^{25} and Cys^{30}-Val^{31}. MrVIB is a potent and selective Na_V channel inhibitor with a preference for the neuronal $Na_V1.8$ subtype and skeletal muscle $Na_V1.4$ subtype over other Na_V channel subtypes. In rat dorsal root ganglion (DRG) neurons, MrVIB blocks TTX-resistant $Na_V1.8$ ($IC_{50} \approx 98$ nM) >100-fold more potently than $Na_V1.9$ ($IC_{50} \approx 1.1$ μM), and about 10-fold more potently than TTX-sensitive $Na_V1.1$, $Na_V1.2$, $Na_V1.3$, $Na_V1.6$ and Nav1.7 channels ($IC_{50} \approx 1$ μM).[105] For Na_V subtypes expressed heterologously in *Xenopus* oocytes, MrVIB inhibits human $Na_V1.8$ channels ($IC_{50} \approx 102$ nM) with a 10-fold higher potency than for rat $Na_V1.2$ and $Na_V1.3$ and human $Na_V1.5$ and $Na_V1.7$ ($IC_{50} \approx 1$ μM).[105] Another study reported selective inhibition of $Na_V1.4$ with an IC_{50} of 222 nM.[105] When tested *in vivo*, MrVIB showed a significant reduction of allodynia and hyperalgesia in rat pain models after intrathecal injection.[105] Synthetic MrVIB also has potent and long-lasting antinociceptive activity in local anaesthetic and postincision allodynia assays.[106] Importantly, MrVIB did not show significant motor side effects and it has a greater therapeutic index than non-selective Na_V channel antagonists such as lidocaine (lignocaine).[105]

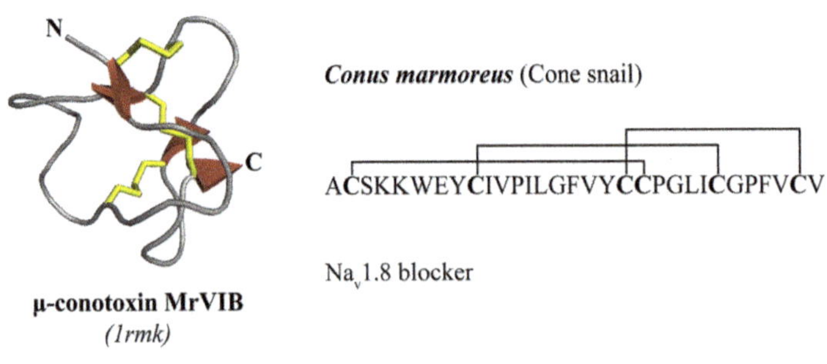

Figure 7.3 μ-Conotoxin MrVIB. The tertiary and primary structures of *Conus marmoreus* MrVIB are shown in the left and right panels, respectively. The three-dimensional structure of the toxin (PDB accession code 1RMK, with β-sheets and disulfide bonds coloured red and yellow, respectively) reveals an ICK fold.

7.2.3.2 α-conotoxin Vc1A

The α-conotoxin family represents the largest and most diverse group of conopeptides characterised to date. They are selective antagonists of nicotinic acetylcholine receptors (nAChRs) expressed in muscle and neuronal cells, and are considered to be one of the best sources of potential leads for nAChR-based therapeutics.[67,107]

α-Conotoxin Vc1A (hereafter called Vc1a) is a small endogenous 16-residue peptide initially identified at the nucleic acid level by a PCR-RACE screen of cDNA from the venom duct of the cone snail *Conus victoria*.[108] It contains two disulfide bonds (I–III, II–IV) that form two loops containing four and seven amino acid residues with the pattern (1)CC(4)C(7) (Figure 7.4).[109] Vc1a possesses 4-hydroxyproline and γ-carboxyglutamate PTMs at positions 6 and 14 respectively, as well as being C-terminally amidated.[110] The main secondary structural elements are a small α-helix spanning Pro^{6}–Asp^{11} and a type I β-turn at the N-terminus.[109]

Synthetic Vc1.1 without the PTMs at positions 6 and 14 originally received attention because, contrary to its endogenous isoform, it demonstrated pain relief and suppressed the vascular responses to sensory nerve C-fibre activation in rat,[108] as well as antagonistic effect in unmyelinated human axons.[111]

Initially it was thought that Vc1.1 was primarily selective for α3-containing nAchR subtypes (α3β2 $IC_{50} \approx 7.3$ μM, and α3β4 $IC_{50} \approx 4.2$ μM).[109] However, it has since been reported that Vc1.1 has 100-fold higher affinity for the α9α10 nAchR subtype, a potential therapeutic target for pain management.[112–114] In order to explain Vc1.1 potency and selectivity towards the α9α10 nAchR subtype, SAR investigations were performed.[115] An extensive panel of point mutants revealed that the proline residues have key roles in maintenance of the active conformation (Pro^{13}) and receptor binding (Pro^{6}). Also, the importance

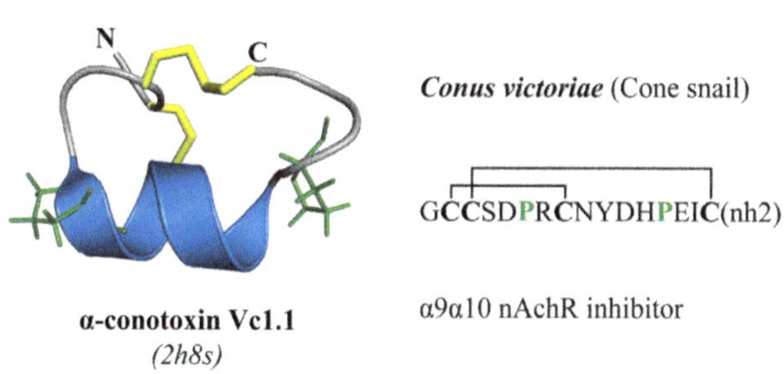

Figure 7.4 α-Conotoxin Vc1.1. Left panel shows the three-dimensional fold (PDB accession code 2H8S) of Vc1.1, a synthetic variant of wild-type conotoxin Vc1A. Disulfide bonds and α-helix are coloured yellow and blue, respectively. The side chains of pharmacophore residues are highlighted in green. Contrary to its native isoform, Vc1.1 inhibits nicotinic acetylcholine receptor (nAChR) with high potency. Vc1.1 contains a C-terminal amidation (nh2) but lacks the two other native PTMs (hydroxyproline and γ-carboxylic glutamic acid).

of residues 5–7 as well as 11–15 defined, as in most α-conotoxins, a conserved hydrophobic patch in the first loop and a more variable cluster in the second loop responsible for binding and subtype selectivity respectively.

Finally, experiments performed *in vivo* in two models of peripheral neuropathy of the rat sciatic nerve demonstrated that Vc1.1 was not only able to induce analgesia when injected intramuscularly near the site of injury but also accelerated the recovery of damaged neurons.[116] To overcome the generally short biological half-lives specific to α-conotoxins as well as their poor activity when administered orally, Clark *et al.* designed a cyclic Vc1.1 mimetic that shows improved *in vivo* stability, enhanced activity and specificity, as well as oral availability.[117]

7.2.3.3 Psalmotoxin 1

Acid-sensing ion channels (ASICs) are proton-gated sodium channels formed by the homomeric or heteromeric association of three subunits (ASIC1a, 1b, 2a, 2b, 3, or 4). ASICs are members of the epithelial sodium channel/degenerin family of ion channels but they are distinguished by their restriction to chordates, their predominantly neuronal distribution, and their activation by decreases in extracellular pH.[118] ASICs have been demonstrated to play crucial functions in nociceptive pathways and are thus considered potential therapeutic targets for the treatment of chronic inflammatory pain.[119–121]

The first toxin discovered and characterised as an inhibitor of ASICs is psalmotoxin 1 (π-theraphotoxin-Pc1a or PcTx1), a 40-residue peptide isolated from the venom of the South American tarantula *Psalmopoeus cambridgei.*[122] The toxin is undergoing preclinical trials for treatment of chronic pain under the name THA 901 by Theralpha (www.theralpha.com). Several three-dimensional solution structures have been determined for this toxin.[123,124] The three disulfide bonds form an ICK scaffold and the most detailed structure available reveals a predominant β-hairpin (from Leu^{21} to Lys^{35}) composed of two antiparallel β-strands (Leu^{21}–Trp^{24} and Val^{32}–Lys^{35}) (Figure 7.5).[124] Minor secondary structures are also present with a 3_{10} helix in the second loop (His^{14}–Asp^{16}), as well as a β-turn in the third loop (Cys^{18}–Leu^{21}). Wild-type, synthetic, and recombinant forms of PcTx1 rapidly and reversibly block ASIC1a homomers with an IC_{50} of ~0.5 nM.[122,124] In contrast, PcTx1 does not inhibit homomeric ASIC1b or ASIC2a channels, nor heteromeric ASIC1a/ASIC3 or ASIC1a/ASIC2a channels. In addition, the voltage-gated potassium (K_V) channels $K_V2.1$, $K_V2.2$, $K_V4.2$, and $K_V4.3$ are insensitive to PcTx1.[122] The selectivity of the toxin is supported by its ability to efficiently inhibit only the more slowly inactivating native proton-gated currents in rat DRGs ($IC_{50} = 0.7$ nM).[122] When tested *in vivo*, intrathecal PcTx1 was shown to be more efficient than morphine in rodent models of acute pain.[125]

Several different studies that used mutagenesis and *in silico* docking approaches were used to define the PcTx1 pharmacophore.[124,126,127] It has been suggested that PcTx1 may bind to the acidic pocket of ASIC1a *via* its β-hairpin structure in loop 4, and more precisely through the interaction of Trp^{24}, Arg^{26},

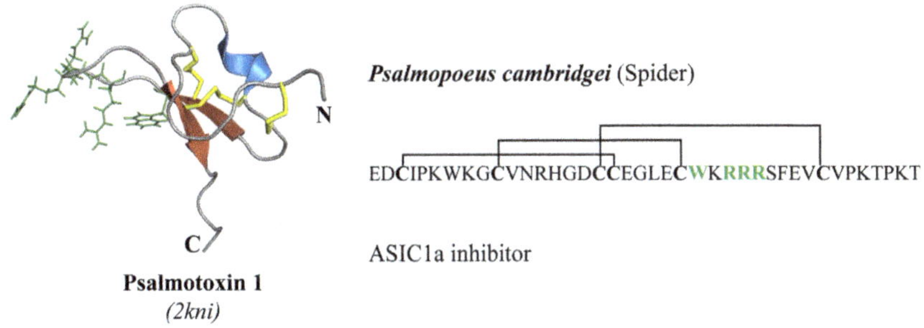

Figure 7.5 Psalmotoxin 1 (π-theraphotoxin-Pc1a). Psalmotoxin 1 was isolated from the venom of the Trinidad chevron tarantula *Psalmopoeus cambridgei*. Its three-dimensional structure (left panel; PDB accession code 2KNI) reveals a 3_{10}-helix (blue), β-strands (red) and three disulfide bonds (yellow). The side chains of the pharmacophore residues responsible for activity on ASIC1a are highlighted in green. The amino acid sequence of the toxin (right panel) shows the typical disulfide framework found in the ICK motif. Pharmacophore residues are highlighted in bold green font.

Arg^{27} and Arg^{28} residues (the co-substitution of Trp^{24} and Arg^{27} to Ala led to a 150-fold decrease of inhibition). Presumably, the arginine residues are able to mimic the persistent activation by protons that trigger the steady-state desensitisation of ASIC1a. Recently, Baconguis I. *et al.* described the X-ray crystal structures of the chicken ASIC1a complex with PcTx1 at high and low pH.[128] They confirm the importance of the arginine-rich hairpin that makes polar interactions in the acidic pocket of the channel, and particularly the involvement of Arg^{26} and Arg^{27} in the establishment of hydrogen bonds with residues located in the thumb (also anchored by the aromatic interaction of Trp^{24}) and palm domains of adjacent channel subunits respectively. Two different expression systems (*Drosophila melanogaster* S2 cells and *Escherichia coli*)[123,124] have been used to produce recombinant toxin, with the latter system producing considerably higher yields.[124]

7.2.3.4 Mambalgin-1 and -2

Diochot *et al.* recently described two new three-finger toxins isolated from the venom of the black mamba *Dendroaspis polylepis polylepis*.[129] The 57-residue mambalgin-1 and mambalgin-2 isopeptides cause potent, rapid, and reversible analgesic effects by inhibiting ASICs expressed in central and peripheral neurons. They contain eight cysteine residues that form four intramolecular disulfide bonds with I–III, II–IV, V–VI and VII–VIII connectivity. The two isoforms differ by a single residue at position 4. A homology model of mambalgin-1 suggests that the toxin forms a triple-stranded β-sheet that connects loops II and III, as well as a short double-stranded β-sheet forming loop I. These three loops emerging from the disulfide-rich core of mambalgin-1 enable its classification as a three-finger toxin.[130]

Mambalgin inhibits ASIC subtypes in both the central and peripheral pain pathways. These toxins block rat homomeric ASIC1a, heteromeric ASIC1a/ASIC2a (naloxone-insensitive analgesia) and heteromeric ASIC1a/ASIC2b subtypes, targets that are abundantly expressed in the CNS.[131–134] Mambalgins also inhibit human homomeric ASIC1b and heteromeric ASIC1a/ASIC1b channels specifically expressed by sensory neurons.[135] These peptides have been also proposed to act as gating modifier toxins by binding to the closed and/or inactivated state of the channels and thereby modifying their affinity for protons. Consequently they inhibit native ASIC currents by reducing the response of the channels to acidic pH. These peptides are analgesic in rodent models of acute and inflammatory pain, with potency similar to that of morphine, but unlike other three-finger toxins they do not cause behavioural neurotoxicity upon central injection in mice.

Theralpha is currently conducting preclinical trials with mambalgin-1 under the name THA 904.

7.2.3.5 APETx2

A selective inhibitor of ASIC3 known as APETx2 was isolated by activity-guided fractionation of venom from the sea anemone *Anthopleura elegantissima.*[136] Theralpha is currently conducting preclinical trials with APETx2 (under the name THA 902) for treatment of chronic inflammatory pain. This toxin has received particular interest as ASIC3 is considered to be a potential analgesic target being predominantly expressed in nociceptors.[137,138]

APETx2 is a highly cationic peptide consisting of 42 residues. The six cysteine residues are linked to form three disulfide bonds with I–V, II–IV, III–VI connectivity; the compact disulfide-rich core has the sequence pattern (3)C(1)C(13)C(9)C(6)CC(4) (Figure 7.6).[136] The three-dimensional structure of APETx2 determined using NMR spectroscopy reveals a central four-stranded β-sheet linked together by type I and type-II″ β-turns; the N- and C-termini, as well as a long loop including residues from Phe[15] to Thr[27], emerge from one end of this disulfide-rich core.[139] APETx2 selectively blocks homomeric rat ASIC3 channels expressed in *Xenopus* oocytes with an IC_{50} of 63 nM; it is 3-fold less active on homomeric human ASIC3 channels transfected in COS cells ($IC_{50} = 175$ nM).[136] APETx2 also inhibits heteromeric ASIC2b/3 channels ($IC_{50} = 117$ nM) and to a lesser extent heteromeric ASIC1a/3 channels ($IC_{50} = 2$ μM) and ASIC1b/3 channels ($IC_{50} = 0.9$ μM) when transfected in COS cells.[136] APETx2 has very little activity towards Na_V and K_V channels, with the highest activity reported against $Na_V1.8$ ($IC_{50} \approx 2.6$ μM).[140]

With its unusual target specificity and selectivity compared to other venoms peptides,[141] APETx2 is a valuable pharmacological tool for the study of ASICs and a promising template for the design of a pain therapeutics directed against ASIC3.

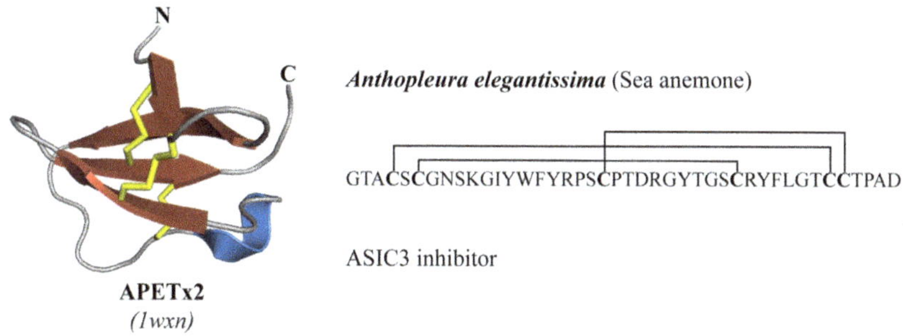

Figure 7.6 Toxin APETx2. Left panel shows the three-dimensional structure (PDB accession code 1WXN) of APETx2 secreted from nematocysts of the sea anemone *Anthopleura elegantissima.* Turns, β-strands and disulfide bonds are coloured blue, red and yellow, respectively. APETx2 displays a cysteine scaffold defined by a C-C-C-C-CC pattern with I–V, II–IV, III–VI disulfide connectivity (shown as black lines in the primary structure on the right panel). APETx2 is a highly selective inhibitor of acid-sensing ion channel 3 (ASIC3).

7.2.3.6 Hannalgesin

Hannalgesin, also called CM-11 or long neurotoxin OH-55, was isolated from the venom of the world's longest venomous snake, the king cobra *Ophiophagus hannah,*[142] as well as cloning *via* a venom-gland cDNA library.[143] Hannalgesin is an α-neurotoxin containing 72 residues and 5 disulfide bonds (the connectivity of the cysteines has been inferred by homology as I–III, II–VI, IV–V, VII–VIII, IX–X). To our knowledge, the three-dimensional structure of the toxin has not been published to date, but the cysteine pattern suggests that hannalgesin would adopt a three-finger toxin fold. Hannalgesin produces a dose-dependent, long-lasting analgesia with 2700-fold higher potency than morphine when assessed with the hot-plate test in rats.[144] However, the same study revealed that this neurotoxin could be toxic for rodents by causing neurological deficits and proconvulsant effects. The biochemical mechanism of action of hannalgesin is not yet fully elucidated. Nevertheless, it has been shown that the toxin significantly increases nitric oxide synthase activity, and the analgesic effect of hannalgesin is strongly reduced by inhibition of this enzyme.[145] Hannalgesin also causes relaxation of precontracted rat annococcygeus muscle.[146] Together, these observations suggest that hannalgesin causes analgesia by an unknown mechanism involving nitric oxide release that may stimulate guanylate cyclase and thus formation of GMPc. In addition, the analgesic effect of the neurotoxin was blocked by naloxone, suggesting its potential interaction with opioid receptors.[144] Recently, Kini *et al.* claimed that they identified the pharmacophore of hannalgesin,[130] using a 'proline bracket' method,[147] located in the C-terminal end of the peptide (unpublished data). They also reported the synthesis of a short peptide from this region of hannalgesin that showed selective analgesia *in vivo* without neurotoxicity (patented data).[148] The short active peptide is currently undergoing preclinical trials with Theralpha under the name THA 903.

7.2.4 Venom Peptides Targeting Nociceptive Receptors

7.2.4.1 $GABA_B$ Receptor Agonists

7.2.4.1.1 α-Conotoxin RgIA. $Ca_V2.1$ channels in peripheral neurons are a proven target for the treatment of chronic neuropathic pain.[149–151] Recent studies have revealed that some α-conotoxins such as Vc1.1 (discussed previously), RgIA, PeIa and AuIB might mediate their potent analgesic effect by acting as agonists of presynaptic $GABA_B$ receptors, thereby leading to an indirect inhibition of $Ca_V2.1$ channels and a consequent decrease in neurotransmitter release.[43,152–155] Although the precise mechanism of their analgesic effects has not been fully determined, it has been suggested that these toxins interact with the $GABA_B$ receptor *via* an allosteric binding site.[114]

RgIA contains 13 residues and, like other α-conotoxins, it contains four cysteine residues with a I–III, II–IV connectivity (Figure 7.7). The three-dimensional structure is comprised of a type I β-turn followed by a 3_{10} helix and a type VIII β-turn.[156]

7.2.4.2 Natriuretic Peptide Receptor Ligand

7.2.4.2.1 Taipan Natriuretic Peptide TNP-c. Natriuretic peptides (NPs) are found in many animals, where they exert hypotensive and vasodilator activities.[157,158] Recent studies have suggested that natriuretic peptides may also be involved in the regulation of pain sensitivity, although the biochemical mechanisms have not yet been fully elucidated.[159–161] Natriuretic peptides contain a single disulfide bond leading to a ring-shaped structure extended by a C-terminal tail. Three natriuretic peptides (TNP-a, TNP-b and TNP-c) have been isolated from the venom of the taipan snake *Oxyuranus microlepidotus*.[162] They are highly similar to other NP family members within the loop region but contain an unusual C-terminal tail. TNP-c, the only active isoform long of 39 amino acids, has a 17-membered ring structure prolonged by a 14-amino-acid C-terminal tail. TNP-c is able to trigger full relaxation of precontracted rat aorta through type-C natriuretic peptide receptor (NPR-C) and type-A guanylate cyclase (GC-A) mediated mechanisms.[162]

7.2.4.3 Sodium Channel Peptides

Several peptide toxins from the venom of theraphosid spiders (commonly known as tarantulas) have useful properties that make them interesting candidates for the design of new analgesics.[62] First, like most spider toxins these peptides contain an ICK motif (like Prialt) that confers thermodynamic stability and resistance to proteases. Second, these toxins act as allosteric modulators or ‘gating modifiers’;[98] thus, they are more likely to exhibit subtype

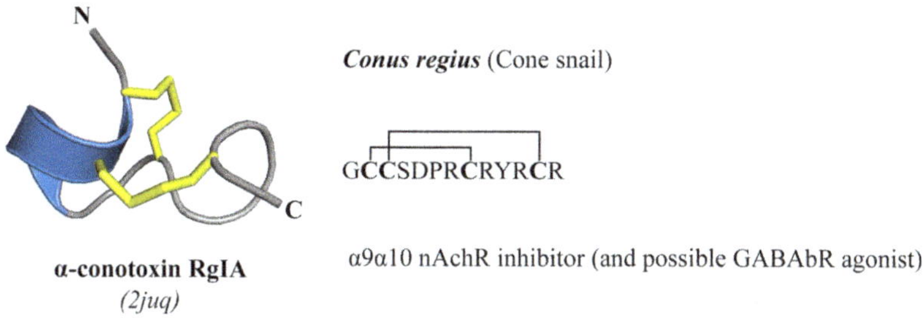

Figure 7.7 α-Conotoxin RgIA. Left panel shows the tertiary structure of α-conotoxin RgIA isolated from *Conus regius*. Helices and disulfide bonds are coloured blue and yellow, respectively. RgIA is a potent inhibitor of the α9α10 nicotinic acetylcholine receptor (nAChR) with agonistic activity on the $GABA_B$ receptor. The disulfide connectivities are shown as black lines in the primary structure on the right.

selectivity than small molecules that bind to the highly conserved pore of the channel.

7.2.4.3.1 ProTx-II. To date, the most potent blocker of human $Na_V1.7$ ($IC_{50} \approx 0.3$ nM) is ProTx-II isolated from the venom of the green velvet tarantula *Thrixopelma pruriens* (see ArachnoServer entry at www.arachnoserver.org/toxincard.html?id = 414).[62] This 30-residue peptide contains 6 cysteine residues arranged in an ICK motif. The disulfide bond framework provides a rigid peptide backbone that forms four loops.[163] ProTx-II is classified as a gating modifier because of its ability to shift the voltage dependence of activation to more positive potentials. It has been proposed that it inhibits activation and inactivation of human $Na_V1.7$, which is highly expressed in nociceptive neurons[95,100] by interacting with the positively charged S4 segments in the voltage sensors of channel domains II and IV that sense membrane depolarisation and induce channel gating (and thus triggers action potentials).[164] ProTx-II also inhibits other sodium channel subtypes ($Na_V1.2$, $Na_V1.3$, $Na_V1.5$, $Na_V1.6$ and $Na_V1.8$) but with an approximate 100-fold lower potency than $Na_V1.7$. Although the binding sites of the toxin on Na_V channels still remain to be precisely determined, it has been suggested that ProTx-II acts by trapping the voltage sensor of Na_V channel domain II in the resting state, impeding outward gating movement of the IIS4 transmembrane segment of the channel and reduces maximum activation of sodium conductance.[165] ProTx-II itself shows no potential as a therapeutic as it is lethal to rats when injected intravenously at 1.0 mg kg^{-1} (although doses of 0.01 and 0.1 mg kg^{-1} are well tolerated) or by intrathecal administration at 0.1 mg/kg. Moreover, the toxin was not efficacious in rodent models of acute and inflammatory pain when administered intravenously at 0.01 or 0.1 mg kg^{-1} or intrathecally at 0.001 or 0.01 mg kg^{-1}. The latter intrathecal dose transiently reduced muscle tone and impaired motor function even though systemic levels at 4 hours after toxin administration were below the level of detection (3 nM).

7.2.4.3.2 Huwentoxin-IV. Huwentoxin-IV (HWTX-IV) is also a gating modifier of Na_V channels (see ArachnoServer entry at www.arachnoserver.org/toxincard.html?id=332). This 35-residue peptide isolated from the venom of the Chinese bird spider *Haplopelma schmidti* has a compact three-dimensional architecture stabilised by three disulfide bonds that form an ICK motif (Figure 7.8).[166] HWTX-IV potently blocks human $Na_V1.7$ ($IC_{50} \sim 23$ nM) by interacting with the voltage sensor in domain II (DII).[167] It traps the DII voltage sensor in the closed state and inhibits channel activation by interacting with five critical residues located in the S1-S2 linker (Glu[753]), the extracellular part of S3 (Glu[811], Leu[814], Asp[816]), and the S3-S4 linker (Glu[818]) of DII.[166–168] It has also been proposed that the positively charged Arg[26] residue in the fourth intercystine loop of HWTX-IV may play an important role in the channel binding. In contrast with ProTx-II, HWTX-IV does not partition into artificial phospholipid bilayers.[169]

7.2.4.3.3 μ-Conotoxin PIIIA. In contrast to the gating modifiers from spider venoms, the μ-conotoxins block the pore of Na_V channels (see ArachnoServer entry at *www.arachnoserver.org/toxincard.html?id=332*). The sequence of μ-conotoxin PIIIA was deduced from mRNA isolated from the cells lining the venom duct of *Conus purpurascens*.[170] This 22-residue peptide is highly basic and contains three post-translationally modified residues (one pyroglutamate and two hydroxyprolines at positions 1, 8, and 18 respectively, as well as an amidated C-terminus). The CC-C-C-CC cysteine framework, with I–IV, II–V, III–VI connectivity, is the same as found in many μ-conotoxins (Figure 7.9). The disulfide bonds form three distinct

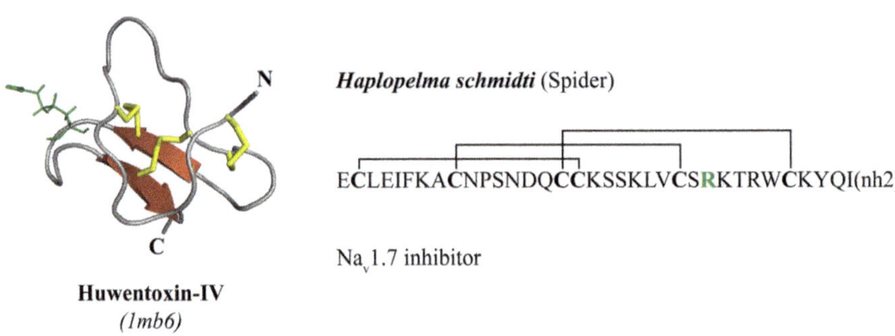

Figure 7.8 Huwentoxin-IV (μ-theraphotoxin-Hh2a). Left panel shows the three-dimensional structure of the spider huwentoxin-IV (PDB accession code 1MB6). Disulfide bonds and β-strands are coloured yellow and red, respectively. The side chain of the functionally important arginine 26 residue is highlighted in green. The right panel shows the amino acid sequence of the toxin highlighting the disulfide connectivity (black lines) and functionally important arginine residue (bold green font). 'nh2' indicates C-terminal amidation.

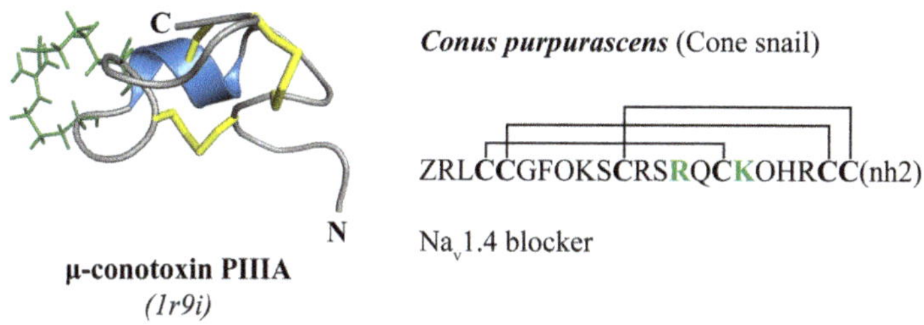

Figure 7.9 μ-Conotoxin PIIIA. Left panel shows the three-dimensional structure of conotoxin PIIIA (PDB accession code 1R9I), a selective blocker of Na$_V$1.4 isolated from the cone snail *Conus purpurascens*. Disulfide bonds and helices are coloured yellow and blue, respectively. The side chains of pharmacophore residues are highlighted in green. The right panel shows the amino acid sequence of the toxin, highlighting the disulfide connectivities (black lines), pharmacophore residues (bold green font), and PTMs (nh2, C-terminal amidation; O, hydroxyproline; Z, pyroglutamic acid).

loops and constrict the overall globular shape of PIIIA, where consecutive turns are found over the N-terminal end with a distorted helix from Ser^{13} up to the C-terminus.[171] PIIIA specifically inhibits tetrodotoxin-sensitive Na$_V$ channels. Although the toxin strongly inhibits the skeletal muscle Na$_V$1.4 channel (IC$_{50}$ ≈ 41 nM), it also blocks the neuronal Na$_V$1.2 subtype with lower potency (IC$_{50}$ ≈ 690 nM).[172] PIIIA-mediated inhibition occurs *via* binding to the so-called sodium channel neurotoxin receptor site 1 which blocks flow of sodium ions through the pore.[173] Residues Arg^{14} and Lys^{17} (and to a lesser extent Arg^{12} and Arg^{20}) face deep into the vestibule of the channel pore.[174] With Arg^{14} positioned in the permeation pathway, PIIIA causes steric and/or electrostatic occlusion of the channel pore.[175] Replacement of His^{19} by Gln increases the selectivity of PIIIA for Na$_V$1.4 over Na$_V$1.2,[175] suggesting that His^{19}, which is highly conserved among μ-conotoxins, increases selectivity for the neuronal subtype. Interestingly, a recent study demonstrated that PIIIA isoforms with different cysteine connectivity (*i.e.*, without changes in the primary structure) can be more potent blockers of Na$_V$1.4.[176]

7.3 Future Prospects and Outlook

As a consequence of their ability to efficiently and selectively target specific subtypes of receptors, transporters or ion channels in mammals, venom peptides have been recognised as particularly useful probes for *in vitro* and *in vivo* pharmacological studies, as well as an unrivalled source of potential candidates and molecular templates for therapeutic applications. As an example, the 500 species within the genus *Conus* have been estimated to elaborate more than 50 000 conopeptides, of which only ~0.1% have been pharmacologically characterised to date. Scorpions comprise about 1,750 extant species that are thought to produce ~100,000 unique peptides.[177]

Spiders (~44,000 species) might produce more than 10 million active peptides.[178] As a consequence of this substantial molecular diversity, and also because a vast majority of this subclass of natural products has been designed to act on the nervous systems of prey or predators, venom peptides have attracted great interest for the development of therapeutic agents dedicated to the modulation of nociception.

Increasing efforts to identify and optimise toxin pharmacophores, coupled with the development of peptidomimetics, will provide novel molecules with improved therapeutic indices that might provide new drugs for the treatment of pain.

Acknowledgement

V.L acknowledges the provision of an Institute for Molecular Bioscience (IMB) Postgraduate Award and support from the Australian National Health and Medical Research Council (NHMRC Program Grant 569927). GFK acknowledges support from the NHMRC (Project Grants APP1012338 and APP1034958) and the Australian Research Council (Discovery Grant DP110103129).

References

1. B. G. Fry, K. Roelants, D. E. Champagne, H. Scheib, J. D. Tyndall, G. F. King, T. J. Nevalainen, J. A. Norman, R. J. Lewis, R. S. Norton, C. Renjifo and R. C. de la Vega, *Ann. Rev. Genom. Hum. Genet.*, 2009, **10**, 483–511.
2. Y. Shimizu, *Ann. Rev. Microbiol.*, 1996, **50**, 431–465.
3. A. D. Cembella, *Phycologia*, 2003, **42**, 420–447.
4. G. P. Rossini and P. Hess, *EXS*, 2010, **100**, 65–122.
5. N. Vilarino, *FEBS J.*, 2008, **275**, 6075–6081.
6. O. M. Pulido, *Marine Drugs*, 2008, **6**, 180–219.
7. L. G. Costa, G. Giordano and E. M. Faustman, *Neurotoxicology*, 2010, **31**, 409–423.
8. H. J. Dominguez, B. Paz, A. H. Daranas, M. Norte, J. M. Franco and J. J. Fernandez, *Toxicon*, 2010, **56**, 191–217.
9. T. Yasumoto, *Proc. Jpn. Acad., Ser. B*, 2005, **81**, 43–51.
10. K. C. Nicolaou, M. O. Frederick and R. J. Aversa, *Angew. Chem. Int. Ed.*, 2008, **47**, 7182–7225.
11. A. Caillaud, P. de la Iglesia, H. T. Darius, S. Pauillac, K. Aligizaki, S. Fraga, M. Chinain and J. Diogene, *Marine Drugs*, 2010, **8**, 1838–1907.
12. M. Wiese, P. M. D'Agostino, T. K. Mihali, M. C. Moffitt and B. A. Neilan, *Marine Drugs*, 2010, **8**, 2185–2211.
13. R. Chau, J. A. Kalaitzis and B. A. Neilan, *Aquat. Toxicol.*, 2011, **104**, 61–72.

14. M. A. Jordan and L. Wilson, *Nat. Rev. Cancer*, 2004, **4**, 253–265.
15. S. M. Johnson, B. E. Saint John and A. P. Dine, *Surg. Infect. (Larchmt)*, 2008, **9**, 205–213.
16. H. Hameed, M. Hameed and P. J. Christo, *Curr. Pain. Headache Rep.*, 2010, **14**, 96–104.
17. S. S. Vansal and D. R. Feller, *Biochem. Pharmacol.*, 1999, **58**, 807–810.
18. S. C. Armstrong and K. L. Cozza, *Psychosomatics*, 2003, **44**, 515–520.
19. M. E. Wall and M. C. Wani, *Cancer Res.*, 1995, **55**, 753–760.
20. J. Y. Wick, *Consult. Pharm.*, 2012, **27**, 322–329.
21. P. Cos, L. Maes, A. Vlietinck and L. Pieters, *Planta Med.*, 2008, **74**, 1323–1337.
22. L. W. Qi, E. H. Liu, C. Chu, Y. B. Peng, H. X. Cai and P. Li, *Curr. Top. Med. Chem.*, 2010, **10**, 434–457.
23. M. I. Damaj, M. Fei-Yin, M. Dukat, W. Glassco, R. A. Glennon and B. R. Martin, *J. Pharmacol. Exp. Ther.*, 1998, **284**, 1058–1065.
24. J. R. Traynor, *Br. J. Anaesth.*, 1998, **81**, 69–76.
25. I. M. Khan, H. Buerkle, P. Taylor and T. L. Yaksh, *Neuropharmacology*, 1998, **37**, 1515–1525.
26. F. Bosmans, C. Maertens, F. Verdonck and J. Tytgat, *FEBS Lett.*, 2004, **577**, 245–248.
27. K. Angerer, *Innovation*, 2011, **24**, 353–369.
28. A. S. Tucker, *Front. Oral Biol.*, 2010, **14**, 21–31.
29. S. Dutertre, A. H. Jin, Q. Kaas, A. Jones, P. F. Alewood and R. J. Lewis, *Mol. Cell Proteomics*, 2013, **12**, 312–329.
30. R. J. Lewis and M. L. Garcia, *Nat. Rev. Drug Discov.*, 2003, **2**, 790–802.
31. A. Gomes, P. Bhattacharjee, R. Mishra, A. K. Biswas, S. C. Dasgupta and B. Giri, *Indian J. Exp. Biol.*, 2010, **48**, 93–103.
32. G. F. King, *Expert Opin. Biol. Ther.*, 2011, **11**, 1469–1484.
33. M. A. Hayashi and A. C. Camargo, *Toxicon*, 2005, **45**, 1163–1170.
34. N. Marsh and V. Williams, *Toxicon*, 2005, **45**, 1171–1181.
35. A. C. Camargo, D. Ianzer, J. R. Guerreiro and S. M. Serrano, *Toxicon*, 2012, **59**, 516–523.
36. G. M. Nicholson, M. J. Little and L. C. Birinyi-Strachan, *Toxicon*, 2004, **43**, 587–599.
37. S. Mouhat, N. Andreotti, B. Jouirou and J. M. Sabatier, *Curr. Pharm. Des.*, 2008, **14**, 2503–2518.
38. R. S. Norton and S. I. McDonough, *Curr. Pharm. Des.*, 2008, **14**, 2480–2491.
39. R. J. Lewis, *IUBMB Life*, 2004, **56**, 89–93.
40. O. B. R. de, A. B. Pizzo, A. C. Fontana, O. G. C. R. de, J. Coutinho-Netto and W. F. Dos Santos, *Eur. J. Pharmacol.*, 2004, **493**, 1–17.
41. R. J. Lewis, *Toxicon*, 2012, **59**, 524–528.
42. K. Nareoja and J. Nasman, *Acta Physiol. (Oxf.)*, 2012, **204**, 186–201.

43. D. J. Adams, B. Callaghan and G. Berecki, *Br. J. Pharmacol.*, 2012, **166**, 486–500.
44. A. W. Bernheimer and B. Rudy, *Biochim. Biophys. Acta*, 1986, **864**, 123–141.
45. M. Arcisio-Miranda, M. P. dos Santos Cabrera, K. Konno, M. Rangel and J. Procopio, *Toxicon*, 2008, **51**, 736–745.
46. W. C. Wimley, *Adv. Exp. Med. Biol.*, 2010, **677**, 14–23.
47. A. Luch, *EXS*, 2010, **100**, 293–315.
48. J. Tibballs, A. A. Yanagihara, H. C. Turner and K. Winkel, *Inflamm. Allergy Drug Targets*, 2011, **10**, 438–446.
49. M. E. Byrnes, *Methods Mol. Biol. (Clifton, N.J.)*, 1994, **36**, 37–52.
50. J. J. Calvete, P. Juarez and L. Sanz, *J. Mass Spectrom.*, 2007, **42**, 1405–1414.
51. P. Favreau, L. Menin, S. Michalet, F. Perret, O. Cheneval, M. Stöcklin, P. Bulet and R. Stöcklin, *Toxicon*, 2006, **47**, 676–687.
52. J. R. Prashanth, R. J. Lewis and S. Dutertre, *Toxicon*, 2012, **60**, 470–477.
53. I. Vetter, J. L. Davis, L. D. Rash, R. Anangi, M. Mobli, P. F. Alewood, R. J. Lewis and G. F. King, *Amino Acids*, 2011, **40**, 15–28.
54. P. E. Dawson, T. W. Muir, I. Clark-Lewis and S. B. Kent, *Science*, 1994, **266**, 776–779.
55. S. Khafagy el and M. Morishita, *Advanced Drug Delivery Reviews*, 2012, **64**, 531–539.
56. N. Svensen, J. G. Walton and M. Bradley, *Trends in Pharmacological Sciences*, 2012, **33**, 186–192.
57. G. P. Miljanich, *Curr. Med. Chem.*, 2004, **11**, 3029–3040.
58. M. McIntosh, L. J. Cruz, M. W. Hunkapiller, W. R. Gray and B. M. Olivera, *Arch. Biochem. Biophys.*, 1982, **218**, 329–334.
59. B. M. Olivera, W. R. Gray, R. Zeikus, J. M. McIntosh, J. Varga, J. Rivier, V. de Santos and L. J. Cruz, *Science*, 1985, **230**, 1338–1343.
60. B. M. Olivera, L. J. Cruz, V. de Santos, G. W. LeCheminant, D. Griffin, R. Zeikus, J. M. McIntosh, R. Galyean, J. Varga and W. R. Gray, *et al.*, *Biochemistry*, 1987, **26**, 2086–2090.
61. D. Chung, S. Gaur, J. R. Bell, J. Ramachandran and L. Nadasdi, *Int. J. Peptide Protein Res.*, 1995, **46**, 320–325.
62. N. J. Saez, S. Senff, J. E. Jensen, S. Y. Er, V. Herzig, L. D. Rash and G. F. King, *Toxins*, 2010, **2**, 2851–2871.
63. N. L. Daly and D. J. Craik, *Curr. Opin. Chem. Biol.*, 2011, **15**, 362–368.
64. V. J. Basus, L. Nadasdi, J. Ramachandran and G. P. Miljanich, *FEBS Lett.*, 1995, **370**, 163–169.
65. K. J. Nielsen, L. Thomas, R. J. Lewis, P. F. Alewood and D. J. Craik, *J. Mol. Biol.*, 1996, **263**, 297–310.
66. B. M. Olivera, G. P. Miljanich, J. Ramachandran and M. E. Adams, *Annu. Rev. Biochem.*, 1994, **63**, 823–867.
67. R. J. Lewis, S. Dutertre, I. Vetter and M. J. Christie, *Pharmacol. Rev.*, 2012, **64**, 259–298.

68. P. T. Ellinor, J. F. Zhang, W. A. Horne and R. W. Tsien, *Nature*, 1994, **372**, 272–275.
69. S. Kaneko, C. B. Cooper, N. Nishioka, H. Yamasaki, A. Suzuki, S. E. Jarvis, A. Akaike, M. Satoh and G. W. Zamponi, *J. Neurosci.*, 2002, **22**, 82–92.
70. S. I. McDonough, L. M. Boland, I. M. Mintz and B. P. Bean, *J. Gen. Physiol.*, 2002, **119**, 313–328.
71. K. Sato, N. G. Park, T. Kohno, T. Maeda, J. I. Kim, R. Kato and M. Takahashi, *Biochem. Biophys. Res. Comm.*, 1993, **194**, 1292–1296.
72. J. I. Kim, M. Takahashi, A. Ogura, T. Kohno, Y. Kudo and K. Sato, *J. Biol. Chem.*, 1994, **269**, 23876–23878.
73. M. J. Lew, J. P. Flinn, P. K. Pallaghy, R. Murphy, S. L. Whorlow, C. E. Wright, R. S. Norton and J. A. Angus, *J. Biol. Chem.*, 1997, **272**, 12014–12023.
74. J. P. Flinn, P. K. Pallaghy, M. J. Lew, R. Murphy, J. A. Angus and R. S. Norton, *Eur. J. Biochem.*, 1999, **262**, 447–455.
75. R. J. Lewis, K. J. Nielsen, D. J. Craik, M. L. Loughnan, D. A. Adams, I. A. Sharpe, T. Luchian, D. J. Adams, T. Bond, L. Thomas, A. Jones, J. L. Matheson, R. Drinkwater, P. R. Andrews and P. F. Alewood, *J. Biol. Chem.*, 2000, **275**, 35335–35344.
76. K. J. Nielsen, D. Adams, L. Thomas, T. Bond, P. F. Alewood, D. J. Craik and R. J. Lewis, *J. Mol. Biol.*, 1999, **289**, 1405–1421.
77. R. D. Penn and J. A. Paice, *Pain*, 2000, **85**, 291–296.
78. K. J. Nielsen, T. Schroeder and R. Lewis, *J. Mol. Recog.*, 2000, **13**, 55–70.
79. D. J. Adams, A. B. Smith, C. I. Schroeder, T. Yasuda and R. J. Lewis, *J. Biol. Chem.*, 2003, **278**, 4057–4062.
80. C. E. Wright, A. D. Robertson, S. L. Whorlow and J. A. Angus, *Br. J. Pharmacol.*, 2000, **131**, 1325–1336.
81. M. T. Smith, P. J. Cabot, F. B. Ross, A. D. Robertson and R. J. Lewis, *Pain*, 2002, **96**, 119–127.
82. D. A. Scott, C. E. Wright and J. A. Angus, *Eur. J. Pharmacol.*, 2002, **451**, 279–286.
83. A. G. Craig, T. Norberg, D. Griffin, C. Hoeger, M. Akhtar, K. Schmidt, W. Low, J. Dykert, E. Richelson, V. Navarro, J. Mazella, M. Watkins, D. Hillyard, J. Imperial, L. J. Cruz and B. M. Olivera, *J. Biol. Chem.*, 1999, **274**, 13752–13759.
84. A. G. Craig, M. Park, W. H. Fischer, J. Kang, P. Compain and F. Piller, *Toxicon*, 2001, **39**, 809–815.
85. U. Westerlind and T. Norberg, *Carbohydrate Res.*, 2006, **341**, 9–18.
86. J. W. Allen, K. Hofer, D. McCumber, J. D. Wagstaff, R. T. Layer, R. T. McCabe and T. L. Yaksh, *Anesth. Analg.*, 2007, **104**, 1505–1513, table of contents.
87. J. M. McIntosh, G. O. Corpuz, R. T. Layer, J. E. Garrett, J. D. Wagstaff, G. Bulaj, A. Vyazovkina, D. Yoshikami, L. J. Cruz and B. M. Olivera, *J. Biol. Chem.*, 2000, **275**, 32391–32397.

88. K. P. Nilsson, E. S. Lovelace, C. E. Caesar, N. Tynngard, P. F. Alewood, H. M. Johansson, I. A. Sharpe, R. J. Lewis, N. L. Daly and D. J. Craik, *Biopolymers*, 2005, **80**, 815–823.
89. I. A. Sharpe, J. Gehrmann, M. L. Loughnan, L. Thomas, D. A. Adams, A. Atkins, E. Palant, D. J. Craik, D. J. Adams, P. F. Alewood and R. J. Lewis, *Nat. Neurosci.*, 2001, **4**, 902–907.
90. S. Furst, *Brain Res. Bull.*, 1999, **48**, 129–141.
91. R. Freeman, *CNS Spectrums*, 2005, **10**, 698–706.
92. A. W. Goddard, S. G. Ball, J. Martinez, M. J. Robinson, C. R. Yang, J. M. Russell and A. Shekhar, *Depression Anxiety*, 2010, **27**, 339–350.
93. I. A. Sharpe, E. Palant, C. I. Schroeder, D. M. Kaye, D. J. Adams, P. F. Alewood and R. J. Lewis, *J. Biol. Chem.*, 2003, **278**, 40317–40323.
94. A. Brust, E. Palant, D. E. Croker, B. Colless, R. Drinkwater, B. Patterson, C. I. Schroeder, D. Wilson, C. K. Nielsen, M. T. Smith, D. Alewood, P. F. Alewood and R. J. Lewis, *J Med. Chem.*, 2009, **52**, 6991–7002.
95. M. Liu and J. N. Wood, *Pain Med. (Malden, Mass.)*, 2011, **12 Suppl 3**, S93–99.
96. S. R. Levinson, S. Luo and M. A. Henry, *Muscle Nerve*, 2012, **46**, 155–165.
97. W. A. Catterall, S. Cestele, V. Yarov-Yarovoy, F. H. Yu, K. Konoki and T. Scheuer, *Toxicon*, 2007, **49**, 124–141.
98. J. K. Klint, S. Senff, D. B. Rupasinghe, S. Y. Er, V. Herzig, G. M. Nicholson and G. F. King, *Toxicon*, 2012, **60**, 478–491.
99. S. England and D. Rawson, *Future Med. Chem.*, 2010, **2**, 775–790.
100. J. J. Clare, *Expert Opin. Invest. Drugs*, 2010, **19**, 45–62.
101. M. Fainzilber, R. van der Schors, J. C. Lodder, K. W. Li, W. P. Geraerts and K. S. Kits, *Biochemistry*, 1995, **34**, 5364–5371.
102. J. M. McIntosh, A. Hasson, M. E. Spira, W. R. Gray, W. Li, M. Marsh, D. R. Hillyard and B. M. Olivera, *J. Biol. Chem.*, 1995, **270**, 16796–16802.
103. N. L. Daly, J. A. Ekberg, L. Thomas, D. J. Adams, R. J. Lewis and D. J. Craik, *J. Biol. Chem.*, 2004, **279**, 25774–25782.
104. A. D. de Araujo, B. Callaghan, S. T. Nevin, N. L. Daly, D. J. Craik, M. Moretta, G. Hopping, M. J. Christie, D. J. Adams and P. F. Alewood, *Angew. Chem. Int. Ed.*, 2011, **50**, 6527–6529.
105. J. Ekberg, A. Jayamanne, C. W. Vaughan, S. Aslan, L. Thomas, J. Mould, R. Drinkwater, M. D. Baker, B. Abrahamsen, J. N. Wood, D. J. Adams, M. J. Christie and R. J. Lewis, *Proc. Natl. Acad. Sci. U. S. A.*, 2006, **103**, 17030–17035.
106. G. Bulaj, M. M. Zhang, B. R. Green, B. Fiedler, R. T. Layer, S. Wei, J. S. Nielsen, S. J. Low, B. D. Klein, J. D. Wagstaff, L. Chicoine, T. P. Harty, H. Terlau, D. Yoshikami and B. M. Olivera, *Biochemistry*, 2006, **45**, 7404–7414.
107. J. M. McIntosh, A. D. Santos and B. M. Olivera, *Annu. Rev. Biochem.*, 1999, **68**, 59–88.

108. D. W. Sandall, N. Satkunanathan, D. A. Keays, M. A. Polidano, X. Liping, V. Pham, J. G. Down, Z. Khalil, B. G. Livett and K. R. Gayler, *Biochemistry*, 2003, **42**, 6904–6911.
109. R. J. Clark, H. Fischer, S. T. Nevin, D. J. Adams and D. J. Craik, *J. Biol. Chem.*, 2006, **281**, 23254–23263.
110. J. A. Jakubowski, D. A. Keays, W. P. Kelley, D. W. Sandall, J. P. Bingham, B. G. Livett, K. R. Gayler and J. V. Sweedler, *J. Mass Spectrom.*, 2004, **39**, 548–557.
111. P. M. Lang, R. Burgstahler, R. V. Haberberger, W. Sippel and P. Grafe, *Neuroreport*, 2005, **16**, 479–483.
112. M. Vincler, S. Wittenauer, R. Parker, M. Ellison, B. M. Olivera and J. M. McIntosh, *Proc. Natl. Acad. Sci. U. S. A.*, 2006, **103**, 17880–17884.
113. S. T. Nevin, R. J. Clark, H. Klimis, M. J. Christie, D. J. Craik and D. J. Adams, *Mol. Pharmacol.*, 2007, **72**, 1406–1410.
114. J. M. McIntosh, N. Absalom, M. Chebib, A. B. Elgoyhen and M. Vincler, *Biochem. Pharmacol.*, 2009, **78**, 693–702.
115. R. Halai, R. J. Clark, S. T. Nevin, J. E. Jensen, D. J. Adams and D. J. Craik, *J. Biol. Chem.*, 2009, **284**, 20275–20284.
116. N. Satkunanathan, B. Livett, K. Gayler, D. Sandall, J. Down and Z. Khalil, *Brain Res.*, 2005, **1059**, 149–158.
117. R. J. Clark, J. Jensen, S. T. Nevin, B. P. Callaghan, D. J. Adams and D. J. Craik, *Angew. Chem. Int. Ed.*, 2010, **49**, 6545–6548.
118. G. R. Dube, A. Elagoz and H. Mangat, *Curr. Pharmaceut. Des.*, 2009, **15**, 1750–1766.
119. N. G. Jones, R. Slater, H. Cadiou, P. McNaughton and S. B. McMahon, *J. Neurosci.*, 2004, **24**, 10974–10979.
120. K. A. Sluka, O. C. Winter and J. A. Wemmie, *Curr. Opin. Drug Discov. Dev.*, 2009, **12**, 693–704.
121. E. Deval, X. Gasull, J. Noel, M. Salinas, A. Baron, S. Diochot and E. Lingueglia, *Pharmacol. Ther.*, 2010, **128**, 549–558.
122. P. Escoubas, J. R. De Weille, A. Lecoq, S. Diochot, R. Waldmann, G. Champigny, D. Moinier, A. Menez and M. Lazdunski, *J. Biol. Chem.*, 2000, **275**, 25116–25121.
123. P. Escoubas, C. Bernard, G. Lambeau, M. Lazdunski and H. Darbon, *Protein Sci.*, 2003, **12**, 1332–1343.
124. N. J. Saez, M. Mobli, M. Bieri, I. R. Chassagnon, A. K. Malde, R. Gamsjaeger, A. E. Mark, P. R. Gooley, L. D. Rash and G. F. King, *Mol. Pharmacol.*, 2011, **80**, 796–808.
125. M. Mazzuca, C. Heurteaux, A. Alloui, S. Diochot, A. Baron, N. Voilley, N. Blondeau, P. Escoubas, A. Gelot, A. Cupo, A. Zimmer, A. M. Zimmer, A. Eschalier and M. Lazdunski, *Nat. Neurosci.*, 2007, **10**, 943–945.
126. Y. J. Qadri, B. K. Berdiev, Y. Song, H. L. Lippton, C. M. Fuller and D. J. Benos, *J. Biol. Chem.*, 2009, **284**, 17625–17633.
127. F. Pietra, *J. Chem. Inf. Model.*, 2009, **49**, 972–977.

128. I. Baconguis and E. Gouaux, *Nature*, 2012, **489**, 400–405.
129. S. Diochot, A. Baron, M. Salinas, D. Douguet, S. Scarzello, A. S. Dabert-Gay, D. Debayle, V. Friend, A. Alloui, M. Lazdunski and E. Lingueglia, *Nature*, 2012, **490**, 552–555.
130. R. M. Kini and R. Doley, *Toxicon*, 2010, **56**, 855–867.
131. E. Lingueglia, J. R. de Weille, F. Bassilana, C. Heurteaux, H. Sakai, R. Waldmann and M. Lazdunski, *J. Biol. Chem.*, 1997, **272**, 29778–29783.
132. R. Waldmann, G. Champigny, F. Bassilana, C. Heurteaux and M. Lazdunski, *Nature*, 1997, **386**, 173–177.
133. C. C. Askwith, J. A. Wemmie, M. P. Price, T. Rokhlina and M. J. Welsh, *J. Biol. Chem.*, 2004, **279**, 18296–18305.
134. A. Baron, N. Voilley, M. Lazdunski and E. Lingueglia, *J. Neurosci.*, 2008, **28**, 1498–1508.
135. C. C. Chen, S. England, A. N. Akopian and J. N. Wood, *Proc. Natl. Acad. Sci. U. S. A.*, 1998, **95**, 10240–10245.
136. S. Diochot, A. Baron, L. D. Rash, E. Deval, P. Escoubas, S. Scarzello, M. Salinas and M. Lazdunski, *EMBO J.*, 2004, **23**, 1516–1525.
137. F. L. Yuan, F. H. Chen, W. G. Lu and X. Li, *Mol. Biol. Rep.*, 2010, **37**, 3233–3238.
138. W. L. Wu, C. F. Cheng, W. H. Sun, C. W. Wong and C. C. Chen, *Pharmacol. Ther.*, 2012, **134**, 127–138.
139. B. Chagot, P. Escoubas, S. Diochot, C. Bernard, M. Lazdunski and H. Darbon, *Protein Sci.*, 2005, **14**, 2003–2010.
140. M. G. Blanchard, L. D. Rash and S. Kellenberger, *Br. J. Pharmacol.*, 2012, **165**, 2167–2177.
141. K. Shiomi, *Toxicon*, 2009, **54**, 1112–1118.
142. J. C. Song, M. C. M. Xiong, Y. Wang and W. Pu, X., *Toxicon*, 1994, **32**, 537–538.
143. Y. Y. He, W. H. Lee and Y. Zhang, *Toxicon*, 2004, **44**, 295–303.
144. X. C. Pu, P. T. Wong and P. Gopalakrishnakone, *Toxicon*, 1995, **33**, 1425–1431.
145. P. Gopalakrishnakone, *Toxicon*, 1996, **34**, 622–622.
146. X. C. Pu, P. Gopalakrishnakone and P. T. H. Wong, *Toxicon*, 1997, **35**, 501–501.
147. R. M. Kini, *Toxicon*, 1998, **36**, 1659–1670.
148. *US Pat.*, 6 613 745, 2003.
149. T. P. Snutch, *NeuroRx*, 2005, **2**, 662–670.
150. T. Pexton, T. Moeller-Bertram, J. M. Schilling and M. S. Wallace, *Expert Opin. Invest. Drugs*, 2011, **20**, 1277–1284.
151. S. Vink and P. Alewood, *Br. J. Pharmacol.*, 2012, **167**, 970–989.
152. B. Callaghan, A. Haythornthwaite, G. Berecki, R. J. Clark, D. J. Craik and D. J. Adams, *J. Neurosci.*, 2008, **28**, 10943–10951.
153. N. L. Daly, B. Callaghan, R. J. Clark, S. T. Nevin, D. J. Adams and D. J. Craik, *J. Biol. Chem.*, 2011, **286**, 10233–10237.

154. H. Klimis, D. J. Adams, B. Callaghan, S. Nevin, P. F. Alewood, C. W. Vaughan, C. A. Mozar and M. J. Christie, *Pain*, 2011, **152**, 259–266.
155. B. Callaghan and D. J. Adams, *Channels (Austin, Tex.)*, 2010, **4**, 51–54.
156. R. J. Clark, N. L. Daly, R. Halai, S. T. Nevin, D. J. Adams and D. J. Craik, *FEBS Lett.*, 2008, **582**, 597–602.
157. K. S. Misono, J. S. Philo, T. Arakawa, C. M. Ogata, Y. Qiu, H. Ogawa and H. S. Young, *FEBS J.*, 2011, **278**, 1818–1829.
158. S. Vink, A. H. Jin, K. J. Poth, G. A. Head and P. F. Alewood, *Toxicon*, 2012, **59**, 434–445.
159. A. Schmidtko, W. Gao, P. Konig, S. Heine, R. Motterlini, P. Ruth, J. Schlossmann, D. Koesling, E. Niederberger, I. Tegeder, A. Friebe and G. Geisslinger, *J. Neurosci.*, 2008, **28**, 8568–8576.
160. F. X. Zhang, X. J. Liu, L. Q. Gong, J. R. Yao, K. C. Li, Z. Y. Li, L. B. Lin, Y. J. Lu, H. S. Xiao, L. Bao, X. H. Zhang and X. Zhang, *J. Neurosci.*, 2010, **30**, 10927–10938.
161. S. Heine, S. Michalakis, W. Kallenborn-Gerhardt, R. Lu, H. Y. Lim, J. Weiland, D. Del Turco, T. Deller, I. Tegeder, M. Biel, G. Geisslinger and A. Schmidtko, *J. Neurosci.*, 2011, **31**, 11184–11192.
162. B. G. Fry, J. C. Wickramaratana, S. Lemme, A. Beuve, D. Garbers, W. C. Hodgson and P. Alewood, *Biochem. Biophys. Res. Comm.*, 2005, **327**, 1011–1015.
163. R. E. Middleton, V. A. Warren, R. L. Kraus, J. C. Hwang, C. J. Liu, G. Dai, R. M. Brochu, M. G. Kohler, Y. D. Gao, V. M. Garsky, M. J. Bogusky, J. T. Mehl, C. J. Cohen and M. M. Smith, *Biochemistry*, 2002, **41**, 14734–14747.
164. Y. Xiao, K. Blumenthal, J. O. Jackson, 2nd, S. Liang and T. R. Cummins, *Mol. Pharmacol.*, 2010, **78**, 1124–1134.
165. S. Sokolov, R. L. Kraus, T. Scheuer and W. A. Catterall, *Mol. Pharmacol.*, 2008, **73**, 1020–1028.
166. K. Peng, Q. Shu, Z. Liu and S. Liang, *J. Biol. Chem.*, 2002, **277**, 47564–47571.
167. Y. Xiao, J. O. Jackson, 2nd, S. Liang and T. R. Cummins, *J. Biol. Chem.*, 2011, **286**, 27301–27310.
168. Y. Xiao, J. P. Bingham, W. Zhu, E. Moczydlowski, S. Liang and T. R. Cummins, *J. Biol. Chem.*, 2008, **283**, 27300–27313.
169. J. J. Smith, T. R. Cummins, S. Alphy and K. M. Blumenthal, *J. Biol. Chem.*, 2007, **282**, 12687–12697.
170. K. J. Shon, B. M. Olivera, M. Watkins, R. B. Jacobsen, W. R. Gray, C. Z. Floresca, L. J. Cruz, D. R. Hillyard, A. Brink, H. Terlau and D. Yoshikami, *J. Neurosci.*, 1998, **18**, 4473–4481.
171. K. J. Nielsen, M. Watson, D. J. Adams, A. K. Hammarstrom, P. W. Gage, J. M. Hill, D. J. Craik, L. Thomas, D. Adams, P. F. Alewood and R. J. Lewis, *J. Biol. Chem.*, 2002, **277**, 27247–27255.

172. P. Safo, T. Rosenbaum, A. Shcherbatko, D. Y. Choi, E. Han, J. J. Toledo-Aral, B. M. Olivera, P. Brehm and G. Mandel, *J. Neurosci.*, 2000, **20**, 76–80.
173. W. A. Catterall, A. L. Goldin and S. G. Waxman, *Pharmacol. Rev.*, 2005, **57**, 397–409.
174. J. R. McArthur, G. Singh, M. L. O'Mara, D. McMaster, V. Ostroumov, D. P. Tieleman and R. J. French, *Mol. Pharmacol.*, 2011, **80**, 219–227.
175. J. R. McArthur, V. Ostroumov, A. Al-Sabi, D. McMaster and R. J. French, *Biochemistry*, 2011, **50**, 116–124.
176. A. A. Tietze, D. Tietze, O. Ohlenschlager, E. Leipold, F. Ullrich, T. Kuhl, A. Mischo, G. Buntkowsky, M. Gorlach, S. H. Heinemann and D. Imhof, *Angew. Chem. Int. Ed.*, 2012, **51**, 4058–4061.
177. P. T. Tan, K. N. Srinivasan, S. H. Seah, J. L. Koh, T. W. Tan, S. Ranganathan and V. Brusic, *Journal of Molecular Graphics & Modelling*, 2005, **24**, 17–24.
178. N. I. Platnick, *The World Spider Catalog, Version 13.5.*, http://research.amnh.org/iz/spiders/catalog. DOI: 10.5531/db.iz.0001.

CHAPTER 8

Non-Oral Treatments for Pain

ANDY PIKE

Pfizer Neusentis, Portway Building, Granta Park, Cambridge CB21 6GS, UK
Email: andy.pike@pfizer.com

8.1 Introduction

For most analgesic applications, outside a professional care environment, delivery of drugs as solid oral formulations is usually the preferred choice, allowing manufacturers to employ well-established, generally low-cost, technologies as well as offering convenience and independence from medical practitioners for the patient. However, the management of pain can be a complex process requiring medication with different modes of action, tolerability profiles and, potentially, means of delivery. Additionally, the range of distinct pharmacological classes is limited, there are class-associated issues such as gastrointestinal bleeding effects with non-steroidal anti-inflammatories (NSAIDs) or constipation and respiratory depression with opioids, and many efficacious compounds are not amenable to dosing as solid oral formulations. Given the challenges of developing effective new pharmacological classes, overcoming these issues as far as possible with novel alternative formulations represents a significant medical and commercial opportunity in pain management.

It has been suggested by Moore that an ideal analgesic for acute pain should have the following attributes, many of which would be equally desirable in a chronic pain therapy: 'a rapid onset of action, act over an extended period of time, reduce awareness of pain quickly and minimise interruption by pain, be well tolerated and produce analgesia over a wide range of pain types in different patient populations'.[1] Depending on the properties of the compound oral

RSC Drug Discovery Series No. 35
Pain Therapeutics: Current and Future Treatment Paradigms
Edited by Charlotte Allerton

Published by the Royal Society of Chemistry, www.rsc.org

delivery, while generally convenient, may not always give the best overall balance of these attributes. There are a number of circumstances of use and drug property considerations which may contribute to oral administration being either impractical, undesirable, or unachievable and require the development of a non-oral means of delivery. These circumstances would include:

- a drug which is efficacious but has poor oral bioavailability following administration as a solid oral dosage form
- pharmacokinetic/pharmacodynamic considerations, such as the desire for a short time to onset that make the delay in time to effect following a solid oral dose suboptimal, *e.g.* breakthrough pain episodes in cancer
- attempts to achieve a local effect to maximise efficacy or avoid systemic adverse effects
- patient capability/behaviour considerations such as being unable to swallow oral medication due to incapacity or illness, a desire to reduce abuse liability, or simply patient preference for a particular delivery route
- the need to move more easily from a fixed dosing regimen to a more flexible regimen, based on patient need for immediate pain relief, while still providing a high degree of control of the total dose available, *e.g.* intrathecal morphine pump for severe pain.

Which of these factors is the priority to address, in combination with the properties of the drug under consideration, will guide the choice of alternative dosing route. Alternative routes of administration of analgesics include parenteral, which might also include local neural blocks, topical/transdermal and transmucosal systems. Many common analgesics have been formulated using one or more of these methods to address specific issues in regard to medical use or aspects of compound limitations in pharmacokinetics, pharmacodynamics and safety.

The aim of this chapter is discuss some alternatives to solid oral dose formulations, their major advantages and limitations, and give some examples of their application to existing marketed analgesics.

8.2 Non-Oral Administration Options

The following section describes some of the factors influencing compound delivery by common non-oral routes, including parenteral, topical/transdermal and transmucosal options.

8.2.1 Injectable Analgesics

Analgesics may be administered by bolus injection or infusion, including the intravenous (IV), subcutaneous and intramuscular routes. While each option may offer different advantages and limitations the common requirements include solubility in an appropriate vehicle, acceptable injection volume, injection site tolerability, and sterility, in addition to demonstrating the systemic safety of the drug itself. Regardless of mechanism or compound, delivering analgesia by injection can offer several advantages over oral dosing including:

- the potential for rapid systemic delivery of drug, especially with intravenous bolus or infusion, hence minimising time to onset of analgesia
- more predictable and controlled drug exposure
- the ability to administer drug to a patient who is incapacitated, *e.g.* in a perioperative or emergency situation
- local administration of nerve blocks or into wound sites.

Typically, to dose a compound by the IV or intramuscular routes requires a trained healthcare professional and for the majority of patients this would not be their preferred choice, especially if pharmacokinetic considerations mean that frequent dosing is required. However, self-administration of subcutaneous drug doses, usually after training by a healthcare professional, is relatively common, with the obvious example of insulin. Therefore, if long-acting antibody-based therapeutics were to become available, such as anti-NGF, then occasional dosing by subcutaneous injection may become viable for chronic pain control.[2–4]

8.2.2 Topical and Transdermal Drug Analgesics

Application of substances to the skin to relieve pain has a long history in the form of various herbal preparations, and more recently chemically defined rubefacients. Rubifacients, which are sometimes termed as counter-irritants, are a variety of substances that give a comforting feeling of warmth. For discussion of exact definition and description see the Bandolier website 'Topical analgesics introduction'.[5] These herbal and rubefacient compounds are still commonly used, although systematic reviews of the literature have suggested there is little evidence for their efficacy in the treatment of acute pain and they compare poorly to NSAIDs in chronic pain states.[6,7] For drugs with more defined pharmacology delivered *via* the skin there are two distinct types, topical and transdermal formulations.

Topical formulations are defined as those designed for dermal application to produce a local effect in the skin or adjacent tissue to the site of application, with the potential benefit of minimising gastrointestinal or systemic effects. For analgesia this will require penetration of the skin to reach the site of action. Topical application can also bypass the hepatic first-pass effect on orally administered drugs, and hence may improve bioavailability.

By contrast, transdermal formulations are defined as those designed for dermal application with the aim of delivering drug to the systemic circulation in order to mediate its pharmacological effect. The potential advantages of this route of administration include: bypass of the first-pass effect, and hence improved bioavailability; the opportunity to provide a reservoir of drug to prolong absorption for an extended period, giving extended effects and improving compliance; and minimising peak-to-trough variation in drug plasma levels. Additionally, depending on the specific drug delivery system and pharmacokinetics, transdermal systems can allow the flexibility of terminating drug administration by simply removing the system from the skin.

The delivery challenges may be minimised by the use of potent compounds to reduce dose requirements, however, this will not in itself make a drug suitable for percutaneous delivery. The challenges of delivering drugs across the skin can be loosely grouped into:

- **physiological factors**: these include the nature of the barrier and the physicochemical limitations it imposes, with the potential for additional variability depending on skin condition, *e.g.* as a result of age, ethnicity, pathology, site of application or environmental interactions, such as temperature
- **formulation/drug factors**, including the vehicle, and any permeability enhancers, the effect of any device employed and the need to avoid local irritability.

8.2.2.1 Physiological Barriers to Drug Permeability Across the Skin

One of the principle functions of the skin is to provide a protective interface between the body and the external environment, a major part of which is the exclusion of harmful agents in the form of infectious organisms or exogenous chemical compounds. As such it presents a significant barrier to percutaneous drug delivery.

The skin is structured in three layers, from the deepest to the surface, termed the hypodermis, dermis and epidermis. The skin barrier to external challenges is the epidermis, which is itself multilayered, with dividing keratinocytes present in the stratum basale, proximal to the dermis, which gradually rise up through the layers to eventually replace surface cells lost by desquamation. Keratinocytes are metabolically active cells and carry out some processes potentially relevant to drug metabolism, such as *N*-acetylation and CYP24 mediated metabolism of vitamin D.[8] As they pass through the layers of the epidermis the keratin content of the keratinocyte cells progressively increases and cytoplasmic organelles and the cell nucleus are lost, though some biochemical capacity may still be retained.[9] At the end of this process the cells, now termed corneocytes, are incorporated into the outermost layer of the epidermis, known as the stratum corneum, where the cells are embedded in a highly structured lipid matrix composed of ceramides, cholesterol and non-essential fatty acids.[10,11] The surface of the stratum corneum is typically at an acidic pH of around 4.8 (sometimes referred to as the 'acid mantle'), with a gradient up to about 6.8 at the stratum basale.[12] While generally being no more than 10–40 μm, or 15–20 cells, thick, the ability of the skin to act as a barrier to the absorption of drugs is thought to largely reflect the properties of the stratum corneum.

As with oral administration, the physicochemical characteristics of a compound will be the major determinant of percutaneous absorption potential, perhaps even more so given the lack of other confounding processes, such as active transport, in the skin. However, unlike oral drug space, where data on thousands of molecules are available to analyse trends and generate physicochemical guidelines for optimal absorption, *e.g.* the 'rule-of-5', the percutaneous absorption of relatively few compounds has been studied in any detail.[13] This likely reflects the fact that for most indications there is little drive to target topical/transdermal delivery ahead of oral and that current compounds are largely formulations of pre-existing drugs. It is therefore challenging to draw generalised conclusions based on the existing data. However, it seems likely that the same parameters influencing oral absorption will be relevant, albeit optimal values are likely to be different given the contrasting environment and different characteristics of the barriers. Choy *et al.*, in a recent review of non-oral drugs, concluded that drugs delivered transdermally fell within more stringent criteria, though again this may be more a reflection of what has been attempted, rather than what is possible, in this space.[14]

Given the lipid matrix in the stratum corneum, it might be supposed that lipophilicity will be an important factor influencing percutaneous absorption. For example, comparisons of related salicylates and NSAIDs suggest an optimal log P value between 2 and 3.[15] Other physicochemical limitations may be imposed by the skin pH and resulting ionisation state of the compound. Studies with acidic molecules have suggested that ionised compounds will penetrate the stratum corneum barrier but at rates much lower than the unionised form; however, recent preclinical data with azelaic acid showed higher permeability at pH 4.9 compared to p 3.9, suggesting that other factors may dominate.[16] Such examples serve to highlight the questionable applicability of conclusions based on single compounds, and series of analogues, to novel molecules. The correlation of molecular weight, log K_{oct}, hydrogen bond donor and acceptor count, melting point and water solubility, on the absorption of three groups of chemicals defined as agricultural, steroid and other organic compounds found that it was only in the steroid group that any significant correlations were observed.[17] Ultimately the rate and extent of percutaneous absorption is likely to be the result of a complex interaction between compound, vehicle and excipients, as well as with the skin itself and hence it may be difficult to reduce to simple physicochemical guidelines.

The evidence for the influence of population and environmental factors such as age, ethnicity, skin condition, site of application and temperature is mixed and relatively limited in scope. For example, a review of the use of transdermal drugs in older people concluded that, despite changes in skin structure with age, there is no consistent evidence for changes in transdermal absorption with increasing age.[18] However, it has been proposed that changes in skin structure with age may disproportionately affect compounds with specific physicochemical properties, *e.g.* hydrophilic compounds, due to changes in the lipid content and hydration state.[19] On the other hand increased temperature, either ambient or induced by exercise or fever, can increase the rate of transdermal absorption of several drugs.[20,21] The current evidence for all these factors has been summarised in a recent review, concluding that the combination of these factors could be significant and cautioning against overgeneralisation when developing transdermal drug delivery systems.[22] While the factors contributing to variability in transdermal and oral drug delivery may be different, a comparison of the two concluded, while acknowledging significant limitations in the data, that the variability was more dependent on the specific compound than the route of delivery and that neither route was inherently more vulnerable to high variability.[23]

8.2.2.2 Formulation Factors

To some degree the challenges of percutaneous absorption may be overcome by appropriate formulation, and the use of excipients either to alter the permeability of the stratum corneum or provide additional solubilisation of the drug.

There are many marketed topical formulations, and often multiple formulations of a given compound, including gels, creams, ointments, *etc*. Many of the terms used for dermally applied products are slightly ambiguous; however, a classification system for what defines a gel, cream, ointment, *etc*., has been suggested by Buhse *et al.*[24] Such formulations have the benefit of relative simplicity and low cost, although they come at the price of a lower level of control of the dose applied and reproducibility of the application.

In order to deliver a more controlled dose of drug *via* dermal application, transdermal drug delivery systems (TDDS) were introduced. The first TDDS were introduced in the late 1970s and early 1980s, including products such as Transderm-Scop (transdermal scopolamine for motion sickness) and Catapres-TTS, a transdermal delivery system for the antihypertensive clonidine. In the United States to date approximately 20 transdermal products have been approved, including analgesics, and other familiar products such as nicotine patches for smoking cessation.[25]

At the most simple level a TDDS will typically consist of a release liner to protect the adhesive surface before use, an adhesive to attach to skin, a drug reservoir or matrix containing drug with any necessary excipients, and a backing layer which protects the exposed surface. However, the use of a delivery device allows increasing levels of sophistication to be incorporated in order to improve or control the absorption process. Prausnitz and Langer suggest, in a review of transdermal drugs marketed to date and the future potential of TDDS, that the technologies group loosely into three generations.[26] First-generation products are those which simply apply drug to the skin but do not actively aid permeability and may therefore be limited in application by the physicochemical properties of the compound. The second generation products include a means of modifying skin permeability, whether by means of chemical excipients or physical processes such as ultrasound or iontophoresis. Finally, third-generation products actively break the stratum corneum barrier, for example by using microneedles or ablation techniques, and may substantially extend the range of molecules which could be delivered.

The potential advantages of transdermal drug delivery need to be weighed against increased costs associated with complex systems, issues arising from patch application such as skin irritation, and the inherent suitability of the drug for cutaneous absorption.

An additional issue which may arise with the increasing complexity of drug delivery devices is increased potential for failure of system components and subsequent impact on efficacy and safety. The IONSYS™ fentanyl HCl patch perhaps illustrates both the potential of sophisticated transdermal delivery systems and the issues which could arise in a complex device. IONSYS™ was developed using an iontophoresis-based technology to control transdermal absorption and allow patient-controlled analgesia with fentanyl, while avoiding the need for venous access. The systemic delivery of up to six 40 μg doses of fentanyl per hour was reported to provide analgesia equivalent to repeated IV bolus dosing of morphine, with a similar incidence of adverse events.[27] Marketing approval was granted in 2006 for control of moderate to severe pain in the hospital setting. However, in 2008 the European Medicines Agency (EMA) recommended suspension of the marketing authorisation after the manufacturer detected corrosion of a component which could have resulted in self-activation of the system and potential overdose, and the authorisation has subsequently been allowed to expire.[28,29] In 2010 the rights to the IONSYS™ device were acquired by Incline Therapeutics who, as of 2012, were still continuing to develop the product.[30]

8.2.3 Transmuscosal Drug Delivery

Transmucosal drug delivery includes the nasal, buccal/sublingual, vaginal, ocular and rectal routes. Mucosal surfaces essentially consist of layers of epithelial cells, although their properties can vary significantly, in different organs and even different regions of the mouth, depending on factors such as the thickness of cell layers, junction properties and the degree of keratinization.[31] The nasal and buccal/sublingual mucosal surfaces are generally reported to have permeability intermediate between skin and intestinal mucosa, therefore potentially offering an absorption advantage over topical/transdermal administration, and often rich vasculature.[32,33] Hence mucosal surfaces potentially offer an easier route for drug absorption than skin, while still maintaining the advantage of transdermal delivery in bypassing first-pass effects. Also, as in the topical/transdermal route, there is the possibility of using excipients to improve compound permeability, though the mechanisms by which this is achieved may be different.[34]

The drawbacks of transmucosal delivery are surface area and accessibility, and often retention of the dose at the site of absorption is one of the biggest challenges. However, assuming this limitation can be overcome with appropriate formulation technology, prolonged residency at the site of absorption and more controlled delivery might be obtained, though dose sizes may be more limited.

A review of the clinical pharmacokinetics of various opioids, benzodiazepines and anti-migraine drugs concluded that it was generally possible to achieve more rapid t_{max} and higher bioavailability with the intranasal route compared to oral administration.[35] The literature suggests that similar advantages can be obtained *via* the oral mucosa.[36]

The potential benefits of the transmucosal route must be offset against increased costs resulting from complexity of the formulation and therefore a significant medical need is required. However, the potential for very rapid absorption offered by transmucosal delivery is a particularly attractive feature for some pain indications.

8.3 Compound Examples

The following section provides some examples of non-oral formulations of pain medications which have been developed to address particular medical requirements, or else mitigate pharmacokinetic, pharmacodynamic or tolerability issues of existing compounds. Because of the number of products available this list is not comprehensive and avoids discussion of non-oral combinations for simplicity.

8.3.1 Paracetamol (Acetaminophen)

Paracetamol (known as acetaminophen in the United States), was discovered in the late 19th century and first marketed by Sterling Winthrop in the 1950s. It remains one of the most widely used analgesic and antipyretic drugs clinically and over the counter. Despite its long history and widespread use, the pharmacology of paracetamol remains poorly understood.[37] Doses of paracetamol are typically quite high, in the 500–1000 mg range, but taken orally paracetamol has good bioavailability, in the range 65–90%, and a plasma half-life of approximately 2–2.5 hours.[38] The reported time to onset of efficacy following oral dosing is typically reported in the range 10–60 minutes with effects lasting for 4–6 hours.[39–41] In addition to solid oral dose formulations, paracetamol has been commercially formulated for administration by both IV and rectal routes of administration. Given its high oral bioavailability and already quite rapid onset of action, the scope for improvement of over oral dosing appears to be modest. However, requirements of emergency administration and perioperative care have driven the development of alternate formulations.

Perhaps the biggest challenges in developing alternative formulations of paracetamol are its relatively high dose, typically up to 1000 mg, in combination with modest aqueous solubility. The first IV formulation of paracetamol employed a water-soluble prodrug, propacetamol, marketed as Pro-Dafalgan®, which is rapidly and quantitatively hydrolysed to the parent compound by plasma esterases (Figure 8.1). A study of time to onset of analgesia following administration of 2 g of propacetamol (equivalent to 1 g of paracetamol) by IV bolus or infusion, following third molar dental surgery, showed a reduction in time to onset of analgesia of 3–5 minutes compared to

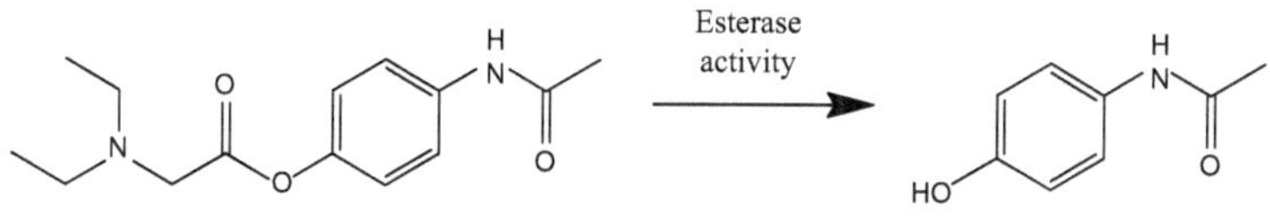

Figure 8.1 Conversion of propacetamol to paracetamol (acetaminophen) by plasma esterases.

11 minutes for an equivalent 1 g oral dose of paracetamol.[40] Pharmacokinetic comparison of propacetamol to subsequently introduced formulations of IV paracetamol demonstrate that very similar C_{max}, area under the curve (AUC), and t_{max} are achieved for equivalent parent dose; however, the adverse-event profile of the prodrug was worse, with more pain at the injection site and other treatment emergent adverse events.[42]

Subsequently an IV formulation of paracetamol, an aqueous solution with the brand name Perfalgan®, was introduced in Europe by Bristol Myers Squibb from 2001. The same formulation was launched in January 2011 in the United States by Cadence Pharmaceuticals under the trade name Ofirmev®, indicated for the management of mild to moderate pain and for use in conjunction with opioids for severe pain and reduction of fever. These IV formulations of paracetamol require a 15-minute, 100 mL infusion, as drug concentration is limited to 10 mg mL^{-1} because of the compound's modest aqueous solubility, and hence are targeted to the perioperative situation.

The comparative pharmacokinetics of IV and oral paracetamol suggest that there should be little overall difference in efficacy between the two routes. This conclusion is supported by systematic Cochrane reviews of the efficacy of IV paracetamol or propacetamol and single-dose oral analgesics, including paracetamol, using the criterion of at least 50% pain relief over 4 or 6 hours.[43,44] While some caution needs to be exercised, given the heterogeneity of the studies analysed, a number needed to treat (NNT) of 4.0 for combined IV paracetamol and propacetamol, *versus* 3.6 for a comparable oral dose (1 g or equivalent) suggests that efficacy is approximately equivalent by the two routes.

Rectal formulations of paracetamol are also relatively commonly administered in the hospital environment, especially in paediatric applications where paracetamol is recommended in World Health Organization (WHO) guidelines as a first line treatment.[45] The pharmacokinetics of rectally administered paracetamol are reported to show prolonged and irregular absorption profiles. This may contribute to the conclusion, in a review of studies of rectal and IV paracetamol, that a single rectal administration required relatively high doses of 40–60 mg kg^{-1} for efficacy.[46] This contrasts to the recommended single IV doses in the paediatric population of 10–15 mg kg^{-1}, lower than that used in adults, reflecting differences in paracetamol pharmacokinetics between age groups.[47]

8.3.2 Lidocaine

Lidocaine was discovered in the 1940s, approved by the United States Food and Drug Administration (FDA) in 1948 and first marketed under the name Xylocaine®.[48] Like its predecessors, procaine and tetracaine, lidocaine blocks signal conduction in neurons by non-selective inhibition of the voltage-gated sodium channels (Na_V) responsible for signal propagation in the neuronal cell membrane. Lidocaine is used as a local and regional anaesthetic, and systemically as a class Ib antiarrhythmic drug.[49] However, it has a fairly narrow range of safe and efficacious blood concentrations (2–5 $\mu g\, mL^{-1}$).[50] Exceeding these concentrations systemically can result in central nervous system (CNS) disturbances, including lightheadedness, dizziness, drowsiness, confusion and potentially seizures, as well as cardiovascular toxicity, manifesting as progressive heart block, reduced cardiac contraction, hypotension and asystole.[51]

Given its established local anaesthetic properties, a local topical application of lidocaine is an attractive option to minimise systemic exposure while achieving efficacy. A variety of gels, ointments and creams are available containing lidocaine, or related topical anaesthetics, as an active ingredient, both over the counter and by prescription. Although it is largely safe and effective, a FDA public health advisory note has been issued detailing circumstances in which systemically toxic levels of a topical anaesthetics were achieved, with fatal consequences.[52] Further guidance was issued highlighting the potential dangers, including application of such products over large areas, formulation strength and various environmental factors that can affect absorption such as covering the area or applying heat.[53] Therefore a more controlled means of delivering the dose would be assumed to be beneficial to safety, especially in a chronic use setting.

The Lidoderm® patch (variously referred to in the literature as a patch or plaster, generally depending on author location) was approved in the United States in 1999 for the treatment of postherpetic neuralgia (licensed for development and commercialisation by Hind Healthcare to the Japanese company Teikoku and marketed under licence in the United States by Endo Pharmaceuticals), and has subsequently been marketed in Europe under the brand name Versatis® (Grunethal GmbH) since 2007. The Lidoderm® patch 5% consists of a 10 cm × 14 cm plaster adhesive hydrogel containing 700 mg (5% w/v) applied to a non-woven polyester felt backing and covered with a polyethylene terephthalate film release liner which is removed before application to the skin.[54] In 2011 Endo Pharmaceuticals reported $825.2 million in sales for Lidoderm® and Grunethal GmbH a further €68 million for Versatis®.

Lidoderm became the first product approved by the FDA for the treatment of postherpetic neuralgia and was subsequently approved by the EMA for this condition. A summary of the available efficacy data in this indication, along with a description of the pharmacokinetics properties and tolerability, have been published in a review article.[55] Given the safety requirement to reduce systemic exposure it is notable that the reported systemic C_{max} concentration produced by the patch was only 0.052 μg mL^{-1}, which is substantially below the reported therapeutic blood concentration of 2–5 μg mL^{-1}.[50,55] A recent review of the efficacy of the Lidoderm® patch highlighted a number of reports suggesting that it may have efficacy in other, largely neuropathic pain conditions, though acknowledging that many of the studies supporting this were case reports, small sample sizes or open label and will require more rigorous follow-up to fully substantiate their findings.[56]

8.3.3 Bupivacaine

Another of the amino-amide class of anaesthetics, bupivacaine shares some common pharmacology with lidocaine, and related local anaesthetics such as articaine, mepivacaine and prilocaine, with both CNS and cardiac-related issues being observed if excessive systemic exposure occurs.[57] It is widely available and can be used as a local anaesthetic for infiltration, nerve block and neuraxial anesthesia, and for postsurgical analgesia *via* injection into wound sites. However, bupivacaine has reduced metabolic clearance compared to other agents and therefore the potential for systemic accumulation may be higher.[58]

Exparel, developed by Pacira Pharmaceutical, combines bupivacaine with their Depofoam Multivesicular liposome technology, was approved by the FDA in 2011 and was designed to improve the duration and extent of analgesia in postsurgical pain.[59] A review of 10 trials using liposomal bupivacaine in wound infiltration situations suggested at least equivalence to standard formulations of bupivacaine hydrochloride, although statistically significant differences were achieved in only 5 of 17 treatment arms.[60] The interpretation is complicated in some studies by a lack of directly comparable dose levels, although it should be noted that this would not necessarily equate with equivalent exposure. However, secondary efficacy measures, including avoidance of, time to first use, and total postoperative consumption of opioid rescue medication, as well patient/care provider satisfaction with postoperative analgesia, also supported an improved profile. A review of cardiac effects showed a benign profile for liposomal bupivacaine; however, this did not significantly differentiate from bupivacaine hydrochloride formulations.[61]

The formulation technology used in Exparel gives the potential for an unusual drug–drug interaction in that other compounds may affect the release rate of bupivacaine from the liposomal formulation.[62]

8.3.4 Capsaicin

Capsaicin (trans-8-methyl-*N*-vanillyl-6-nonenamide) is familiar to many as the major component of chilli peppers responsible for the ‘burning’ sensation generated when ingested. Capsaicin was first isolated, and shown to be the causative agent of the burning sensation produced by chilli peppers, in the 19th century. As recently as 1997 the pharmacological mechanism behind the chilli sensation was demonstrated to result from a specific effect of capsaicin on the transient receptor potential cation channel subfamily V member 1 (TRPV1) receptor.[63]

The mechanism of topical capsaicin application has been described as 'defunctionalisation', rather than acute desensitization of the TRPV1 receptor, though it is unclear how this relates to dose.[64] This proposed defunctionalisation-based mechanism would be unlikely to be tolerable at a systemic level and capsaicin has additional limitations as an oral drug, including unpleasant gastrointestinal side effects and poor pharmacokinetics. The use of topical application could potentially mitigate the pharmacokinetic and gastrointestinal tolerability issues and allow sufficiently high local concentrations to be achieved to gain a therapeutic effect, although local irritation and the potential for contact with sensitive areas remain an issue. Multiple brands of topical capsaicin formulations are available, either by prescription or over the counter in some markets, typically containing less than 0.1% capsaicin and applied by multiple daily applications over a period of days or weeks. However, the handling issues may make long-term use of such formulations problematic.

In 2010 the Qutenza® patch (a.k.a. NGX-4010) from NeurogesX (licensed to Astellas Pharma in Europe) was approved for use in neuropathic pain associated with postherpetic neuralgia. Qutenza® consists of a high (8%) concentration capsaicin patch formulation applied for a period of 30–90 minutes (recommended 60 minutes) by a medical professional in conjunction with a local anaesthetic to alleviate the acute nociceptive effects of capsaicin. The single treatment is claimed to produce a response for a period of up to 12 weeks.[65]

With regard to the mode of action it is unclear if this is consistent between repeated low-dose application and the single high exposure obtained with Qutenza®. Interestingly, a systematic Cochrane review of topical capsaicin for neuropathic pain suggested a NNT of 6.6 with low-dose (0.075%) topical formulations compared to 12 for Qutenza®.[66] However, the authors note that the overall amount of data available for analysis was limited, only two studies for Qutenza®, and that there was considerable heterogeneity in endpoints for the low-dose formulation studies. It was also noted, and has been remarked on by other authors, that the potential for application site pain/irritation/skin reactions resulting from capsaicin application is a potential source of functional unblinding.[66,67]

Sales of the Qutenza® patch have been poor, with NeurogesX reporting sales of $1.0 million and a further $0.8 million from the Astellas Pharma partnership in 4th quarter 2011.[68] The poor uptake is perhaps due to the relatively complex and time-consuming application process and the need for healthcare practitioners to have product available ready for use. In March 2012 the company announced they would cease activities relating to the commercialisation of the Qutenza® patch to focus resources on the development of a follow-up product referred to as NGX-1998. In contrast to the patch technology used with Qutenza®, NGX-1998 is a liquid formulation, topically applied using a sponge applicator, containing 20% capsaicin designed to treat pain associated with neuropathic pain conditions. It has been reported that, unlike Qutenza®, NGX-1998 does not require pretreatment with a topical anaesthetic and that the application period is much shorter, which may increase convenience for both patient and physician. The liquid formulation/applicator combination may also be easier to apply irrespective of location on the body.

8.3.5 Diclofenac

Diclofenac is an example of the NSAIDs class, which mediate their effect through the inhibition of the cyclooxygenase 1 (COX-1) and cyclooxygenase 2 (COX-2) enzymes. Perhaps the most significant issue associated with the chronic use of oral NSAIDs is the potential to produce gastrointestinal ulcers and episodes of bleeding which is thought to result, at least in part, from their COX-1/2 inhibitory action.[69] While there are some differences in incidence reported between specific agents, this is likely to be at least partially due to differences in dose and local concentrations, rather than intrinsic tolerability.[70] One attempt to improve tolerability led to the development of the COX-2 selective class of 'coxib' compounds, including celecoxib, rofecoxib and valdecoxib (although rofecoxib and valdecoxib have since been withdrawn from the market due to cardiovascular safety concerns), to retain the anti-inflammatory properties while reducing the anti-clotting effects thought to be principally mediated *via* COX-1. However, an alternative approach to minimising gastrointestinal side effects of NSAIDs is to avoid significant local exposure in the gastrointestinal tract by using the topical route of administration. Evans *et al.* concluded that, after adjusting for confounding factors such as concomitant oral NSAIDs and/or ulcer-healing drugs, topically applied NSAIDs were not significantly associated with gastrointestinal bleeding.[71]

The market for topical NSAIDs historically has seen a significant geographical split, with no products approved in the United States until 2007, perhaps due to questions about efficacy and speculation that any effects observed simply resulted from the systemic exposure obtained. In 2007 the FDA approved Voltaren® gel for osteoarthritic joint pain, to be followed by additional diclofenac formulations including other gel formulations, the topical solution Pennsaid®, and the Flector® patch (Table 8.1).[72–75]

The question of relative exposure at the site of action compared to systemic levels following topical *versus* oral diclofenac administration may still be somewhat equivocal, though the published data tend to support the hypothesis that absolute concentrations in subcutaneous fat and muscle at the application site are at least comparable using the topical formulations studied.[76,77] It should be noted that these studies used different topical formulations and plasma/tissue sampling methods, which may contribute to the differences in findings. However, comparison of the clinical pharmacokinetic parameters reported following multiple doses of various topical formulations compared to similar doses of oral diclofenac support substantially reduced systemic exposure compared to a comparable oral dose (Table 8.1). While local

Table 8.1 Comparison of pharmacokinetic parameters following oral dosing of diclofenac sodium *versus* topical formulations of diclofenac.

			Plasma pharmacokinetic parameters[a]					
Product name	*Production description*	*Dosing*	*Dosing day*	*Plasma* C_{max} *(ng/mL)*	*Plasma* t_{max} *(h)*	*Plasma* $t_{1/2}$ *(h)*	*Plasma AUC (0–t) (ng h* mL^{-1}*)*[b]	*References*
None	Diclofenac sodium50 mg enteric-coated tablets	3 times/day (=150 mg diclofenac sodium/day)	7	2270 ± 778	6.5 (1–14)	NS	3890 ± 1710	69
Voltaren	Diclofenac sodium topical gel 1%	4 × 4 g per day (=160 mg diclofenac sodium/day)	7	15 ± 7.3	14 (0–24)	NS	233 ± 128	70
Pennsaid	Diclofenac sodium 1.5% w/w topical solution	40 drops (~1.2 mL) to each knee, 4 times/day (~154 mg diclofenac sodium/day)	7	19.4 ± 9.3	4.0 ± 6.5	79.0 ± 38.1	695.4 ± 348.9	71
Flector	Diclofenac epolamine topical patch 1.3%	Patch (10 × 14 cm) applied 2 times/day (=360 mg of diclofenac epolamine/day)	8	17.4 ± 13.5	5.4 ± 3.7	26.4 ± 12.4	119.3 ± 75.7	72

[a]Parameters are reported as mean ± standard deviation or median (range).
[b]AUC values reported are for 0–24 hours on the day of dosing, except Flector for which AUC is reported 0–12 hours.

concentrations are not often reported a review of 19 published clinical trials concluded that topical diclofenac consistently demonstrated efficacy, consistent with sufficient local exposure being achieved to obtain pharmacological effects.[78] This conclusion is more broadly supported by systematic reviews encompassing all topically applied NSAIDs, though it is interesting to note that indomethacin and benzydamine were found to be not significantly better than placebo.[71,72,79,80]

8.3.6 Fentanyl

First synthesised at Janssen Pharmaceutica in 1960, fentanyl is a μ-opoid agonist approximately 100 times more potent, when compared by analgesically equivalent IV doses, than morphine.[81,82] It has become a very significant product in the management of acute pain, though its potential as an oral drug is limited by high clearance. However, the high potency and permeability of fentanyl make it an excellent candidate for alternative dose routes and various transdermal and transmucosal formulations have been introduced.[83] The contrasting properties of the transdermal and transmucosal formulations provide examples of different non-oral approaches being used, with the same compound, to address the differing requirements of chronic and episodic pain states.

The first fentanyl transdermal patch introduced was Duragesic® (a.k.a. Durogesic), by Ortho McNeil Janssen, and was approved in the United States and Europe in the 1990s for the management of severe chronic pain. It is a good example of the commercial opportunity of using non-oral delivery technology to address a specific medical need and compound limitations, in this case the effective delivery of a highly efficacious compound with poor oral pharmacokinetics over an extended period, with sales reaching $2 billion in 2004 before loss of exclusivity in January 2005.[84] Since then a number of transdermal fentanyl-based products have been introduced. The Duragesic® patch system, based on a drug-containing reservoir and an adhesive formulation, provides continuous delivery of fentanyl for up to 72 hours with a range of doses from 12.5 to 100 $\mu g\, h^{-1}$ provided by different surface areas of patch (5–40 cm^2), the appropriate dose being estimated by means of conversion from various doses of commonly used opiates. The pharmacokinetic properties show a prolonged absorption phase with C_{max} concentrations after a single patch application occurring after 17–48 hours, which is consistent with the reported time to onset of action.[85–87] The delay is proposed to result from time taken for the drug to build up in the skin at the adhesion site, which is consistent with a prolonged apparent half-life following patch removal of 13–25 hours compared with 3.1–7.9 hours following an intravenous bolus dose.[87,88] This prolonged half-life after patch removal will act to cover periods after patch exchange.

Most issues with use of the fentanyl patch are related to the drug itself and therefore largely independent of the formulation. Patch-specific issues include a warning regarding the effect of heating the patch, which may increase drug absorption rates, and the requirement to ensure safe disposal of used patches to avoid accidental exposure or abuse.

Breakthrough pain is a condition common in cancer patients who can experience sudden onset of extremely severe episodic pain, above background levels, which is not controlled by their maintenance medication, and may last up to an hour.[89] Given the severity and episodic nature of these attacks, in contrast to chronic pain states, time to onset of meaningful pain relief is a critical consideration, and a short duration of action can actually be beneficial to avoid prolonged drug effects at the end of the episode. Fentanyl has good efficacy and a short duration of action and various transmucosal formulations have been developed targeting a reduction in time to onset.

The first of these products was the Actiq® lozenge (or 'lollipop'), developed by Anesta and FDA approved in 1998 (and subsequently acquired by Cephalon). The Actiq® lozenge demonstrated good efficacy and tolerability and had good acceptance by patients, reaching peak sales of $612 million in 2006.[90,91] Crucially, in comparison to a transdermal patch formulation, the transmuscoal Actiq® lozenge was rapidly absorbed, reaching t_{max} after less than 1 hour (Table 8.2). Nevertheless, some aspects of Actiq®'s performance were still considered suboptimal, including some degree of use-dependent release rate, *e.g.* levels of saliva and how the patient positioned the lozenge, as well as the taste.[92–94]

Further transmucosal delivery products have been developed, and approved within the last 5 years, which may improve some of these characteristics, including nasal sprays, *e.g.* Instanyl® and PecFent®, and sublingual/buccal oral transmucosal products, *e.g.* Abstral® (sublingual tablet), Effentora® (buccal tablet) and Onsolis® (mucoadhesive buccal film).

Generally these products have been shown to provide short times to onset of action and give comparable C_{max} values, which may be limited by opioid side

Table 8.2 Comparison of C_{max} and t_{max} of fentanyl following administration as the Duragesic® transdermal patch or Actiq® transmucosal lozenge.

		Plasma pharmacokinetic parameters		
Product	*Dosing*	*Plasma C_{max} (ng/mL)*	*Plasma t_{max} (h)*	*References*
Duragesic® patch[a]	100 mg h^{-1} (highest single patch dose). Applied for up to 72 h/patch	3.36 ± 1.28	29.9 ± 13.7	90
Actiq® lozenge[b]	1600 mg lozenge (highest dose available)	2.51 (23)	0.33 (0.33–824)	91

[a]Pharmacokinetic parameters reported as mean ± standard deviation.
[b]Pharmacokinetic parameters reported as mean (coefficient of variation) and median range for C_{max} and t_{max} respectively.

effect liabilities, though total exposures do vary, possibly in relation to the differences in total exposure times.[92] However, relatively few data are available making direct comparison of the pharmacokinetics and pharmacodynamics of these products. An open-label comparison of the Instanyl® spray to the Actiq® lozenge has been published, demonstrating that the intranasal administration gave a shorter median time to meaningful pain relief, a higher proportion of patients achieving meaningful relief 5 and 10 minutes after dosing, as well as a greater degree of patient preference for the spray over the lozenge.[95] Future head-to-head comparisons may guide the appropriate route for non-oral formulations of other mechanisms.

8.3.7 Buprenorphine

Buprenorphine is a synthetic opioid analgesic which functions as a partial agonist of the μ-opioid receptor and has been in use for around 30 years. Because of its high first-pass metabolism buprenorphine is not viable as an oral drug and was therefore originally developed in the 1980s as sublingual and injectable formulations. However, like fentanyl, buprenorphine has pharmacological and physicochemical characteristics which make it a good candidate for transdermal delivery, with low dose requirements of 0.5–2 mg and high lipophilicity.

Two patch formulations of buprenorphine are currently marketed: Transtec®, marketed by Grünenthal, and Butrans®, marketed by the Purdue/Mundipharma/Napp group.

Significant geographic differences exist in the adoption of transdermal buprenorphine products, with Transtec® first marketed in Switzerland and Germany in 2001, and shortly after in the rest of Europe with sales of €90 million reported in 2011.[96] Approval for the Butrans® patch was given to Purdue Pharma L.P. by the FDA as recently as 2010.

The literature suggest that the Butrans patch can obtain equivalent efficacy outcomes compared to sublingual dosing in osteoarthritis patients, and also reduce common opioid side effects such as nausea, dizziness and vomiting.[97] It has also been reported to effectively replace the codeine component of co-codamol when administered with oral paracetamol to osteoarthritis patients, and result in less use of ibuprofen as rescue medication.[98]

8.4 Conclusions

The significant unmet medical need in pain and the relatively limited range of distinct pharmacological mechanisms available for its control have combined to drive a need to address the shortcomings of existing analgesic drugs through novel, non-oral formulations.

The drawbacks associated with cost of goods and complexity of non-oral formulations will inevitably mean they are not always employed as the formulation approach of choice. However, the application of non-oral delivery technologies has been used to address pharmacodynamic, pharmacokinetic and safety aspects of many pain medications and resulted in some significant medical and commercial successes. In contrast, the relatively poor uptake of products such as the Qutenza® patch suggests that convenience is still a significant driver of acceptance by both medical professionals and patients and needs to be one of the key criteria for the design of complex delivery systems.

Given the success of existing non-oral products in optimising the performance of existing drugs and mechanisms, it is unclear in which direction this field may move in future. Some products have suggested the potential for improved patient-controlled analgesia, though the requirements for such technology to be extremely robust have also been highlighted, and the potential for drug combinations, often used in pain management, in single non-oral products has so far been relatively ignored. However, given the difficulties in bringing any effective new analgesic mechanisms to market and the high medical need for such treatments, the best application of non-oral formulations of the future may remain compensating for the shortcomings of compounds with novel mechanisms of action and the potential to deliver effective pain relief.

Abbreviations and Definitions

AUC	area under the concentration *versus* time curve
C_{max}	maximum concentration reached following dosing
IV	intravenous
log K_{oct}	log of the octanol:water partition coefficient (K_{oct})
t_{max}	time at which C_{max} occurs
% w/v	concentration term defined by (weight of solute in grams/volume of solvent in millilitres) × 100

References

1. N. D. Moore, *Acute Pain*, 2009, **11**, 129–137.
2. V. Kumar and B. A. Mahal, *J. Pain Res.*, 2012, **5**, 279–287.
3. K. Haddley, *Drugs Future*, 2010, **35**, 373–378.
4. A. Cattaneo, *Curr. Opin. Mol. Ther.*, 2010, **12**, 94–106.
5. Bandolier website: www.medicine.ox.ac.uk/bandolier/.
6. L. Mason, R. A. Moore, J. E. Edwards, H. J. McQuay, S. Derry and P. J. Wiffen, *BMJ*, 2004, **328**, 995–997.
7. P. Matthews, S. Derry, R. A. Moore and H. J. McQuay, *Cochrane Database Syst. Rev.*, 2009.
8. J. M. Baron and H. F. Merk, *Curr. Opin. Allergy Clin. Immunol.*, 2001, **1**, 287–291.
9. P. M. Elias, *J. Invest. Dermatol.*, 2004, **122**, xxxvi–xxxix.
10. J. A. Bouwstra and G. S. Gooris, *Open Dermatol. J.*, 2010, 10–13.
11. P. M. Elias, *J. Invest. Dermatol.*, 2012, **132**, 2131–2133.
12. J. W. Fluhr and P. M. Elias, *Exog. Dermatol.*, 2002, **1**, 163–175.

13. C. A. Lipinski, F. Lombardo, B. W. Dominy and P. J. Feeney, *Adv. Drug Delivery Rev.*, 1997, **23**, 3–25.
14. Y.-B. Choy and M. R. Prausnitz, *Pharm. Res.*, 2011, **28**, 943–948.
15. T. Yano, A. Nakagawa, M. Tsuji and K. Noda, *Life Sci.*, 1986, **39**, 1043–1050.
16. J. Swarbrick, G. Lee, J. Brom and N. P. Gensmantel, *J. Pharm. Sci.*, 1984, **73**, 1352–1355.
17. A. Alikhan, S. Farahmand and H. I. Maibach, *J. Appl. Toxicol.*, 2009, **29**, 590–596.
18. L.-Z. Kaestli, A.-F. Wasilewski-Rasca, P. Bonnabry and N. Vogt-Ferrier, *Drugs Aging*, 2008, **25**, 269–280.
19. K. V. Roskos, H. I. Maibach and R. H. Guy, *J. Pharmacokinet. Biopharm.*, 1989, **17**, 617–630.
20. T. F. Barkve, K. Langseth-Manrique, J. E. Bredesen and K. Gjesdal, *Am. Heart J.*, 1986, **112**, 537–541.
21. J. Vanakoski, T. Seppala, E. Sievi and E. Lunell, *Clin. Pharmacol. Ther.*, 1996, **60**, 308–315.
22. I. Singh and A. P. Morris, *Int. J. Pharm. Invest.*, 2011, **1**, 4–9.
23. J. Levin and H. Maibach, *J. Pharm. Sci.*, 2012, **101**, 4293–4307.
24. L. Buhse, R. Kolinski, B. Westenberger, A. Wokovich, J. Spencer, C. W. Chen, S. Turujman, M. Gautam-Basak, G. J. Kang, A. Kibbe, *et al.*, *Int. J. Pharm.*, 2005, **295**, 101–112.
25. K. S. Paudel, M. Milewski, C. L. Swadley, N. K. Brogden, P. Ghosh and A. L. Stinchcomb, *Ther. Delivery*, 2010, **1**, 109–131.
26. M. R. Prausnitz and R. Langer, *Nat. Biotechnol.*, 2008, **26**, 1261–1268.
27. E. R. Viscusi, L. Reynolds, F. Chung, L. E. Atkinson and S. Khanna, *JAMA, J. Am. Med. Assoc.*, 2004, **291**, 1333–1341.
28. European Medicines Agency, *Press release*, 2008. www.emea.europa.eu/ema/index.jsp?curl=pages/news_and_events/news/2009/11/news_detail_000249.jsp&mid=WC0b01ac058004d5c1.
29. European Medicines Agency, *Public statement*, 2011. www.emea.europa.eu/docs/en_GB/document_library/Public_statement/2011/01/WC500101495.pdf.
30. Thomson Reuters Cortellis Database, *Incline Therapeutics Announces Acquisition of IONSYS(TM) and $43 million Series A Financing.*
31. C. A. Squier and M. J. Kremer, *J. Natl. Cancer Inst. Monogr.*, 2001, **29**, 7–15.
32. G. Campisi, C. Paderni, R. Saccone, O. Di Fede, A. Wolff and L. I. Giannola, *Curr. Pharm. Des.*, 2010, **16**, 641–652.
33. R. Bhati and R. K. Nagrajan, *Int. J. Pharm. Sci. Res.*, 2012, **3**, 659–681.
34. J. A. Nicolazzo, B. L. Reed and B. C. Finnin, *J. Controlled Release*, 2005, **105**, 1–15.
35. N. M. L. Veldhorst-Janssen, A. A. A. Fiddelers, P.-H. M. van der Kuy, C. Neef and M. A. E. Marcus, *Clin. Ther.*, 2009, **31**, 2954–2987.

36. H. Zhang, J. Zhang and J. B. Streisand, *Clin. Pharmacokinet.*, 2002, **41**, 661–680.
37. K. Toussaint, X. C. Yang, M. A. Zielinski, K. L. Reigle, S. D. Sacavage, S. Nagar and R. B. Raffa, *J. Clin. Pharm. Ther.*, 2010, **35**, 617–638.
38. J. A. Forrest, J. A. Clements and L. F. Prescott, *Clin. Pharmacokinet.*, 1982, **7**, 93–107.
39. P. L. Moller, S. E. Norholt, H. E. Ganry, J. H. Insuasty, F. G. Vincent, L. A. Skoglund and S. Sindet-Pedersen, *J. Clin. Pharmacol.*, 2000, **40**, 370–378.
40. P. L. Moller, S. Sindet-Pedersen, C. T. Petersen, G. I. Juhl, A. Dillenschneider and L. A. Skoglund, *Br. J. Anaesth.*, 2005, **94**, 642–648.
41. S. Daniels, S. Reader, P. Berry and M. Goulder, *Eur. J. Clin. Pharmacol.*, 2009, **65**, 343–353.
42. B. Flouvat, A. Leneveu, S. Fitoussi, B. Delhotal-Landes and A. Gendron, *Int. J. Clin. Pharmacol. Ther.*, 2004, **42**, 50–57.
43. A. Tzortzopoulou, E. D. McNicol, M. S. Cepeda, M. B. D. Francia, T. Farhat and R. Schumann, *Cochrane Database Syst. Rev.*, 2011, **10**.
44. R. A. Moore, S. Derry, H. J. McQuay and P. J. Wiffen, *Cochrane Database Syst. Rev.*, 2011, **9**.
45. World Health Organization, 2012. *WHO Guidelines on the Pharmacological Treatment of Persisting Pain in Children with Medical Illnesses.* WHO, Geneva.
46. J. Romsing, S. Moiniche and J. B. Dahl, *Br. J. Anaesth.*, 2002, **88**, 215–226.
47. B. Bannwarth and F. Pehourcq, *Drugs*, 2003, **63**, 5–13.
48. T. Gordh, T. E. Gordh and K. Lindqvist, *Anesthesiol. Clin.*, 2010, **113**, 1433–1437.
49. J. L. Anderson, *Clin. Ther.*, 1984, **6**, 125–141.
50. B. J. Campbell, M. Rowbotham, P. Stitzlein Davies, P. I. Jacob and N. L. Benowitz, *J. Pharm. Sci.*, 2002, **91**, 1343–1350.
51. C. P. Denaro and N. L. Benowitz, *Med. Toxicol. Adverse Drug Exper.*, 1989, **4**, 412–428.
52. United States Food and Drug Administration, *Postmarket Drug Safety Information for Patients and Providers*, 2007. Public Health Advisory: Life-Threatening Side Effects with the Use of Skin Products Containing Numbing Ingredients for Cosmetic Procedures.
53. U.S. Food and Drug Administration, FDA News Release, January 16 2009, *FDA Alerts Public about Danger of Skin Numbing Products: Serious and life-threatening risks associated with improper use*.
54. Electronic Medicines Compendium, Last visited September 2012, Versatis 5% Medicated Plaster, www.medicines.org.uk/emc/medicine/19291.
55. K. P. Garnock-Jones and G. M. Keating, *Drugs*, 2009, **69**, 2149–2165.
56. G. Mick and G. Correa-Illanes, *Curr. Med. Res. Opin.*, 2012, **28**, 937–951.
57. M. Naguib, M. M. A. Magboul, A. H. Samarkandi and M. Attia, *Drug Saf.*, 1998, **18**, 221–250.
58. M. Cusato, T. Niebel, S. Bettinelli and M. Regazzi, *Eur. J. Pain, Suppl.*, 2011, **5**, 471–475.

59. M. S. Angst and D. R. Drover, *Clin. Pharmacokinet.*, 2006, **45**, 1153–1176.
60. S. D. Bergese, S. Ramamoorthy, G. Patou, K. Bramlett, S. R. Gorfine and K. A. Candiotti, *J. Pain Res.*, 2012, **5**, 107–116.
61. S. D. Bergese, E. Onel, M. Morren and J. Morganroth, *Reg. Anesth. Pain Med.*, 2012, **37**, 145–151.
62. Pacira Pharmaceutical, 2012. *Exparel Prescribing Information*, Dated 23 January 2012.
63. M. J. Caterina, M. A. Schumacher, M. Tominaga, T. A. Rosen, J. D. Levine and D. Julius, *Nature*, 1997, **389**, 816–824.
64. F. Touska, L. Marsakova, J. Teisinger and V. Vlachova, *Curr. Pharm. Biotechnol.*, 2011, **12**, 122–129.
65. M. Wallace and M. Pappagallo, *Expert Rev. Neurother.*, 2011, **11**, 15–27.
66. S. Derry, R. Lloyd, R. A. Moore and H. J. McQuay, *Cochrane Database Syst. Rev.*, 2009, **4**.
67. W. Y. Zhang and W. P. A. Li, *Eur. J. Clin. Pharmacol.*, 1994, **46**, 517–522.
68. NeurogesX, Press release, 2012. *NeurogesX Reports Fourth Quarter and Year-End 2011 Results.*
69. B. Cryer, *Curr. Opin. Gastroenterol.*, 2001, **17**, 503–512.
70. D. Henry, L. L.-Y. Lim, L. A. G. Rodriguez, S. P. Gutthann, J. L. Carson, M. Giffin, R. Savage, R. Logan, Y. Moride, *et al.*, *BMJ*, 1996, **312**, 1563–1566.
71. J. M. Evans, A. D. McMahon, M. M. McGilchrist, G. White, F. E. Murray, D. G. McDevitt and T. M. MacDonald, *BMJ (Clin. Res. Ed.)*, 1995, **311**, 22–26.
72. J.-L. Kienzler, M. Gold and F. Nollevaux, *J. Clin. Pharmacol.*, 2009, **50**, 50–61.
73. Covidien, 2010. *Pennsaid® prescribing information.*
74. Endo Pharmaceuticals, 2009. *Voltaren® Gel Prescribing information.*
75. A. Assandri, S. Canali and C. Giachetti, *Drugs Exp. Clin. Res.*, 1993, **19**, 89–95.
76. M. Brunner, P. Dehghanyar, B. Seigfried, W. Martin, G. Menke and M. Muller, *Br. J. Clin. Pharmacol.*, 2005, **60**, 573–577.
77. S. Miyatake, H. Ichiyama, E. Kondo and K. Yasuda, *Br. J. Clin. Pharmacol.*, 2009, **67**, 125–129.
78. J. Zacher, R. Altman, N. Bellamy, P. Bruehlmann, J. Da Silva, E. Huskisson and R. S. Taylor, *Curr. Med. Res. Opin.*, 2008, **24**, 925–950.
79. T. Massey, S. Derry, R. A. Moore and H. J. McQuay, in *Cochrane Database Syst. Rev.* 2010, **6**.
80. R. A. Moore, M. R. Tramer, D. Carroll, P. J. Wiffen and H. J. Mcquay, *BMJ*, 1998, **316**, 333–338.
81. T. H. Stanley, *J. Pain Symptom Manage.*, 1992, **7**, S3–7.
82. A. E. Patanwala, S. M. Keim and E. Brian .L., *Ann. Pharmacother.*, 2010, **44**, 1800–1809.
83. S. D. Roy and G. L. Flynn, *Pharm. Res.*, 1989, **6**, 825–832.
84. Johnson & Johnson, 2004. Securities and Exchange Commission, Form 10-K.

85. G. K. Gourlay, S. R. Kowalski, J. L. Plummer, D. A. Cherry, P. Gaukroger and M. J. Cousins, *Pain*, 1989, **37**, 193–202.
86. G. K. Gourlay, S. R. Kowalski, J. L. Plummer, D. A. Cherry, S. M. Szekely, L. E. Mather, H. Owen and M. J. Cousins, *Pain*, 1990, **40**, 21–28.
87. W. Jeal and P. Benfield, *Drugs*, 1997, **53**, 109–138.
88. P. W. Peng and A. N. Sandler, *Anesthesiol. Clin.*, 1999, **90**, 576–599.
89. S. Randive and V. Mehta, *Curr. Clin. Pharmacol.*, 2012, **7**, 116–120.
90. G. W. Hanks, M. Nugent, C. M. B. Higgs and M. A. Busch, *Palliat. Med*, 2004, **18**, 698–704.
91. Thomson Reuters Cortellis Database.
92. F. Elsner, G. Zeppetella, J. Porta-Sales and I. Tagarro, *Clin. Drug Invest.*, 2011, **31**, 605–618.
93. Janssen, 2009 (revised July 2012). DURAGESIC® (Fentanyl Transdermal System); highlights of prescribing information.
94. Cephalon Inc., Revised December 2011. ACTIQ® (fentanyl citrate) oral transmucosal lozenge; highlights of prescribing information.
95. S. Mercadante, L. Radbruch, A. Davies, P. Poulain, T. Sitte, P. Perkins, T. Colberg and M. A. Camba, *Curr. Med. Res. Opin.*, 2009, **25**, 2805–2815.
96. Grunenthal, *Grunenthal Annual Report 2011*, Grunenthal, 2012.
97. I. G. V. James, C. M. O'Brien and C. J. McDonald, *J. Pain Symptom Manage.*, 2010, **40**, 266–278.
98. P. G. Conaghan, C. M. O'Brien, M. Wilson and J. P. Schofield, *Osteoarthritis Cartilage*, 2011, **19**, 930–938.

CHAPTER 9

Affective and Cognitive Modulation of Pain

WEREDESELAM M. OLANGO AND DAVID P. FINN*

Pharmacology and Therapeutics, School of Medicine, NCBES Neuroscience Cluster and Centre for Pain Research, University Road, National University of Ireland, Galway, Ireland
*Email: david.finn@nuigalway.ie

9.1 Introduction

It is now widely acknowledged that the intensity and severity of perceived pain does not necessarily correlate linearly with the degree of tissue damage/injury/inflammation or the intensity of nociceptive activity. In this respect, the importance of context and the modulation of pain by emotion is now widely recognised. In particular, stress, fear and anxiety exert potent, but complex, modulatory influences on pain.[1–6] Research suggests that the interaction between emotion and degree of arousal can influence the impact on pain. Positive emotions generally inhibit pain (regardless of arousal level), whereas negative emotions with low to moderate arousal enhance pain and negative emotions with high arousal inhibit pain.[2,7–12] Hence, differential levels of valence and arousal may determine whether an aversive stimulus induces hyperalgesia or analgesia.

Over the past 20 years, there has been growing interest in the relationship and interaction between persistent pain and psychiatric disorders. Anxiety and depression share a complex reciprocal relationship with pain, whereby perception of pain and chronic pain complaints are often greater in patients with these psychiatric disorders, in turn showing a higher prevalence in chronic

RSC Drug Discovery Series No. 35
Pain Therapeutics: Current and Future Treatment Paradigms
Edited by Charlotte Allerton

Published by the Royal Society of Chemistry, www.rsc.org

pain patients.[1,13–16] Clinical studies have shown that anxiety is associated with an increased frequency of chronic pain complaints.[1,17–21] Correlational studies also indicate that anxiety is related to increased pain reports in high trait anxious subjects.[8,22] For example, research indicates that post-traumatic stress disorder (PTSD) is frequently accompanied by acute pain episodes as well as chronic musculoskeletal pain.[23–25] It is also the case that chronic pain patients are more likely to present with depression (21.7% *versus* 10.0%) or anxiety disorders (35.1% *versus* 18.1%) when compared to the general population.[26] Fishbain *et al.* (1986) found generalised anxiety disorder to be the most prevalent of the anxiety disorders in patients with chronic pain, seen in approximately 15% of chronic pain subjects. Including generalised anxiety disorder, adjustment disorder with anxious mood, obsessive compulsive disorder (OCD), PTSD and agoraphobia, the prevalence of clinical anxiety among the population with chronic pain may be as high as 60%.[27] The co-occurrence of anxiety and/or depression with chronic pain amplifies the negative effects of each alone, often complicating the treatment and resulting in poor outcome.[1,13,14,23]

In addition, clinically, stress has a major impact on pain perception. For example, stressful events worsen the symptoms perceived by patients suffering from chronic pain.[28–31] Stress amplifies nociception in irritable bowel syndrome (IBS), headaches and abdominal pain.[32–34] Painful medical syndromes such as chronic shoulder/neck pain syndrome,[35] the complex regional pain syndrome[36] and fibromyalgia (FM)[37] are all strongly associated with or precipitated by stress.

The goal of this chapter is to provide a comprehensive review of research which has investigated emotional and cognitive modulation of pain, with a major emphasis on the impact of stress on pain. The chapter deals primarily with the phenomena of stress-induced analgesia (SIA) and stress-induced hyperalgesia (SIH), reviewing rodent and human models used to study these phenomena and then discussing what is known about the neurobiological mechanisms mediating them. Concepts of cognitive modulation of pain, including the influence of attention and distraction-induced analgesia (DIA), are also introduced in the context of their relevance to the modulation of pain by stress and emotion. The chapter concludes with some perspectives on the implications and complications that emotional and cognitive factors present for the current and future treatment of pain.

9.2 Stress-Induced Analgesia

There is now a large body of evidence demonstrating that stress or fear can induce a potent form of endogenous analgesia which is adaptive and evolutionarily preserved.[4] Acute, highly arousing, intense stress or fear usually suppresses pain in rodents[38–44] and in humans.[2,45–51] Both conditioned (CS) and unconditioned (US) stressors have been shown to induce analgesia, phenomena referred to herein as fear-induced (or fear-conditioned) analgesia (FCA) and SIA, respectively. Exposing rodents directly to the US (such as brief foot shock) results in an active coping 'fight or flight' response and hypoalgesia on standard tests of pain reactivity, and is known as SIA.[52–54] Exposing a rat to a non-aversive stimulus or context that was previously paired with an aversive stimulus (US) (such as foot shock) results in conditioned somatomotor defensive responses including hypoalgesia (FCA).[55–58] Hence, FCA utilises the principle of Pavlovian conditioning and is characterised by a robust decrease in nociceptive behaviour in rodent models which, at its peak, can suppress pain-related behaviour by as much as 90%.[58,59] SIA/FCA is thus a potent endogenous pain suppression response that occurs during or following exposure to unconditioned or conditioned stressful or fearful stimuli. Exposure to such aversive environments is thought to result in recruitment and activation of the intrinsic descending inhibitory pain pathway that projects from the amygdala to the midbrain periaqueductal grey (PAG) and the brainstem rostroventromedial medulla (RVM), terminating in the dorsal horn of the spinal cord (DHSC).[60] Fear is an immediate alarm reaction to present/imminent/expected threat (*i.e.* there is certain expectation) characterised by impulses to escape and a surge of sympathetic arousal which mobilises the organism to take action or assume defensive behaviour (fight or flight response).[61] In unconditioned or conditioned fear associated with exposure to a noxious/aversive stimulus, the rats typically exhibit intense fear-related behaviour (freezing and 22 kHz ultrasonic vocalisation) and hypoalgesia. The perceptual–defensive–recuperative model proposes that inhibition of pain under such conditions has essential survival value because pain-related responses may compromise performance or interfere with other defensive responses, such as escape.[38]

9.2.1 Rodent and Human Models of Stress-Induced Analgesia

Experimental studies investigating the interaction between stress/aversion/fear and pain usually involve the combination of one or more stressors with one or more pain tests/models. Models of SIA/FCA play a key role in dissecting the exact mechanisms underpinning this form of potent endogenous analgesia and identifying novel potential therapeutic targets for pain and stress-related disorders. As has been reviewed in detail by Butler and Finn,[4] numerous different aversive stimuli (which can be conditioned or unconditioned) have been used to model SIA/FCA (see Table 9.1). These include exposure to a mild aversive environment such as exposure to a novel arena, elevated plus maze, or aversive stimuli that may have a noxious component (such as footshock). Some aversive stimuli are predominately psychological in nature (*e.g.* social isolation, restraint, contextual fear) while others may be predominantly physical in nature (*e.g.* forced swimming, cold water immersion, or predators such as biting mice, biting insects, cats, or snake). These aversive stimuli are paired with noxious stimuli, such as the chemical irritant formalin, radiant heat, tail pinches, and intracutaneous electrical current, constituting pain models which are used to assess the level of nociception. In the case of models that employ conditioned stress in rodents (*i.e.* to model FCA), the unconditioned stressor (commonly footshock) is paired with the conditioned stimulus (such as a context, light or

Table 9.1 Summary of rodent and human models used to study SIA/FCA.

Species	*Stressors*	*Painful stimuli*	*References*
Rodents	Context associated with footshocks	Formalin, radiant heat	188,331,334
	Social learning of fear to biting insects	Insect bites	335
	Footshocks	Formalin, radiant heat	70,336
	Continuous cold swim	Radiant heat	337
	Intermittent cold swim	Radiant heat	108
	Haemorrhagic shock	Formalin	338
	Restraint stress	Nitroglycerin	339
	Elevated plus maze	Radiant heat	340
	Exposure to predators	Radiant heat	77
	Social isolation	Radiant heat	341
	Defeat experience	Radiant heat	342
	Predator odour	Radiant heat	134
	Social learning of fear to biting insects	Insect bites	335
	Social conflict	Attack by intruders	343
	Intraperitoneal injection of hypertonic saline	Radiant heat	344
	Food restriction	Radiant heat	345
Humans	Green light previously paired to mental arithmetic + noise	Intracutaneous electric current	45
	Auditory stimulus previously paired to mental arithmetic + noise	Intracutaneous electric current	104
	Spider phobics exposed to spiders	Intracutaneous electric current	346
	Re-exposure to handshocks	Radiant heat	2
	War veterans with PTSD viewing combat video	Radiant heat	347
	Re-exposure to footshocks	Flexion reflex	47
	Transcutaneous electrical nerve stimulation	Radiant heat	348
	Peripheral electrical stimulation	Dental pain	349
	Transcutaneous electrical nerve stimulation	Tourniquet ischaemia	350
	Handshock	Radiant heat	2

Adapted (with modification) from Butler and Finn, 2009.[4]

auditory cue) through Pavlovian conditioning. Human research to understand the impact of anxiety/fear on pain perception has commonly involved induction of anxiety or fear on healthy volunteers with subsequent pain testing.

Inheritance of SIA in mice, studied using a selective breeding technique which artificially selects genes to change trait-relevant gene frequencies in an experimental population, has helped to advance our understanding of the genetic basis of variability that exists in pain response. Randomly mated mice were tested for nociception following swim stress, and postswim latencies on the hot plate were used to select progenitors of low (LA) and high (HA) analgesia lines.[62] Significant divergence in SIA magnitude between HA and LA mice was achieved in the first generation of selection, arguing for both the high heritability and the involvement of few gene loci in mediation of this trait.[63] Indeed, Mogil *et al.* identified a sex-specific locus on chromosome 8 that is significantly associated with non-opioid SIA in female mice but not in males.[64] However, complete abolition of SIA required more than 21 generations of selective breeding, indicating that SIA is very well conserved.[65] Selective breeding for high and low swim SIA was associated with high and low levorphanol-induced analgesia, respectively.[66,67] Such breeding also exerted a striking influence on the magnitude of the opioid-mediated type of footshock analgesia[68] and on stimulation-produced analgesia,[69] but had no effect on non-opioid SIA,[68] suggesting a high degree of common genetic determination in opioid and swim SIA. The data also suggest that individual differences in analgesic responsiveness to opiate drugs could result from genetically determined individual differences in endogenous pain inhibitory mechanisms.[66]

9.2.2 Neurobiological Mechanisms Mediating Stress-Induced Analgesia

Stress/fear inhibits or enhances pain sensation by activating neural pathways in the brain and spinal cord that engage various neurochemicals. The underlying mechanisms are likely to be dependent on the recruitment of neural substrates that are important in the descending inhibitory/facilitatory pain pathways, including the amygdala, PAG, the RVM and the DHSC. Opioid and non-opioid forms of emotional modulation of pain have been described and will be discussed here.

9.2.2.1 Role of Opioids in SIA

Opioids have been the mainstay of pain treatment for thousands of years, and they remain so today. After the discovery of opioid receptors in the early 1970s, stimulation-produced analgesia and its naloxone reversibility strongly pointed to the existence of endogenous opioids.[70] Opioids such as heroin and morphine exert their effects by mimicking naturally occurring substances, called endogenous opioid peptides or endorphins. The endogenous opioid system is made up of three main opioid G-protein coupled receptor (GPCR) families, delta (δ), kappa (κ) and mu (μ) which are all G-protein coupled, and the endogenous ligands such as dynorphins, enkephalins, endorphins and endomorphins.[71] Among other things, the diverse functions of this system include modulation of pain and anxiety/fear.[72,73]

In general, two major types of endogenous pain inhibitory systems are commonly believed to exist: opioid and non-opioid mediated. A large body of evidence suggests a role for the opioid system in SIA/FCA.[40,74–76] Factors determining opioid dependency of SIA/FCA include warm (opioid) *versus* cold (non-opioid) swim stress,[77] forepaw (opioid) *versus* hindpaw (non-opioid) foot shock,[78] intermittent (opioid) *versus* continuous (non-opioid) foot shock,[79] and uncontrollable (opioids)[80] *versus* controllable stress (non-opioids).[81]

Pharmacological studies have shown that systemic or intracerebral administration of opioid receptor antagonists attenuates SIA/FCA in rats,[70,82,83] whereas enhancing endogenous opioid levels using catabolic enzyme (enkephalinase) inhibitors enhanced SIA in rats, in a μ-opioid receptor dependent manner.[84,85] The use of transgenic mice revealed an important role of β-endorphin over met-enkephalin[86,87] and the necessity of μ- or δ-opioid receptors for the expression of opioid-mediated SIA.[88]

The opioidergic system at different levels in the central nervous system (CNS) has been implicated in SIA/FCA. In the PAG, social conflict-induced SIA was associated with decreased β-endorphin-like immunoreactivity.[89] Blocking μ-opioid receptors in the vlPAG with naltrexone attenuated both SIA and FCA in rats.[90,91] In contrast, an infusion of morphine into the nucleus accumbens, but not into the caudate-putamen or prefrontal cortex, impaired the acquisition of FCA in rats on the hotplate test in a dose-dependent manner and was reversed by systemic or intra-accumbal infusion of naloxone.[92] Microinjection of morphine into the amygdala also impaired FCA in rats exposed to a hotplate apparatus in a dose-dependent, naloxone-sensitive manner.[93] In addition, expression of FCA in the tail-flick test after pairing auditory CS and footshock was dependent on μ-opioid receptors in the RVM, but not δ- or κ-opioid receptors.[94] Further support for the involvement of μ-opioid receptors in the RVM in FCA arises from a study where intra-RVM pretreatment with antisense oligodeoxynucleotides directed against μ-opioid receptor mRNA attenuated FCA.[95] Moreover, the κ-opioid receptor agonist, U69593, injected directly into the RVM, completely blocked expression of FCA.[96] These results suggest differential roles of these receptors in FCA. A role for opioid signalling in the hypothalamus and thalamus in SIA has also been demonstrated,[97] and an opioid-mediated pituitary mechanism subserving conditioned analgesia has also been described.[98] However, shock-induced analgesia was seen in both hypophysectomised and sham-operated rats and was sensitive to naloxone, implicating opioids of CNS rather than pituitary origin in its mediation.[99]

Spinal μ-, κ- and δ-opioid receptors have all been shown to mediate SIA/FCA.[100–102] Indeed, combined blockade of all three classes of opioid receptor in the spinal cord antagonised all of the non-opioid forms of SIA, suggesting that all forms of SIA may in fact be dependent on opioid receptor activation at the level of the spinal cord.[100]

In humans, analgesia to nociceptive flexion reflex following footshock stress was reversed by naloxone, providing evidence for involvement of endogenous opioids in SIA.[47,103] Study of SIA in war veterans with PTSD demonstrated that re-exposure to a stimulus resembling the original traumatic event increased pain tolerance to a thermal stimulus, an effect attenuated by naloxone.[49] In addition, in a classical conditioning paradigm combining auditory stimulus with mental arithmetic plus white noise, conditional analgesia was, at least partially, mediated by the endogenous opioid system.[104] It is believed that exposure to conditioned or unconditioned stressors increases levels of endogenous opioids in different brain regions and in the spinal cord, particularly at the level of PAG, RVM and DHSC, and that these in turn act on opioid receptors to produce analgesia.[105]

9.2.2.2 *Role of the HPA Axis in SIA*

The hypothalamo–pituitary–adrenal (HPA) axis is made up of the hypothalamus, the pituitary gland and the adrenal glands and represents the major hormone system responsible for maintenance of homeostasis following exposure to stress.[106,107] In work by Bodnar and colleagues, sham-operated rats exhibited profound analgesia following initial cold water swim stress, while hypophysectomised rats did not display any cold water swim stress-induced operant escape shifts and significantly less cold water swim SIA, indicating the involvement of the pituitary gland in the mediation of cold water swim SIA.[108] Functional blockade of the hypophyseal–adrenocortical system, produced by systemic administration of hydrocortisone or dexamethasone or by implantation of dexamethasone above the paraventricular nucleus of the hypothalamus, resulted in reductions in SIA.[109–111] In addition, both hypophysectomy and dexamethasone administration blocked the analgesic consequences of inescapable shock.[112]

It has been suggested that selective breeding for high and low swim SIA modifies the extent of HPA axis involvement in the generation of SIA. Thus, dexamethasone reduced swim SIA in HA mice, but not in LA mice.[113] The chronic pretreatment with metyrapone potentiated cold swim SIA, an effect which was reversed by naloxone, suggesting that the corticosteroid modulation (of the HPA axis) may have a role in regulating SIA, and interacts with opioid-mediated pain-inhibiting systems.[114] SIA which is blocked by dexamethasone and hypophysectomy is also blocked by adrenalectomy and reinstated with corticosterone therapy.[112] Moreover, adrenalectomy prevented haemorrhagic shock-induced analgesia, which was reversed by corticosterone administration.[115] However, FCA was not accompanied by changes in plasma corticosterone levels in rats.[116]

9.2.2.3 *Role of GABA in SIA*

GABAergic neurons constitute the major mode of inhibitory transmission throughout the CNS. The inhibitory neurotransmitter, GABA, acts on two distinct types of receptor: $GABA_A$ receptors, which are ligand-gated ion channels, and $GABA_B$ receptors, which are G-protein-coupled receptors.[117,118] GABA activity at postsynaptic $GABA_A$ receptors mediates hyperpolarisation by an increasing Cl^- flux. The $GABA_A$ receptor–ionophore complex also contains modulatory receptor sites for the benzodiazepines and barbiturates.[119] $GABA_B$ receptors are located pre- and postsynaptically, and exert their effects by inhibiting voltage-gated calcium channels (thus reducing transmitter release) and by opening potassium channels (thus reducing postsynaptic excitability).[117,120,121] GABAergic neurotransmission plays a key role in supraspinal modulation of pain[72,122,123] and fear.[124–127]

A number of studies have shown a role for GABAergic involvement in FCA[74,128–130] and SIA.[131–133] The effect of GABA receptor agonists/antagonists on SIA is dependent on the dose of drug used, the type of stress and the receptor subtype targeted. Muscimol, a $GABA_A$ receptor agonist, suppressed psychological SIA at low dose but tended to potentiate it at high dose, potentiated swim SIA dose-dependently, and did not affect footshock SIA.[131] Both bicuculline, a $GABA_A$ receptor antagonist, and picrotoxin, a Cl^- channel blocker, dose-dependently suppressed psychological and footshock SIA with little or no effect on swim SIA.[131] Baclofen, a $GABA_B$ receptor agonist, had no influence on SIA, while, CGP 35348, a $GABA_B$ receptor antagonist, dose-dependently blocked footshock SIA without affecting either psychological or swim SIA.[131] Thus, GABAergic signalling plays an important role in the production of SIA with different forms of stress engaging the two receptor subtypes differentially. In male mice, a 30 s exposure to 2-propylthietane, the major component of weasel odour (the odour of a predator) elicited analgesia on the hotplate test which was partially sensitive to the $GABA_A$ receptor antagonist bicuculline.[134] The reduction in formalin-induced nociception following exposure to an aversively conditioned environment was sensitive to the benzodiazepine midazolam,[59,128,135] an effect reversed by the benzodiazepine binding site antagonist flumazenil.[128] Hypoalgesia on the tail-flick test induced by transfer stress was also slightly reduced by diazepam.[136] However, unlike the formalin test,[128,135] on tail-flick test the benzodiazepine diazepam did not affect conditioned analgesia to footshock.[136]

Studies investigating neural substrates involved in GABAergic modulation of FCA have revealed some of the discrete brain regions involved. For example, intra-amygdala and intra-PAG administration of benzodiazepines prevented the expression of FCA in rats.[59,129] A microdialysis study in our laboratory revealed that GABA levels in the basolateral amygdala (BLA), but not in the dorsolateral (dl)PAG, were significantly lower in rats expressing FCA, suggesting that reduced GABAergic tone in the BLA may facilitate expression of FCA.[127] Supporting this hypothesis, a subsequent study demonstrated that enhancing GABAergic tone in the BLA *via* intra-BLA administration of the $GABA_A$ receptor agonist muscimol prevented the full expression of FCA.[130] It has been suggested that non-opioid SIA induced by cold water swim stress might be related to alterations in the rates of general ligand–receptor interactions including $GABA_A$/benzodiazepine system in the cerebral cortex, hippocampus and cerebellum during stress.[137] A role for $GABA_A$ receptors in the spinal cord was shown in cold water swim SIA in mice, as indicated by attenuation following intrathecal administration of the antagonists picrotoxin and biculline.[132] In human volunteers, the analgesic effects of a repetitive stress induced by anticipation of pain (noxious footshock) were attenuated by diazepam.[138] Thus, both enhancing and suppressing GABAergic transmission appear to result in either potentiation or attenuation of SIA/FCA, depending primarily on the type of stress. It appears that following unconditioned stress

GABAergic transmission largely enhances analgesia while following conditioned stress it attenuates analgesia.

9.2.2.4 Role of Glutamate in SIA

Glutamate, the major excitatory neurotransmitter in the mammalian CNS, fulfils a vitally important role in information transfer and synaptic plasticity within the CNS. Exposure to aversive stimuli is characterised by alterations in extracellular levels of glutamate[139,140] and a perturbation of glutamatergic transmission is implicated in the affective symptoms of several psychiatric disorders, including anxiety and depression.[73,141] Glutamate is also very important in the sensitisation of dorsal horn neurons which underlies long-term, painful states[142,143] and is involved in mechanisms of descending pain modulation.[144,145]

A role for glutamate in FCA[146] and SIA[134,147–149] has been demonstrated. A non-opioid form of SIA induced by forced walking combined with the formalin test was sensitive to the *N*-methyl-D-aspartate (NMDA) receptor antagonist LY-235959, albeit in an age-dependent manner.[149] In contrast to a 30 minute exposure to biting flies, which elicited opioid-mediated analgesia, a 5 minute exposure to biting flies elicited analgesia that was antagonised by the competitive NMDA antagonist NPC 1262, though this effect was only seen in male mice, indicating a glutamatergic SIA which is sex-specific.[148] A 30 second exposure to odour of a predator, an ethologically relevant threatening stimulus, elicited analgesia that was blocked by NPC 1262.[147] It was shown that NMDA receptors in the BLA participate in FCA, as bilateral infusion of the NMDA receptor antagonist *d,l*-2-amino-5-phosphonovaleric acid into the rat BLA attenuated analgesia to tone and context which were previously paired with footshock.[146] It appears that antagonising the glutamatergic system at the NMDA receptor attenuates SIA/FCA which implies that acute fear or stress engages the glutamatergic system to enhance descending inhibitory pain pathway activation. Recent evidence also suggests a key role for metabotropic mGluR5 receptors in the dlPAG in mediating unconditioned SIA in rats, *via* a mechanism which involves mobilisation of the endocannabinoid 2-arachidonylglycerol.[150] The role of the endocannabinoid system in SIA and FCA is discussed in more detail in Section 9.2.2.6 below.

9.2.2.5 Role of Monoamines in SIA

The monoamines noradrenaline (norepinephrine), serotonin and dopamine remain the most intensively investigated transmitters implicated in regulation of mood and descending control of pain. The discovery of multiple classes of receptors with divergent roles in the mediation of descending inhibition and descending facilitation has contributed to our expanding knowledge of mechanisms of endogenous analgesia and facilitation of pain.[72] Anxiogenic and other stressful stimuli activate monoaminergic projections that heavily innervate corticolimbic regions involved in integrating the response to anxiety and are accompanied by emotional and autonomic manifestations of fear behaviour.[73]

Serotonin. Systemic administration of quipazine, an agonist at 5-HT_{2A} and 5-HT_3 receptors, enhanced antinociception in rats following stress, and BC-105, an antagonist at 5-HT_{2A} and 5-HT_{2C} receptors, blocked the increase in stress-induced tail-flick latency.[151] Systemic administration of the 5-HT_{1A} receptor agonist 8-OH-DPAT had biphasic effects on open-arm confinement-induced antinociception, antagonising the response at the lowest dose and enhancing it at the highest dose in mice.[152] Forepaw, but not hindpaw, footshock SIA in rats was significantly attenuated by depletion of spinal serotonin or by combined depletion of spinal serotonin and noradrenaline.[153] In addition, exposure to predator odour produced analgesia that was partially sensitive to 8-OH-DPAT.[147]

Despite the well-established role of the amygdala in stress/fear and pain, bilateral injections of 8-OH-DPAT or NAN-190 (another 5-HT_{1A} receptor antagonist) into the amygdala did not alter open-arm confinement-induced antinociception.[154] Repeated footshock-induced analgesia assessed by latency to tail withdrawal from heat was associated with increased frontal cortex serotonin turnover,[155] while FCA was accompanied by reduced tissue serotonin levels in the cerebellum.[156]

Noradrenaline. Clonidine, an α_2-adrenoceptor agonist, potentiated cold water stress-induced analgesia in a synergistic or additive manner in rats.[157] However, Snow and colleagues showed that systemically administered clonidine markedly decreased SIA in rats and phenoxybenzamine, an α_1-adrenoceptor antagonist, increased the peak and duration of SIA.[151] The α_2-adrenoceptor antagonist yohimbine also potentiated cold water swim SIA on the hotplate and tail-flick tests.[157] Furthermore, acute, but not chronic, pretreatment with the tricyclic antidepressant desipramine potentiated cold water swim SIA as measured by the hotplate test.[158] However, FCA was not associated with changes in noradrenaline.[116,156]

At the level of the spinal cord, a role for serotonin and noradrenaline has been demonstrated in forepaw footshock SIA.[153] The analgesia induced by acute exposure to cold water swim stress was shown to covary with levels of brain noradrenaline and is reduced by lesions of the locus coeruleus.[157] These data implicate a role for the noradrenergic receptor system in SIA/FCA though there are equivocal data as to whether enhanced or subdued noradrenergic activity favours SIA/FCA.

Dopamine. When dopaminergic tone was increased with systemic administration of apomorphine (agonist at dopamine D1 and dopamine D2 receptors), a stress-induced increase in rat tail-flick latency was markedly attenuated, whereas, systemic blockade of D2 receptors with haloperidol enhanced SIA.[151] However, hypoalgesia on the rat tail-flick test induced by transfer stress and conditioned analgesia upon re-exposure to an aversive context were both blocked by chlorpromazine, antagonist of D2 receptor.[136] Repeated footshock-induced analgesia was associated with decreased hypothalamic dopamine turnover in rats.[155] Tissue levels of 3,4-dihydroxyphenylacetic acid and the

3,4-dihydroxyphenylacetic acid:dopamine ratio were significantly increased in the PAG while levels of dopamine in the thalamus were decreased in rats expressing FCA.[116] In addition, FCA was accompanied by increased homovanillic acid (HVA):dopamine (DA) ratio in the cerebellum.[156] Overall, both enhancing and inhibiting monoaminergic signalling seem to enhance or reduce SIA/FCA depending on the type of stress and pain models used.

9.2.2.6 Role of the Endocannabinoid System in SIA

The endogenous cannabinoid (endocannabinoid) system consists of the CB_1 and CB_2 receptors, the naturally occurring endogenous ligands (endocannabinoids) that act at these receptors (the two best-characterised being anandamide [AEA] and 2-arachidonoyl glycerol [2-AG]), and enzymes involved in the synthesis and degradation of the endocannabinoids (the latter including fatty acid amide hydrolase [FAAH] which catabolises AEA and monoacylglycerol lipase [MAGL] which catabolises 2-AG). Recent years have seen an explosion of research into this novel signalling system and its role in pain[159–162] and in stress/fear/anxiety.[163–168] CB_1 receptors, the predominant cannabinoid receptor subtype present in the brain,[169] are located throughout brain regions that are key components of the descending inhibitory/facilitatory pain pathway and stress/fear/anxiety circuitry. Both painful stimuli[170] and stress/fear[171,172] have been shown to increase levels of endocannabinoids in relevant brain regions. Taken together, these findings suggest an overlap between neural substrates and pathways involved in cannabinoid-mediated modulation of pain and aversion (fear, anxiety) and implicate the endocanabinoid system as an important common denominator. Furthermore, cannabinoids activate the descending PAG–RVM pathway *via* GABA-mediated disinhibition.[173]

To date, a number of studies have shown a role for the endocannabinoid system in endogenous analgesia, including SIA/FCA. Electrical stimulation-induced analgesia in the dlPAG was blocked by rimonabant and occurred concomitantly with the release of endogenous anandamide.[170] Evidence from pharmacological studies, employing both systemic and intracerebral administration of drugs, supports a role for the endocannabinoids in SIA. Intraperitoneal administration of the CB_1 receptor antagonist rimonbant or its analogue AM251, but not a CB_2 receptor antagonist, abolished a non-opioid form of SIA in rats, assessed with the rat tail-flick test after foot shock as an unconditioned stimulus.[171] Enhancement of cannabinoid SIA was also observed following systemic injection of FAAH inhibitors (AA-5-HT and palmitoyltrifluoromethylketone).[174] The use of transgenic mice lacking the CB_1 receptor further implicates a role for endocannabinoids in SIA.[175] CB_1 receptor knockout mice did not exhibit antinociception following forced swim and showed a decrease in the immobility induced by the previous exposure to electric footshock.[175] Similarly, in a model that combines the formalin test of tonic persistent pain with contextual fear in rats, systemic administration of rimonabant attenuated the fear-induced suppression of formalin-evoked nocicpetive behaviour, providing the first evidence for a role of CB_1 receptors in mediating FCA.[176] Subsequent work revealed enhanced cannabinoid-mediated FCA in rats following systemic administration of the FAAH inhibitor URB597.[177,178] It has been proposed that the endocannabinoid system acts in concert with the endogenous opioid system to mediate the expression of FCA[178] in a similar manner to the functional association between the endogenous opioid and cannabinoid systems during pain modulation.[179,180] Moreover, studies conducted in mice suggest interactions between the endocannabinoid system and the cholecystokininergic system (particularly CCK2 receptors) are important for expression of unconditioned SIA.[181]

A role for the endocanabinoid system in the PAG,[171,174] BLA[182] and RVM[174] and in spinal cord [183] in SIA has been demonstrated. Microinjection of rimonabant into the PAG[171] and BLA[182] but not central nucleus of amygdala (where CB_1 receptors are largely absent) prevented non-opioid SIA induced by continuous footshock. However, neither the FAAH inhibitor URB597 nor the MAGL inhibitor URB602 enhanced SIA when administered into the BLA, at doses that potentiated SIA following administration into the dorsolateral PAG.[182] Moreover, blockade of CB_1 receptors in the RVM attenuated cannabinoid SIA. whereas pharmacological inhibition of FAAH enhanced SIA in a CB_1-dependent manner.[174] The rapid increase of 2-AG/AEA in the midbrain PAG after stress indicates that endocannabinoid release might be responsible for SIA.[171] A recent study by Hohmann and colleagues demonstrates that footshock-induced increases in 2-AG concentrations in the dlPAG and subsequent expression of SIA are likely due to mGluR5-mediated stimulation of the 2-AG synthesising enzyme diacylglycerol lipase-α.[150] Unlike SIA, FCA was not prevented by intra-BLA rimonabant,[156,184] suggesting a differential role of CB_1 receptors in the BLA in FCA and SIA. However, like SIA, we have shown that FCA in rats is fully attenuated by injection of the CB_1 receptor antagonist/inverse agonist rimonabant into the dlPAG and is associated with increased tissue concentrations of anandamide in this region (see Figure 9.1).[185] We have also shown that microinjection of the FAAH inhibitor URB597 directly into the ventral hippocampus results in enhanced FCA in rats, an effect which is blocked by coadministration of rimonabant.[186]

Figure 9.2 provides a summary of the key neurobiological mechanisms that are believed to mediate both SIA and SIH.

9.3 Distraction-Induced Analgesia

Reductions in pain perception can also be achieved with stimuli that are not aversive in nature. Thus, exposure to non-stressful stimuli that command attention can result in a so-called DIA. Of course attention is likely to play an important role in SIA also. In fact, SIA could be a form of DIA, the stress being a distractor. In experiments employing painful electrical stimulation applied to the ankle of arachnophobes exposed to a spider cue, high anxiety was associated with reduced pain ratings, but this effect was lost when differences in subjective attention were corrected, suggesting the importance of attention in SIA/FCA.[51] Attention towards and away from painful stimuli has been shown

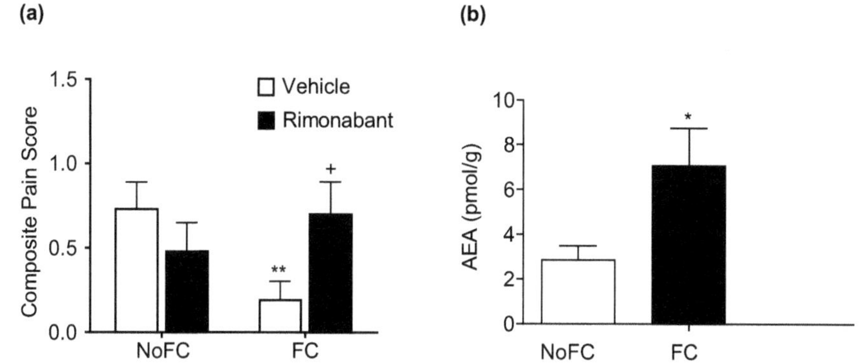

Figure 9.1 (a) The effect of intradorsolateral PAG administration of the CB_1 receptor antagonist/inverse agonist rimonabant (0.4 nmol/0.2 μL) on fear-conditioned analgesia in male Lister-Hooded rats. Formalin-evoked nociceptive behaviour was assessed during a 15 minute re-exposure to an observation chamber paired 24 hours previously with footshock. (b) The effect of re-exposure to an observation chamber paired 24 hours previously with footshock on tissue concentrations of the endocannabinoid anandamide (AEA) in the dorsolateral PAG. $^{**}p<0.01$, $^{*}p<0.05$ *versus* NoFC-Vehicle; $^{+}p<0.05$ *versus* FC-Vehicle. Data expressed as mean ± SEM (n = 5–9). FC, fear-conditioned; NoFC, non-fear-conditioned. Data previously published in Olango *et al.* (2012).[189]

to have different effects on perception of pain. From a clinical point of view, understanding the mechanism involved in DIA is vital as it is devoid of an aversive component and thus has a direct therapeutic implication. Indeed, techniques utilising the principle of DIA are currently used clinically to treat pain. Examples of distracting stimuli used so far to induce analgesia include counting objects,[187] light centrifugal rotation,[188a] listening to music,[188b] virtual reality environment,[189–196] viewing picture of romantic partner[197] and affective pictures.[198] In humans, the use of a virtual reality environment resulted in a marked reduction in subjective pain complaints during painful clinical procedures including dental procedures[195] and dressing of paediatric burns.[189,190,195] Brain regions thought to be involved and identified using imaging techniques in clinical studies of DIA include the prefrontal cortex, anterior cingulate cortex (ACC), amygdala, PAG and RVM.[199–202] An important role for the monoaminergic system in attentional control has been demonstrated in clinical studies.[203] However, a detailed understanding of the neurochemical mechanisms mediating DIA has proven elusive owing to the paucity of animal models of DIA. Our laboratory has reported on a novel putative model of DIA in rats whereby exposure to a novel object significantly reduces formalin-evoked nociceptive behaviour.[204] Exposure to a novel arena or to another formalin-injected rat had little/no effect on nociceptive behaviour, while exposure to the novel object reduced formalin-evoked nociceptive behaviour for up to 20 minutes (see Figure 9.3). This reduction in nociceptive behaviour was accompanied by a significant reduction in serotonin and

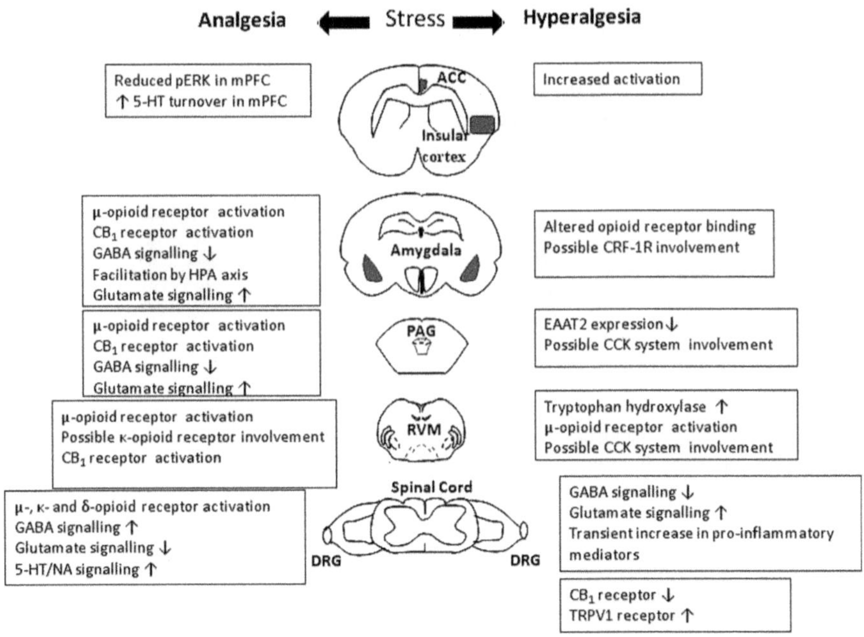

Figure 9.2 Summary of some of the key sites and neurobiological mechanisms thought to mediate stress-induced analgesia (SIA) and stress-induced hyperalgesia (SIH). mPFC, medial prefrontal cortex; PAG, periaqueductal grey; RVM, rostroventromedial medulla; DRG, dorsal root ganglia; pERK, phosphorylated extracellular signal regulated kinase; 5-HT, 5-hydroxytryptamine; NA, noradrenaline; GABA, gamma-aminobutyric acid; CRF-R1, corticotropin releasing factor receptor subtype 1; EAAT2, excitatory amino acid transporter 2; CCK, cholecystokinin; TRPV1, transient receptor potential vanilloid 1.

dopamine metabolites in the medial prefrontal cortex.[204] However, there is a paucity of studies investigating neurochemistry of non-aversive DIA.

9.4 Stress-Induced Hyperalgesia

Stress/anxiety does not invariably suppress pain; it can also enhance nociception and exacerbate pain in a phenomenon referred to herein as SIH. Evidence suggests that depending on the type of stress and pain model employed, stress/anxiety can enhance pain in rodents[205–211] and in humans.[2,8,50,212–217] In general, more sustained but less intense stress in rodents[205] or moderately arousing state/trait anxiety in humans[2,218,219] is often associated with enhanced pain through SIH. Less intense stress results in anxiety, which is a future-orientated emotion characterised by negative affect and apprehensive anticipation of potential threats (uncertain expectation).[2] This uncertainty results in hypervigilance, uneasiness and somatic tension leading to increased alertness and scrutiny towards the environment that enhances sensory receptivity

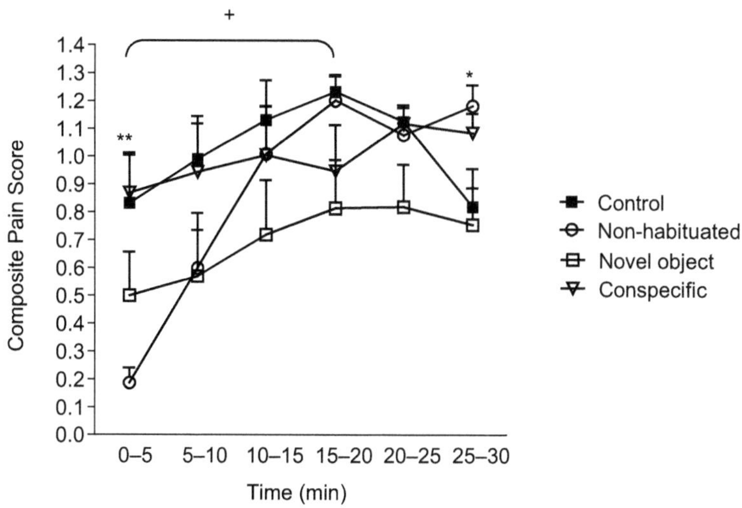

Figure 9.3 Effect of distractors on formalin-evoked nociceptive behaviour in rats. Formalin-evoked nociceptive behaviour was assessed in male Lister-Hooded rats during exposure to one of three different stimuli: a novel arena (Non-habituated), novel object, or another formalin-injected rat separated by a Perspex partition (Conspecific). $^{+}P<0.05$ for novel object *versus* control from 0–20 mins. $^{**}P<0.01$ for non-habituated. *versus* control at 0–5 mins; $^{*}P<0.05$ for non-habituated *versus* control at 25–30 mins. Data are means ± SEM (n = 6–8).
Data previously published in Ford *et al.* (2008).[204]

contributing to hyperalgesia. In other words, in the absence of sufficient information as to the nature of threat or when the animal is not expecting an identifiable grave danger (*e.g.*, being restrained) animals might have difficulty elaborating an orientated response. In this case, increased sensitivity to painful stimuli may represent a more adaptive response than analgesia, enabling the organism to detect any possible threat as early as possible. This hypothesis is in agreement with the general adaptive model of injury-related behaviour,[220] which suggests that when there is a high probability of injury, a fear state (active defensive response) is elicited that inhibits pain; in contrast, when the probability of injury is low, an anxiety state (a passive defensive response) and hyperalgesia result. The same principle was shown to apply in humans where anxiety/fear induced by an aversive event produces analgesia[11,47,221], but anxiety occurring in the absence of knowledge regarding a forthcoming event is accompanied by hyperalgesia.[2,44,217] It has been suggested that the anticipation of pain activates brain regions in close proximity to brain regions activated by pain itself.[222]

9.4.1 Rodent and Human Models of Stress-Induced Hyperalgesia

A number of animal models have been devised to study SIH (Table 9.2). Aversive stimuli or environments used to induce SIH include brief and mildly

Table 9.2 Summary of rodent and human models used to study SIH.

Species	*Stressor/source of anxiety*	*Painful stimulus*	*Ref.*
Rodents	Social threat	Formalin/radiant heat/ acetic acid injection	253,351
	Water avoidance	Colorectal distension/ thermal and mechanical stimuli	206,230,293
	Restraint stress	Rectal distension/TF/ hot tail immersion/ TMJ formalin/HP	228,242,250,260,278,352–354
	Holding/novelty	Tail shock/mechanical stimuli	208,223
	Rotational stress	Formalin	355
	Prenatal stress (restraint)	Formalin	356–358
	Vibration	TF	224,359
	Foot or tail shock	Tail immersion/ colorectal distension/ TF/ tail shock	360–362
	Forced swim	Thermal stimuli, mechanical stimuli, formalin, HP, carrageenan	209–211,302,363,364
	Noise stress	Bradykinin, adrenaline	240,365
	Exposure to cold	Mechanical stimuli/ footshock/capsaicin/ Freund's adjuvant	241,246,247,249
	REM sleep deprivation	Mechanical stimuli	303
Humans	Patients with mood disorders	Electrical pain/ thermal stimuli/ cold pressor pain	19,366–368
	Preoperative state anxiety	Postoperative analgesic consumption	21,235,236
	High and low trait anxiety	Thermal stimuli/ mechanical stimuli	8, 216
	Footshock	Radiant heat/ lower level of shock	215
	Threatening pictures	Immersing arms into hot water	217
	Experimentally induced anxiety (pain-relevant negative information)	Radiant heat/ cold pressor/foot shock / ischaemic arm pain/ mechanical pressure	2,8,50,212–214,369

TF, tail-flick test; HP, hotplate test, TMJ, temporomandibular joint.

aversive but innocuous stimuli such as novelty, vibration and holding.[223,224] Other aversive stimuli used are usually applied repeatedly over a prolonged period of time with the intention of inducing a mild form of stress and hypervigilance. The use of stressors such as forced swimming as a model of SIH has a second advantage since immobility induced by inescapable swim stress is thought to be a model of human depression[225] and patients with depression have increased pain sensitivity.[226] Thus, such models may have significant clinical validity.

Visceral hyperalgesia is commonly observed in patients with conditions such as IBS and interstitial cystitis. Psychological stress is widely believed to play a major role in the precipitation or exacerbation of IBS.[227] Response to colonic/bladder distension after a single or repeated stress has been used to induce visceral hyperalgesia.[206,228–232]

As in the case of animal models, stressors used to model SIA/FCA and SIH in humans can overlap, albeit with some differences with respect to how they are applied. Studies of SIH in humans focus on assessing pain responding in individuals with existing anxiety (trait) or experimentally induced anxiety. To study whether relevant, but not irrelevant, anxiety intensifies pain, volunteers were exposed to laboratory-induced general anxiety or pain-specific anxiety through instructions/warnings.[8,50,213,214,233] Some authors, instead of inducing anxiety in the laboratory, classified healthy subjects from the population into high and low anxiety (trait anxiety) or high or low anxiety sensitivity (a fear of anxiety-related symptoms) using standard anxiety scales (such as the Taylor Manifest Anxiety Scale) and assessed their response to painful stimuli.[216] Subjects with high anxiety showed enhanced pain responding compared with those with low anxiety. A number of studies have examined the responses of patients with mood disorders to painful stimuli in the laboratory (Table 9.2). Moreover, studies on patients undergoing elective surgical procedures suggest that psychological variables such as preoperative anxiety can serve as predictors for postoperative pain response.[21,234–239]

It should be noted that pain perception is variable across individuals and this could be a possible confounding factor when studying the interaction between anxiety and pain in humans. As ethical guidelines require that patients are aware they are in a 'pain' experiment, experimental factors such as human attention and anticipation of a noxious stimulus can confound the results. Furthermore, though fear/anxiety may be induced experimentally, the subjects are reassured that no real danger would happen to them, unlike animal studies where subjects are almost certainly unaware of the nature, severity and potential consequences of the stimulus/threat they are about to experience.

A number of stressors that have been used to produce SIH have also been used to produce SIA; one determining factor appears to be the repetitiveness of the stressor. Repetitive stress favours the induction of hyperalgesia in rodents. Hence, repeated exposure to loud sound,[240] cold environment,[241] restraint[242] or swim stress[210] potentiates pain response. In general, models that induce SIH involve chronic exposure to the stressor (days or weeks rather than minutes), implying that SIH may be the result of more chronic psychological stress rather than acute/physical stress as in the case of SIA. Studies have also suggested that rat models of chronic pain rather than acute pain are associated with elevated pain behaviours in the presence of an aversive stimulus.[208] Children who experience the chronic stress of recurrent abdominal pain display SIH to the cold pressor test.[243] However, the effect of chronic *versus* acute stress on pain perception is not consistent, as acute stress can also produce hyperalgesia and prolonged stress may evoke analgesia. For example, the earliest studies by Vidal and colleagues showed that acute exposure to emotionally arousing non-noxious stress, such as inescapable holding, novel environments or vibration, was shown to produce hyperalgesia, albeit in an immediate and transient fashion,[207,223,224] and exposure to chronic unpredictable stress was found to be analgesic.[244,245]

Paradigms which use chronic stress as opposed to acute stress, however, result in hyperalgesia that usually lasts longer than that induced by acute stress and seems to mimic the human chronic pain condition associated with chronic anxiety-related disorders. For example, repeated cold stress was shown to facilitate the response to noxious stimuli for up to 3 days after the last exposure to cold stress.[241,246–249] Other well-characterised forms of chronic stress paradigms used to model SIH include chronic restraint stress[250–252] and repeated forced swim stress,[209,210] which induce long-lasting hyperalgesia, the hyperalgesia at times lasting 28 days after the cessation of the chronic stress.

It has also been proposed that the severity of the stressor differentially modulates pain sensitivity, such that more severe stressors evoke SIA, whereas less severe stressors evoke SIH. Supporting this proposal, in a model of social threat, acute stress evoked by the potential for actual physical aggression produced analgesia, whereas limiting physical contact between animals eliminated the acute/intense stress and perhaps instead triggered psychological stress/anxiety from the mere presence of an unfamiliar stranger male, thereby producing SIH.[253] In addition, thermal pain reactivity is inhibited in rats after exposure to very severe shock, whereas hyperalgesia is observed after low to moderate-intensity shock. Maier and colleagues showed that rats subjected to inescapable noxious shocks exhibited long-term analgesia whereas rats receiving an identical number of shocks but which were able to escape did not show analgesia.[44] In humans, the use of mild shock (5 mA) which was brief and aversive, but only mildly painful, resulted in hyperalgesia.[215] In contrast, a higher intensity of shock (12 mA and 70 mA) resulted in analgesia.[47,215] Thus, exposure to mild stress repeatedly might create uncertainty about future threats. Increased alertness and early detection of threats, including nociceptive stimuli, would then seem an appropriate adaptive response.

The notion that differential levels of severity may determine whether an aversive stimulus induces hyperalgesia or analgesia is consistent with theories of attentional influences on pain modulation by anxiety/fear which suggest that stimuli that are severe enough to take the attention away from the pain suppresses pain and less intense stimuli increase the attention to the nociceptive stimulus, enhancing pain perception.[51,254,255] Whereas hyperalgesia may occur when the anticipatory anxiety is directed towards the pain itself,[233,256–258] analgesia may occur when anxiety is directed towards a stressor that shifts the attention from the pain.[103,259] However, this is not always the case as fear/anxiety irrelevant to pain can both enhance and suppress pain depending on the intensity.[2]

Anxiety-related modulation of pain may also be influenced by the type of nociceptive stimulus used to evoke pain responding. While anxiety demonstrated no association with pain intensity during cold stimulation, a significant hypersensitivity was found during heat stimulation in humans.[216] Bradesi *et al.* (2005) described a differential modulatory role of repeated psychological water avoidance stress on visceral *versus* somatic nociception in rats. Chronic water avoidance stress resulted in a transient somatic antinociceptive response but sustained visceral hyperalgesia.[206] Similarly, restraint stress reduced the duration of lick/guard responses to thermal stimulation (analgesic effect) while also increasing sensitivity to thermal stimulation (hyperalgesic effect), as assessed by learned escape responses.[260] In summary, the types of stressor, its intensity and duration, as well as the type of the pain model/test used influence the direction of pain modulation by stress, and affect the potency of analgesic or hyperalgesic effect. The literature suggests that the stress-regulatory circuit activated by a particular stressor is crucially dependent on stimulus attributes.[261]

The influence of genetic factors on anxiety-related hyperalgesia in rodents has been examined by assessing pain behaviour across inbred rat strains. The use of inbred strains makes it easier to dissociate genetic and environmental components of a trait because the animals within each strain are theoretically identical in their genotype. Using strains of rats with genetic differences in anxiety traits, studies have shown that the threshold for nociceptive response also appeared to be subject to genetic variation.[262] Fecho and colleagues demonstrated significant strain differences in pain sensitivity in two inbred strains, Lewis and Fischer, and the outbred Sprague-Dawley (SD) strain, which elicit different responses in tests of fear or anxiety.[263] At baseline, Fisher rats were the most sensitive to mechanical stimulation (with von Frey monofilament testing) and the least sensitive to noxious heat pain (the Hargreaves radiant heat test). Following intraplantar administration of carrageenan, Lewis rats showed the least, and Fisher rats showed the highest, thermal hyperalgesia and mechanical allodynia/hyperalgesia.[263] In another study, Lewis rats displayed higher levels of avoidance of different types of anxiogenic stimuli[264] and enhanced hyperalgesia in the formalin test, compared with spontaneously hypertensive rat counterparts.[265] The authors suggested a possibility of differences in the gene *Tac1r* between these two strains, a gene that encodes for the substance P receptor neurokinin (NK)1 receptor, known to influence both pain and anxiety.[265]

Wistar-Kyoto (WKY) rats display a heightened level of anxiety-related behaviour in response to behavioural tests of stress and anxiety such as the acoustic startle, open field and elevated plus maze tests compared with other strains including SD, Wistar, spontaneously hypertensive and Fisher rats (Figure 9.4).[266–269] These rats also exhibit neurochemical differences in response to stress and anxiety tests compared with SD and Lewis rats.[266,270] Interestingly, WKY rats also exhibit reduced hotplate response latencies and enhanced formalin-evoked nociceptive behaviour, compared with SD rats (Figure 9.4).[266] In addition, WKY rats exhibit an exaggerated visceromotor response to innocuous colorectal distension compared with low/moderate-anxiety

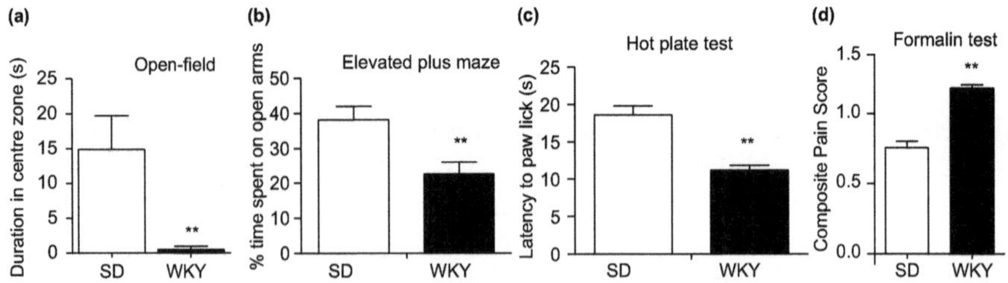

Figure 9.4 Compared with Sprague-Dawley (SD) controls, Wistar-Kyoto (WKY) rats display an anxiogenic phenotype as evidenced by (a) the reduction in time spent in the centre zone of an open-field arena and (b) reduced time spent on the open arms of the elevated plus maze. (c) WKY rats demonstrate a reduction in the latency to hindpaw lick in the hotplate test and (d) enhanced formalin-evoked nociceptive behaviour, indicating hyperalgesia to noxious thermal and chemical somatic stimuli. $^{**}P<0.01$ *versus* SD. Data are means ± SEM (n = 12).

Fisher and Sprague–Dawley rats.[262] Using a peripheral nerve injury model, it has been shown that depression-like behaviour in WKY rats is associated with exacerbated mechanical allodynia compared to Wistar rats.[271] Moreover, water avoidance stress resulted in augmented urinary bladder hyperalgesia in WKY rats.[230] Chronic stress effects were only observed in high-anxiety rats, suggesting a genetic component in SIH.[230] Thus, rats with high anxiety state showed increased sensitivity to colonic[262,272–274] or urinary bladder distension [230] and somatic[266,271] noxious stimuli. The availability of multiple models is useful as even if a particular neurobiological mechanism underlies the altered pain perception in one animal model, alternative mechanisms may underlie this response in the other models. To date, work on genes associated with anxiety–pain comorbidity is lacking. It has been reported that in patients with surgical pain, polymorphisms in three prespecified pain–mood candidate genes (catechol-*O*-methyl transferase, serotonin transporter and brain-derived neurotrophic factor) were not associated with late postoperative change in mood or with a pain–gene interaction on mood.[275]

9.4.2 Neurobiological Mechanisms Mediating Exacerbation of Pain Responding by Stress or Anxiety

9.4.2.1 Role of Opioids in SIH

It was reported that long-lasting delayed swim stress-induced thermal hyperalgesia was prevented by systemic administration of μ-opioid receptor antagonists, suggesting that endogenous opioid systems may also paradoxically play a role in SIH.[276] Prestress treatment with low doses of the opioid receptor antagonist naloxone during exposure to swim stress prevented the development of SIH, indicating that activation of μ-opioid receptors during forced swim stress might be required for the induction of hyperalgesia.[276] It has been shown that illness induced by lithium or lipopolysaccharides results in hyperalgesia in the rat formalin and tail-flick tests[101] which was blocked by naloxone.[277] The attenuation of both hyperalgesia and analgesia with naloxone might be explained by the role opioid receptors play in both descending facilitation and inhibition of pain.

In Long Evans rats, stress-induced increased sensitivity to thermal stimulation after restraint stress was attenuated and enhanced following morphine and naloxone systemic administration, respectively. The authors suggested that the endogenous opioid system mediates effects of stress on these responses by enhancing stress-induced hyporeflexia and opposing SIH.[278] In contrast to the enhanced antinociception following administration of U-50488H (a κ-opioid receptor agonist) observed in repeated cold-stressed mice, the antinociception induced by DAMGO (μ-opioid receptor agonist) or morphine was reduced, suggesting that repeated cold-stressed mice are hyposensitive to supraspinal μ-opioid receptor-mediated antinociception, whereas their antinociception through κ-opioid receptors is increased.[247] In addition, it was demonstrated that repeatedly restrained rats displayed decreased responses to morphine, requiring an increased dose to show the classic antinociceptive effect on the temporomandibular joint formalin test.[242] Opioid-mediated novelty-induced analgesia on tail-flick latencies [279] was not seen in repeatedly restrained rats,[280] suggesting that repeated restraint stress induces an alteration in the nociceptive response, perhaps as a result of altered levels or release of endogenous opioids in these animals. In line with this idea, previous studies showed that chronic restraint stress could modify the activity of the opioid systems.[281] For example, the density of opioid receptors in the repeatedly restrained rats decreased significantly in CNS structures such as the spinal cord, frontal cortex and hippocampus.[282] Indeed, Omiya *et al.* (2000) showed that hypofunction of the supraspinal μ-opioid receptor may explain the hyperalgesic effect of repeated cold stress in mice.[247]

Endogenous opioids are released in the CNS in response to noxious or aversive stimuli.[105] Thus, frequent release of endogenous opioids as a consequence of repeated exposure to stressors could lead to overactivation and desensitisation and/or downregulation of opioid receptors, resulting in a tolerance to the analgesic effects of endogenous opioids, which might be implicated in the hyperalgesia and reduced response to morphine observed after repeated swim stress. Though it is well known that exogenous opioids induce analgesia, hyperalgesia can also occur following opioid administration in rats[283,284] and in humans.[285] Exposure to stress in rats after prior pain or fentanyl experiences 2 weeks earlier resulted in hyperalgesia.[208] Thus, exposure to increased levels of endogenous or exogenous opioids could have either analgesic or hyperalgesic effects on subsequent pain response depending on the frequency, duration of exposure and prior experience. However, it should be noted that activity of opioid receptors is strongly coupled to modulation of the inhibitory amino acid neurotransmitter GABA and the excitatory amino acid neurotransmitter glutamate.[286] For example, tolerance to the analgesic effects of opioids is associated with hyperalgesia[287] and increased activity of NMDA receptors.[288] Thus, alterations in opioid and NMDA receptor function could be implicated in the development and maintenance of SIH.

9.4.2.2 Role of the HPA Axis in SIH

Using a model of intermittent sound stress in rats, it was suggested that stress-induced enhancement of hyperalgesia requires activity in the sympathoadrenal system (*via* release of adrenaline) and HPA axis (*via* release of corticosterone).[240,289] Hypophysectomy potentiated inescapable holding-induced hyperalgesia, but attenuated novelty-induced hyperalgesia.[223] However, dexamethasone, which is known to block the stress-induced release of adrenocorticotropic hormone (ACTH) and endorphin from the anterior lobe of the pituitary did not affect novelty-induced hyperalgesia but enhanced hyperalgesia induced by holding.[223] Thus, hypophyseal factors, not affected by dexamethasone and originating from the pituitary, may participate in novelty-induced hyperalgesia, whereas analgesic mediators originating in the anterior pituitary (*e.g.* opioids) appear to counteract the holding-induced hyperalgesia.

Corticotropin releasing factor (CRF) is a hypothalamic peptide that stimulates the synthesis and release of ACTH and β-endorphin from the pituitary and interacts with CRF receptors, subtype 1 (CRF_1 receptor) and/or subtype 2 (CRF_2 receptor).[290,291] It acts both peripherally, as a hormone, and centrally on various brain regions that mediate the central response to stress (*i.e.* hypothalamus, amygdala, locus coeruleus, dorsal raphe nucleus, and hippocampus). CRF is reported to play an important role in stress-induced visceral hyperalgesia.[292] Peripheral administration of the selective CRF_1 receptor antagonist CP-154526, prior to water avoidance stress, prevented delayed stress-induced colonic hyperalgesia.[293] In addition, systemic administration of the selective CRF_1 receptor antagonist, NBI 35965, abolished stress-induced visceral hyperalgesia.[294] In rats, central administration of CRF mimics the effect of restraint stress in increasing the number of abdominal contractions to rectal distension (visceral hyperalgesia),[295] and the hyperalgesic effect of such stress was blocked by central administration of the CRF antagonist α-helical CRF9–41. It has also been reported that CRF_1 receptors are involved in stress-induced visceral hyperalgesia in a rat model of neonatal stress.[296]

In humans, peripherally administered CRF decreased perception thresholds and increased intensity ratings in response to rectal distension in healthy volunteers.[297] Another study in humans has also shown that α-helical CRF significantly reduces the abdominal pain evoked by electrical stimulation in patients with IBS.[298]

9.4.2.3 Role of GABA in SIH

Exposure to stress alters GABAergic transmission in the CNS, which in turn regulates the stress response.[299,300] Stress is also thought to decrease the activity of the $GABA_A$ receptor complex, an effect mimicked by the *in vivo* administration of different inhibitors of GABAergic transmission and antagonised by anxiolytic benzodiazepines.[301]

In forced swim-stressed rats, spinal GABA release was reduced in association with behavioural hyperalgesia, both of which were prevented by prestress treatment with diazepam, a positive modulator of $GABA_A$ receptors.[302]

The antihyperalgesic effect of diazepam was blocked by flumazenil, a selective antagonist of benzodiazepine binding sites, suggesting the involvement of $GABA_A$ receptors. In the same experiment, prestress treatment with diazepam blocked pain-induced c-Fos expression in laminae I–VI of the ipsilateral lumbar dorsal horn in a flumazenil-sensitive manner. This finding indicates an involvement of spinal GABA receptors in SIH. In addition, it has been reported that diazepam abolishes the immediate and transient thermal and mechanical hyperalgesia observed after brief exposure to non-noxious stressors such as a novel environment[223] or 15 minute exposure to restraint/vibration stress.[224] However, diazepam did not affect inescapable holding-induced hyperalgesia.[224] Thus, as is the case with SIA, the role of GABA receptors in SIH depends on the types of stressors involved, which in turn is likely to impact on GABAergic tone in key neural substrates involved in stress–pain interactions.

9.4.2.4 Role of Glutamate in SIH

Mechanical hypersensitivity induced by REM sleep disturbance in a rat model of nerve ligation-induced neuropathic pain and sham controls was attenuated by spinal administration of an mGluR5 antagonist or an NMDA receptor antagonist.[303] In nerve-injured animals, the antihyperalgesic effect was most pronounced with the mGluR5 antagonist. This work suggests a role for glutamate receptors in SIH and that mechanical hypersensitivity following REM sleep disturbance and peripheral nerve injury share common spinal mechanisms involving mGluR5.[303] Administration of the NMDA receptor antagonist BN2572, prior to innocuous stress, completely prevented SIH in rats with prior pain and fentanyl experiences, suggesting that a sustained NMDA receptor blockade is required to counteract the activation of NMDA-dependent pronociceptive systems induced by innocuous stress in pain and prior opioid-experienced rats.[208] In addition, the NMDA receptor antagonist ketamine, at a dose that did not alter rat behaviour in the hotplate test in non-stressed rats, prevented and reversed SIH, suggesting the involvement of NMDA receptors in both the initiation and maintenance of this phenomenon.[276]

Several studies have found that stressors such as restraint and forced swimming induce glutamate release,[304,305] and in rats subjected to repeated cold stress the release of glutamate evoked by capsaicin was markedly increased in the DHSC.[249] This could lead to compensatory changes in the NMDA receptor system, perhaps interfering with the descending pain pathway. In addition, NMDA receptor antagonists also prevent tolerance to morphine analgesia induced by repeated social defeat stress[306] and hyperalgesia induced by repeated administration of an opioid receptor agonist.[307]

9.4.2.5 Role of Monoamines in SIH

Serotonin. It has been reported that a cold stress paradigm that decreased the levels of both serotonin and 5-hydroxyindoleacetic acid (5-HIAA) in the hypothalamus, midbrain, thalamus, pons and medulla of rats[308] induced hyperalgesia in mice which was suppressed by the systemic administration of 5-hydroxytryptophan, a precursor of serotonin, and by L-3,4-dihydroxyphenylalanine, a precursor of catecholamines.[309] Administration of fluoxetine, a selective serotonin reuptake inhibitor, attenuated chronic restraint stress-induced hyperalgesia in the formalin test in rats.[310] In addition, forced swim-induced hyperalgesia was completely blocked by acute pretreatment with tryptophan, a precursor of serotonin, and long-term pretreatment with clomipramine and fluoxetine, an effect not attributed to their analgesic properties.[209] Restraint SIH to thermal stimuli was associated with increased pERK immunoreactivity in neurons of the RVM, three-quarters of which are serotonergic in nature.[252] In the same study, levels of tryptophan hydroxylase were significantly increased in the RVM. These data suggest that the hyperalgesia after chronic/subchronic stress might be mediated by changes in the activity of the central serotoninergic system. Moreover, WKY rats exhibited SIH which was inversely correlated with serotonin and 5-HIAA levels in the hypothalamus.[266]

The effect of swim stress on serotonin release in the brain is regionally specific and bidirectional. Microdialysis studies in freely moving rats have shown increased serotonin release in several brain regions, especially in the raphe magnus, following short-lived forced swimming.[311–313] On the other hand, prolonged forced swimming (*e.g.* 30 minutes) diminishes the efflux of serotonin in the amygdala and lateral septum.[314] Moreover, swim stress produced decreases in extracellular serotonin levels in the forebrain targets of the dorsal raphe nucleus, a major serotonergic nucleus.[312] In cold-stressed rats, the levels of both serotonin and 5-HIAA decreased in the hypothalamus, thalamus, midbrain, pons and medulla.[248] Current treatments for painful conditions such as temporomandibular disorders that show increased stress, depression, anxiety and somatisation[315,316] utilise drugs, such as fluoxetine, that increase the levels or activity of serotonin.[317] Thus, potentially changes in the activity of central serotoninergic systems might explain, at least in part, the bidirectional changes in nociception (analgesia and hyperalgesia) seen after different stress conditions.

Noradrenaline. Cerebral release of noradrenaline during stress has been implicated as a factor in the expression of anxiety,[318] which in turn enhances pain sensitivity in human subjects.[10,218] In addition, repeated stress is associated with increased noradrenaline turnover.[319] Clonidine, an α_2-adrenoceptor agonist and inhibitor of the synaptic release of noradrenaline, was shown to block vibration-induced hyperalgesia, indicating that enhanced noradrenergic system activity is involved in this form of SIH.[224] However, pretreatment with milnacipran, a dual serotonin/noradrenaline uptake inhibitor, reversed repeated forced swim stress-induced muscle hyperalgesia without modifying prestress muscle nociception, suggesting that enhanced central noradrenaline and/or serotonin reverses this phenomenon.[211] Furthermore, chronic restraint SIH to thermal stimuli applied to the tail was associated with significantly decreased pERK immunoreactivity in the rat locus coeruleus, a major nucleus of noradrenaline in the brain.[252]

Dopamine. At present, direct experimental evidence implicating a role of dopamine in SIH is lacking. However, exposure to acute stressors increases activation of dopaminergic neurons within the ventral tegmental area (VTA)[320,321] and release of dopamine in a number of brain regions, including the nucleus accumbens.[322,323] Rats exposed to chronic unavoidable stress exhibit decreased dopamine efflux in the nucleus accumbens shell, which is associated with a reduced activity of dopaminergic neurons.[324] Indeed, there is a disruption of both serotonergic and dopaminergic function that occurs within the nucleus accumbens following chronic stress, the impact on dopamine outlasting that of serotonin.[325] Fibromyalgia, a clinical disorder characterised by disturbance of emotion and nociception, is thought to embody a hypodopaminergic state. A decrease in the concentration of dopamine metabolites (along with those of serotonin and noradrenaline) in patients with fibromyalgia has also been shown.[326,327] Future work investigating the role of the dopaminergic system in SIH is needed to improve our understanding.

9.4.2.6 Role of the Endocannabinoid System in SIH

Despite the well-established role of the endocannabinoid system in stress, anxiety and pain,[328] very few studies have investigated the role of endocannabinoids in SIH, with the only studies to date focusing on visceral hyperalgesia. Intraperitoneal administration of the CB_1 receptor agonist arachidonyl-2-chloro ethylamine (ACEA) significantly diminished the enhanced visceromotor reflex to colorectal distension and also attenuated changes in electromyogram response in rats stressed by partial restraint, whereas the CB_1 receptor antagonist/inverse agonist (rimonabant) had the opposite effect.[329] In the same experiment, a stress-induced upregulation of colon CB_1 receptors was demonstrated. In another study, visceral motor response increased significantly in water-avoidance stressed rats, indicating hyperalgesia.[330] Treatment of water-avoidance stressed rats with the cannabinoid receptor agonist WIN 55,212-2, prevented the development of visceral hyperalgesia. In the same experiment, levels of anandamide in the dorsal root ganglia of stressed rats were increased; while CB_1 receptor expression was decreased.[330] These results suggest that endocannabinoid signalling through CB_1 may play an important role in stress-induced visceral hyperalgesia. More work is needed to understand the role of endocannabinoids in other forms of SIH.

9.5 Conclusions and Implications for the Treatment of Pain

Clearly, the reciprocal interactions between stress and pain are highly complex and present both challenges and opportunities for the improved treatment of pain and its comorbidity with stress-related psychiatric disorders. The ability of humans to engage potent forms of endogenous analgesia including SIA, FCA and DIA has clinical potential. A thorough understanding of these forms of potent endogenous analgesia and the neurochemical and molecular mechanisms underpinning them could allow for targeting/harnessing of those mechanisms for therapeutic gain. For obvious reasons, induction of SIA/FCA *per se* is not likely to be a useful or ethically valid means of treating pain clinically. However, if it were possible to engage the same potent analgesic mechanisms that underpin SIA/FCA, but avoid induction of stress/anxiety/fear, then this could be a useful therapeutic approach. As Harris and Westbrook described, fear is sufficient, but not necessary for FCA,[74] and other evidence also exists supporting the notion that it is possible to isolate the aversive aspect from the analgesia[331] and *vice versa*.[156,332] Expression of SIA might also be useful as a predictor for development of psychological disorders. For example, SIA sensitivity in abused women has been used as a marker for the future development of PTSD.[333] Furthermore, SIA-based tests aimed at investigating the functional integrity of endogenous analgesic systems may be useful as a means to better understand the pathophysiology of pain and anxiety disorders and also as potential diagnostic markers of these disorders. Clearly, attentional modulation of pain through DIA has already found a place in clinical practice as reviewed above. Increased understanding of the neurobiological mechanism mediating DIA should, however, facilitate improvements in DIA approaches and their efficacy. In addition, improved understanding of the neurochemical and receptor mechanisms involved in DIA should allow for a more rational design of patient treatment plans which might incorporate pharmacological approaches alongside DIA with the aim of achieving synergistic analgesic efficacy.

On the other hand, exacerbation of pain by stress, and comorbidity of pain with stress-related psychiatric disorders including anxiety and depression represent significant clinical challenges. Increased prevalence of catastrophising in these patient groups adds further to the challenge. Once again, a multidisciplinary approach combining psychological interventions such as cognitive behavioural therapy (CBT) with pharmacological treatment is likely to result in the best treatment outcomes for patients. Our ever-increasing understanding of overlap and interactions that exist between the neural substrates and neurochemical mechanisms that regulate pain and mood means that it may be possible to develop new drugs which can treat both pain and co-occurring anxiety/depression. Indeed, the current use of drugs such as pregabalin, amitriptyline and duloxetine for the treatment of both pain and anxiety/depression illustrates the close associations that exist between pain and psychiatric disorders and suggests that novel drugs with improved efficacy and fewer adverse effects may eventually emerge from research focused on understanding pain–affect interactions.

Acknowledgements

Funding from Science Foundation Ireland (10/IN.1/B2976 and 05/YI2/B686) and the Irish Research Council for Science, Engineering and Technology is gratefully acknowledged. We thank Ms Fiona McGowan for her assistance with formatting the chapter.

References

1. G. J. Asmundson and J. Katz, *Depression Anxiety*, 2009, **26**, 888.
2. J. L. Rhudy and M. W. Meagher, *Pain*, 2000, **84**, 65.
3. J. L. Rhudy and M. W. Meagher, *Curr. Opin. Psychiatr.*, 2001, **14**, 241.
4. R. K. Butler and D. P. Finn, *Prog. Neurobiol.*, 2009, **88**, 184.
5. G. K. Ford and D. P. Finn, *Pain*, 2008, **140**, 3.
6. K. Wiech and I. Tracey, *NeuroImage*, 2009, **47**, 987.
7. M. de Wied and M. N. Verbaten, *Pain*, 2001, **90**, 163.
8. M. J. Dougher, *J. Abnorm. Psychol.*, 1979, **88**, 137.
9. M. W. Meagher, R. C. Arnau and J. L. Rhudy, *Psychosom. Med.*, 2001, **63**, 79.
10. J. L. Rhudy and M. W. Meagher, *Pain*, 2003, **104**, 617.
11. J. L. Rhudy and M. W. Meagher, *Pain Med.*, 2003, **4**, 244.
12. J. L. Rhudy and M. W. Meagher, *J. Pain*, 2001, **2**, 57.
13. M. J. Bair, R. L. Robinson, W. Katon and K. Kroenke, *Arch. Intern. Med.*, 2003, **163**, 2433.
14. R. Lieb, G. Meinlschmidt and R. Araya, *Psychosom. Med.*, 2007, **69**, 860.
15. C. H. Dominick, F. M. Blyth and M. K. Nicholas, *Pain*, 2012, **153**, 293.
16. G. Gambassi, *Arch. Gerontol. Geriatr.*, 2009, **49 Suppl 1**, 103.
17. T. M. Palermo and D. Drotar, *J. Pediatr. Psychol.*, 1996, **21**, 683.
18. J. H. Atkinson, M. A. Slater, T. L. Patterson, I. Grant and S. R. Garfin, *Pain*, 1991, **45**, 111.
19. R. H. Dworkin, W. C. Clark and J. D. Lipsitz, *Psychiatry Res.*, 1995, **56**, 173.
20. S. Lautenbacher, J. Spernal, W. Schreiber and J.-C. Krieg, *Psychosom. Med.*, 1999, **61**, 822.
21. Z. N. Kain, F. Sevarino, G. M. Alexander, S. Pincus and L. C. Mayes, *J. Psychosom. Res.*, 2000, **49**, 417.
22. R. M. Malow, J. A. West and P. B. Sutker, *J Abnorm. Psychol.*, 1987, **96**, 184.
23. G. J. G. Asmundson and J. Katz, *Pain*, 2008, **138**, 247.
24. J. D. Otis, T. M. Keane and R. D. Kerns, *J. Rehabil. Res. Dev.*, 2003, **40**, 397.
25. J. C. Shipherd, M. Keyes, T. Jovanovic, D. J. Ready, D. Baltzell, V. Worley, V. Gordon-Brown, C. Hayslett and E. Duncan, *J. Rehabil. Res. Dev.*, 2007, **44**, 153.
26. L. A. McWilliams, B. J. Cox and M. W. Enns, *Pain*, 2003, **106**, 127.
27. D. A. Fishbain, M. Goldberg, B. R. Meagher, R. Steele and H. Rosomoff, *Pain*, 1986, **26**, 181.
28. A. Zautra, B. Parrish, C. Van Puymbroeck, H. Tennen, M. Davis, J. Reich and M. Irwin, *J. Behav. Med.*, 2007, **30**, 187.
29. C. Zaza and N. Baine, *J. Pain Symptom Manage.*, 2002, **24**, 526.
30. D. A. Fishbain, B. Cole, R. B. Cutler, J. Lewis, H. L. Rosomoff and R. S. Rosomoff, *Pain Med.*, 2006, **7**, 509.

31. R. Conrad, G. Schilling, C. Bausch, J. Nadstawek, H. C. Wartenberg, I. Wegener, F. Geiser, K. Imbierowicz and R. Liedtke, *Pain*, 2007, **133**, 197.
32. E. J. Bennett, C. C. Tennant, C. Piesse, C.-A. Badcock and J. E. Kellow, *Gut*, 1998, **43**, 256.
33. C. C.-M. Boey and K.-L. Goh, *J. Gastroenterol. Hepatol.*, 2002, **17**, 1250.
34. G. Alfven, V. Östberg and A. Hjern, *J. Psychosom. Res.*, 2008, **65**, 381.
35. K. B. Nilsen, T. Sand, R. H. Westgaard, L. J. Stovner, L. R. White, R. Bang Leistad, G. Helde and M. Rø, *Eur. J. Pain*, 2007, **11**, 743.
36. L. A. Grande, J. D. Loeser, J. Ozuna, A. Ashleigh and A. Samii, *Pain*, 2004, **110**, 495.
37. B. Van Houdenhove and P. Luyten, *Acta Neurol. Belg.*, 2006, **106**, 149.
38. R. C. Bolles and M. S. Fanselow, *Behav. Brain Sci.*, 1980, **3**, 291.
39. R. J. Bodnar, D. D. Kelly, M. Brutus and M. Glusman, *Neurosci. Biobehav. Rev.*, 1980, **4**, 87.
40. A. I. Basbaum and H. L. Fields, *Annu. Rev. Neurosci.*, 1984, **7**, 309.
41. M. S. Fanselow, *Ann. N. Y. Acad. Sci.*, 1986, **467**, 40.
42. L. R. Watkins and D. J. Mayer, *Ann. N. Y. Acad. Sci.*, 1986, **467**, 273.
43. A. H. Lichtman and M. S. Fanselow, *Brain Res.*, 1990, **533**, 91.
44. S. F. Maier, R. C. Drugan and J. W. Grau, *Pain*, 1982, **12**, 47.
45. H. Flor and S. M. Grusser, *Eur. J. Pain*, 1999, **3**, 317.
46. R. M. Malow, *Pain*, 1981, **11**, 397.
47. J. C. Willer, H. Dehen and J. Cambier, *Science*, 1981, **212**, 689.
48. M. N. Janal, E. W. D. Colt, W. C. Clark and M. Glusman, *Pain*, 1984, **19**, 13.
49. R. K. Pitman, B. A. van der Kolk, S. P. Orr and M. S. Greenberg, *Arch. Gen. Psychiatry*, 1990, **47**, 541.
50. M. Al Absi and P. D. Rokke, *Pain*, 1991, **46**, 43.
51. S. A. Janssen and A. Arntz, *Pain*, 1996, **66**, 145.
52. M. S. Fanselow, *Behav. Neurosci.*, 1984, **98**, 79.
53. J. W. Grau, *Behav. Neurosci.*, 1984, **98**, 278.
54. S. F. Maier, *Behav. Neurosci.*, 1989, **103**, 131.
55. J. Iwata and J. E. LeDoux, *Behav. Neurosci.*, 1988, **102**, 66.
56. J. LeDoux, J. Iwata, P. Cicchetti and D. Reis, *J. Neurosci.*, 1988, **8**, 2517.
57. F. J. Helmstetter and P. S. Bellgowan, *Brain Res.*, 1993, **612**, 253.
58. D. Finn, S. Beckett, D. Richardson, D. Kendall, C. Marsden and V. Chapman, *Eur. J. Neurosci.*, 2004, **20**, 848.
59. J. A. Harris and R. F. Westbrook, *Behav. Neurosci.*, 1995, **109**, 295.
60. R. K. Butler and D. P. Finn, *Prog. Neurobiol.*, 2009, **88**, 184.
61. D. H. Barlow, B. F. Chorpita and J. Turovsky, *Nebr. Symp. Motiv.*, 1996, **43**, 251.
62. I. Panocka, P. Marek and B. Sadowski, *Brain Res.*, 1986, **397**, 152.
63. W. F. Sternberg and J. F. Mogil, 2001, **15**, 229.
64. J. S. Mogil, S. P. Richards, L. A. O'Toole, M. L. Helms, S. R. Mitchell, B. Kest and J. K. Belknap, *J. Neurosci.*, 1997, **17**, 7995.

65. J. S. Mogil, W. F. Sternberg, P. Marek, B. Sadowski, J. K. Belknap and J. C. Liebeskind, *Proc. Natl. Acad. Sci. U. S. A.*, 1996, **93**, 3048.
66. P. Marek, J. S. Mogil, J. K. Belknap, B. Sadowski and J. C. Liebeskind, *Brain Res.*, 1993, **608**, 353.
67. K. Lutfy, B. Sadowski, P. Marek, I.-S. Kwon, J. F. W. Keana and E. Weber, *Pharmacol. Biochem. Behav.*, 1996, **54**, 495.
68. P. Marek, I. Panocka and B. Sadowski, *Pain*, 1987, **29**, 393.
69. P. Marek, R. Yirmiya, I. Panocka and J. C. Liebeskind, *Brain Res.*, 1989, **489**, 182.
70. H. Akil, D. J. Mayer and J. C. Liebeskind, *Science*, 1976, **191**, 961.
71. H. Akil, S. J. Watson, E. Young, M. E. Lewis, H. Khachaturian and J. M. Walker, *Annu. Rev. Neurosci.*, 1984, **7**, 223.
72. M. Millan, *Prog. Neurobiol.*, 2002, **66**, 355.
73. M. J. Millan, *Prog. Neurobiol.*, 2003, **70**, 83.
74. J. A. Harris and R. F. Westbrook, *Psychopharmacology (Berl)*, 1994, **115**, 65.
75. H. Akil, E. Young, J. M. Walker and S. J. Watson, *Ann. N. Y. Acad. Sci.*, 1986, **467**, 140.
76. Z. Amit and Z. H. Galina, *Physiol. Rev.*, 1986, **66**, 1091.
77. M. Kavaliers and K. P. Ossenkopp, *Pain*, 1988, **32**, 223.
78. L. R. Watkins, D. A. Cobelli and D. J. Mayer, *Brain Res.*, 1982, **245**, 97.
79. J. W. Lewis, J. T. Cannon and J. C. Liebeskind, *Science*, 1980, **208**, 623.
80. S. A. Janssen and A. Arntz, *Behav. Res. Ther.*, 1997, **35**, 823.
81. S. A. Janssen and A. Arntz, *Behav. Res. Ther.*, 1999, **37**, 77.
82. M. S. Fanselow, D. J. Calcagnetti and F. J. Helmstetter, *J. Pharmacol. Exp. Ther.*, 1989, **250**, 825.
83. S. L. Hart, H. Slusarczyk and T. W. Smith, *Eur. J. Pharmacol.*, 1983, **95**, 283.
84. R. E. Chipkin, M. B. Latranyi and L. C. Iorio, *Life Sci.*, 1982, **31**, 1189.
85. R. Greenberg and E. H. O'Keefe, *Life Sci.*, 1982, **31**, 1185.
86. B. L. Kieffer, *Trends Pharmacol. Sci.*, 1999, **20**, 19.
87. M. Rubinstein, J. S. Mogil, M. Japón, E. C. Chan, R. G. Allen and M. J. Low, *Proc. Natl. Acad. Sci. U. S. A.*, 1996, **93**, 3995.
88. C. Contet, C. Gaveriaux-Ruff, A. Matifas, C. Caradec, M.-F. Champy and B. L. Kieffer, *Neuropsychopharmacology*, 2005, **31**, 1733.
89. P. Kulling, H. R. Frischknecht, A. Pasi, P. G. Waser and B. Siegfried, *Brain Res.*, 1988, **450**, 237.
90. F. J. Helmstetter and J. Landeira-Fernandez, *Brain Res.*, 1990, **537**, 88.
91. C. P. Wiedenmayer and G. A. Barr, *Behav. Neurosci.*, 2000, **114**, 125.
92. R. F. Westbrook, J. A. Harris, A. J. Good and G. Paxinos, *Q. J. Exp. Psychol. B*, 1992, **45**, 99.
93. A. J. Good and R. F. Westbrook, *Behav. Neurosci.*, 1995, **109**, 631.
94. H. Foo and F. J. Helmstetter, *Pain*, 1999, **83**, 427.
95. H. Foo and F. J. Helmstetter, *Mol. Brain Res.*, 2000, **76**, 282.
96. H. Foo and F. J. Helmstetter, *NeuroReport*, 2000, **11**, 3349.
97. A. Kurumaji, M. Takashima and H. Shibuya, *Peptides*, 1987, **8**, 355.

98. M. Gaiardi, M. Bartoletti, C. Gubellini, A. Bacchi and M. Babbini, *Pain*, 1983, **17**, 83.
99. J. W. Lewis, E. H. Chudler, J. T. Cannon and J. C. Liebeskind, *Proc. West Pharmacol. Soc.*, 1981, **24**, 323.
100. L. R. Watkins, E. P. Wiertelak and S. F. Maier, *Brain Res.*, 1992, **582**, 1.
101. E. P. Wiertelak, H. Y. T. Yang, K. Mooney-Heiberger, S. F. Maier and L. R. Watkins, *Brain Res.*, 1994, **634**, 214.
102. L. Menendez, F. Andres-Trelles, A. Hidalgo and A. Baamonde, *Brain Res.*, 1993, **611**, 264.
103. J. C. Willer and D. Albe-Fessard, *Brain Res.*, 1980, **198**, 419.
104. H. Flor, N. Birbaumer, R. Schulz, S. M. Grüsser and R. F. Mucha, *Eur. J. Pain*, 2002, **6**, 395.
105. A. L. Curtis, N. T. Bello and R. J. Valentino, *J. Neurosci.*, 2001, **21**, RC152.
106. J. P. Herman, H. Figueiredo, N. K. Mueller, Y. Ulrich-Lai, M. M. Ostrander, D. C. Choi and W. E. Cullinan, *Front. Neuroendocrinol.*, 2003, **24**, 151.
107. C. Tsigos and G. P. Chrousos, *J. Psychosom. Res.*, 2002, **53**, 865.
108. R. J. Bodnar, M. Glusman, M. Brutus, A. Spiaggia and D. D. Kelly, *Physiol. Behav.*, 1979, **23**, 53.
109. A. A. Filaretov, A. I. Bogdanov and N. I. Yarushkina, *Neurosci. Behav. Physiol.*, 1996, **26**, 572.
110. S. Mousa, C. H. Miller, Jr. and D. Couri, *Psychopharmacology (Berl)*, 1983, **79**, 199.
111. A. A. Filaretov, A. I. Bogdanov and N. I. Iarushkina, *Fiziol. Zh. im. I. M. Sechenova*, 1995, **81**, 40.
112. A. J. MacLennan, R. C. Drugan, R. L. Hyson, S. F. Maier, J. t. Madden and J. D. Barchas, *J. Comp. Physiol. Psychol.*, 1982, **96**, 904.
113. I. Panocka, B. Sadowski and P. Marek, *Physiol. Behav.*, 1987, **40**, 597.
114. S. Mousa, C. H. Miller, Jr. and D. Couri, *Neuroendocrinology*, 1981, **33**, 317.
115. T. Fukuda, C. Nishimoto, M. Miyabe and H. Toyooka, *J. Anesth.*, 2007, **21**, 348.
116. D. Finn, M. Jhaveri, S. Beckett, A. Madjd, D. Kendall, C. Marsden and V. Chapman, *Neuroscience*, 2006, **138**, 1309.
117. N. G. Bowery and T. G. Smart, *Br. J. Pharmacol.*, 2006, **147**, S109.
118. N. G. Bowery, *Annu. Rev. Pharmacol. Toxicol.*, 1993, **33**, 109.
119. R. W. Olsen, J. Yang, R. G. King, A. Dilber, G. B. Stauber and R. W. Ransom, *Life Sci.*, 1986, **39**, 1969.
120. J. G. Richards and H. Mohler, *Neuropharmacology*, 1984, **23**, 233.
121. B. Bettler, K. Kaupmann, J. Mosbacher and M. Gassmann, *Physiol. Rev.*, 2004, **84**, 835.
122. S. J. Enna and K. E. McCarson, *Adv. Pharmacol.*, 2006, **54**, 1.
123. K. Rea, M. Roche and D. P. Finn, *Br. J. Pharmacol.*, 2007, **152**, 633.
124. D. J. Berlau and J. L. McGaugh, *Neurobiol. Learn. Mem.*, 2006, **86**, 123.
125. M. Davis and K. M. Myers, *Biol. Psychiat.*, 2002, **52**, 998.

126. D. Paré, G. Quirk and J. Ledoux, *J. Neurophysiol.*, 2004, **92**, 1.
127. K. Rea, Y. Lang and D. P. Finn, *J. Pain*, 2009, **10**, 1088.
128. J. A. Harris and R. F. Westbrook, *Brain Res. Bull.*, 1996, **39**, 227.
129. F. J. Helmstetter, *Pharmacol. Biochem. Behav.*, 1993, **44**, 433.
130. K. Rea, M. Roche and D. P. Finn, *J. Pain*, 2011, **12**, 712.
131. S. Tokuyama, M. Takahashi and H. Kaneto, *Jpn. J. Pharmacol.*, 1992, **60**, 105.
132. P. Killian, B. B. Holmes, A. E. Takemori, P. S. Portoghese and J. M. Fujimoto, *J. Pharmacol. Exp. Ther.*, 1995, **274**, 730.
133. A. J. Houston, J. C. Wong and I. S. Ebenezer, *Methods Find. Exp. Clin. Pharmacol.*, 1997, **19**, 167.
134. M. Kavaliers and E. Choleris, *Brain Res.*, 1997, **768**, 30.
135. M. Fanselow and F. Helmstetter, *Behav. Neurosci.*, 1988, **102**, 233.
136. B. Jakoubek, *Physiol. Bohemoslov.*, 1984, **33**, 171.
137. M. Amitani, Y. Umetani, R. Hosoi, K. Kobayashi, K. Abe and O. Inoue, *Synapse*, 2005, **58**, 23.
138. J. C. Willer and M. Ernst, *Brain Res.*, 1986, **362**, 398.
139. M. Karreman and B. Moghaddam, *Brain Res.*, 1996, **716**, 180.
140. W. Timmerman, G. Cisci, A. Nap, J. B. de Vries and B. H. C. Westerink, *Brain Res.*, 1999, **833**, 150.
141. B. Moghaddam, *Biol. Psychiat.*, 2002, **51**, 775.
142. J. Lerma, A. V. Paternain, A. Rodríguez-Moreno and J. C. López-García, *Physiol. Rev.*, 2001, **81**, 971.
143. W. J. Martin, A. B. Malmberg and A. I. Basbaum, *J. Neurosci.*, 2001, **21**, 5321.
144. L. R. Watkins, E. P. Wiertelak, L. E. Goehler, K. Mooney-Heiberger, J. Martinez, L. Furness, K. P. Smith and S. F. Maier, *Brain Res.*, 1994, **639**, 283.
145. H. Vanegas, *Neurosci. Lett.*, 2004, **361**, 225.
146. H. J. Lee, J.-S. Choi, T. H. Brown and J. J. Kim, *J. Neurosci.*, 2001, **21**, 4116.
147. M. Kavaliers, D. D. Colwell and T. S. Perrot-Sinal, *Brain Res.*, 1997, **766**, 11.
148. M. Kavaliers, D. D. Colwell and E. Choleris, *Pain*, 1998, **77**, 163.
149. K. Onodera, S. Sakurada, S. Furuta, A. Yonezawa, T. Hayashi, I. Honma and S. Miyazaki, *Drugs Exp. Clin. Res.*, 2001, **27**, 193.
150. L. C. Gregg, K. M. Jung, J. M. Spradley, R. Nyilas, R. L. Suplita, 2nd, A. Zimmer, M. Watanabe, K. Mackie, I. Katona, D. Piomelli and A. G. Hohmann, *J. Neurosci.*, 2012, **32**, 9457.
151. A. E. Snow, S. M. Tucker and W. L. Dewey, *Pharmacol. Biochem. Behav.*, 1982, **16**, 47.
152. R. L. Nunes-de-Souza, A. Canto-de-Souza, M. da-Costa, R. V. Fornari, F. G. Graeff and I. R. Pela, *Psychopharmacology (Berl)*, 2000, **150**, 300.
153. L. R. Watkins, J. N. Johannessen, I. B. Kinscheck and D. J. Mayer, *Brain Res.*, 1984, **290**, 107.

154. R. L. Nunes-de-Souza, A. Canto-de-Souza, M. da-Costa, R. V. Fornari, F. G. Graeff and I. R. Pelá, *Psychopharmacology*, 2000, **150**, 300.
155. J. A. Rosecrans, S. E. Robinson, J. H. Johnson, D. J. Mokler and J. S. Hong, *Brain Res.*, 1986, **382**, 71.
156. M. Roche, E. O'Connor, C. Diskin and D. Finn, *Eur. J. Neurosci.*, 2007, **26**, 2643.
157. R. J. Bodnar, K. P. Merrigan and E. Sperber, *Pharmacol. Biochem. Behav.*, 1983, **19**, 447.
158. R. J. Bodnar, P. E. Mann and E. A. Stone, *Pharmacol. Biochem. Behav.*, 1985, **23**, 749.
159. K. Rea, M. Roche and D. P. Finn, *Br. J. Pharmacol.*, 2007, **152**, 633.
160. J. Guindon and A. G. Hohmann, CNS Neurol, *Disord.: Drug Targets*, 2009, **8**, 403.
161. M. D. Jhaveri, D. Richardson and V. Chapman, *Br. J. Pharmacol.*, 2007, **152**, 624.
162. V. Di Marzo, *Nat. Rev. Drug Discovery*, 2008, **7**, 438.
163. J. P. Chhatwal and K. J. Ressler, *CNS Spectr.*, 2007, **12**, 211.
164. P. Lafenetre, F. Chaouloff and G. Marsicano, *Pharmacol. Res.*, 2007, **56**, 367.
165. E. M. Marco and M. P. Viveros, *Mini-Rev. Med. Chem.*, 2009, **9**, 1407.
166. F. A. Moreira and B. Lutz, *Addict. Biol.*, 2008, **13**, 196.
167. D. P. Finn, *Immunobiology*, 2009.
168. B. B. Gorzalka, M. N. Hill and C. J. Hillard, *Neurosci. Biobehav. Rev.*, 2008, **32**, 1152.
169. M. Herkenham, A. Lynn, M. Johnson, L. Melvin, B. de Costa and K. Rice, *J. Neurosci.*, 1991, **11**, 563.
170. J. M. Walker, S. M. Huang, N. M. Strangman, K. Tsou and M. C. Sanudo-Pena, *Proc. Natl. Acad. Sci. U. S. A.*, 1999, **96**, 12198.
171. A. Hohmann, R. Suplita, N. Bolton, M. Neely, D. Fegley, R. Mangieri, J. Krey, J. Walker, P. Holmes, J. Crystal, A. Duranti, A. Tontini, M. Mor, G. Tarzia and D. Piomelli, *Nature*, 2005, **435**, 1108.
172. G. Marsicano, C. Wotjak, S. Azad, T. Bisogno, G. Rammes, M. Cascio, H. Hermann, J. Tang, C. Hofmann, W. Zieglgänsberger, V. Di Marzo and B. Lutz, *Nature*, 2002, **418**, 530.
173. B. Szabo and E. Schlicker, *Handb. Exp. Pharmacol.*, 2005, 327.
174. R. Suplita, J. Farthing, T. Gutierrez and A. Hohmann, *Neuropharmacology*, 2005, **49**, 1201.
175. O. Valverde, C. Ledent, F. Beslot, M. Parmentier and B. P. Roques, *Eur. J. Neurosci.*, 2000, **12**, 533.
176. D. P. Finn, S. R. Beckett, D. Richardson, D. A. Kendall, C. A. Marsden and V. Chapman, *Eur. J. Neurosci.*, 2004, **20**, 848.
177. R. K. Butler, G. K. Ford, M. Hogan, M. Roche, K. M. Doyle, J. P. Kelly, D. A. Kendall, V. Chapman and D. P. Finn, *J. Psychopharmacol.*, 2012, **26**, 83.
178. R. K. Butler, K. Rea, Y. Lang, A. M. Gavin and D. P. Finn, *Pain*, 2008, **140**, 491.

179. D. L. Cichewicz and E. A. McCarthy, *J. Pharmacol. Exp. Ther.*, 2003, **304**, 1010.
180. S. P. Welch and M. Eads, *Brain Res.*, 1999, **848**, 183.
181. K. Kurrikoff, J. Inno, T. Matsui and E. Vasar, *Eur. J. Neurosci.*, 2008, **27**, 2147.
182. K. Connell, N. Bolton, D. Olsen, D. Piomelli and A. G. Hohmann, *Neurosci. Lett.*, 2006, **397**, 180.
183. R. L. Suplita, T. Gutierrez, D. Fegley, D. Piomelli and A. G. Hohmann, *Neuropharmacology*, 2006, **50**, 372.
184. M. Roche, P. Johnston, O. N. Mhuircheartaigh, W. M. Olango, K. Mackie and D. P. Finn, *Eur. J. Pain*, 2010, **14**, 487.
185. W. M. Olango, M. Roche, G. K. Ford, B. Harhen and D. P. Finn, *Br. J. Pharmacol.*, 2012, **165**, 2549.
186. G. K. Ford, S. Kieran, K. Dolan, B. Harhen and D. P. Finn, *Pain*, 2011, **152**, 2495.
187. L. K. Zeltzer, M. J. Dolgin, S. LeBaron and C. LeBaron, *Pediatrics*, 1991, **88**, 34.
188. (a) R. L. Hayes, G. J. Bennett, P. G. Newlon and D. J. Mayer, *Brain Res.*, 1978, **155**, 69; (b) S. Fowler-Kerry and J. R. Lander, *Pain*, 1987, **30**, 169.
189. D. Das, K. Grimmer, A. Sparnon, S. McRae and B. Thomas, *BMC Pediatrics*, 2005, **5**, 1.
190. Y. S. Schmitt, H. G. Hoffman, D. K. Blough, D. R. Patterson, M. P. Jensen, M. Soltani, G. J. Carrougher, D. Nakamura and S. R. Sharar, *Burns*, 2011, **37**, 61.
191. K. M. Malloy and L. S. Milling, *Clin. Psychol. Rev.*, 2010, **30**, 1011.
192. K. Sato, S. Fukumori, T. Matsusaki, T. Maruo, S. Ishikawa, H. Nishie, K. Takata, H. Mizuhara, S. Mizobuchi, H. Nakatsuka, M. Matsumi, A. Gofuku, M. Yokoyama and K. Morita, *Pain Med.*, 2010, **11**, 622.
193. L. M. Dahlquist, K. E. Weiss, E. F. Law, S. Sil, L. J. Herbert, S. B. Horn, K. Wohlheiter and C. S. Ackerman, *J. Pediatr. Psychol.*, 2010, **35**, 617.
194. H. Hoffman, G. Chambers, W. Meyer, L. Arceneaux, W. Russell, E. Seibel, T. Richards, S. Sharar and D. Patterson, *Ann. Behav. Med.*, 2011, **41**, 183.
195. H. G. Hoffman, D. R. Patterson, G. J. Carrougher and S. R. Sharar, *Clin. J. Pain*, 2001, **17**, 229.
196. B. Kipping, S. Rodger, K. Miller and R. M. Kimble, *Burns*, 2012, **38**, 650.
197. J. Younger, A. Aron, S. Parke, N. Chatterjee and S. Mackey, *PLoS ONE*, 2010, **5**, e13309.
198. M. de Tommaso, R. Calabrese, E. Vecchio, V. De Vito Francesco, G. Lancioni and P. Livrea, *Int. J. Psychophysiol.*, 2009, **74**, 139.
199. P. Petrovic and M. Ingvar, *Pain*, 2002, **95**, 1.
200. I. Tracey, A. Ploghaus, J. S. Gati, S. Clare, S. Smith, R. S. Menon and P. M. Matthews, *J. Neurosci.*, 2002, **22**, 2748.
201. M. Valet, T. Sprenger, H. Boecker, F. Willoch, E. Rummeny, B. Conrad, P. Erhard and T. R. Tolle, *Pain*, 2004, **109**, 399.

202. A. L. H. Buffington, C. A. Hanlon and M. J. McKeown, *Pain*, 2005, **113**, 172.
203. K. E. Scholes, B. J. Harrison, B. V. O'Neill, S. Leung, R. J. Croft, A. Pipingas, K. L. Phan and P. J. Nathan, *Neuropsychopharmacology*, 2006, **32**, 1600.
204. G. K. Ford, O. Moriarty, B. E. McGuire and D. P. Finn, *Eur. J. Pain*, 2008, **12**, 970.
205. H. Imbe, Y. Iwai-Liao and E. Senba, *Front. Biosci.*, 2006, **11**, 2179.
206. S. Bradesi, I. Schwetz, H. S. Ennes, C. M. R. Lamy, G. Ohning, M. Fanselow, C. Pothoulakis, J. A. McRoberts and E. A. Mayer, *Am. J. Physiol-Gastr. L.*, 2005, **289**, G42.
207. C. Vidal and J. Jacob, *Ann. N. Y. Acad. Sci.*, 1986, **467**, 73.
208. C. Rivat, E. Laboureyras, J.-P. Laulin, C. Le Roy, P. Richebe and G. Simonnet, *Neuropsychopharmacology*, 2007, **32**, 2217.
209. L. Quintero, M. Moreno, C. Avila, J. Arcaya, W. Maixner and H. Suarez-Roca, *Pharmacol. Biochem. Behav.*, 2000, **67**, 449.
210. L. Quintero, M. C. Cuesta, J. A. Silva, J. L. Arcaya, L. Pinerua-Suhaibar, W. Maixner and H. Suarez-Roca, *Brain Res.*, 2003, **965**, 259.
211. H. Suarez-Roca, L. Quintero, J. L. Arcaya, W. Maixner and S. G. Rao, *Physiol. Behav.*, 2006, **88**, 82.
212. R. Schumacher and M. Velden, *Percept. Mot. Skills*, 1984, **58**, 339.
213. M. Weisenberg, O. Aviram, Y. Wolf and N. Raphaeli, *Pain*, 1984, **20**, 371.
214. A. Cornwall and D. C. Donderi, *Pain*, 1988, **35**, 105.
215. A. E. Williams and J. L. Rhudy, *J. Pain*, 2007, **8**, 598.
216. T. Thompson, E. Keogh, C. C. French and R. Davis, *Pain*, 2008, **134**, 187.
217. J. L. Rhudy, P. M. Dubbert, J. D. Parker, R. S. Burke and A. E. Williams, *Pain Med.*, 2006, **7**, 483.
218. L. E. Carter, D. W. McNeil, K. E. Vowles, J. T. Sorrell, C. L. Turk, B. J. Ries and D. R. Hopko, *Pain Res. Manag.*, 2002, **7**, 21.
219. J. E. James and D. Hardardottir, *Br. J. Health Psychol.*, 2002, **7**, 149.
220. E. T. Walters, *Int. Rev. Neurobiol.*, 1994, **36**, 325.
221. H. Beecher, *JAMA*, 1969, **209**, 1080.
222. A. Ploghaus, I. Tracey, J. S. Gati, S. Clare, R. S. Menon, P. M. Matthews and J. N. P. Rawlins, *Science*, 1999, **284**, 1979.
223. C. Vidal and J. Jacob, *Neurosci. Lett.*, 1982, **32**, 75.
224. E. Jorum, *Pain*, 1988, **32**, 349.
225. R. D. Porsolt, M. Le Pichon and M. Jalfre, *Nature*, 1977, **266**, 730.
226. B. A. Arnow, E. M. Hunkeler, C. M. Blasey, J. Lee, M. J. Constantino, B. Fireman, H. C. Kraemer, R. Dea, R. Robinson and C. Hayward, *Psychosom. Med.*, 2006, **68**, 262.
227. M. M. Delvaux, *Can. J. Gastroenterol.*, 1999, **13 Suppl A**, 32A.
228. S. Bradesi, H. Eutamene, R. Garcia-Villar, J. Fioramonti and L. BuÉNo, *Neurogastroenterol. Motil.*, 2002, **14**, 75.
229. A. Ait-Belgnaoui, S. Bradesi, J. Fioramonti, V. Theodorou and L. Bueno, *Pain*, 2005, **113**, 141.
230. M. T. Robbins, J. DeBerry and T. J. Ness, *Physiol. Behav.*, 2007, **91**, 544.

231. M. Gue, C. Bonbonne, J. Fioramonti, J. More, C. Del Rio-Lacheze, C. Comera and L. Bueno, *Am. J. Physiol-Gastr. L.*, 1997, **272**, G84.
232. M. Toulouse, A. M. Coelho, J. Fioramonti, A. Lecci, C. Maggi and L. Buéno, *Br. J. Pharmacol.*, 2000, **129**, 193.
233. F. Benedetti, M. Amanzio, C. Casadio, A. Oliaro and G. Maggi, *Pain*, 1997, **71**, 135.
234. A. Martinez-Urrutia, *J. Consult. Clin. Psychol.*, 1975, **43**, 437.
235. L. E. Scott, G. A. Clum and J. B. Peoples, *Pain*, 1983, **15**, 283.
236. T. N. Wise, W. A. Hall and O. Wong, *J. Psychosom. Res.*, 1978, **22**, 513.
237. M. Papaioannou, P. Skapinakis, D. Damigos, V. Mavreas, G. Broumas and A. Palgimesi, *Pain Med.*, 2009, **10**, 1452.
238. F. Vaughn, H. Wichowski and G. Bosworth, *AORN*, 2007, **85**, 589.
239. H. Y. V. Ip, A. Abrishami, P. W. H. Peng, J. Wong and F. Chung, *Anesthesiology*, 2009, **111**, 657.
240. S. G. Khasar, O. A. Dina, P. G. Green and J. D. Levine, *J. Pain*, 2009, **10**, 1073.
241. M. Satoh, Y. Kuraishi and M. Kawamura, *Pain*, 1992, **49**, 273.
242. G. H. Gameiro, A. da Silva Andrade, M. de Castro, L. F. Pereira, C. H. Tambeli and M. C. Ferraz de Arruda Veiga, *Pharmacol. Biochem. Behav.*, 2005, **82**, 338.
243. L. M. Dufton, B. Konik, R. Colletti, C. Stanger, M. Boyer, S. Morrow and B. E. Compas, *Pain*, 2008, **136**, 38.
244. F. Pinto-Ribeiro, A. Almeida, J. M. Pêgo, J. Cerqueira and N. Sousa, *Neurosci. Lett.*, 2004, **359**, 73.
245. M. F. Pignatiello, G. A. Olson, A. J. Kastin, R. H. Ehrensing, J. H. McLean and R. D. Olson, *Pharmacol. Biochem. Behav.*, 1989, **32**, 737.
246. C. Kawanishi, M. Fukuda, R. Tamura, H. Nishijo and T. Ono, *Physiol. Behav.*, 1997, **62**, 849.
247. Y. Omiya, K. Goto, A. Ishige and Y. Komatsu, *Pharmacol. Biochem. Behav.*, 2000, **65**, 261.
248. T. Hata, T. Kita, E. Itoh and A. Kawabata, *J. Auton. Pharmacol.*, 1988, **8**, 45.
249. K. Okano, M. Ueda, Y. Kuraishi and M. Satoh, *Neurosci. Res.*, 1997, **29**, 319.
250. I. L. da Silva Torres, S. N. S. Cucco, M. Bassani, M. S. Duarte, P. P. Silveira, A. P. Vasconcellos, A. S. Tabajara, G. Dantas, F. U. Fontella, C. Dalmaz and M. B. C. Ferreira, *Neurosci. Res.*, 2003, **45**, 277.
251. I. L. da Silva Torres, C. D. Bonan, L. Crema, M. De Leon Nunes, A. M. O. Battastini, J. J. F. Sarkis, C. Dalmaz and M. B. C. Ferreira, *Eur. J. Pharmacol.*, 2003, **481**, 197.
252. H. Imbe, S. Murakami, K. Okamoto, Y. Iwai-Liao and E. Senba, *Pain*, 2004, **112**, 361.
253. D. J. Langford, A. H. Tuttle, C. Briscoe, C. Harvey-Lewis, I. Baran, P. Gleeson, D. B. Fischer, M. Buonora, W. F. Sternberg and J. S. Mogil, *J. Pain*, 2011, **12**, 125.

254. A. Arntz, *Behav. Res. Ther.*, 1991, **29**, 41.
255. A. Arntz, L. Dreessen and P. De Jong, *Pain*, 1994, **56**, 307.
256. N. Sawamoto, M. Honda, T. Okada, T. Hanakawa, M. Kanda, H. Fukuyama, J. Konishi and H. Shibasaki, *J. Neurosci.*, 2000, **20**, 7438.
257. T. Koyama, J. G. McHaffie, P. J. Laurienti and R. C. Coghill, *Proc. Natl. Acad. Sci. U. S. A.*, 2005, **102**, 12950.
258. J. R. Keltner, A. Furst, C. Fan, R. Redfern, B. Inglis and H. L. Fields, *J. Neurosci.*, 2006, **26**, 4437.
259. G. W. Terman, M. J. Morgan and J. C. Liebeskind, *Brain Res.*, 1986, **372**, 167.
260. C. D. King, D. P. Devine, C. J. Vierck, J. Rodgers and R. P. Yezierski, *Brain Res.*, 2003, **987**, 214.
261. J. P. Herman and W. E. Cullinan, *Trends Neurosci.*, 1997, **20**, 78.
262. W. D. Gunter, J. D. Shepard, R. D. Foreman, D. A. Myers and Beverley, *Physiol. Behav.*, 2000, **69**, 379.
263. K. Fecho, A. G. Nackley, Y. Wu and W. Maixner, *Physiol. Behav.*, 2005, **85**, 177.
264. A. Ramos, O. Berton, P. Mormède and F. Chaouloff, *Behav. Brain Res.*, 1997, **85**, 57.
265. A. Ramos, A. L. Kangerski, P. F. Basso, J. E. Da Silva Santos, J. Assreuy, L. F. Vendruscolo and R. N. Takahashi, *Behav. Brain Res.*, 2002, **129**, 113.
266. N. N. Burke, E. Hayes, P. Calpin, D. M. Kerr, O. Moriarty, D. P. Finn and M. Roche, *Neuroscience*, 2010, **171**, 1300.
267. J. R. Glowa and C. T. Hansen, *Behav. Genet.*, 1994, **24**, 79.
268. W. P. Paré, *Physiol. Behav.*, 1992, **51**, 1051.
269. C. Gentsch, M. Lichtsteiner and H. Feer, *Behav. Brain Res.*, 1987, **25**, 101.
270. M. C. Pardon, G. G. Gould, A. Garcia, L. Phillips, M. C. Cook, S. A. Miller, P. A. Mason and D. A. Morilak, *Neuroscience*, 2002, **115**, 229.
271. Q. Zeng, S. Wang, G. Lim, L. Yang, J. Mao, B. Sung, Y. Chang, J.-A. Lim, G. Guo and J. Mao, *Brain Res.*, 2008, **1200**, 27.
272. B. Greenwood-Van Meerveld, A. C. Johnson, R. D. Foreman and B. Linderoth, *Auton. Neurosci.*, 2005, **122**, 69.
273. S. M. O'Mahony, D. C. Bulmer, A. M. Coelho, P. Fitzgerald, C. Bongiovanni, K. Lee, W. Winchester, T. G. Dinan and J. F. Cryan, *Neurogastroenterol. Motil.*, 2010, **22**, 573.
274. S. M. Gibney, R. D. Gosselin, T. G. Dinan and J. F. Cryan, *Neuroscience*, 2010, **165**, 675.
275. M. Max, T. Wu, S. Atlas, R. Edwards, J. Haythornthwaite, A. Bollettino, H. Hipp, C. McKnight, I. Osman, E. Crawford, M. Pao, J. Nejim, A. Kingman, D. Aisen, M. Scully, R. Keller, D. Goldman and I. Belfer, *Mol. Pain*, 2006, **2**, 14.
276. H. Suarez-Roca, J. A. Silva, J. L. Arcaya, L. Quintero, W. Maixner and L. Pinerua-Shuhaibar, *Behav. Brain Res.*, 2006, **167**, 205.
277. G. P. McNally, I. N. Johnston and R. F. Westbrook, *Behav. Neurosci.*, 2000, **114**, 1183.

278. C. D. King, D. P. Devine, C. J. Vierck, A. Mauderli and R. P. Yezierski, *Neuroscience*, 2007, **147**, 174.
279. B. Siegfried, C. A. Netto and I. Izquierdo, *Behav. Neurosci.*, 1987, **101**, 436.
280. I. L. S. Torres, A. P. Vasconcellos, S. N. Silveira Cucco and C. Dalmaz, *Braz. J. Med. Biol. Res.*, 2001, **34**, 241.
281. G. Drolet, É. C. Dumont, I. Gosselin, R. Kinkead, S. Laforest and J.-F. Trottier, *Prog. Neuro-Psychopharmacol. Biol. Psychiatry*, 2001, **25**, 729.
282. G. Dantas, I. L. D. S. Torres, L. M. Crema, D. R. Lara and C. Dalmaz, *Neurochem. Res.*, 2005, **30**, 1.
283. J.-P. Laulin, A. Larcher, E. Célèrier, M. L. Moal and G. Simonnet, *Eur. J. Neurosci.*, 1998, **10**, 782.
284. J.-P. Laulin, P. Maurette, J.-B. Corcuff, C. Rivat, M. Chauvin and G. Simonnet, *Anesth. Analg.*, 2002, **94**, 1263.
285. M. S. Angst, W. Koppert, I. Pahl, D. J. Clark and M. Schmelz, *Pain*, 2003, **106**, 49.
286. M. J. Christie, M. Connor, C. W. Vaughan, S. L. Ingram and E. E. Bagley, *Clin. Exp. Pharmacol. Physiol.*, 2000, **27**, 520.
287. D. J. Mayer, J. Mao, J. Holt and D. D. Price, *Proc. Natl. Acad. Sci.*, 1999, **96**, 7731.
288. K. A. Trujillo and H. Akil, *Science*, 1991, **251**, 85.
289. S. G. Khasar, J. Burkham, O. A. Dina, A. S. Brown, O. Bogen, N. Alessandri-Haber, P. G. Green, D. B. Reichling and J. D. Levine, *J. Neurosci.*, 2008, **28**, 5721.
290. T. L. Bale and W. W. Vale, *Annu. Rev. Pharmacol. Toxicol.*, 2004, **44**, 525.
291. M. H. Perrin and W. W. Vale, *Ann. N. Y. Acad. Sci.*, 1999, **885**, 312.
292. Y. Taché, V. Martinez, L. Wang and M. Million, *Br. J. Pharmacol.*, 2004, **141**, 1321.
293. I. Schwetz, S. Bradesi, J. A. McRoberts, M. Sablad, J. C. Miller, H. Zhou, G. Ohning and E. A. Mayer, *Am. J. Physiol-Gastr. L.*, 2004, **286**, G683.
294. M. Million, D. E. Grigoriadis, S. Sullivan, P. D. Crowe, J. A. McRoberts, H. Zhou, P. R. Saunders, C. Maillot, E. A. Mayer and Y. Taché, *Brain Res.*, 2003, **985**, 32.
295. M. Gue, C. Del Rio-Lacheze, H. Eutamene, V. Theodorou, J. Fioramonti and L. Bueno, *Neurogastroenterol. Motil.*, 1997, **9**, 271.
296. I. Schwetz, J. A. McRoberts, S. V. Coutinho, S. Bradesi, G. Gale, M. Fanselow, M. Million, G. Ohning, Y. Taché, P. M. Plotsky and E. A. Mayer, *Am. J. Physiol-Gastr. L.*, 2005, **289**, G704.
297. T. Lembo, V. Plourde, Z. Shui, S. Fullerton, H. Mertz, Y. Tache, B. Sytnik, J. Munakata and E. Mayer, *Neurogastroenterol. Motil.*, 1996, **8**, 9.
298. Y. Sagami, Y. Shimada, J. Tayama, T. Nomura, M. Satake, Y. Endo, T. Shoji, K. Karahashi, M. Hongo and S. Fukudo, *Gut*, 2004, **53**, 958.
299. J. M. Verkuyl, S. E. Hemby and M. Joëls, *Eur. J. Neurosci.*, 2004, **20**, 1665.
300. J. M. Verkuyl, H. Karst and M. Joëls, *Eur. J. Neurosci.*, 2005, **21**, 113.

301. G. Biggio, A. Concas, M. G. Corda, O. Giorgi, E. Sanna and M. Serra, *Pharmacol. Ther.*, 1990, **48**, 121.
302. H. Suarez-Roca, L. Leal, J. A. Silva, L. Pinerua-Shuhaibar and L. Quintero, *Behav. Brain Res.*, 2008, **189**, 159.
303. H. Wei, W. Zhao, Y.-X. Wang and A. Pertovaara, *Brain Res.*, 2007, **1148**, 105.
304. M. Engelmann, G. Wolf and T. F. W. Horn, *Neurosci. Lett.*, 2002, **324**, 252.
305. F. U. Fontella, D. A. Vendite, A. S. Tabajara, L. O. Porciúncula, I. L. d. S. Torres, F. M. Jardim, L. Martini, D. O. Souza, C. A. Netto and C. Dalmaz, *Neurochem. Res.*, 2004, **29**, 1703.
306. I. V. Belozertseva and A. Y. Bespalov, *Naunyn-Schmiedeberg's Arch. Pharmacol.*, 1998, **358**, 270.
307. J. Mao, D. D. Price, J. Lu and D. J. Mayer, *Neurosci. Lett.*, 1998, **250**, 193.
308. T. Hata, E. Itoh and A. Kawabata, *Jpn. J. Pharmacol.*, 1991, **56**, 101.
309. H. Ohara, M. Kawamura, A. Namimatsu, T. Miura, R. Yoneda and T. Hata, *Jpn. J. Pharmacol.*, 1991, **57**, 243.
310. G. H. Gameiro, P. H. Gameiro, A. da Silva Andrade, L. F. Pereira, M. T. Arthuri, F. K. Marcondes and M. C. F. de Arruda Veiga, *Physiol. Behav.*, 2006, **87**, 643.
311. D. Hellhammer, J. Hingtgen, S. Wade, P. Shea and M. Aprison, *Psychosom. Med.*, 1983, **45**, 115.
312. A. Adell, J. M. Casanovas and F. Artigas, *Neuropharmacology*, 1997, **36**, 735.
313. M. Ikeda and T. Nagatsu, *Naunyn-Schmiedebergs Arch. Pharmacol.*, 1985, **331**, 23.
314. L. G. Kirby, A. R. Allen and I. Lucki, *Brain Res.*, 1995, **682**, 189.
315. R. Gatchel, J. Garofalo, E. Ellis and C. Holt, *J. Am. Dent. Assoc.*, 1996, **127**, 1365.
316. D. A. Jones, G. B. Rollman and R. I. Brooke, *Pain*, 1997, **72**, 171.
317. P. E. Stokes and A. Holtz, *Clin. Ther.*, **19**, 1135.
318. M. Tanaka, M. Yoshida, H. Emoto and H. Ishii, *Eur. J. Pharmacol.*, 2000, **405**, 397.
319. A. Tsuda, M. Tanaka, Y. Ida, S. Tsujimaru, I. Ushijima and N. Nagasaki, *Pharmacol. Biochem. Behav.*, 1986, **24**, 115.
320. M. J. Bannon, P. J. Elliott, J. E. Alpert, M. Goedert, S. D. Iversen and L. L. Iversen, *Nature*, 1983, **306**, 791.
321. P. W. Kalivas and R. Abhold, *Brain Res.*, 1987, **414**, 339.
322. M. Bertolucci-D'Angio, A. Serrano and B. Scatton, *J. Neurosci. Methods*, 1990, **34**, 135.
323. Y.-L. Wu, M. Yoshida, H. Emoto and M. Tanaka, *Neurosci. Lett.*, 1999, **275**, 69.
324. C. Gambarana, F. Masi, A. Tagliamonte, S. Scheggi, O. Ghiglieri and M. Graziella De Montis, *J. Neurochem.*, 1999, **72**, 2039.
325. P. B. Wood, *Med. Hypotheses*, 2004, **62**, 420.

326. E. Legangneux, J. J. Mora, O. Spreux-Varoquaux, I. Thorin, M. Herrou, G. Alvado and C. Gomeni, *Rheumatology*, 2001, **40**, 290.
327. I. J. Russell, H. Vaeroy, M. Javors and F. Nyberg, *Arthritis Rheum.*, 1992, **35**, 550.
328. D. P. Finn, *Immunobiology*, 2010, **215**, 629.
329. L. Shen, X. J. Yang, W. Qian and X. H. Hou, *J. Neurogastroenterol. Motil.*, 2010, **16**, 281.
330. S. Hong, J. Fan, E. S. Kemmerer, S. Evans, Y. Li and J. W. Wiley, *Gut*, 2009, **58**, 202.
331. F. Helmstetter and M. Fanselow, *Physiol. Behav.*, 1987, **39**, 501.
332. I. B. Kinscheck, L. R. Watkins and D. J. Mayer, *Brain Res.*, 1984, **298**, 33.
333. P. Nishith, M. G. Griffin and T. L. Poth, *Biol. Psychiatry.*, 2002, **51**, 867.
334. W. T. Chance, G. M. Krynock and J. A. Rosecrans, *Pain*, 1978, **4**, 243.
335. M. Kavaliers, E. Choleris and D. D. Colwell, *Behav. Neurosci.*, 2001, **115**, 661.
336. J. T. Madden, H. Akil, R. L. Patrick and J. D. Barchas, *Nature*, 1977, **265**, 358.
337. R. J. Bodnar, D. D. Kelly, A. Spiaggia, C. Ehrenberg and M. Glusman, *Pharmacol. Biochem. Behav.*, 1978, **8**, 667.
338. T. Fukuda, C. Nishimoto, M. Miyabe and H. Toyooka, *Anesth. Analg.*, 2001, **93**, 424.
339. A. Costa, A. Smeraldi, C. Tassorelli, R. Greco and G. Nappi, *Neurosci. Lett.*, 2005, **383**, 7.
340. C. Lee and R. J. Rodgers, *Psychopharmacology (Berl)*, 1990, **102**, 507.
341. S. Puglisi-Allegra, C. Castellano and A. Oliverio, *Brain Res.*, 1982, **252**, 373.
342. R. J. Rodgers and J. I. Randall, *Physiol. Behav.*, 1987, **41**, 279.
343. R. J. Rodgers and C. A. Hendrie, *Physiol. Behav.*, 1983, **30**, 775.
344. D. M. Wright and D. W. Lincoln, *Physiol. Behav.*, 1985, **34**, 691.
345. C. H. Wideman, H. M. Murphy and S. B. McCartney, *Peptides*, 1996, **17**, 63.
346. S. A. Janssen and A. Arntz, *Pain*, 1996, **66**, 145.
347. B. A. van der Kolk, M. S. Greenberg, S. P. Orr and R. K. Pitman, *Psychopharmacol. Bull.*, 1989, **25**, 417.
348. S. Marchand, M. C. Bushnell and G. H. Duncan, *Clin. J. Pain*, 1991, **7**, 122.
349. S. M. Abdulhameed, R. J. Feigal, J. D. Rudney and K. C. Kajander, *Anesth. Prog.*, 1989, **36**, 52.
350. D. M. Walsh, C. Liggett, D. Baxter and J. M. Allen, *Pain*, 1995, **61**, 39.
351. J. Andre, B. Zeau, M. Pohl, Cesselin, Francois, Benoliel, Jean-Jacques and C. Becker, *J. Neurosci.*, 2005, **25**, 7896.
352. L. Bardin, N. Malfetes, A. Newman-Tancredi and R. Depoortère, *Behav. Brain Res.*, 2009, **205**, 360.
353. A. Dhir, S. S. V. Padi, P. S. Naidu and S. K. Kulkarni, *Eur. J. Pharmacol.*, 2006, **535**, 192.

354. G. D. Gamaro, M. H. Xavier, J. D. Denardin, J. A. Pilger, D. R. Ely, M. B. C. Ferreira and C. Dalmaz, *Physiol. Behav.*, 1998, **63**, 693.
355. S. Boccalon, B. Scaggiante and L. Perissin, *Life Sci.*, 2006, **78**, 1225.
356. I. P. Butkevich and E. A. Vershinina, *Brain Res.*, 2001, **915**, 88.
357. I. P. Butkevich, G. A. Barr, V. A. Mikhailenko and V. A. Otellin, *Neurosci. Lett.*, 2006, **403**, 222.
358. I. P. Butkevich, G. A. Barr and E. A. Vershinina, *Eur. J. Pain*, 2007, **11**, 888.
359. A. J. Devall, Z.-W. Liu and T. A. Lovick, *Psychoneuroendocrinology*, 2009, **34**, 587.
360. G.-J. Geerse, L. C. A. van Gurp, V. M. Wiegant and R. Stam, *Behav. Brain Res.*, 2006, **175**, 9.
361. K. Tyler, S. Moriceau, R. M. Sullivan and B. Greenwood-van Meerveld, *Neurogastroenterol. Motil.*, 2007, **19**, 761.
362. T. E. King, E. D. Crown, A. N. Sieve, R. L. Joynes, J. W. Grau and M. W. Meagher, *Behav. Brain Res.*, 1999, **100**, 33.
363. H. Imbe, K. Okamoto, T. Donishi, E. Senba and A. Kimura, *Brain Res.*, 2010, **1329**, 103.
364. G. A. S. Metz, M. E. Schwab and H. Welzl, *Physiol. Behav.*, **72**, 29.
365. S. G. Khasar, P. G. Green and J. D. Levine, *Pain*, 2005, **116**, 79.
366. G. Adler and W. F. Gattaz, *Biol. Psychiat.*, 1993, **34**, 687.
367. M. W. Otto, M. J. Dougher and R. A. Yeo, *J. Nerv. Ment. Dis.*, 1989, **177**, 210.
368. N. G. Ward, V. L. Bloom, S. Dworkin, J. Fawcett, N. Narasimhachari and R. O. Friedel, *J. Clin. Psychiatry*, 1982, **43**, 32.
369. F. Benedetti, M. Amanzio, S. Vighetti and G. Asteggiano, *J. Neurosci.*, 2006, **26**, 12014.

CHAPTER 10

Pain Management Companies

DAVID S. REYNOLDS

Pfizer Neusentis, Portway Building, Granta Park, Cambridge CB21 6GS, UK
Email: David.Reynolds@pfizer.com

10.1 Introduction

The global pain market in 2010 was \$22 billion, making it one of the largest therapeutic areas by sales. Furthermore the market has demonstrated year-on-year growth of approximately 3% in recent years.[1] Pain is predominantly a symptom of other disorders, although it can also be a disease state in its own right, which means that several different pathophysiological processes likely underlie similar clinical symptoms. This heterogeneity of causes means that a range of different mechanisms of drugs are required to effectively treat the different classes of pain. Currently the main drug classes are the non-steroidal anti-inflammatory drugs (NSAIDs, *e.g.* ibuprofen, naproxen, diclofenac and celecoxib), opioids (*e.g.* morphine, oxycodone), antiepileptics (*e.g.* carbamazepine, pregabalin), antidepressants (*e.g.* duloxetine) and local anaesthetics (*e.g.* lidocaine). Each of these classes contributes a significant share of the market, usually with a small number of products generating the majority of the sales. Likewise it is a small number of companies that account for the majority of the value of the overall pain market. Pfizer has the number 1 pain franchise by sales with two major products (Lyrica® and Celebrex®) and a range of smaller products, followed by Purdue and then Eli Lilly each with a single large product. The top five companies are completed by Johnson & Johnson and Novartis, both with several smaller products.[1] These five companies along with two other pain-focused companies, Grünenthal and Endo, are described in the following sections of this chapter. Although 76% of the total prescription pain

RSC Drug Discovery Series No. 35
Pain Therapeutics: Current and Future Treatment Paradigms
Edited by Charlotte Allerton

Published by the Royal Society of Chemistry, www.rsc.org

market by sales in 2010 is accounted for by just five companies and a relatively short list of products, the picture is completely different if viewed in terms of number of days of therapy. By this measure over-the-counter (OTC) analgesics, such as paracetamol (acetaminophen), aspirin and ibuprofen, are by far the highest-volume pain drugs. A short overview of this category of medicines is given at the end of the chapter.

An interesting feature of the branded pain market is that the intellectual property protecting the product is often not on the molecule itself, but rather the formulation. Products such as OxyContin® (oxycodone extended release) and Lidoderm® patch (transdermal lidocaine) contain generic active drugs, and yet can still garner blockbuster sales (defined as annual sales >$1 billion) despite this. Over the last 15 years few novel mechanisms have proved superior to existing drugs and therefore many companies have focused on reformulation and combination strategies in an attempt to maintain their pain franchises. Although this strategy has proved effective, the marketplace is changing and it will be increasingly difficult to derive high sales from this approach. Reformulation approaches can offer convenience and improved performance either on the safety or duration-of-action front, but they do not fundamentally improve the efficacy or change the potential for adverse effects. In an increasingly cost-conscious marketplace the payers, be they insurance companies, national health services or the patients themselves, are demanding improved product performance in order to justify premium pricing. That said, the combination of the high unmet need (especially for neuropathic pain conditions), large numbers of patients and paucity of novel mechanisms for physicians to use means that any new therapy that has reasonable benefits over existing treatments is likely to have rapid uptake and potentially disrupt current market dynamics. Despite this fact there are almost no new mechanism drugs in late-stage clinical development (*i.e.* having demonstrated proof of concept in Phase 2a trials) and a relatively small number in Phase 1 or 2a. The majority of novel early mechanisms under investigation are selective ion-channel modulators.[2] The discovery of new channel subtypes over the last decade or so, *e.g.* transient receptor potential (TRP) families, twin pore potassium channels, has improved our understanding of pain neurotransmission, such that these targets are now ripe for utilisation as novel pain therapies.[3] Additionally, improvements in high-throughput electrophysiological screening capabilities for ion-channel targets has made them easier targets for pharmacological researchers.

10.2 Pfizer

Pfizer Inc. is the world's largest pharmaceutical company by revenues. It was founded in 1849 in New York City as a manufacturer of fine chemicals and it slowly evolved into a pharmaceutical company from the 1940s onwards. It is still headquartered in New York. It researches, develops and manufactures medicines and vaccines in a wide range of indications including immunology, inflammation, cardiovascular, oncology, neuroscience and pain.[4] Pfizer's path to be the largest pharmaceutical company largely occurred *via* a series of major acquisitions starting with Warner-Lambert in 2000, Pharmacia in 2003 and then Wyeth in 2009. Pfizer's original research sites in Groton, Connecticut, USA and Sandwich, Kent, UK have discovered and developed a number of Pfizer's biggest-selling drugs, which include amilodopine (Norvasc®) for hypertension, sildenafil (Viagra®) for erectile dysfunction and maraviroc (Selzentry®) for HIV. In contrast, Pfizer's marketed pain portfolio has largely come through acquisitions, with the exception of eletriptan (Relpax®) for the treatment of migraine which was discovered at the Sandwich site, albeit as a 'me too' drug following the success of GlaxoSmithKline's sumatriptan (Imigran®) (Table 10.1).

The Warner-Lambert acquisition kick-started Pfizer's major presence in the pain market through gabapentin (Neurontin®) and subsequently pregabalin (Lyrica®). Both drugs bind to the $\alpha 2\delta$ accessory subunit of calcium channels and are the only two drugs in this class.[5] At the time gabapentin was approved and marketed as an adjunctive medication for the control of partial seizures in epilepsy; however it was the U.S. Food and Drug Authority (FDA) approval for postherpetic neuralgia (PHN) in 2003 that brought a shift in the sales towards pain indications. Although PHN makes up a relatively small proportion of the number of patients with neuropathic pain, there is high medical need and, certainly at the time, a very limited number of treatment options. The approval for one type of neuropathic pain led to increasing use of gabapentin for other neuropathic pain conditions, such as painful diabetic neuropathy, *via* off-label usage (*i.e.* at the physician's discretion for use in conditions not covered by the drug label). Pfizer capitalised on its gabapentin franchise by launching pregabalin in 2005 just as the gabapentin patents expired at the end of 2004.[6] Lyrica has shown sustained growth since launch, reaching sales of $4.06 billion in 2011.

Celecoxib (Celebrex®) and valdecoxib (Bextra®, withdrawn in 2005) were added to the pain franchise through the Pharmacia acquisition in 2003.[4] The coxib drug class are all selective cyclooxygenase 2 (COX-2) inhibitors in comparison to other NSAIDs which typically block both COX-1 and COX-2 enzyme subtypes.[7] This selectivity results in equivalent analgesia to NSAIDs but with much reduced gastrointestinal side effects.[8] The familiarity of patients and physicians with NSAIDs was a key factor in the rapid market uptake of the COX-2 drug class from their launch in 1999 as they were perceived as 'super aspirins' and they soon achieved worldwide sales of more than $3 billion by 2002. In September 2004 rofecoxib (Vioxx®, marketed by Merck) was voluntarily withdrawn from the market due to concerns about increased risk of heart attack and stroke associated with long-term, high-dose use.[9] Not surprisingly, this significantly impacted the use of COX-2 drugs and prompted close examination of the cardiovascular safety data for all drugs in this class. In 2005 valdecoxib was withdrawn from the market for similar concerns.[10] However, celecoxib remained available and sales have steadily recovered, at least partly due to heavy marketing activities, exceeding $2.78 billion in 2011.

Pfizer has established the largest pain franchise of any company, with pregabalin and celecoxib both being the market-dominant product in their drug class.

Table 10.1 Key Pfizer pain products and compounds in development.

Product name	*Generic name and mechanism*	*Launch date (US)*	*Comments*
Neurontin®	Gabapentin, $\alpha 2\delta$ ligand	1994	First pain indication approved in 2002; patents expired in 2004
Lyrica®	Pregabalin, $\alpha 2\delta$ ligand	2005	Multiple pain indications added over the years
Relpax®	Eletriptan, serotonin 1B/1D receptor antagonist	2003	Key patent will expire in 2016
Celebrex®	Celecoxib, cyclooxygenase II enzyme inhibitor	1999	Will go off patent in 2014
Bextra®	Valdecoxib, cyclooxygenase II enzyme inhibitor	2003	Withdrawn 2005 following the Vioxx withdrawal
Flector patch®	Diclofenac epolamine (topical)	2008	Only available in the USA
Avinza®	Morphine sulfate (extended release), μ opioid receptor agonist	2002	Protected *via* a formulation patent until 2017
Embeda®	Morphine sulfate/naltrexone hydrochloride (extended release), μ opioid receptor agonist/antagonist	2009	Withdrawn in 2011 due to manufacturing stability issues. Relaunch date unknown
Oxecta®	Oxycodone (immediate release), μ opioid receptor agonist	2012	Formulation including niacin as abuse deterrent ingredient was not approved
Remoxy®	Oxycodone (controlled release), μ opioid receptor agonist	Not launched yet	FDA issued complete response letters in 2008 and 2011. Issues may lie with ORADUR sustained-release gel technology
NGF sequestration	Tanezumab, monoclonal antibody against nerve growth factor	Phase 3	FDA clinical hold lifted in May 2012 and development activities have resumed
Sodium channel blockers	$Na_V1.7$ and $Na_V1.8$ selective blockers	Phase 2	Proof of concept in pain indications yet to be established

The other two major classes of analgesics in today's market are the opioids (*e.g.* morphine, oxycodone) and sodium channel blockers (carbamazepine, lidocaine), neither of which Pfizer had a stake in. In an attempt to gain rapid market entry with opioids Pfizer acquired King Pharmaceuticals in October 2010 for \$3.6 billion.[11] King was a mid-sized pharmaceutical company based in Bristol, Tennessee, USA, with revenues of \$1.7 billion in 2009 and employing over 3000 people.[12] King's largest products were metaxolone (Skelaxin®) a muscle relaxant; injectable adrenaline (EpiPen®) for emergency treatment of allergy reactions including anaphylaxis; and a topical thrombin product (Thrombin-JMC®) for local application during surgery to control bleeding. For pain indications King had two established products: Avinza ® (morphine sulfate extended release) and a topical diclofenac epolamine patch (FlectorPatch®) for acute pain due to sprains, strains and bruises; both had modest revenues of \$131 and £139 million respectively in 2009.[12] These were reasonably attractive products in their own right, but it was the potential of the emerging pain portfolio that was the driver for the acquisition.

The market-dominant opioid product is Purdue's extended-release oxycodone (OxyContin®), which in 2010 had just had a tamper-resistant formulation approved (OxyContin OP®).[13] For Pfizer Embeda® (morphine sulfate/naltrexone hydrochloride), which had been launched in September 2009, was the promising product long term because of its tamper-resistant formulation that could compete directly with OxyContin for market share. Embeda capsules contain pellets of morphine sulfate with a sequestered core of naltrexone hydrochloride. If the capsules are swallowed as intended the morphine is absorbed in the gastrointestinal tract and produces analgesia, but little if any of the naltrexone is released from the core of the pellets which are then excreted in the faeces. If, however, the capsules are tampered with by crushing, chewing or other mechanical means then the naltrexone is also absorbed and antagonises the effects of the morphine, thus preventing abuse of the drug product by opioid-dependent patients or abusers. However, in March 2011 Pfizer voluntarily recalled the product because of stability issues in the manufacturing process.[14] The company have stated that they are working to re-release the product, but no timelines have been stated.[15]

The other technology platform that Pfizer inherited from King for abuse deterrence was the ORADUR® technology which King licensed from Durect in 2005. ORADUR allows the formulation of gel capsules containing the active pharmaceutical ingredient with a high-viscosity matrix component such as sucrose acetate isobutyrate, which releases drug in a controlled manner for 12–24 hours.[16] The high viscosity of the matrix prevents accelerated drug release following crushing or chewing, thus making abuse more difficult; additionally it is difficult to extract the active drug *via* dissolution and the viscosity makes injection impossible. The ORADUR technology has been used to develop Remoxy® (oxyxodone controlled release), initially designed to compete directly with OxyContin. Remoxy was first filed with the FDA in 2008, but received a complete response letter and was not approved. A second filing in 2011 resulted in another complete response letter. Pfizer have continued to work on the manufacturing issues with Remoxy and are likely to resubmit it for approval by the FDA in 2013.[16] Given the 5 year delay in getting this product to market, it is no longer certain that it will be a large commercial success. Purdue's OxyContin has continued to strengthen its market dominance, so it will be a significant challenge for Pfizer to meaningfully impact such an established brand.

The other strategic goal of the King acquisition was also likely to be entry into patch products for pain. Although the US market is relatively under-developed for patch therapies compared to other parts of the globe, particularly Japan, it has undergone rapid growth over the last few years. The leading product by far is the Lidoderm® patch (transdermal lidocaine), marketed by Endo in the United States.[1] Although this patch is only approved for PHN it enjoys significant off-label usage such that it had sales of $1.2 billion in 2012. King had Eladur® (bupivacine patch) in Phase 2 proof-of-concept trials at the time of the acquisition. Bupivacaine is an established sodium channel blocking local anaesthetic like lidocaine that would be well placed to compete with the Lidoderm patch. Eladur was originally developed by Durect, but then licensed to King. The patch technology used, Transdur®, would allow for continuous delivery of bupivacaine for up to 3 days. However, by March 2012 Pfizer had returned the rights to Durect, presumably because of disappointing clinical results.[17]

In contrast to Eladur, the FlectorPatch® is on the market and generating modest sales.[15] The challenge for branded NSAID products is to differentiate them sufficiently from generic oral alternatives and so justify the branded price. Clearly a topical product will likely avoid the gastrointestinal adverse effects of NSAIDs, which will suit some patients for whom this is a particular issue. However, in today's market with considerable payer pressures it will be difficult to grow this product significantly. Additionally, Novartis have increased their sales and marketing efforts around Voltaren® (diclofenac gel) which is a direct competitor.

In looking at the business strategy in 2010 it was clear that the intention of the King Pharmaceuticals acquisition was to drive short-term commercial success with rapid entry into the opioid market. To a large extent that success has not been realised, and moderate commercial returns are now likely over the mid to long term if, and when, the tamper-resistant products are approved. In contrast, the Icagen acquisition of 2011 was very much driven by research and development.[18] Icagen was a small biotechnology company of only 30 or so scientists based in Research Triangle Park, Durham, North Carolina, USA. It was founded in 1992 as an ion-channel focused company and developed high-throughput, high-quality expertise in electrophysiological screening and characterisation of ion-channel modulators discovered through their own small-molecule drug discovery efforts. Over their 20 year history they worked in a number of therapeutic areas and with a range of industrial partners, but pain research was always a key focus. Icagen first entered into a research collaboration with Pfizer in 2007 when the two companies began working on subtype-selective sodium channel modulators.[19] The challenge was to develop a drug with the profound analgesic properties of a local anaesthetic (*e.g.* lidocaine), but without the numbing effects of complete sensory nerve block and not carrying the cardiovascular or central nervous system (CNS) depressant effects due to the important role of sodium channels in those organs. Of the nine channel subtypes (Na_V 1.1–1.9) the 1.7 and 1.8 subtypes have the most literature linking them to pain transmission and so research primarily focused there.[20] Subtype-selective Na_V1.7 blockers is an intensely competitive research area with several large pharma and multiple biotechs all competing to be the first to prove the concept and ultimately get a new drug on the market. The acquisition of Icagen closed in October 2011 with Pfizer paying $56 million to acquire the remaining shares in the company (it already owned 11%).[15] After the deal closed Pfizer retained the Icagen staff and research facility in North Carolina and integrated it into Neusentis, its pain research unit, based in Cambridge, UK.[21]

In summary, Pfizer is well positioned to remain a key player in the pain market for the foreseeable future and it possesses franchise options in most of the key mechanisms.

10.3 Purdue Pharma

Purdue Pharma was founded over 60 years ago by physicians and is currently headquartered in Stamford, Connecticut, USA. It is a privately owned company whose main focus has been pain products since 1991.[22] It is the number 2 pain company by sales, with the vast majority of those carried by its leading opioid analgesic OxyContin® (oxycodone sustained release).[1] In 2011 OxyContin had sales of $3.6 billion, a big driver of which is the tamper-resistant formulation which was first launched in 2010 and has rapidly replaced its previously unprotected formulation. The US patent for the product expires in 2013, which will open the way for generics, and predicted sales will be just $250 million by 2016.[1]

Purdue's only new product of note is a buprenorphine patch (Butrans®) which it licensed from Grünenthal in July 2010 (Section 10.7). In 2011 Butrans was launched in the US market for moderate to severe chronic pain[23] where its main competitor is Johnson & Johnson's Duragesic® fentanyl patch. From publicly available data it appears that Purdue has only two other pain drugs in clinical development. An extended-release formulation of hydrocodone, which should allow once-daily dosing, is in Phase III testing. V-116517 (mechanism unknown) entered into Phase 2 trials for osteoarthritis and PHN in 2012.[24] Whether either of these two potential products will be able to offset the loss in sales in the mid to long term, when OxyContin comes off patent, is uncertain.

10.4 Eli Lilly & Co.

Eli Lilly was founded in 1876; it is headquartered in Indianapolis, Indiana, USA and employs approximately 38 000 people. Its main therapeutic area focus has long been metabolic disease and psychiatry. It was one of the first companies to produce human insulin using DNA recombinant technology and it maintains a strong presence in this market. Lilly was also the first company to launch a new class of antidepressant: a selective serotonin reuptake inhibitor, fluoxetine (Prozac®) in 1988, which sparked a wave of 'me too' products and an explosion in the value of the antidepressant market. Lilly followed up fluoxetine with a mixed serotonin/noradrenaline reuptake inhibitor (SNRI), duloxetine (Cymbalta®), which was first launched for major depressive disorder in 2004. Since then Lilly has added a number of pain indications to the label for duloxetine including diabetic peripheral neuropathy (2004), fibromyalgia (2009) and chronic musculoskeletal pain (2011). The majority of sales and marketing efforts on Cymbalta are targeted to the depression market and therefore that indication is the origin of the bulk of the prescriptions. However, depression and pain are often comorbid[25] and therefore the treatment of pain does constitute a reasonable proportion of its $4.1 billion sales in 2011.[26] Lilly has attempted to protect its pain franchise with additional pipeline assets targeting novel mechanisms. However, both a cannabinoid CB2 receptor agonist (LY282360) and a dual mechanism compound (potentiation of metabotropic glutamate type 2 receptors/ antagonism of cysteinyl leukotriene type 1 receptors; LY2300559) have recently been discontinued. In October 2012 the company website listed only one compound in Phase 2 for migraine prevention and a second in Phase 1 for osteoarthritis.[27] In both cases the mechanism of action has not been disclosed. Lilly will therefore have a declining presence in the pain marketplace; at least until new-mechanism compounds can reach the market towards the tail end of the decade.

10.5 Johnson & Johnson

Johnson & Johnson (J&J) was founded in 1886 selling ready-to-use surgical dressings, and then expanded into other consumer healthcare and pharmaceutical products over the next century. It is headquartered in New Brunswick, New Jersey, USA, and employs approximately 118 000 people. It is probably the most diversified of the large pharmaceutical companies with around 250 subsidiary companies across the globe, 3 of which make major contributions to its pain franchise: Janssen Pharmaceuticals, Ortho-McNeil Pharmaceuticals and ALZA Corporation. J&J has a broad pain portfolio of prescription-only and OTC products. Their substantial OTC division possesses the key brands Tylenol® (paracetamol [acetaminophen]), Motrin® (ibuprofen) and St Joseph® (aspirin). In 2009 J&J's OTC analgesics collectively had revenues of $1.54 billion and made up about one-fifth of their total OTC sales.[28]

J&J's prescription-only analgesics are all different types of opioids and are shown in Table 10.2. Duragesic® (fentanyl patch) had sales of $748 million in 2010[1] despite losing patent protection in 2005. It garnered sales of more than $1 billion in the 2002–2005 period, making it J&J's largest pain product by far. Duragesic® was developed by ALZA, and first launched in the United States in 1991. It releases fentanyl transdermally from the patch at a constant rate over the 72 hour period for which the patch should be worn, and so provides stable opioid exposure for the treatment of moderate to severe chronic pain.[29]

Table 10.2 Key Johnson & Johnson pain products and compounds in development.

Product name	*Generic name and mechanism*	*Launch date (US)*	*Comments*
Duragesic® patch	Fentanyl, short-acting μ-opioid agonist	1991	Manufacturing complexity for patches reduces the number of generic competitors and helps to maintain sales
Nucynta®	Tapentadol, mixed μ-opioid agonist and NET blocker	2009	J&J has commercial rights licensed from Grünenthal to much of the world outside Europe and South America
Ultram®	Tramadol, mixed μ-opioid agonist and NET blocker	1997	Originally sold under licence from Grünenthal
Ultracet®	Tramadol and paracetamol	2001	Marketed by Grünenthal under licence in Europe
Ultram ER®	Tramadol	2005	An extended-release formulation to allow once-daily dosing
Jurnista®	Hydromorphone, μ-opioid agonist	2007 (EU) 2010 (US)	Osmotic controlled-release formulation for once-daily dosing

It is often coprescribed with another opioid for acute use in case of breakthrough pain.[30] From 2005 generic fentanyl patches have been available and have eroded Duragesic sales; however, the complexity of manufacturing that is required for transdermal patches has limited that competition and helped to maintain reasonable revenues.

Nucynta® (tapentadol) is J&J's only analgesic whose patent protection is of the drug itself, rather than the formulation. Tapentadol was licensed from Grünenthal who discovered the drug as a follow-on to tramadol. Grünenthal retained marketing rights in Western Europe and South America, but J&J markets the drug in much of the rest of the world. Tapentadol is described more fully in the Grünenthal section (Section 10.7). J&J is experienced at marketing mixed opioid agonists in that it possesses a trio of tramadol-containing products: Ultram®, Ultram ER® and Ultracet ® (tramadol/paracetamol combination). All three generate relatively modest revenues and it is expected these will decline further now that tapentadol is on the market.[1]

The last of J&J's prescription analgesics is a controlled-release hydromorphone sold as Jurnista® (EU) and Exalgo® (US). Hydromorphone is a semisynthetic μ-opioid receptor agonist with relatively short duration of action. The key feature of the product's formulation is the OROS controlled-release system that was developed by ALZA, a J&J subsidiary. The tablets are coated with a water-permeable jacket with laser-drilled holes in it. Once the tablet is in the gastrointestinal tract water enters the tablet by osmosis and the pressure slowly pushes the drug through the drilled holes. In this way sustained release of the drug is possible, so converting a 4–6 times a day drug like hydromorphone into a once-daily drug.[31]

J&J has two new agents in clinical development for pain indications: fulranumab and JNJ-39439335. Fulranumab is an NGF antibody licensed from Amgen. It is currently in Phase 2 trials and has demonstrated analgesic efficacy in low back pain[32] and osteoarthritis.[33] It was subject to clinical hold by the FDA as a result of the adverse events experienced in the tanezumab clinical trials, which led to the discontinuation of several trials. JNJ-39439335 is a TRPV1 receptor antagonist in Phase 1. This ion-channel target has been of great interest to pharmaceutical companies since it was discovered in 1997,[34] but most companies have discontinued R&D efforts with this target. Phase 1 studies from multiple companies demonstrated sustained and non-tolerated hyperthermia following TRPV1 antagonist administration, which is believed to be mechanism related.[35] The profile of this compound has not been publicly released and therefore it is currently unknown if this compound will display an improved adverse-event profile.

10.6 Novartis

Novartis was formed from the merger of Ciba-Geigy and Sandoz in 1996. Both were Swiss pharmaceutical companies with over 100 years of history and the resultant company remains headquartered in Basel, Switzerland with approximately 100 000 employees. Novartis had combined sales of over $1 billion in 2011 for its pain drugs despite none of them still being covered by patent protection.[1] Its largest product is Voltaren® (diclofenac) was first launched in 1973 and has been generic for many years. Voltaren is no longer sold in the United States (other generics are available), but is marketed in over 140 countries and it is brand loyalty that maintains its sales of just under $800 million in 2011. It is available in non-oral formulations such as Voltaren Ophtha® eye drops and Voltaren gel which was launched in the United States in 2010 and will compete with Pfizer's Flector Patch. Novartis' two other products of note are carbamazepine (Tegretol®; $365 million sales in 2011) and oxcarbazepine (Trileptal®; $263 million sales in 2011), which is a metabolite of carbamazepine. Both drugs are described as sodium channel blockers, although they possess a range of other ion-channel pharmacologies.[36] Carbamazepine has been a mainstay of the epilepsy market since its launch in 1974 and is first line therapy for partial seizures. Although carbamazepine is not approved for neuropathic pain conditions it is widely used off label, despite the fact its efficacy has not been established in placebo-controlled clinical trials in the literature. It is approved for use in trigeminal neuralgia, for which there are a limited number of treatment options and high unmet medical need. The sales of carbamazepine and oxcarbazepine are both in decline due to competition from generics.[6]

Novartis has a limited pain pipeline to renew its pain franchise. Internal R&D efforts have generated SAF312 which is in Phase 2 development for pain and overactive bladder. A Phase 2 trial in dental pain was initiated in 2009 and completed in 2011, although the results have not been released.[37] In 2012 a trial into bladder overactivity due to spinal cord lesions began and is currently recruiting. Novartis also licensed the commercial rights from GW

Pharmaceuticals to Sativex® (nabiximols; cannabinoid agonists) in April 2011 in Middle Eastern, Asian and Australian markets.[38] Sativex buccal spray is currently approved for use in treating spasticity in multiple sclerosis.[39] It is also launched for cancer and neuropathic pain in Canada, although approval in other markets for these indications is uncertain at this time.

10.7 Grünenthal GmbH

The company was founded as Chemie Grünenthal GmbH (later renamed Grünenthal GmbH), by Hermann Wirtz in 1946 and remains owned by the Wirtz family to this day. The headquarters and R&D activities are located in Aachen, Germany. The company has approximately 4500 employees and operates in 26 countries, mostly in Europe and South America. Commercial activities in North America and Asia are *via* partnerships with large pharmaceutical companies with presence in those regions, such as J&J (see Section 10.5). In 2011 its revenues were €947 million.[40] Being privately owned, Grünenthal is able to invest much more heavily into R&D (~25% of revenues) than most publicly listed pharmaceutical companies. Public pharmaceutical companies are beholden to their shareholders and hence are more tightly constrained as to what percentage of profits can be reinvested back into research, typically around 15%.

Until recently Grünenthal developed and marketed a range of products across pain, respiratory, antibacterial, cardiovascular and contraceptive therapeutic areas. However, in the last few years the company has decided to focus on pain products and sold off its portfolio of mostly generic drugs to other companies.

Tramadol hydrochloride (Tramal®, Ultram®) was first launched in 1977 in Europe and is partnered with J&J in the United States and many other markets. Tramadol is structurally related to codeine and is a centrally acting analgesic approved for use in moderate to severe pain. It is a weak μ-opioid receptor agonist and also a weak inhibitor of noradrenaline reuptake transporter.[41] Tramadol is metabolised to *O*-desmethyltramadol which has significantly higher affinity for the μ-opioid receptor.[41] Both pharmacologies contribute to the analgesic profile, although the adverse-event profile is more closely related to that typical of opioid analgesics including dizziness, somnolence, nausea, vomiting and constipation. It is also available in combinations with other analgesics, for example with paracetamol (Ultracet®).

Grünenthal followed the success of tramadol by developing tapentadol (Nucynta® (US), Palexia® (EU)) which was first launched as an immediate-release formulation in the United States in 2008 in partnership with J&J. Tapentadol has similar pharmacology to tramadol in that it is mixed μ-opioid receptor agonist and noradrenaline reuptake transporter inhibitor, although it is more potent than tramadol and hence possesses opioid-like efficacy.[42] It has demonstrated equivalent change in pain intensity scores, improved response rate and nearly a 50% reduction in adverse events leading to discontinuation compared with oxycodone in a pivotal Phase 3 chronic low back pain trial.[43] The key to tapentadol's success has been the launch of the sustained-release formulation in 2010, which includes the tamper-resistant technology INTAC®.[44] Grünenthal developed INTAC internally and it functions in much the same way as that found in OxyContin® or Remoxy®. The fact that tapentadol is the first novel opioid in over 25 years, it possesses a favourable clinical profile and is available with a sustained-release, tamper-resistant formulation, all result in rapidly growing sales (predicted to be $916 million by 2016).[1]

Grünenthal also markets two patch products for the treatment of pain: buprenorphine (Transtec®) and lidocaine (Versatis®). Transtec® was first launched in Europe in 2001 for the treatment of moderate to severe pain. However, it was not until the product was licensed to Purdue that it gained FDA approval and was launched in the United States in 2011. Buprenorphine is a partial μ-opioid receptor agonist[45] and therefore it has a different clinical profile to other opioid patches, such as fentanyl. The partial agonism means that it cannot achieve the same level of efficacy as fentanyl, but also it means it has a lower propensity for adverse effects, particularly respiratory depression,[45] and therefore it can be used in opioid-naive patients. Grünenthal licensed the rights to the lidocaine patch (sold in the United States as Lidoderm®) from Endo for the European market and launched the product as Versatis® in 2007. Although the product is growing rapidly, it only sold $59 million in 2011[40] and so has only a fraction of the sales of Lidoderm.

Grünenthal has two novel drugs in clinical trials for pain, GRT-6005 and GRT-6010, both of which are in Phase 2. Both compounds follow in the footsteps of tapentadol using the strategy of mixing μ-opioid agonism with other pharmacological activities with the goal of maintaining efficacy but reducing or removing the troublesome side effects of classical opioids. In order to be successful in the marketplace they will need to clearly differentiate from tapentadol. GRT-6005 is described as a nociceptin (ORL-1) receptor agonist with additional agonist activity at μ-opioid receptors.[46] The nociceptin receptor is a G-protein coupled receptor whose endogenous agonist is nociceptin (also known as Orphanin FQ).[47] The receptor shares some homology with classical opioid receptors (μ, δ and κ) although most opioid drugs do not bind to the nociceptin receptor. The precise role of this receptor in analgesia is somewhat unclear, as some studies indicate an analgesic effect, whereas others indicate hyperalgesia. GRT-6005 has completed Phase 2 trials in different patient populations including bunionectomy, painful diabetic neuropathy and osteoarthritis of the knee. Currently it is being assessed in a Phase 2 trial in chronic low back pain in elderly patients.[47] The reason for the wide range of Phase 2 trials has not be publicly stated, but is possible that Grünenthal are trying to establish clinically meaningful differentiation from tapentadol and other marketed opioid drugs.

10.8 Endo Health Solutions

Endo Health Solutions Inc was formed *via* a management buyout from DuPont Merck in 1997. It is a privately held company based in Chadds Ford, Pennsylvania, USA. Endo has been diversifying itself in recent years and moving from a pain management company to one that also includes pelvic health and endocrinology. Endo's most successful product is the Lidoderm® patch (lidocaine), which had sales of $1.1 billion in 2010.[1] It also possesses several opioid-containing products including Opana® (oxymorphone), Percocet® (oxycodone/paracetamol combination) and Percodan® (oxycodone/aspirin). Opana ER®, a tamper-resistant formulation of oxymorphone, has recently been launched and is expected to help revive sales of the product after a series of manufacturing issues with Opana negatively impacted it.[48]

Lidocaine has been in use for many years as a local anaesthetic and at lower doses as an antiarrhythmic drug,[49] but it cannot be given systemically for analgesia because of its cardiovascular and CNS depressant adverse effects. However, local application to the skin is viable and able to produce analgesia in peripheral neuropathies.[50] It was developed by Endo and Teikuko, a Japanese company which is a leader in medicated plasters and patches. Lidoderm was launched in the USA in 1999 and has been incredibly successful, especially considering it is only actually approved for postherpetic neuralgia (PHN).[51] The majority of its sales are generated through off-label uses for other peripheral, localised pain indications, such as DPN and low back pain. Lidoderm is due to go off patent in September 2013 and Endo has had protracted interactions with the FDA about the extent of data required by the generics company Watson Pharmaceuticals to launch a competitor product from that time. For oral products a generic competitor is only required to demonstrate bioequivalence (*i.e.* the same drug levels are achieved in the body following administration) compared with the patent-protected product—no efficacy or safety data are required, as they can be referenced to the original drug once patent expiry occurs. However, for products with local actions demonstrating bioequivalence in the systemic circulation is often insufficient as it may not be an accurate indicator of local drug levels at the site of application; in these cases efficacy data must also be provided, which is much more costly to generate. In the end Endo came to a settlement with Watson, such that the companies will work in partnership from 2013 to supply and market Lidoderm®.[52] Launch dates for other Lidoderm® generics are uncertain, but unlikely to occur before 2015.

Endo does not have publicly disclosed pain compounds in development and therefore it is likely to concentrate on its diversified portfolio for future revenue generation.

10.9 Over-the-Counter Analgesics

The prescription medicine pain market is certainly large both in terms of overall sales and number of days of therapy. However, it is important to consider the non-prescription market as well, because the overall volume of product sales is markedly higher even if the average price is much lower. Pain medicines that are available OTC (*i.e.* without prescription) fall into two categories: either only available from pharmacies where the pharmacist can control to whom they are sold, or on the open shelf, meaning they can be sold in any retail outlet,

Table 10.3 Top 10 over-the-counter analgesics by sales in 2010.

Product	*Company*	*Active ingredient*	*% of market*
Advil	Pfizer	Ibuprofen	4.8
Voltaren	Novartis	Diclofenac	4.3
Tylenol	Johnson & Johnson	Paracetamol	4.1
Efferalgan	Bristol Myers Squibb	Paracetamol	3.7
Nurofen	Reckitt Benckiser	Ibuprofen	3.5
Doliprane	Sanofi-Aventis	Paracetamol	3.3
Aspirin	Bayer	Aspirin	2.5
Panadol	GlaxoSmithKline	Paracetamol	2.4
Salonpas (topical)	Hisamitsu	Menthol/Methyl Salicylate	2.0
Aleve	Bayer	Naproxen	2.0

including supermarkets, garages and other locations where controls on sales do not exist. OTC analgesics collectively make up ~15% of global OTC sales across all therapeutic categories and 5 of the top 15 OTC brands are oral analgesics. Paracetamol is the most common active ingredient of these products, with aspirin, diclofenac and ibuprofen being the three other major drugs. Hundreds of companies produce versions of these drugs and they are sold under global brand names (*e.g.* Advil®, Tylenol®), national brands (Anadin®, Doliprane®), as well as pharmacy (*e.g.* Boots ibuprofen) and supermarket chain (*e.g.* Tesco paracetamol) own labels. The top 10 brands globally are shown in Table 10.3. All of these products are maintained as best-sellers by active product improvement formulations (*e.g.* gel capsules, dissolve-in-the-mouth granules, *etc.*), differentiated claims achieved through carefully designed clinical trials (*e.g.* faster onset than similar comparators) and high-profile marketing campaigns. All three of these factors mean than smaller manufacturers struggle to compete with multinational pharmaceutical companies and hence a limited number of companies account for the majority of sales value in the OTC markets.[28] The market dynamics and drivers for OTC products are very different from that of the patent-protected, prescription-only market and are beyond the scope of this chapter.

10.10 Potential New Pain Management Companies

The previous sections describe the key players in today's pain markets; however, patent expirations and launch of new products are likely to change the landscape over the next 5–10 years. Advances in ion-channel screening capabilities have brought a range of ion-channel targets within reach from a practical standpoint within drug discovery. Additionally the emergence of the neurotrophin signalling cascade as an efficacious pathway to target has provided additional impetus. It is too early to tell which mechanisms and which companies pursuing them will be the eventual winners in terms of product launches and revenue generation. A few of the potentially leading companies are described briefly here.

Of the large pharmaceutical companies that do not possess in-line patent-protected pain products, Abbott is currently one of the most active. They have drugs against a range of mechanisms in Phase 2 development including an NGF-blocking antibody (ABT-110), a histamine H3 receptor blocker (ABT-652) and a T-type ($Ca_V3.2$) calcium channel blocker (ABT-639). They are also known to be carrying out research into selective sodium channel blockers.[53] Sanofi and Merck also have active research groups pursuing novel mechanism analgesics.

A whole host of biotech companies and mid-sized pharmaceutical companies are pursing novel formulation or prodrug approaches to existing agents. These approaches carry a much smaller risk in terms of whether they will be effective treatments or not, because they are variations of agents with proven efficacy. The risk here is whether they will be differentiated from the slew of products already out there, in order to meet the payer and regulatory requirements. One of the leading players in this field is Nektar Therapeutics.[54] They have a polymer conjugation technology platform that they have applied to known opioids and local anaesthetics. Polymer conjugation (often a pegylated chain) to a known analgesic creates a prodrug, which is classed as a new chemical entity and hence can be patent protected. The prodrug reduces the rate of release of the active analgesic into the bloodstream, slows down brain penetration and makes physical methods of tampering ineffective. Several agents are currently in Phase 2 proof-of-concept trials, and therefore whether this approach is effective is currently unknown.

There are also a range of small biotechnology companies with ion-channel drug discovery expertise which are pursuing novel mechanism analgesics. Three of note are Convergence Pharmaceuticals,[55] who span out from GlaxoSmithKline with their sodium and calcium channel antagonist programmes, Xenon[56] who are pursuing sodium channels (among other targets) and Zalicus[57] who also have sodium and calcium channel programmes. As yet none of these companies have progressed their compounds beyond Phase 2a proof-of-concept trials. Successful programmes are likely to be partnered with large pharma for late-stage clinical trials and commercialisation.

References

1. Scripps Business Insights™. *The Pain Market Outlook to 2016.*
2. Y. A. Andreev, A. A. Vassilevski and S. A. Kozlov, *Recent Pat. Inflamm. Allergy Drug Discov*, 2012, **6**, 35.
3. E. Colombo, S. Francisconi, L. Faravelli, E. Izzo and P. Pevarello, *Future Med. Chem.*, 2012, **2**, 803.
4. www.pfizer.com.
5. A. J. Thorpe and J. Offord, *Curr. Opin. Investig. Drugs*, 2010, **11**, 235.
6. Datamonitor. *2011 Key Companies: Neuropathic Pain.*
7. L. J. Marnett, *Annu. Rev. Pharmacol. Toxicol.*, **49**, 265.
8. F. E. Silverstein, G. Faich, J. L. Goldstein, L. S. Simon, T. Pincus, A. Whelton, R. Makuch, G. Eisen, N. M. Agrawal, W. F. Stenson,

A. M. Burr, W. W. Zhao, J. D. Kent, J. B. Lefkowith, K. M. Verburg and G. S. Geis, *JAMA*, **284**, 1247.
9. www.pbm.va.gov/vioxx/Dear%20Healthcare%20Professional.pdf.
10. Press Release: www.fda.gov/Drugs/DrugSafety/PostmarketDrugSafety InformationforPatientsandProviders/ucm106148.htm.
11. Press Release: www.pfizer.com/files/investors/presentations/king_investor_deck_101210.pdf.
12. King Pharmaceuticals. *Annual Report 2010*.
13. Press Release: www.fda.gov/NewsEvents/Newsroom/PressAnnouncements/ucm207480.htm.
14. Press Release: www.reuters.com/article/2011/03/16/us-pfizer-recall-idUSTRE72FA3Q20110316.
15. Pfizer. *Annual Report 2011*.
16. www.durect.com/wt/durect/page_name/oradur.
17. Press Release: www.reuters.com/article/2012/03/01/durect-idUSL4E8E16M920120301.
18. Press Release: www.reuters.com/article/2011/07/20/us-pfizer-icagen-idUSTRE76J39L20110720.
19. Icagen. *Annual Report 2007*.
20. M. Liu and J. N. Woods, *Pain Med.*, 2011, **12** Suppl 3, S93.
21. www.neusentis.com.
22. www.purduepharma.com/.
23. www.fda.gov/downloads/Drugs/DrugSafety/UCM219146.pdf.
24. http://clinicaltrials.gov/ct2/results?term=V-116517.
25. J. P. Lepine and M. Briley, *Hum. Psychopharmacol*, 2004, **19** Suppl 1, S3.
26. Eli Lilly. *Annual Report 2011*.
27. www.lilly.com/SiteCollectionDocuments/Pipeline/Clinical%20 Development%20Pipeline/index.html.
28. N. Hall, *Global OTC Analgesics Review*, 2010.
29. www.accessdata.fda.gov/drugsatfda_docs/label/2005/19813s039lbl.pdf.
30. A. H. Vallerand, *Nurs. Clin. North Am.*, 2003, **23**, 435.
31. D. R Guay, *Consult. Pharm.*, 2010, **25**, 816.
32. P. Sanga, K. Karcher, S. Wang, K. Kelly, C. Oh and J. Thipphawong, *J. Pain*, 2011, **12** Suppl 1, 53.
33. S. Vallow, P. Sanga, E. Polverejan, S. Wang, K. Kelly and J. Thipphawong, *J. Pain*, 2011, **12** Suppl 1, 54.
34. M. J. Caterina, M. A. Schumacher, M. Tominga, T. A. Rosen, J. D. Levine and D. Julius, *Nature*, 1997, **389**, 816.
35. N. R. Gavva, *Trends Pharmacol. Sci.*, 2008, **29**, 550.
36. S. M. Grant and D. Faulds, *Drugs*, 1992, **43**, 873.
37. http://clinicaltrials.gov/ct2/results?term=saf312.
38. Press release: www.gwpharm.com/news.aspx.
39. J. Sastre-Garriga, C. Vila, S. Clissold and X. Montalban, *Expert Rev. Neurother*, 2011, **11**, 627.
40. Grünenthal. *Annual Report 2011*.
41. P. Dayer, L. Collart and J. Desmeules, *Drugs*, 1994, **47** Suppl 1, 3.

42. T. M. Tzschentke, T. Christoph, B. Kögel, K. Schiene, H. H. Hennies, W. Englberger, M. Haurand, U. Jahnel, TI. Cremers, E. Friderichs and J. De Vry, *J. Pharmacol. Exp. Ther.*, 2007, **323**, 265.
43. R. Buynak, DY. Shapiro, A. Okamoto, I. van Hove, C. Rauschkolb, A. Steup, B. Lange, C. Lange and M. Etropolski, *Expert Opin. Pharmacother.*, 2010, **11**, 1787.
44. www.intac.grunenthal.com/grt-intac/212000676.jsp.
45. N. Vadivelu and R. L. Hines, *J. Opioid Manag.*, 2007, **3**, 49.
46. L. C. Choiu, Y. Y. Liao, P. C. Fan, P. H. Kuo, C. H. Wang, C. Reimer and E. P. Prinssen, *Curr. Drug Targets*, 2007, **8**, 117.
47. http://clinicaltrials.gov/ct2/results?term = GRT-6005.
48. Press Release: www.reuters.com/article/2012/01/09/us-fda-endo-idUSTRE8081SE20120109.
49. K. A. Collinsworth, S. M. Kalman and D. C. Harrison, *Circulation*, 1974, **50**, 1217.
50. K. P. Garnock-Jones and G. M. Keating, *Drugs*, 2009, **69**, 2149.
51. www.accessdata.fda.gov/drugsatfda_docs/label/2005/020612s007lbl.pdf.
52. Press Release: www.accessdata.fda.gov/drugsatfda_docs/label/2005/020612s007lbl.pdf.
53. Thomson Reuters Partnering Database.
54. www.nektar.com/product_pipeline/all_phases.html.
55. www.convergencepharma.com/.
56. www.xenon-pharma.com/.
57. www.zalicus.com/.

CHAPTER 11

Enabling Next-Generation Pain Therapeutics through Precision Medicine

SERENA SCOLLEN* AND ANNE PHELAN

Pfizer Neusentis, Pfizer Worldwide R&D, Portway Building, Granta Park, Cambridge CB21 6GS, UK
*Email: Serena.Scollen@Pfizer.com

11.1 Precision Medicine

Whether we call it 'stratified', 'personalised' or 'precision' medicine the basic concepts are the same and allude to the fact that researchers in all domains of the healthcare system need to become more innovative in the steps from scientific discoveries to clinical utility and the development of approaches that will result in better healthcare outcomes for patients.

Precision medicine as a concept is very simple and very few companies, academics or funding bodies would dispute the importance of this venture. The benefit to patients of identifying precision approaches in advance rather than downstream of drug use is irrefutable. For pharmaceutical companies, if genomic biomarkers for drug efficacy and/or adverse drug reactions (ADRs) can be identified in advance of administering drugs to large patient populations, then markets can be well defined and the number of clinical trial failures and ADRs reported can be reduced. Huge advances in 'omics' technologies (*e.g.* genomics, proteomics and metabolomics) over the last decade have advanced the understanding of disease mechanisms and enabled precision

RSC Drug Discovery Series No. 35
Pain Therapeutics: Current and Future Treatment Paradigms
Edited by Charlotte Allerton

Published by the Royal Society of Chemistry, www.rsc.org

medicine strategies to take a forefront in drug development. Identifying markers and/or selecting patient populations prior to or during drug development is now a reality. At the time of development of warfarin, an oral anticoagulant drug approved for use in 1954, it was not possible to identify the genomic markers for drug efficacy and ADRs now known (*VKORC1* and *CYP2C9*). However, if this drug had been developed recently, it would have been absolutely possible to identify these markers before drug approval.

It is too simplistic at this point in time to think that we will have full 'personalised medicine' in place in the very near future. This implies that we would be able to personalise or select a drug for a particular individual based upon their 'omic' profile. Costs to generate 'omic' data are becoming less prohibitive but there are still gaps in our ability to interpret, analyse and even store data on this large scale. A generalised interpretation of an 'omic' profile into a meaningful outcome would currently be flawed in most cases, because of the complexity of most disease mechanisms. More likely, over the next decade, medicine will become increasingly 'precise'. Drugs/therapies will be developed for specific subpopulations and the ability to classify an individual into a subpopulation will become more accurate. As our understanding of disease mechanisms at a molecular level advance there will be a rapid increase in the identification of clinically significant patient subgroups (in addition to diagnostic technologies) and development of more targeted and efficient therapies.

The aim of precision medicine is twofold:

- to select the right drug with better efficacy for a specific disease subpopulation and therefore a higher chance of efficacy within an individual
- to avoid giving a patient a therapy that may have a higher likelihood of detrimental or no effect.

These aims are of course the current aims of all prescribing clinicians, but currently on average only about half the patients who receive a given therapy actually benefit from it and many may experience side effects. When there are large knowledge gaps about the disease mechanisms at a molecular level, the trial and error approach with therapy selection for a patient is the only option.

To achieve these precision medicine aims the fundamental strategy is to bridge the gaps in our understanding of the molecular mechanisms underlying a human disease or trait (disease diversity) and identification of the differences between patients. Increased knowledge in these areas will lead to:

- reclassification of diseases
- reclassification of treatment options for individuals
- more informed therapeutic target selection
- patient stratification (including alternative trial designs and companion diagnostics)

To expedite success in the near term, this strategy must be applied at every stage of drug development.

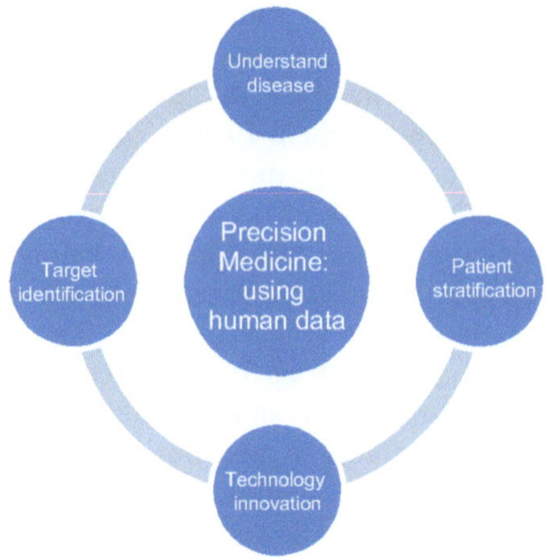

Figure 11.1 Essential components for precision medicine.

These key aspects of precision medicine (Figure 11.1) are discussed in this chapter. A crucial component to aid this process is the continued development of innovative technologies, which, together with the expanded use of human data, will truly enable a new era in pain research.

11.1.1 Target Selection Based on Preclinical Animal Models

Traditionally, within the pharmaceutical industry, target selection and validation for potential new pain therapies follows a fairly standard route. A receptor/ion channel/enzyme is implicated in pain signalling pathways (typically in the literature), or through a perceived understanding of its pharmacological or biophysical role within the cell type of interest (*e.g.* dorsal root ganglion, spinal cord). Recombinant cell-based assays or *ex vivo* rodent tissue preparations are established in order to identify a sufficiently potent and selective tool compound or antibody, to be able to ultimately validate the target through *in vivo* models of pain. Alternatively, knockout mouse or RNAi mediated knockdown are conducted to enable *in vivo* evaluation of the consequences of inhibition or 'loss' of the target of interest. Studies are conducted 'blind' with positive controls and data interpreted accordingly.

This classical route of target validation has proven successful for some mechanistic classes of pain medications, perhaps most notably the opioids, which appear to generate robust efficacy in the majority of animal models of pain tested.[1] However, preclinical models of efficacy have frequently not provided the desired translational outcome in the clinic, at great cost to the industry.

Perhaps the most famous example is the NK1 (substance P) receptor story. The antagonist was shown to attenuate nociceptive responses sensitised by both inflammatory and nerve damage in animal pain models.[2] In humans, the molecule demonstrated suitable systemic and central nervous system (CNS) exposure, receptor occupancy studies showed that the compound gained access to the target,[3] but the compound failed to show any clinical efficacy in a range of clinical pain states. The conclusion drawn for this target was that NK1 is relevant for pain in rodents, but it is not relevant to pain in humans. Now clearly this was a disappointing and expensive discovery, but as an industry we have apparently continued to make the same 'mistakes'. A more recent example is PF-04457845. This was published as a small molecule that selectively inhibits fatty acid amide hydrolase (FAAH) *in vivo* (as determined by activity-based protein profiling), and raises brain anandamide levels for up to 24 hours, resulting in profound cannabinoid receptor-dependent reductions in inflammatory pain, manifest as efficacy in complete Freund's adjuvant (CFA)-based rodent inflammatory pain models.[4] However, only a couple of years later a clinical trial with PF-04457845 failed to induce effective analgesia in patients suffering from pain due to osteoarthritis of the knee.[5] Again, despite a clear biomarker and excellent pharmacological activity, the mechanism did not translate from rodent models to clinical efficacy.

There have been many other subsequent failures in the clinic of compounds which have shown excellent preclinical efficacy across a range of models. The issue is not the technical validity of the assay or the manner in which the studies are conducted, but rather the relevance of preclinical evoked models of pain to complex human painful conditions and the oversimplification of assessing a single parameter in rat such as the withdrawal reflex, with the anticipation that it will translate through to humans.[6] Clearly the time is ripe for change, and with the dawn of new technologies enabling target identification through direct interrogation of human genetic and epigenetic data and validation using human-induced pluripotent stem cell derived samples, we are on the brink of an exciting era in drug discovery.

11.2 Understanding Painful Diseases Through Use of Human Data and Patient Stratification

Diseases involving pain are usually complex and heterogeneous and are often difficult to diagnose. Complex pain disorders can be categorised into acute pain (burn, labour), cancer pain and chronic pain. Chronic pain can further be classified into idiopathic, inflammatory, musculoskeletal, neuropathic, visceral and widespread pain, although some disorders are difficult to classify. Examples of pain disorders and stratification are shown in Table 11.1. A patient may have more than one of these painful conditions at one time (or in their recent medical history). In addition, comorbidities are often present in chronic pain patients, for example depression, anxiety and sleep disturbance. These factors often complicate diagnosis, research, stratification and treatment. Diagnostic classifications evolve as the understanding about the mechanisms of disease evolves.

To understand the underlying genetics and mechanisms of painful diseases, as well as enable patient stratification for therapeutic target selection and trials,

Table 11.1 Complex pain disorders and patient stratification.

Acute pain	*Cancer pain*	*Chronic pain*							
Non-chronic	*Cancer*	*Back pain*	*Idiopathic*	*Inflammatory*	*Miscellaneous*	*Musculoskeletal*	*Neuropathic*	*Visceral*	*Widespread*
Burn pain	Cancer pain	Back pain	Burning mouth syndrome	Chronic pancreatitis	Chronic pain (pain clinic)	Osteoarthritis	Neuropathic pain	Dysmenorrhoea	Fibromyalgia
Delayed-onset muscle soreness		Low back pain	Temporo-mandibular joint disorder	Familial Mediterranean fever	Diabetic foot ulcer pain	Whiplash	Complex regional pain syndrome	Endometriosis	Chronic widespread pain
Labour pain		Lumbar disc disease	Irritable bowel syndrome	Rheumatoid arthritis	Multi-somatoform disorder	McArdle's disease	Phantom limb pain	Epigastric pain syndrome	Multiple regional pain
Postoperative pain		Lumbar root pain	Interstitial cystitis	Ulcerative colitis	Pain in depressed patients		Post-herpetic neuralgia	Functional dyspepsia	
Menstrual pain		Lumbar root pain (postop)	Chronic pelvic pain		Sickle cell disease				
			Postmastectomy pain						
		Sciatica	Vulvodynia						

it is essential to dissect the phenotype of pain into both diagnostic and quantitative subgroups (*e.g.* by measuring pain sensitivity *via* quantitative sensory testing or by functional imaging). It is also crucial to identify different underlying causes that drive painful disease and separate diseases that have been traditionally regarded as a single disease into specific categories.

11.2.1 Genetics

The genetic component of painful disorders, shown by heritability estimates, is moderate (Table 11.2). These estimates confirm that there is a genetic component to these painful disorders. However, calculations are in the main based upon twin studies that have measured the concordance of the painful disease among twin pairs rather than the direct twin–twin correlation of pain ratings. The overlap between developing painful diseases and actual pain intensity is unknown. It is important to consider that identifying genes that underlie the painful disease without measuring pain may lead to uncertainty as to whether the gene is actually involved in pain physiology itself.

A moderate heritability estimate also indicates that other predisposing non-inherited factors, such as environmental triggers, exist. It is also possible that epigenetic factors play a significant role in the development of these disorders. Identifying the underlying genetic predisposing factors may lead to understanding disease pathways and aid with the identification of pain therapeutic targets. It is also important to take other contributing factors into consideration to fully understand the mechanism of disease.

Different strategies are required to identify the genetic component of painful disorders. For common complex pain disorders it is expected that a cumulative effect of rare and common variants will drive genetic susceptibility. For very rare and extreme pain disorders it is expected that very rare variants that have a dramatic effect on phenotype will be causative of the disorder.

11.2.1.1 Common Complex Disorders

The underlying genetic heritability of complex pain disorders is relatively unknown compared to other complex disorders, such as cardiovascular disease or

Table 11.2 Heritability of diseases with painful components.

Disease	*Heritability estimate (%)*	*Reference*
Chronic pain	29	7
Severe chronic pain	44	7
Back pain	34	8[a]
Migraine	45	8[a]
Neck pain	24–58	8[b]
Visceral pain	22–27	8[b]
Rheumatoid arthritis	65/53	8[b]
Chronic widespread pain/fibromyalgia	~50	8[b]

[a]Meta-analysis of reported studies.
[b]Reviewed.

type 2 diabetes. This is attributable to a lack of large-scale pain disorder cohorts that can contribute to consortia and meta-analysis required for both discovery and independent validation essential for identification of common variant susceptibility loci by genome-wide association (GWA) studies and rare variant association by sequencing studies.

For nearly a decade GWA studies have been used successfully to identify susceptibility genes for complex disorders with similar heritability estimates. An advantage of these studies is that there is no prior hypothesis that a particular gene will be associated with the disease.

Experience from GWA of other traits tells us that variants revealed to be associated with a pain trait by GWA may be:

- In or near genes known to cause mendelian traits with aberrant function of a protein relevant to the trait of study. For example, myocardial infarction variants were identified in or near genes known to cause mendelian dyslipidaemias (*LDLR*, *APOB* and *PCSK9*) or established molecular targets for LDL-C lowering therapies (HMGCR).[9]
- In or close to genes known to have a role in the pathophysiology of the disease (variants upstream and within the first intron of *TRPM8* have been associated with migraine without aura).[10] *TRPM8* encodes a cold-activated ion channel that is expressed in sensory neurons, which could be involved in cold allodynia and therefore could have a key role in pain transduction.
- In regions that contain genes that have not previously been implicated in pain or as yet have an unknown function. Using cardiovascular traits again as an example, the *SORT1* locus (chromosome 1p13) has the strongest association with LDL-C of any locus in the genome ($P = 1 \times 10^{-170}$)[11] but before the discovery of this locus, originally in 2008, variants in the region had not been implicated in lipid metabolism. Functional evidence for a novel regulatory pathway for lipoprotein metabolism was identified *via in vitro* and *in vivo* studies. A common non-coding variant rs12740374 was shown to create a C/EBPα binding site that confers increased expression of sortilin-1 and lower plasma LDL-C levels, thus improving cardiovascular health.[9] A tentative association has been made between *ASTN2* and migraine.[10] *ASTN2* is a member of the astrotactin gene family but currently it remains unclear how this ASTN2 could have a role in migraine pathophysiology.
- In stretches of DNA known as gene deserts that are devoid of genes or other types of non-coding DNA. Such a region, 8q24, was found to harbour common variants associated with breast, ovarian, prostate, colon and bladder cancer.[12]

A GWA approach is unlikely to capture all of the underlying heritability of a complex disorder. Large GWA studies for Crohn's disease (discovery panel: meta-analysis of six GWA studies comprising of >6000 cases and >15 000 controls; replication comprising of >15 000 cases, >14 000 and 414 parent–offspring trios) identified 71 independent loci that account for only 23.2% of heritability.[13] So-called 'missing heritability', the remaining genetic variation that causes susceptibility to disease, could be attributable to larger numbers of variants of smaller effect sizes yet to be found, rarer variants with larger effect sizes or gene–gene interactions, among other possibilities.

Since the first report in 2005, in excess of 1000 GWA reports have been published for a diverse range of complex disorders or traits. The only pain GWA reported are for migraine, for which there have been several,[10,14,15,16] and chronic widespread pain (CWP).[17] In addition to these studies, GWA studies for painful diseases such as osteoarthritis have been reported,[18] but have not directly phenotyped pain and so it is unclear currently if the signals identified are pain- or disease-related. GWA studies that have small discovery cohorts (<1000 cases and 1000 controls) or those that lack large independent replication cohorts/substantive meta-analysis have not been considered here because of the high chance of false-positive signals and lack of power due to multiple testing burden.

Diagnostic criteria for migraine and CWP are well established, aiding collection of large cohorts of disease cases and controls for discovery GWA and meta-analysis. The International Classification of Headache Disorders (ICHD) was first published in 1988. Similarly the American College of Rheumatology (ACR) introduced fibromyalgia classification criteria in 1990. These classifications increase recognition of the syndromes and enable not only research but also drug trials.

More frequently than GWA studies, candidate gene association studies have been performed for many biologically plausible pain genes (*COMT*, *OPRM1* and *GCH1* to name a few). As in other fields, there are inconsistent reports of association and confidence in the data is hampered by lack of replication. The studies are often underpowered because of small sample size and are subject to other problems such as confounding from population substructure. The approach is also notoriously subject to publication bias, with negative results failing to enter the public domain, resulting in 'winner's curse'. Current reports of association mostly require replication in large, well-defined case–control cohorts (>1000 cases and >1000 controls) or GWA studies.

It is not yet known if genes will be identified that affect multiple pain disorders or whether genes will be identified specific to a disorder. If genes are associated with multiple disorders, it will be important to understand if they are associated with pain itself or modulators of pain (*e.g.* depression). For development of a pain therapeutic, it is crucial to make this distinction and therefore studies will be required where the patients are stratified into categories *via* deep phenotyping with analysis broadened to encompass a wider phenotype where applicable.

Further GWA studies for pain traits are required that use large case–control cohorts (>1000 cases and >1000 controls) where the cases (and controls) have been precisely phenotyped. Once signals have been independently replicated, new biological hypotheses can be derived. New avenues for pain research may be initiated to characterise these regions, understand the direction of effect and relevance to the phenotype. It is possible that a novel regulatory pathway will be identified for pain *via* GWA of common variants to a disorder and evidence may build that modulation of such pathway may provide a novel therapeutic mechanism. It should also be noted that for some loci identified, such as those in desert regions, it will be more difficult to identify functionality.

11.2.1.2 Rare Variants with an Intermediate Effect Size Contributing to Complex or Rare Phenotypes

With the realisation that GWA was not going to reveal all the genetic susceptibility to a common complex disease *via* the cumulative effect of common variant hypothesis, the focus for many diseases has shifted to rare variants. Due to developing sequencing technology and reducing costs, it has become possible to capture rare variants with a moderate effect size that could be contributing to disease. Study designs include the use of:

- families, to capture mendelian or near-mendelian-type disorders
- phenotype extremes, to capture rare variants enriched in a population with an extreme phenotype

The first approach has been very successful uncovering many new variants for rare phenotypes, but the second has struggled with study power. Despite this, a significant enrichment of variants in genes of the angiotensin pathway was identified to contribute to pain perception in humans.[19]

It would also be possible to design a case–control study (akin to GWA). The cost of sequencing the exome and sample sizes required to cope with multiple testing burden (when testing the burden of rare coding mutations on 20 000 genes) is still out of reach. It has been reported that sample sizes required for gene discovery in exome sequencing studies of complex traits would be >1000 and ideally between 3000 and 10 000.[20]

11.2.1.3 Extremely Rare Disorders

There are severe pain disorders that have extremely rare prevalence in the general population and usually occur in families where there has been a history of consanguinity. Genetic mutations for many of pain disorders have been identified (examples given in Table 11.3) and many have been validated by functional studies showing that the mutation results in aberrant protein function. As an example, *SCN9A* missense mutations can result in loss of function in the sodium ion channel $Na_V1.7$ it encodes, causing a condition known as congenital insensitivity to pain (CIP), whereas gain-of-function mutations in the same gene have been reported to cause paroxysmal extreme pain disorder, inherited erythromelalgia (IEM) and most recently small-fibre neuropathy (http://omim.org/entry/603415).[21–23]

Table 11.3 Examples of pain disorders in which a responsible gene has been identified.

Phenotype	*Gene*	*Protein*	*OMIM gene entry*[a]
Syndromes featuring absence of pain			
Congenital Insensitivity to Pain (CIP)	*SCN9A*	Voltage-gated sodium channel, type IX, alpha subunit (NaV1.7)	603415
Heriditary sensory and autonomic neuropathy 1A/1C (HSANI)	*SPTLC1*	Serine palmitoyltransferase, long chain 1	605712
Heriditary sensory and autonomic neuropathy 1D (HSANI)	*ATL1*	Atlastin GTPase 1	606439
Heriditary sensory and autonomic neuropathy 1E (HSANI)	*DNMT1*	DNA methyltransferase 1	126375
Heriditary sensory and autonomic neuropathy 2A (HSANII)	*HSN2*	Splice variant of *WNK1* gene	605232
Heriditary sensory and autonomic neuropathy 2B (HSANII)	*FAM134B*	Family with sequence similarity 134, member B	613114
Heriditary sensory and autonomic neuropathy 3 (HSAN-III)	*IKBKAP*	Inhibitor of kappa light polypeptide gene enhancer in B cells, kinase complex-associated protein	603722
Heriditary sensory and autonomic neuropathy 4 (HSAN-IV)	*NTRK1*	Neurotrophic tyrosine kinase receptor type 1	191315
Heriditary sensory and autonomic neuropathy 4 and 5 (HSAN-IV/V)	*NGFB*	Nerve growth factor, beta subunit	162030
Syndromes featuring pain			
Familial hemiplegic migraine type 1 (FHM-I)	*CACNA1A*	Voltage-dependent calcium channel, P/Q type, alpha-1A subunit	601011
Familial hemiplegic migraine type 2 (FHM-II)	*ATP1A2*	ATPase, Na+/K+ transporting, alpha-2 polypeptide	182340
Familial hemiplegic migraine type 3 (FHM-III)	*SCN1A*	Voltage-gated sodium channel, neuronal type I, alpha subunit	182389
Migraine with aura	*KCNK18*	Potassium channel, subfamily K, member 18 (TRESK)	613655
Familial Episodic pain syndrome	*TRPA1*	Transient receptor potential cation channel, subfamily A, member 1	604775
Paroxysmal extreme pain disorder (PEPD)	*SCN9A*	Voltage-gated sodium channel, type IX, alpha subunit (NaV1.7)	603415
Inherited Erythermalgia (IEM)	*SCN9A*	Voltage-gated sodium channel, type IX, alpha subunit (NaV1.7)	603415

[a]http://omim.org/entry/
Table adapted from Ref. 24.

These disorders have a mendelian inheritance with one variant/mutation having a dramatic effect on the phenotype of the individual. The application of linkage studies (and now exome sequencing, whole genome sequencing or chip approaches) have successfully identified variants for several painful diseases. There has been success in identifying the variants that are causative of these diseases, due to the availability of well-phenotyped consanguineous families.

11.3 Epigenetics—a New Era in Target Identification

Epigenetic modifications, in concert with the genetic template, act to fundamentally regulate transcriptional activity of all cellular processes in normal tissues, acting as a bridge between genotype and phenotype. There are a growing number of examples where dysregulation of epigenetic processes has been linked directly to diseases such as cancer, asthma, diabetes and a range of neurological disorders.[25] Clearly most traction has been gained in the oncology field where for example, hypermethylation of tumour suppressor gene promoters such as FBX032, MLH1 and RUNX3, results in stable gene-silencing and the resultant activation of pathways culminating in tumorigenesis.[25] Indeed, the four epigenetic therapies approved by the U.S. Food and Drug Authority (FDA) to date have focused on inhibitors of histone deacetylation or DNA methylation for the treatment of blood and lymphoid cancers,[26] with some success. However, as epigenetic mechanisms underwriting transcriptional cascades in complex multifactorial diseases such as pain emerge, so too will our ability to exploit the epigenome for novel therapeutic opportunities. Some early advances have been made in understanding the epigenetic regulation of pain systems; for example, in traumatic injury models of neuropathic pain, at least 10% of the nociceptor transcriptome is altered. The differential expression of particular potassium channel subunits appears to contribute to nociceptor hyperexcitability,[27] as does the downregulation of postsynaptic genes, such as *GAD65*, *GAD67* and *GABAB1* in spinal inhibitory systems;[28,29] however, the full complexity of these systems is still to be elucidated.

11.3.1 Pathway Analysis

With the furious advancement of high-throughput exome sequencing, RNAseq and ChIPseq technologies, we are rapidly building a picture of gene expression and epigenetic signatures of human health and disease. It is increasingly clear, however, that cellular/molecular processes operate within interaction networks, and are governed by their properties. It is therefore imperative that results from large-scale genetic and epigenetic studies should be studied in the context of these networks. Numerous resources of protein interaction data now exist: BioGrid,[30] String,[31] HPRD,[32] *etc*. These can be used for drug target discovery in many ways. One such approach, 'network pharmacology',[33] relies on the topological properties of biological networks. Biological networks are thought to be scale-free; that is, the distribution of node connectivity (a.k.a. degree) is not uniform, with the presence of highly connected hubs being a key feature.[34]

This property provides protection against random failure of any one node, for example through the presence of a loss-of-function single-nucleotide polymorphism (SNP). However, such hub nodes are ideal drug targets and a network pharmacology approach seeks to inactivate as many as necessary in parallel to disrupt the structure of the network and thus its function. This is a blunt approach, where little understanding of the underlying biology is required.

In contrast, other approaches to the mining of data in a network context have been proposed. An approach called 'causal reasoning' has recently been characterised that utilises a database of directed regulatory interactions between molecular entities.[35] Given a set of differentially expressed genes (for example), it identifies upstream regulators whose known regulatory activity correctly explains a significant proportion of the gene expression changes observed, taking into account their directionality. In so doing pathways can recapitulate the key drivers of disease and relevant expression changes identified. These regulators represent points of intervention that can be validated within the laboratory given a suitably amenable assay. Figure 11.2 shows the result of a causal reasoning analysis as applied to 138 genes associated with thermal nociception from an exome sequencing study.[19] The most significant regulator identified in this study, angiotensin II, has strong links to neuropathic pain and its receptor is the subject of a promising drug development program (www.spinifexpharma.com.au/).

11.4 Application of Genetics to Target Identification

To date, human genetic evidence used to identify or develop a drug target has come from the identification of mutations that result in aberrant protein function that causes an extreme pain phenotype. Several genes and mutations have been identified for rare, monogenic disorders of pain (Table 11.3). *SCN9A* is unique in this list as mutations in this gene result in more than one pain state, leading to high interest in the $Na_V1.7$ sodium channel as a drug target, and blockers for $Na_V1.7$ are under development for the treatment of pain. Loss-of-function mutations in *SCN9A* were identified in individuals suffering from a condition known as CIP.[36] These mutations are inherited as a recessive trait. A single mutation in *SCN9A* results in a truncated $Na_V1.7$ protein, resulting in a complete loss of function of the channel. This loss of function has been demonstrated using electrophysiology and mammalian cell lines. The rationale for a therapeutic for pain is based upon replicating the loss-of-function phenotype in clinical trials using a small-molecule inhibitor. Gain-of-function mutations in *SCN9A* that result in IEM[21] and paroxysmal extreme pain disorder (PEPD)[22] are inherited as a dominant trait. These disorders are also very rare and extreme, individuals suffering from episodes of burning pain in their hands and feet or rectum. These mutations were shown in *in vitro* functional studies to confer hyperexcitability to peripheral sensory neurons, contributing to symptom production in IEM/PEPD.[21,22] More recently, mutations in the

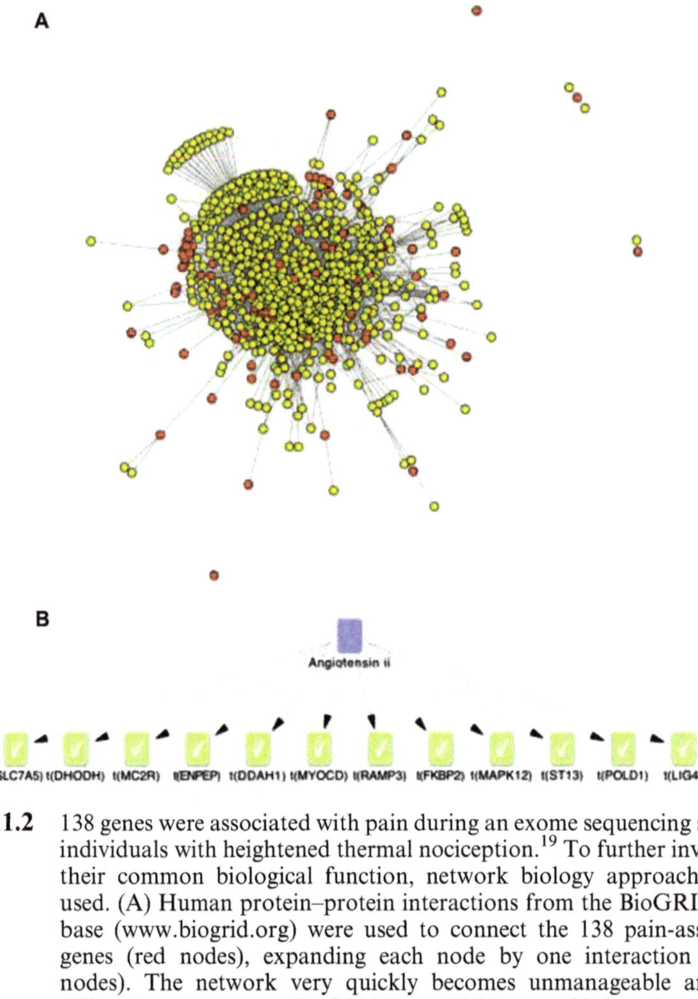

Figure 11.2 138 genes were associated with pain during an exome sequencing study of individuals with heightened thermal nociception.[19] To further investigate their common biological function, network biology approaches were used. (A) Human protein–protein interactions from the BioGRID database (www.biogrid.org) were used to connect the 138 pain-associated genes (red nodes), expanding each node by one interaction (yellow nodes). The network very quickly becomes unmanageable and it is difficult to uncover meaningful biology. (B) More sophisticated approaches such as Causal Reasoning[35] can be used to reveal more subtle interactions present in the dataset. Here, a significant proportion of the 138 genes were identified to be downstream of the regulator angiotensin II.[19]

gene have been reported to be causative of other pain disorders such as a particular subgroup of idiopathic small-fibre neuropathy patients, and again studies have indicated hyperexcitability.[23] Other than these pain disorders, *SCN9A* mutations have been reported in rare familial forms of epilepsy and febrile seizures.[37]

Over the next few years there will be an increase in the number of reported 'pain' genes from both rare extreme phenotype families and large case–control association studies. These new genes may lead to a greater understanding of pain pathways, providing novel therapeutic targets and new therapeutic approaches.

Table 11.4 General guide to interpreting the strength of genetic association data.

Strength of association	*Overall*	*Mendelian studies*	*Candidate gene association studies*	*GWA studies*
No data	Lack of studies on phenotype of interest	Linkage peak over large region	~100 cases and controls	Gene region not well covered on GWA study chips
Low	No evidence of association despite compelling studies		10 000 cases and controls with no clear association Several independent GWA studies on relevant phenotype with no association for genes of interest (ensuring panel includes gene of interest)	
	Poorly designed studies		Several studies of ~100 cases and controls with non-conflicting association	
	Tentative association	Linkage peak for narrow region <5 Mb, LOD>3.0	~1000 cases and controls with inconsistent associations	P-value below but near threshold of 10^{-8}
Medium	Association in several studies	Linkage peak for narrow region <5 Mb, LOD>3.0, with fine mapping data	>1000 cases and controls with replication/meta-analysis, P<applicable threshold (Bonferroni correction)	$P<10^{-8}$
High	Robust statistical genetic evidence, some functional data	As above + sequencing and functional data	>10 000 cases and controls, P<applicable threshold (Bonferroni correction), with weak functional evidence	Several independent GWAS ($P<10^{-8}$), weak functional evidence
	Robust statistical genetic evidence, good functional evidence and phenotype is relevant to indication	As above + sequencing and functional data with high relevance to phenotype	>10 000 cases and controls with strong functional evidence and high relevance to phenotype	Several independent GWAS ($P<10^{-8}$), strong functional evidence including direction of effect and high relevance to phenotype

To consider using human genetic association data for drug target identification, it is paramount that the evidence is compelling, the functional consequence has been characterised and the underlying molecular pathway has been dissected. Guidelines for interpreting genetic association data are given in Table 11.4. Ideally a potential target will only be considered if there is high confidence in the association. Patient phenotyping must be robust and a complete genome-wide coverage approach is essential for both linkage and GWA studies. For GWA studies steps are often taken to fine map a region to identify a causative gene and where several genes are involved, bioinformatic approaches can be applied to identify the involvement of a biological pathway.

For pain phenotypes, high confidence of association (and potential new targets from genetic evidence) is currently mostly limited to mutations/variants identified for rare mendelian disorders and migraine.

The impact of identifying pain genes through GWA studies of common complex diseases will vary depending on the novelty of the molecular pathway. The challenges for setting up a GWA study for pain traits start with being able to access large enough populations with genotype and phenotype data for discovery and replication studies. Following this, it is often necessary to identify actual causal variants. If an entirely new pathway were discovered, it would take at least a decade for a compound to reach clinical trial and longer to have an impact on a large patient population. Having said that, if a compound targeting a pathway were already established, repositioning efforts could expedite this process. If a GWA study identifies genes known to have a role in the pathophysiology of pain then the time to impact and the drug development effort will be reduced. These challenges are in some ways similar to identification of pain genes through rare variant studies. The advantage of rare variant studies is that it may be easier to functionally characterise the affected protein involved (they are more likely to have a dramatic effect), shortening the time to impact a drug discovery programme.

11.5 Technology Innovation—Humanising Drug Development

To understand human disease mechanisms and improve drug development it is crucial that models that are closest to the species and phenotype of interest are used.

In 2006, induced pluripotent stem (iPS) cells were established from mouse embryonic fibroblasts following the induction of four transcriptional factors: Oct3/4, Sox2, Klf4 and c-Myc.[38] This revolutionised research by generating an opportunity to derive cells from patient populations with specific genotype and phenotype data. These cells can be used for disease modelling, screening for new drugs and investigating effects of gene variants on pharmacology of clinical compounds (Figure 11.3).

Much of the knowledge gained to date about the functional consequence of genetic variants has been acquired through the use of cell lines or genetically

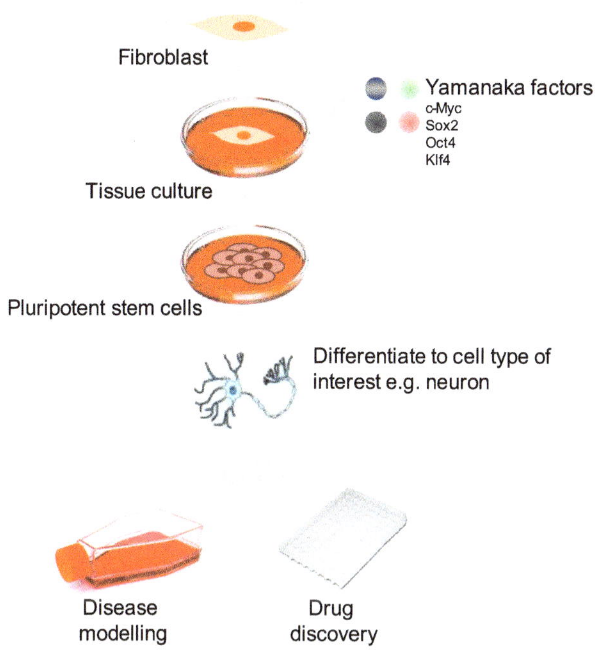

Figure 11.3 Differentiated somatic cells (fibroblasts) can be reprogrammed into induced pluripotent stem cells, which in turn have the potential to differentiate into cell types such as neurons. This supply of genetically identical cells can be used increase the understanding of disease mechanisms as well as drug discovery.

modified mouse models. Transfection of a vector containing the gene/variant of interest into an immortalised human cell line (or rodent primary cell) can provide evidence of a change in the protein function caused by the gene/variant. The limitations are that overexpression can occur, human tissue-relevant proteins can be missing and the genetic background is often unknown. An advantage of using a genetically modified mouse model to study protein function (*e.g.* knockin or knockout) over use of an immortalised cell line is that the relevant tissue can be sourced; however, using this model to directly translate to humans can be flawed due to species divergence, differing genetic architecture and functional redundancy.[39] The advantage of iPS cell approaches is the potential to study the effects of a particular gene mutation on protein function in cells sourced from patients and has the potential to model human physiology. Specific human tissue types (*e.g.* dorsal root ganglion neurons and brain tissue) are difficult to source for the study of pain and so iPS cell models will be of enormous value. Once the cells are generated, extensive phenotypic characterisation of their structure, electrophysiology, transcription, translation and intracellular signalling can be performed. The value of IPS cells in understanding pathophysiology has already been demonstrated in IPS-derived cardiomyocytes, where a range of gene variants underlying cardiac arrhythmia syndromes have been investigated.[40]

As with any scientific approach there are caveats. There will still be variation in primary cells sourced from different patients and it will not always be possible to generalise for the entire disease population from a small number of sourced iPS cells used for modelling the disease. However, the 'omic' profile of donor patients will be available and so the limitations of interpretation can be predefined to extend only to patients with a 'matched' profile. Until iPS cell technology becomes higher throughput it will be most easily applied to mendelian-type disorders where there is a single genetic component having a dramatic effect to phenotype and an early onset of disease, *e.g.* to study the effect of mutations in *SCN9A* that cause loss of function of $Na_V1.7$ resulting in CIP, and gain-of-function mutations that result in IEM. The interpretation for these mendelian disorders will be more easily applied to a disease population as the effect of other background variants or unknown factors within the population are not as pronounced. One difficulty is access to patients, but this will shortly be bypassed by newly developed gene-targeting technologies, such as use of zinc finger nucleases that are being developed to create isogenic iPS cells from any patient, thus reducing the need to access patients with the actual genetic disorder.

It is not currently possible to mimic the environmental influence or the late-onset phenotype of a common complex pain phenotype in a laboratory setting. It will be more difficult to study phenotypes that are complex, involving many genetic factors or many different cell types or interactions. Co-differentiation and co-culture of several different cell types will invariably take longer to develop and this may not be possible in the near term.

As previously noted, it is also possible to use these iPS cells lines as drug screens. Many drugs demonstrate preclinical efficacy but do not perform well in Phase 2 trials due to species-specific differences. The introduction of an iPS cell model into preclinical validation pipelines may substantially lower the risk of failed clinical trials. The development of iPS cell technology will aid the humanisation of drug development. Imminent research will be in a small clinical population but the knowledge of pain mechanisms can quickly be translated to a larger clinical population. In the short term, although yet to be proven, there is a huge potential for iPS cell use to increase the understanding of pain mechanisms. In the longer term, once large numbers of iPS cells are available for specific pain disorders along with comprehensive genotype and phenotype data from the source patient, the study of disease-state modifiers will be possible.

11.6 Patient Stratification

Late-stage attrition has been a problem for drug development over the past decade. Future drug development success depends on the ability to reveal activity for a drug in specific subpopulations rather than in broad populations that may hide the activity. This can be achieved through the identification of biomarkers: genomic, proteomic (measurements taken from biological samples such as proteins in blood) or imaging (such as PET).

Currently for pain disorders, patient stratification has been challenging based upon the lack of measureable biomarkers; however, there are some examples from other disease areas that demonstrate that patient stratification has been successful. With increasing understanding of pain mechanisms, it may soon be possible to give examples within the pain area. To establish patient sub-population selection for targeted drug development it is essential to understand disease phenotypes and classify according to mechanism of disease. However, the application of stratified medicine is only possible where there is evidence that therapeutically relevant strata exist. Unless there is a clear, unambiguous understanding of the mechanism of disease there is the chance that the population selected may not be the right population and this in itself may lead to failed drug studies.

Here, we present three examples of patient stratification approaches. Two examples come from oncology, where tissue for research is accessible and subgroups of patients can often be identified based on tumour subtypes. Pain research is at a disadvantage in this respect: tissue is not readily accessible and subgroups can only be made by the phenotype of the disorder itself through clinical examination or testing at the current time. The third example is a treatment for asthma where patients have been stratified using a protein biomarker.

Crizotinib (Xalkori, Pfizer), an inhibitor of the MET and ALK kinases, is an example where understanding the molecular mechanism of disease and selecting patients based on this understanding was crucial to successful drug development. Crizotinib was developed for patients with lung cancer, but in an unselected population the efficacy of the drug was not obvious. During the development of this drug it was reported that there was a small subset of non-small-cell lung cancer patients whose tumour growth was driven by a particular chromosomal translocation involving the gene encoding ALK.[41] A clinical trial that included patients with this translocation showed that this drug was efficacious (the median disease-free survival with crizotinib was a year compared to just a few months with standard care). Had the trial completed in a non-specific lung cancer cohort, then the non-responders would have likely masked the activity of the drug, resulting in a failed study.

Following the identification of somatic missense mutations in the B-raf protein (a serine/threonine kinase) in a subset of malignant melanomas,[42] specific inhibitors of mutated B-raf protein have been developed as a cancer therapeutic. The B-raf protein plays a role in regulating the MAP kinase/ERKs signalling pathway, which affects cell division, differentiation and secretion. The mutated protein is oncogenic, resulting in overactive downstream signalling causing excessive transcription of genes that promote cell proliferation, leading to tumour genesis. The protein has been found to be mutated in several cancers including colorectal cancer, non-small-cell lung cancer and hairy cell leukaemia. Therapeutics have been designed to target the active confirmation of B-raf with the aim of preventive activation of other proteins and therefore inhibiting downstream signalling leading to inhibition of cell proliferation and potentially apoptosis resulting in tumour regression.

To treat metastatic melanoma, highly selective small-molecule kinase inhibitors have been developed to offset the elevated kinase activity of mutated B-raf proteins (*e.g.* Zelboraf (vemurafenib), Roche). Approximately 70–80% of acquired melanocytic naevi and 40–60% of malignant melanoma contain a *BRAF* mutation, the vast majority of which result in a single amino acid change at codon 600 ($BRAF^{V600E}$).[42,43] Patients with $BRAF^{V600E}$ mutation-positive, inoperable or metastatic melanoma who could benefit from this therapeutic can be identified using a diagnostic test to screen for this mutation. Two clinical trials were performed where patients were selected for inclusion based on these criteria (a Phase 3 study comparing Zelboraf to dacarbazine chemotherapy in 675 patients with previously untreated $\mathrm{BRAF}^{\mathrm{V600E}}$ mutation-positive, inoperable or metastatic melanoma and a Phase 2 study with 132 patients with previously treated $\mathrm{BRAF}^{\mathrm{V600E}}$ mutation-positive, inoperable or metastatic melanoma) and both showed, among other outputs, approximately 50% response rate for tumour shrinkage. Zelboraf is not recommended for use in melanoma patients who lack the $\mathrm{BRAF}^{\mathrm{V600E}}$ mutation.

The third example is the identification of a molecularly defined subpopulation of asthma patients for which a drug, lebrikizumab (Genentech, Roche), could be efficacious. Lebrikizumab is a humanised monoclonal antibody that binds to and inhibits interleukin-13 and is currently in clinical trial for the treatment of asthma that cannot be adequately controlled with inhaled glucocorticoids. A Phase 2 study (219 patients) suggested that serum periostin levels could be used as a protein biomarker.[44] Periostin is a protein that can be induced by interleukin-13 cytokines and becomes elevated in patients with moderate to severe disease. In the aforementioned Phase 2 study, patients with poorly controlled asthma respond well to lebrikizumab but patients with high levels of periostin gain even greater improvement in lung function when given lebrikizumab compared to placebo. If the patients taking lebrikizumab are stratified in the periostin-defined subpopulation, patients with high periostin have a greater improvement in lung function than those categorised as having low periostin, indicating that this biomarker could be used to identify super-responders. Two Phase 3 studies are currently in progress to determine if patients who have high serum periostin levels respond especially well to lebrikizumab and to establish the safety and efficacy of this drug over a longer-term period.

For pain it will be difficult to identify a measure of response or companion diagnostics in the near term, but clearly with a focus to identify potential super-responder subgroups within ongoing trials, precision medicine options may emerge.

11.6.1 Precision Medicine Trial Design

There is a growing acceptance of a clinical development path which entails clinical proof of concept (POC) for a novel drug on a small genetically or phenotypically homogenous patient population whose underlying cause of disease is well understood. Pharmacological success in these small and well-controlled patient cohorts can rapidly lead to increased mechanistic confidence in a target and underwrite the financial commitment required to embark on costly Phase 3 studies.

One such study is exemplified by the recent clinical trial conducted by Xenon Pharmaceuticals,[45] on a small group of patients suffering from the rare genetic disorder IEM. These patients have a gain-of-function mutation in the *SCN9A* gene ($Na_V1.7$), which promotes spontaneous hyperexcitability of the channel and spontaneous pain, characteristic of IEM. The aim of the study was to demonstrate that antagonism of the hyperexcitable $Na_V1.7$ channel could alleviate the pain of IEM, thereby providing POC for a $Na_V1.7$ blocker in the treatment of pain. The study, conducted in just four patients, was reported as a success, thereby validating the $Na_V1.7$ mechanism. The anticipation now is that blockade of the wild-type $Na_V1.7$ channel will also prove efficacious in the treatment of a range of painful conditions where sodium channel activity is regarded as key to the pathophysiology of disease.

11.7 Conclusion

Traditional approaches have been successful in the past and efficacious drugs have been developed without any appreciation of precision medicine. The advantage of preclinical animal models and traditional clinical trial design is that the limitations are well understood and researchers and clinicians are able to interpret data accordingly. However, there is a clear need to expedite drug development success. With increasing examples of successful precision medicine approaches for a range of disease areas, this is clearly an area worthy of exploration on a larger scale. That said, as with any approach, there will be limitations and the approach should be used to complement rather than replace traditional drug discovery. The major limitation to date within the pain field is a lack of large informative and interpretable datasets. Ultimately, as our understanding of disease mechanisms increase so will the success rate of tailored precision medicine approaches.

For improved target selection and to prevent the development of drugs that are non-efficacious in humans, it is imperative that human 'omics' data complements mechanistic evidence of the involvement of a receptor/ion channel/enzyme with pain signalling pathways (typically identified *in vitro* or in animal models). Through increased understanding of the human molecular mechanisms underlying painful disorders, it may be possible to identify novel targets and diagnostics and clinically separate patient populations.

To accelerate generation and interpretation of 'omic' datasets, pre-competitive initiatives are being established (between academic institutions, biotechs and pharmaceutical companies). Over the past few decades new technologies have aided discoveries in human genetics, epigenetics, proteomics and metabolomics that have contributed to our understanding of the molecular mechanisms of certain diseases and patient treatment options.

Finally, investment in precision medicine from research through to clinical trials requires a long-term effort. It took 9 years from the discovery of B-raf mutations to FDA-approved drug. The challenges and approaches taken for each drug discovery programme will vary and the timeframe will not be directly transferable. For example, in the case of $Na_V1.7$, mutations have been identified in a small clinically relevant population implicating the sodium channel as a major player in pain pathways; however, there are no known damaging $Na_V1.7$ mutations in a common pain disorder in this gene to apply a diagnostic approach. The focus of precision medicine for this target has been POC using small phenotypically homogeneous patient populations whose underlying cause of disease is well understood, an obvious, successful and economical route to human target validation.

Acknowledgement

Thanks to Dr Ben Sidders for his expertise on pathway analyses.

References

1. G. T. Whiteside, A. Adedoyin and L. Leventhal, *Neuropharmacology*, 2008, **54**, 767.
2. R. Hill, *TIPS*, 2000, **21**, 244.
3. M. Bergström, R. J. Hargreaves, H. D. Burns, M. R. Goldberg, D. Sciberras, S. A. Reines, *et al.*, *J. Biol. Psychiatr.*, 2004, **55**(10), 1007.
4. K. Ahn, D. S. Johnson, M. Mileni, D. Beidler, J. Z. Long, M. K. McKinney, *et al.*, *Chem. Biol.*, 2009, **16**(4), 411.
5. J. P. Huggins, T. S. Smart, S. Langman, L. Taylor and T. Young, *Pain*, 2012, **153**(9), 1837.
6. J. S. Mogil, *Nat. Rev. Neurosci.*, 2009, **10**, 283.
7. L. J. Hocking, Generation Scotland, A. D. Morris, A. Dominiczak, D. Porteous and B. H. Smith, *Eur. J. Pain*, 2012, **16**(7), 1053.
8. C. Nielsen, G. Knudsen and O. Steingrímsdóttir, *Clin. Genet.*, 2012, **82**(4), 331.
9. K. Musunuru, A. Strong, M. Frank-Kamenetsky, N. E. Lee, T. Ahfeldt, K. V. Sachs, *et al.*, *Nature*, 2010, **466**(7307), 714.
10. T. Freilinger, V. Anttila, B. de Vries, R. Malik, M. Kallela, G. M. Terwindt, *et al.*, *Nat. Genet.*, 2012, **44**(7), 777.
11. T. M. Teslovich, K. Musunuru, A. V. Smith, A. C. Edmondson, I. M. Stylianou, M. Koseki, *et al.*, *Nature*, 2010, **466**(7307), 707.
12. M. Ghoussaini, H. Song, T. Koessler, A. A. Al Olama, Z. Kote-Jarai, K. E. Driver, *et al.*, *J. Natl. Cancer Inst.*, 2008, **100**(13), 962.
13. A. Franke, D. P. McGovern, J. C. Barrett, K. Wang, G. L. Radford-Smith, T. Ahmad, *et al.*, *Nat. Genet.*, 2010, **42**(12), 1118.
14. V. Anttila, H. Stefansson, M. Kallela, U. Todt, G. M. Terwindt, M. S. Calafato, *et al.*, *Nat. Genet*, 2010, **42**(10), 869.

15. D. I. Chasman, M. Schürks, V. Anttila, B. de Vries, U. Schminke, L. J. Launer, *et al.*, *Nat. Genet.*, 2011, **43**(7), 695.
16. L. Ligthart, B. de Vries, A. V. Smith, M. A. Ikram, N. Amin, J. J. Hottenga, *et al.*, *Eur. J. Hum. Genet.*, 2011, **19**(8), 901.
17. M. J. Peters, L. Broer, H. L. Willemen, G. Eiriksdottir, L. J. Hocking, K. L. Holliday, *et al.*, *Ann. Rheum. Dis.*, 2013, **72**(3), 427.
18. A. G. Day-Williams, L. Southam, K. Panoutsopoulou, N. W. Rayner, T. Esko, K. Estrada, *et al.*, *Am. J. Hum.Genet.*, 2011, **89**(3), 446.
19. F. M. K. Williams, S. Scollen, D. Cao, Y Memari., C. L. Hyde, B. Zhang, *et al.*, *PLoS Genet.*, 2012, **8**(12), e1003095.
20. S. Kathiresan and D. Srivastava, *Cell*, 2012, **148**(6), 1242.
21. S. D. Dib-Hajj, A. M. Rush, T. R. Cummins, F. M. Hisama, S. Novella, L. Tyrrell, *et al.*, *Brain*, 2005, **128**(Pt 8), 1847.
22. C. R. Fertleman, M. D. Baker, K. A. Parker, S. Moffatt, F. V. Elmslie, *et al.*, *Neuron*, 2006, **52**, 767.
23. C. G. Faber, J. G. J. Hoeijmakers, H.-S. Ahn, X. Cheng, C. Han, J.-S. Choi, *et al.*, *Ann. Neurol.*, 2012, **71**, 26.
24. J. S. Mogil, *Trends Genet.*, 2012, **28**(6), 258–66.
25. P. A. Jones and S. B. Baylin, *Cell*, 2007, **128**, 683.
26. G. S. Mack., *Nat. Biotech.*, 2010, **28**, 1259.
27. L.-Y. Chien, J.-K. Cheng, D. Chu, C.-F. Cheng and M.-L. Tsaur, *J. Neurosci.*, 2007, **27**, 9855.
28. J. G. Meisner, A. D. Marsh and D. R. March, *J. Neurotrauma*, 2010, **27**, 729.
29. X. L. Wang, Q. Zhang, Y. Z. Zhang, Y. T. Liu, R. Dong, Q. J. Wang and Y. X. Guo, *Neurosci. Lett.*, 2011, **490**, 112.
30. C. Stark, B. J. Breitkreutz, A. Chatr-Aryamontri, L. Boucher, R. Oughtred, M. S. Livstone, *et al.*, *Nucleic Acids Res.*, 2011, **39**(database issue), D698.
31. D. Szklarczyk, A. Franceschini, M. Kuhn, M. Simonovic, A. Roth, P. Minguez, *et al.*, *Nucleic Acids Res.*, 2011, **39**(database issue), D561.
32. T. S. Keshava Prasad, R. Goel, K. Kandasamy, S. Keerthikumar, S. Kumar, S. Mathivanan, *et al.*, *Nucleic Acids Res.*, 2009, **37**(database issue), D767.
33. A. L. Hopkins, *Nat. Chem. Biol.*, 2008, **4**(11), 682–90.
34. A. L. Barabási and Z. N. Oltvai., *Nat. Rev. Genet.*, 2004, **5**(2), 101–13.
35. L. Chindelevitch, D. Ziemek, A. Enayetallah, R. Randhawa, B. Sidders, C. Brockel, *et al.*, *Nature*, 2006, **444**(7121), 894.
36. J. J. Cox, F. Reimann, A. K. Nicholas, G. Thornton, E. Roberts, K. Springell, *et al.*, *Nature*, 2006, **14;444**(7121), 894.
37. N. A. Singh, C. Pappas, E. J. Dahle, L. R. Claes, T. H. Pruess, P. de Jonghe, *et al.*, *plos genet.*, 2009, **5**(9), e1000649.
38. K. Takahashi and S. Yamanaka, *Cell*, 2006, **126**, 663.
39. I. Barbaric, G. Miller and T. N. Dear, *Brief Funct Genomic Proteomic*, 2007, **6**(2), 91.

40. M. Hoekstra, C. L. Mummery, A. A. Wilde, C. R. Bezzina and A. O. Verkerk, *Front. Physiol.*, 2012, **3**, 346.
41. M. Soda, Y. L. Choi, M. Enomoto, S. Takada, Y. Yamashita, S. Ishikawa, *et al.*, *Nature*, 2007, **448**(7153), 561.
42. H. Davies, G. R. Bignell, C. Cox, P. Stephens, S. Edkins, S. Clegg, *et al.*, *Nature*, 2002, **417**(6892), 949.
43. P. M. Pollock, U. L. Harper, K. S. Hansen, L. M. Yudt, M. Stark, C. M. Robbins, *et al.*, *Nat. Genet.*, 2003, **33**(1), 19.
44. J. Corren, R. F. Lemanske, N. A. Hanania, P. E. Korenblat, M. V. Parsey, J. R. Arron, *et al.*, *N. Engl. J. Med.*, 2011, **365**(12), 1088.
45. Y. P. Goldberg, N. Price, R. Namdari, C. J. Cohen, M. H. Lamers, C. Winters, *et al.*, *Pain*, 2012, **153**(1), 80.

CHAPTER 12

Case Study: Discovery and Development of Pregabalin (Lyrica®)

ANDREW J. THORPE*[a] AND LLOYD E. KNAPP[b]

Pfizer Inc., [a]Primary Care U.S. Medical Affairs, New York, NY, USA; [b]Primary Care Clinical Sciences, Groton, CT, USA
*Email: andrew.thorpe@pfizer.com

12.1 Preclinical Studies Leading to the Discovery of Pregabalin

Pregabalin originated from a research programme led by Dr Richard Silverman of Northwestern University.[1] Dr Silverman sought to develop a series of 4-amino-3-alkylbutanoic acids as inactivators of γ-aminobutyric acid aminotransferase (GABA-AT).[1,2] This is the enzyme responsible for the breakdown of γ-aminobutyric acid (GABA). GABA is the principle inhibitory neurotransmitter found in the brain and reduction in brain concentrations of this important neurotransmitter can elicit seizures. However, while found to be substrates for this enzyme this series of compounds were also found to be activators of L-glutamic acid decarboxylase (GAD), an enzyme that catalyses the synthesis of GABA from its precursor.[1–3] In the series of compounds investigated, racemic 3-aminomethyl-5-methylhexanoic acid was found to be a weak activator of GAD.[3] However, it was also found to prevent seizures in a mouse model of epilepsy.[3]

RSC Drug Discovery Series No. 35
Pain Therapeutics: Current and Future Treatment Paradigms
Edited by Charlotte Allerton

Published by the Royal Society of Chemistry, www.rsc.org

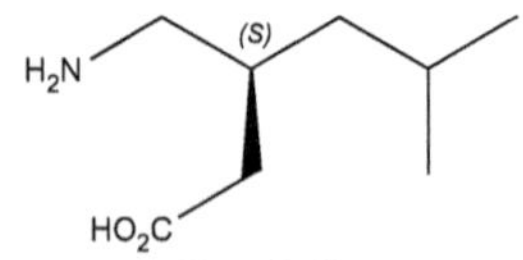

Pregabalin

Figure 12.1 Chemical structure of pregabalin.

Subsequent work demonstrated that the (*S*)-enantiomer ((*S*)-3-aminomethyl-5-methylhexanoic acid, pregabalin, Lyrica®) possessed robust antiepileptic activity (Figure 12.1).[4]

The finding of antiepileptic activity in a preclinical model ultimately catalysed what turned out to be a significant investment by many researchers in aiming to delineate both the mechanism of action and pharmacologic activity of this substituted γ-amino acid. Pregabalin has been extensively studied across many preclinical models including models of anxiety,[5] epilepsy[6] and across a number of pain models.[7–9] Activity in these models helped define the pharmacological activity of pregabalin.

Therefore, with the above information in mind, there is the question of how pregabalin actually produces its pharmacological effect.

12.1.1 Studies to Delineate the Proposed Mechanism of Action of Pregabalin

Despite possessing antiseizure activity it was determined that the activation effect on GAD was not related to anticonvulsant activity.[1] In addition, pregabalin does not bind to $GABA_A$, $GABA_B$ or benzodiazepine receptors.[10] With respect to other potential binding sites and drug targets, pregabalin does not bind or interact with certain other known drug targets for anxiolytic, antiepileptic and analgesic activity (*e.g.* pregabalin, has no activity at opioid receptors and does not modify cyclooxygenase activity).[11] As opposed to the mechanism of action of several well-known drug classes, pregabalin does not inhibit the transport of serotonin or noradrenaline (norepinephrine).[11,12] However, studies revealed high affinity of pregabalin for the $\alpha_2\delta$ protein.[13]

12.1.2 Role and Function of the $\alpha_2\delta$ Protein

The $\alpha_2\delta$ protein is an auxiliary protein associated with voltage-gated calcium channels.[14,15] Its expression with other proteins forms a fully functional calcium channel complex (Figure 12.2). It is known that coexpression of the $\alpha_2\delta$ protein leads to an increase in the number of functional calcium channels at the cell surface and coexpression of the $\alpha_2\delta$ subunit also furnishes fully optimised channel kinetics and also increased stabilisation of channel.[14,15] The $\alpha_2\delta$ protein consists of two heavily glycosylated amino acid domains, which are formed following post-translation modification.[14] Linked through two disulfide bonds, the mature complex consists of a membrane-bound δ portion with a bound,

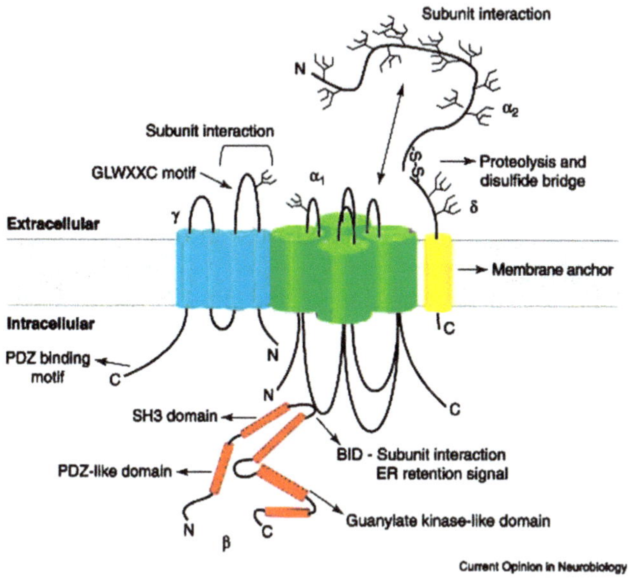

Figure 12.2 Features and structural components of the auxiliary subunits of the voltage-gated calcium channel. Studies have shown that pregabalin binds to the α_2 portion of the $\alpha_2\delta$ protein.[10]
Reprinted from Curr. Opin. Neurobiol. **13**, J. Arikkath & K. P. Campbell, Auxiliary subunits: essential components of the voltage-gated calcium channel complex, 298–307. Copyright (2003), with kind permission from Elsevier.

extracellular α_2 region that extends outward from the cell surface.[10,14] $\alpha_2\delta$ Subunits consist of a number of structural domains, one of which consists of a peptide sequence that is homologous to the type A von Willebrand motif. Von Willebrand motifs are regions of a protein known to bind with other proteins.

Four genes encode the known subtypes of the $\alpha_2\delta$ protein.[10,16,17] It has been established that pregabalin binds only to types 1 and 2. A series of studies support the conclusion that the pharmacological activity of pregabalin is driven through binding to the $\alpha_2\delta$ type 1 protein.[10,18–20] This included a set of experiments using genetically modified mice. Such studies have been invaluable in probing the mechanism of action of pregabalin and supporting the hypothesis that the $\alpha_2\delta$ type 1 protein is the site of action of pregabalin.[21–23]

From point mutation studies, it is known that amino-acid mutation in the extracellular α_2 region of the $\alpha_2\delta$ type 1 protein reduces pregabalin binding.[21] A mutant mouse was generated with an arginine-to-alanine mutation at amino acid 217 of the $\alpha_2\delta$ type 1 protein (referred to as a R217A mutation). This mutation reduces pregabalin binding to $\alpha_2\delta$ type 1 and also reduces pregabalin binding in regions of the central nervous system that are known to express the $\alpha_2\delta$ type 1 protein.[21] When pregabalin was administered to R217A mice, the pain-relieving effect was reduced while that of morphine was unaffected by this mutation. This indicates that other modes of analgesic action were preserved.[23] This effect was also replicated in the Vogel conflict model of anxiety where pregabalin produced robust anxiolytic effects in wild-type mice but was inactive in R217A mice at doses up to 320 mg/kg.[22] These results strongly support the hypothesis that binding of pregabalin to $\alpha_2\delta$ type 1 protein is required for analgesic and anxiolytic-like pharmacology in mouse models and also that $\alpha_2\delta$ binding is a primary mechanism of pregabalin action. Although not presented, structure–activity relationships also strongly suggest that the pharmacology of pregabalin is mediated through binding to the $\alpha_2\delta$ protein.[19,24]

In order to resolve the question of the possible role of the $\alpha_2\delta$ type 2 protein in the pharmacological action of pregabalin, a second strain of mutated mice was developed where point mutation of the $\alpha_2\delta$ type 2 protein (arginine at position 279 replaced by alanine, or R279A) rendered $\alpha_2\delta$-2 (R279A)-containing receptors insensitive to binding pregabalin.[22] In contrast to the findings with the R217A mice, pregabalin produced robust anxiolytic-like activity in wild-type mice and also in the $\alpha_2\delta$-2(R279A) mutant mice, thus showing that the $\alpha_2\delta$ type 2 protein was not responsible for the pharmacology of pregabalin in this model.

In conclusion, the totality of data indicates that the anxiolytic-like effects and hyperalgesic effects of pregabalin are likely to be mediated by pregabalin drug binding at the $\alpha_2\delta$ type 1 subunit. Further evidence that the $\alpha_2\delta$ type 1 subunit plays an important role in the development of allodynia comes from observation that mice overexpressing the $\alpha_2\delta$ type1 subunit exhibit an allodynia phenotype.[25] Following injury, upregulation of $\alpha_2\delta$ type 1 mRNA and protein in dorsal root ganglia (DRG) and/or spinal cord is seen as reported in a number of studies.[26–28]

Other studies have sought to further understand the functional consequences of binding. In general, it could be considered that several main avenues of exploration have pursued the functional effect of binding of pregabalin to the $\alpha_2\delta$ protein: modulation of the release of excitatory neurotransmitters from *in vitro* preparations; effect on calcium channel conduction; trafficking of $\alpha_2\delta$ subunits and modulation of the action of endogenous pain pathways.

12.1.3 Modulation of the Release of Excitatory Neurotransmitters

As neurotransmitters are known to play an important role in maintaining the delicate balance of the CNS, numerous experiments have explored the effect of pregabalin on modifying release of neurotransmitters under a variety of experimental conditions. In summary, these experiments have shown that pregabalin subtly reduces release of excitatory neurotransmitters such as substance P and glutamate.[10,29] It is quite likely that this pharmacology underlies its action in pain and in seizure models.

Pregabalin has been shown to reduce calcium-dependent release of the neurotransmitter glutamate in a variety of *in vitro*/*ex vivo* systems.[10] Suppression of glutamate release is not seen in all systems,[30,31] but the majority of evidence demonstrates that pregabalin effectively reduces the release of glutamate under a variety of induced conditions. In an *in vivo* setting, it has been shown through microdialysis in the spinal cord of the rat that pregabalin reduces formalin-induced release of glutamate in the dorsal horn.[32] Similarly, pretreatment with pregabalin in the chronic constriction injury (CCI) model also reduced the elevation of glutamate in the spinal cord after injection of formalin into the hindpaw. Most recently, in neocortical slices from transgenic mice having the R217A point mutation of the $\alpha_2\delta$ type 1 protein, pregabalin did not affect potassium-evoked glutamate release yet inhibited this release in wild-type mice.[33] These results show that pregabalin attenuated stimulus-evoked glutamate release in rodent neocortical slices and that the $\alpha_2\delta$ type 1 subunit of voltage-gated calcium channels appears to mediate this effect (thus complementing the results seen with the R217A mutants in models of pain and anxiety).

Pregabalin has also been studied in *in vitro* systems for effect on the release of radiolabelled monoamine neurotransmitters (noradrenaline, dopamine and serotonin).[34–37] Across these studies pregabalin was shown to reduce the release of potassium-stimulated evoked tritiated-neurotransmitter release from human neocortex[34,37] and from rat neocortical slices.[35,36]

Despite the studies that demonstrated an effect on various neurotransmitters, elucidating the effect of pregabalin on calcium channel currents has been rather more difficult to show conclusively.[38] A more recent study has provided evidence that pregabalin does reduce channel-mediated currents and reduces the amplitude of excitatory postsynaptic currents.[38]

12.1.4 Trafficking of $\alpha_2\delta$ Type 1 Subunits

A compelling subhypothesis to the overarching $\alpha_2\delta$ mechanism of pregabalin's action has emerged. This hypothesis proposes that pregabalin inhibits the development of allodynia through arresting the trafficking of $\alpha_2\delta$ type 1 protein from the cell body of DRG neurons to synaptic boutons in the dorsal horn of the spinal cord.[39]

In a series of experiments it was shown that $\alpha_2\delta$ type 1 immunoreactivity in spinal cord sections at the level of ligation in the sciatic nerve ligation (SNL) model was increased in lamina II and lamina I, respectively, indicating that ligation led to increased levels of protein in the spinal cord. It was also shown that this increase was not due to increased mRNA for the protein. Immunogold electronmicroscopy also showed that the elevated $\alpha_2\delta$ type 1 was virtually confined to excitatory (glutamatergic) rather than inhibitory (GABA/glycinergic) synapses. Overall, these results showed that $\alpha_2\delta$ type 1 protein levels were increased in excitatory presynaptic terminals in the dorsal horn. As with other studies, pregabalin reduced allodynia to cold and mechanical allodynia in the SNL model. Furthermore, pregabalin strongly reduced ligation-stimulated increase of $\alpha_2\delta$ type 1 protein in the dorsal horn at the level of ligation. This reduction was not restricted to the dorsal horn but was also observed in the axons of DRG neurons. Electron microscopy showed that the $\alpha_2\delta$ type 1 in the axon was associated with structures that are involved in protein trafficking and

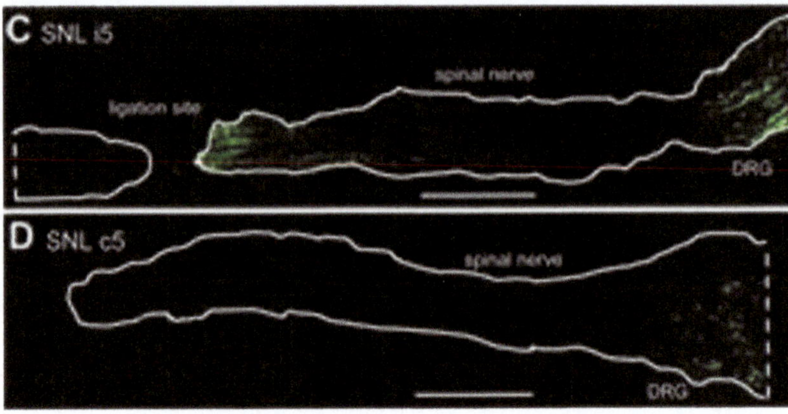

Figure 12.3 Increase of $\alpha_2\delta$-1 protein (over time) in dorsal root axons and accumulation proximal to the ligation site in the SNL model.
Republished with permission of Society for Neuroscience, from C. S. Bauer, *et al.*, The increased trafficking of the calcium channel subunit α_2-δ-1 to presynaptic terminals in neuropathic pain is inhibited by the α_2-δ ligand pregabalin, *J. Neurosci.*, 2009, **29**, 4076–4088.

immunofluorescence determined that there was stronger staining for $\alpha_2\delta$ type 1 protein at the site proximal to the site of ligation (Figure 12.3).

It was subsequently found that chronic application of pregabalin for 3.5 days reduced the accumulation of $\alpha_2\delta$ type 1 protein at the site proximal to the site of ligation of the nerve by 24–32%. This data provided strong evidence that the mechanism of action of pregabalin in models of neuropathic pain is driven, at least in part, by inhibiting trafficking of the $\alpha_2\delta$ type 1 protein from its site of synthesis in the body of the DRG to its target site on the plasma membrane of presynaptic terminals in the deep layers of the dorsal horn. This effect would therefore reduce the level of functional calcium channels at the synaptic junction on excitatory neurons and effectively lead to reduced release of excitatory neurotransmitters.

12.1.5 Enhancement of the Action of Endogenous Pain Pathways

A recent avenue of research has explored the effect of pregabalin at supraspinal sites.[8] Various pathways and mechanisms at the level of the spinal cord and supraspinal sites likely underlie neuropathic pain. Both behavioural and neurochemical evidence has been provided demonstrating pregabalin may have a supraspinal mechanism of action. It has been shown that intracerebroventricular administration of pregabalin afforded complete reversal of thermal and mechanical hypersensitivity in the partial SNL model in the mouse. This study also provided evidence that the effects of pregabalin was manifested through activating the descending noradrenergic pathway which originates in the locus coeruleus and terminating in the lumbar spinal cord; mice pretreated with

6-hydroxydopamine (6-OHDA, which depletes central noradrenaline levels) showed delayed and weakened effect of pregabalin administered by both intraperitoneal and intracerebroventricular routes. An effect of pregabalin on descending facilitatory pathways has been supported in other work.[40]

12.1.6 Mechanism of Action—Summary

Overall, the totality of data indicates that the preclinical actions of pregabalin are highly likely to be mediated through interaction with the $\alpha_2\delta$ type 1 protein. Effect of that interaction appears to result in several subsequent possible mechanisms that may all play a role in contributing to the observed pharmacology of pregabalin in specific conditions.

12.2 Clinical Development of Pregabalin

Pregabalin (Lyrica®) has been shown to be a medicine useful for a variety of clinical indications in an extensive clinical development programme. The following describes selected elements of pregabalin's clinical development. In addition, pregabalin's development, specifically in neuropathic pain and fibromyalgia, reflected emerging science in these areas and, therefore, evolving regulatory agency advice and guidance.

12.2.1 Atypical Development Path

The clinical development of pregabalin followed a typical path during Phase 1 (*e.g.* tolerability and safety studies, dose-ranging studies, single- and multiple-dose pharmacokinetic [PK] studies and dose proportionality studies). During Phases 2–3 however, the development path was rather atypical in that not only were these development phases integrated into one, but also that multiple, full-development programmes were pursued in several disease states simultaneously—neuropathic pain, anxiety, and epilepsy, and then expanded from there to include fibromyalgia and other potential indications. The pregabalin approach to full development culminated in initial marketing authorisations/approvals of pregabalin in the European Union (EU; peripheral neuropathic pain and adjunctive therapy for adult patients with partial onset seizures) and in the United States of America (neuropathic pain associated with diabetic peripheral neuropathy, postherpetic neuralgia and adjunctive therapy for adult patients with partial onset seizures)[12] in mid- and late 2004, respectively. Additional indications followed these initial approvals as briefly described in the following sections. Pregabalin has received regulatory approvals in over 130 countries worldwide.

12.2.2 Early Clinical Studies and Emerging Profile

The earliest clinical studies for pregabalin were clinical pharmacology, Phase 1, PK, dose-ranging, tolerability and safety studies, which were initiated in 1996. A predictable, reliable PK profile began to emerge from the early data, with greater than 90% oral absorption being achieved with the oral solid formulation.[41] The predominant route of elimination from the body was determined to be through renal excretion of unchanged drug, as pregabalin was essentially not metabolised.[41] Among other parameters, including safety and tolerability, these studies established a projected dose range that could be considered in subsequent clinical efficacy and safety trials, and that was 150–600 mg/day. Pregabalin doses up to 900 mg/day were studied, with the latter dose being less well-tolerated than lower doses due to a higher incidence and duration of reported adverse effects in some subjects (*e.g.* CNS effects with dizziness and somnolence common reports).[41] The projected dose range of 150–600 mg/day, therefore, was used in Phase 2–3 clinical efficacy and safety studies, and turned out to be a very accurate projection of the clinical dose range in these subsequent studies.

Clinical pharmacology studies also characterised pharmacokinetic parameters with two different dosing regimens of pregabalin: twice or three times a day. The plasma half-life determined in human healthy volunteers is approximately 6 hours,[41] and this would conventionally support dosing three times daily (TID). A decision was made to investigate twice-daily (BID) dosing as well in some of the early clinical efficacy and safety studies, and also to progress investigation of dosing three times a day. Twice-daily dosing eventually became the standard dosing regimen in latter stages of Phase 3–4 clinical development.

Clinical pharmacology work also revealed another aspect of the emerging PK profile. Pregabalin has a linear, predictable plasma concentration profile across doses studied up to 900 mg/day h with either BID or TID dosing.[41] This was reinforced via positive results in the Phase 2–3 clinical studies. Data indicated a low potential for pregabalin to have pharmacokinetic drug–drug interactions (PK DDI) and pregabalin was shown to have a lack of significant effect *via* induction or inhibition of major cytochrome P450 enzymes that are common in metabolism of other drugs.[41,42]

Early Phase 2 studies were initiated in pain (acute dental pain) and epilepsy populations in 1997, and in anxiety in 1998.[43,44] Acute dental pain and the shorter treatment duration Phase 2 epilepsy and anxiety studies were intended to provide an early indication of clinical efficacy on which to base decisions for the larger, later stage trials. Positive results in these studies in all three areas of investigation, therefore, supported further studies and development.

The formation of a talented team to execute clinical development programmes was critical to success. As in this case, the then planned multiple Phase 2/3 full-development clinical programmes required multiple disciplines to be highly engaged and interactive—clinicians (including neurologists, psychiatrists, pain specialists, generalists), clinical pharmacologists, statisticians, study monitors, regulatory experts, pharmacists, formulation specialists, and so on—was a key to success of the multi-indication programme.

12.2.3 Full Clinical Development for Neuropathic Pain, Generalised Anxiety Disorder and Epilepsy

Full development of pregabalin began in earnest in neuropathic pain (painful diabetic peripheral neuropathy [DPN] and postherpetic neuralgia [PHN], epilepsy [partial onset seizures] and anxiety [generalised anxiety disorder and social anxiety disorder; note that in the United States, Lyrica is not indicated for either anxiety condition] in 1998, with results of all of these early Phase 2–3 studies then available in 1999 and subsequently reported.[44–58] An example of the efficacy observed in one study (one of the pivotal studies supporting the U.S. registration for DPN) is presented in Figure 12.4.[46]

The results from early clinical pharmacology and Phase 2 clinical studies were well founded, as reflected in the robust, positive results in these Phase 2–3 studies. Results also demonstrated the efficacy and safety of BID or TID dosing. The basis was thus established to continue full development in all three areas, with emphasis switching to BID dosing in latter pregabalin development.

It is, we believe, at least interesting to reflect on this time period to better appreciate the environment and procedures that were developed which enabled the rapid execution of this development programme. The high energy and the high degree of constant activity appeared unprecedented, given the multiplicity of parallel programmes and the rapidity with which the programmes were able to be progressed. Studies were being designed, initiated, and providing results on a frequent basis. Standardisation of protocol elements, database management, programming/output, and other related activities into what one might refer to as integrated protocol modules, now a wider industry practice, was derived and was a notable success factor during this time. Standardisation enabled very rapid turnaround in data generation and decision-making for subsequent studies.

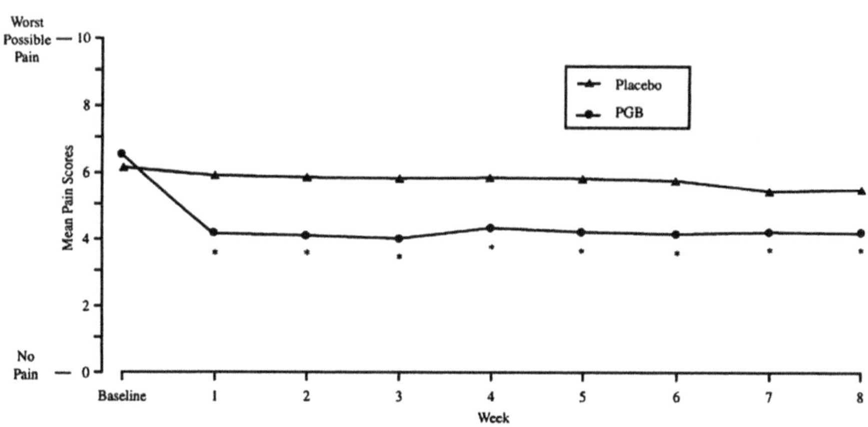

Figure 12.4 A selected example of the efficacy of pregabalin in a randomised, double-blind, placebo-controlled, 8 week trial evaluating the effectiveness of pregabalin in alleviating pain associated with DPN.
Taken from J. Rosenstock, M. Tuchman, L. LaMoreaux & U. Sharma, Pregabalin for the treatment of painful diabetic peripheral neuropathy: a double-blind, placebo-controlled trial. *Pain*, **110**, 628–638 (2004). This figure has been reproduced with permission of the International Association for the Study of Pain (IASP). The figure may NOT be reproduced for any other purpose without permission.

How Phases 2 and 3 were combined is worthy of comment. Traditionally a sequential approach would have been taken, that is, dose–response studies would have been conducted initially, usually with a smaller sample size and a number of doses explored. Subsequently, progression to pivotal efficacy studies with a generally larger sample size and either a targeted dose or a more limited number of doses would have been undertaken.

With pregabalin, there was little distinction between Phase 2 and Phase 3 studies and these developmental phases were essentially combined. While this approach entailed greater inherent strategic risk; reductions in development time, time to submission, time to approval and market launch were substantially shorter, and benefits to patients realised all the sooner. This worked out quite well for pregabalin with one comment. The combined Phase 2–3 program focused on the upper and lower doses of the projected range (150–600 mg/day), and while upper and lower doses were accurately projected, the CNS tolerability profile suggested that additional doses in between 150 or 600 mg/day would be useful to investigate and characterise from both an efficacy and safety standpoint. Earlier Phase 2–3 studies investigated fixed doses at the lower and upper anticipated range (*i.e.* 150 and 600 mg/day) and later studies investigated fixed doses intermediate to 150 and 600 mg/day, as well as flexible dose designs where the pregabalin dose was optimised based on individual subject response and tolerability in the range of 150–600 mg/day. The assessment of additional doses was done in subsequent studies to provide a more complete characterisation of the safety and tolerability profile by dose. In retrospect, adding more doses to the initial studies that combined Phase 2 and 3 might have achieved even greater efficiencies than the step-wise, iterative approach in the pregabalin programme.

While the primary focus of this chapter is on the development of pregabalin for the treatment of pain conditions, it is interesting to note the positive clinical response to pregabalin across what may appear to be dissimilar conditions—neuropathic pain, fibromyalgia, epilepsy, and anxiety. Pregabalin was really the first drug to be formally studied in a wide variety of conditions in adequate-sized clinical studies that would provide statistical power for the comparisons of interest. It is not actually known why pregabalin is effective across these conditions, but one would assume that a common or sufficiently similar pathophysiology underlies and links these conditions and that the $\alpha_2\delta$ type 1 protein is at some point, relevant in that pathophysiology.

Pregabalin's clinical development was impacted in the latter stages of work leading to initial submissions for the original indications due to a partial clinical hold related to an animal toxicology finding in mice.[59] While it is certainly not unprecedented to have animal findings emerge during a development programme, there was limited experience with this particular finding both within the industry, in Pfizer, as well as in the regulatory agency environments. Several years of intense and rigorous scientific investigation followed to characterise the mechanism of the mouse toxicology finding such that the partial clinical hold was lifted and approvals of specific indications granted. The animal finding in mice is reflected in current product labelling for pregabalin.[12]

For neuropathic pain, DPN and PHN were selected as representative conditions of neuropathic pain, and the clinical programmes were conducted reflecting the state of regulatory advice from the EU and U.S. regulatory authorities at the time. One challenge for any programme in neuropathic pain at the time was the evolving understanding for aetiologies and pathophysiology for neuropathic pain, and whether these could be grouped or considered as separate, distinct conditions. For example, one could group various conditions into peripheral or central neuropathic pain depending upon the location of the primary insult to the nervous system. Alternatively, each neuropathic pain condition could be considered separately. Regulatory guidelines and advice evolved in this environment as the state of knowledge expanded and neuropathic pain became more familiar to the various regulatory agencies. It is also interesting to note that regulatory agencies addressed labelling for neuropathic pain indications differently in the EU, the United States and (as further discussed below) Japan. In the United States, pregabalin is currently labelled specifically for neuropathic pain associated with DPN and PHN, as well as neuropathic pain associated with spinal cord injury,[12] whereas in the EU[60] and Japan, DPN and PHN are subsumed under a general label of peripheral neuropathic pain (possibly reflecting differing regulatory philosophies as the state of knowledge regarding neuropathic pain evolved). Further, a discussion paper considers the potential extrapolation of efficacy demonstrated in a given pain condition being reasonably applied to another pain condition or conditions.[61] The workshop that preceded the discussion paper and the publication itself were contributed to by representatives of pain research experts from academia and from regulatory bodies—in particular, the U.S. Food and Drug Authority (FDA).[61] In summary, two studies underwrote the U.S. registration and label for pregabalin in neuropathic pain associated with DPN[45,46] and three studies for PHN.[48–50] Since then, pregabalin also received U.S. approval for neuropathic pain associated with spinal cord injury.[12,62,63]

For the pain programme, several, significant challenges were encountered during the course of the clinical development programme. The first was the duration of double-blind treatment. Double-blind treatment duration in registration studies for neuropathic pain ranged from 5 to 13 weeks. The initial feedback from regulatory agencies on the proposed duration of double-blind treatment in planned studies which ranged from 5 to 8 weeks depending upon the specific study, indicated general agreement with this proposed duration of treatment. However, it was following results of the early Phase 2–3 studies that the regulatory agency advice began to evolve, with both the EU and U.S. agencies subsequently requiring 12 week treatment duration studies. The rationale for this requested treatment duration from both agencies was that 12 week clinical studies could better represent treatment for a drug to be used chronically in clinical practice. It was further clarified that 12 week treatment duration meant 12 weeks on fixed dose and could not include any dose escalation or titration period. As such, shorter-duration studies came to be considered as 'supportive' and not pivotal in the regulatory review processes. While the regulatory standard is now understood to be 12 week duration studies, as we learn more about clinical studies of chronic pain conditions, the authors wonder if there may be an opportunity in future programmes to discuss a pivotal role for studies of shorter duration than 12 weeks. For example, might two studies be of 12 weeks duration at fixed dose and then other studies of shorter duration depending upon the weekly profile or pattern of pain reduction in the intervening weeks up to 12 weeks duration? For pregabalin, the effect in pain reduction is relatively rapid (in some patients as early as 1 week[45,46,64]) in that one observes near maximal effects within the first 2–4 weeks of treatment, which is generally consistent with that then observed with 12 weeks of treatment in the same study. It would also appear that the near maximal treatment effect in approximately 2–4 weeks includes any dose escalation to the final randomised dose and that one may not need to wait to observe the effect at fixed randomised dose in all instances.

The second and third significant challenges are inter-related and linked to the first challenge above; these are maintenance of placebo treatment for 12 weeks, or longer as may be needed for dose escalation to the fixed randomised dose, and imputation methods for subjects when not completing the study. Maintenance of study subjects randomised to placebo in pain studies that typically require moderate to severe levels of pain to demonstrate the pain effect become increasingly challenging to retain the longer the study. So when one requires a 12 week study, and initiation of treatment requires time, the overall time in study can become quite long for subjects, particularly those randomised to placebo and left in persistent pain. Imputation in statistical methods has received increasing discussion during this time of pregabalin's development, wherein assumptions are needed to be made regarding individual subject data that does not exist for those subjects who did not complete the study. Last-observation-carried-forward (LOCF) analyses were often commonly used before this time. However, as any approach has certain limitations, the LOCF approach had the added criticism of potentially assigning a 'good outcome' in pain response to a subject who was not able to complete the 12 week study. The FDA requested more conservative landmark analyses where a subject is assigned a 'poor outcome' in pain response. One approach suggested by the FDA used baseline-observation-carried-forward (BOCF), where pretreatment baseline pain data is imputed and used in the efficacy analysis for subjects who do not complete the study. This then becomes 'no change' for these subjects, perhaps a conservative estimate. For a drug in development that would have discontinuations during the study, as might be more the case with CNS-active drugs like pregabalin, the result is that a larger sample size is required for powering a study for BOCF analysis. A report by the U.S. National Science Academy on the prevention and treatment of missing data in clinical trials emphasises avoidance of missing data where possible in the conception and design of a given clinical study, and the use of various imputation methods and sensitivity analyses to providing the optimum context of the study data in contrast to reliance on a single method missing data imputation and analysis.[65]

Another challenge realised during the program was the globalisation of studies. Not so many years ago, much of clinical research was conducted with a focus on U.S. and western European centres. Given the dynamics noted above—more studies, larger/longer studies, and increased cost of studies, in addition to competitive factors of many sponsors conducting research in the same population—other countries around the globe have been added to provide research centres. This has its own set of challenges such as in language, culture, time differences, travel, and connectivity to geographically dispersed investigators, and general clinical trial research experience.

In addition, increasing placebo response was another issue during the programme. This factor has previously affected many areas of research, most notably psychiatry (*e.g.* anxiety and depression studies, for example). The impact of increasing placebo response translates to a smaller placebo-corrected or placebo-subtracted treatment effect (delta), greater variability in general, and a much higher proportion of studies not meeting their primary efficacy endpoint. Mitigation of the large placebo response comes in the form of large sample sizes to power the trial to show an effect, resulting in increased time, resources, subjects and costs. This is true for pregabalin in neuropathic pain, even with an overwhelming evidence of efficacy from prior studies. The reasons for a large placebo response are not known but possible causes include the inclusion of an increased number of trial countries (and therefore variability) less experienced (new) pain research sites, changes in research population over time (particularly post-marketing), availability of proven therapies, heightened patient expectation and less rigorous approaches on the part of investigators and study coordinators to pain research (again, possibly related to expectations of efficacy for a proven drug). Note that attempting to solve the issue of placebo response in research remains quite active, with discussions continuing in scientific/academic/regulatory venues and meetings, and in the trial and error mitigations currently employed in studies being conducted.

The overall result of past and contemporary challenges and mitigation approaches in the clinical development of pain drugs, is that the clinical development programme for potential new medicines, not unlike pregabalin, may require larger studies, longer studies, and studies with increasing resource demand. This could include an increase in the number of investigators, training to ensure standard approaches in pain research across geographical and cultural differences that may exist, number and instruction of potential pain subjects, development time and costs.

12.2.4 Development for Fibromyalgia

Development of pregabalin for the treatment of fibromyalgia also started late in 1999. Clinical development of new medicines for fibromyalgia also faced the challenge of evolving knowledge in medicine as well as in the regulatory environment, with understanding of emerging science in this area informing subsequent clinical development as well as regulatory agency guidance. Pregabalin fibromyalgia development proceeded with results from a major study reported in 2005.[66] This study demonstrated efficacy in patients with fibromyalgia and provided dose information that enabled discussions with the regulatory agencies

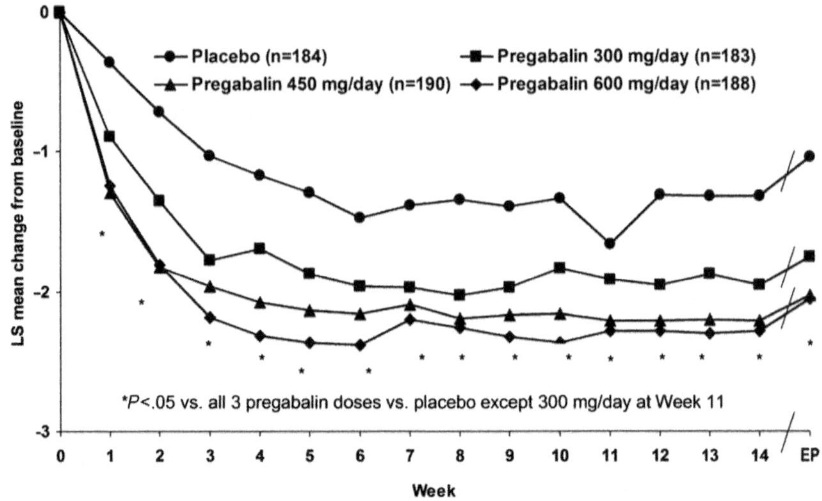

Figure 12.5 Weekly and endpoint mean pain scores in a 14-week evaluation of pregabalin *versus* placebo in patients with fibromyalgia.
Reprinted from M. Arnold Lesley, *et al.*, A 14-week, randomized, double-blinded, placebo-controlled monotherapy trial of pregabalin in patients with fibromyalgia, *J. Pain*, 2008, **9**, 792–805. Copyright (2008), with permission from Elsevier.

and which informed subsequent clinical studies in the programme. In all, several fibromyalgia studies were conducted[64,66,67] and two supported registration in the United States,[64,67] with the results from one study comparing the efficacy and safety of pregabalin monotherapy in reducing pain in fibromyalgia patients when compared to placebo shown in Figure 12.5.[64] Pregabalin doses in fibromyalgia were ultimately determined to be 300–450 mg/day, which differed slightly from doses in other approved indications.

It is also interesting to note that with the evolving nature of the U.S. regulatory landscape, a 6 month treatment duration, placebo-controlled study was requested of pharmaceutical sponsors at the time. The request for data from a 6 month controlled study in fibromyalgia was new and had not previously been requested of sponsors. The pregabalin study was initiated and was successful in its efficacy endpoints, representing one of the two 'pivotal' studies reflected in U.S. labelling.[67] Pfizer selected a 6 month randomised withdrawal design study due to some of the considerations noted above (*i.e.* variability and placebo response) that might confound such a study, particularly a study with a 6 month treatment duration. The randomised withdrawal design has been reported as a potential design to consider addressing issues of placebo response, and was the first that we are aware of in such a large and potentially pivotal study. With success in the United States, Japan initiated a study with pregabalin in Japanese patients with fibromyalgia, which ultimately was also successful, and which supported registration in Japan for this indication.

12.2.5 Japanese Development for Neuropathic Pain and Fibromyalgia

Despite the very significant challenges of placebo response and potential concerns for additional cultural variation in pain reporting, development of pregabalin for neuropathic pain and fibromyalgia was conducted in Japan. A study was conducted to address tolerability and PK of pregabalin in Japanese subjects with doses up to 600 mg/day. Initial considerations in Japanese development of pregabalin were that subjects may tolerate pregabalin 600 mg/day less well than Western subjects; however, this was not the case as the Japanese subjects had a similar tolerability profile to Western subjects. Acceptable tolerability of the full clinical dose range from the West would then enable a study to be conducted in Japan which would in essence bridge to the results of a comparably designed study previously conducted in the West. Neuropathic pain, particularly DPN and following spinal cord injury, and fibromyalgia studies, had not been conducted in Japan by any sponsor. To the Japanese team's credit, the studies conducted in Japan were highly successful and a tremendous accomplishment given the unprecedented nature of the programme in this country. The clinical team in Japan applied learnings from Western studies regarding the importance of accurate pain diagnosis, limiting variability to pain response, placebo response in general, and training provided even for experienced investigators. The overall result was ultimately positive studies and approval of pregabalin for the indications of neuropathic pain (which includes peripheral and central neuropathic pain conditions) as well as for fibromyalgia, all of which were unprecedented accomplishments in pharmaceutical development in Japan.

12.2.6 Clinical Development—Summary

The clinical development of pregabalin is a remarkable programme and story and has resulted in a very robust set of clinical data which has supported the registration and marketing of pregabalin for a number of indications worldwide. While some of the learnings are unique to pregabalin, there is also much to learn from the programme that can be applied to future programmes in pain development as well as other indications. Most importantly, many, many patients have benefited from pregabalin's clinical usage, with an estimated exposure of greater than 19 million patient-years to date.[11]

References

1. R. B. Silverman, From basic science to blockbuster drug: the discovery of Lyrica, *Angew. Chem. Int. Ed. Engl.*, 2008, **47**, 3500–3504.
2. R. Andruszkiewicz and R. B. Silverman, 4-Amino-3-alkylbutanoic acids as substrates for gamma-aminobutyric acid aminotransferase, *J. Biol. Chem.*, 1990, **265**, 22288–22291.

3. R. B. Silverman, R. Andruszkiewicz, S. M. Nanavati, C. P. Taylor and M. G. Vartanian, 3-Alkyl-4-aminobutyric acids: the first class of anticonvulsant agents that activates L-glutamic acid decarboxylase, *J. Med.. Chem.*, 1991, **34**, 2295–2298.
4. P.-w. Yuen, G. D. Kanter, C. P. Taylor and M. G. Vartanian, Enantioselective synthesis of PD144723: a potent stereospecific anticonvulsant, *Bioorg. Med. Chem. Lett.*, 1994, **4**, 823–826.
5. M. J. Field, R. J. Oles and L. Singh, Pregabalin may represent a novel class of anxiolytic agents with a broad spectrum of activity, *Br. J. Pharmacol.*, 2001, **132**, 1–4.
6. M. G. Vartanian, *et al.*, Activity profile of pregabalin in rodent models of epilepsy and ataxia, *Epilepsy Res.*, 2006, **68**, 189–205.
7. Y. Cao, H. Wang, C.-Y. Chiang, J. O. Dostrovsky and B. J. Sessle, Pregabalin suppresses nociceptive behavior and central sensitization in a rat trigeminal neuropathic pain model, *J. Pain*, 2013, **14**, 193–204.
8. Y. Takeuchi, K. Takasu, H. Ono and M. Tanabe, Pregabalin, S-(+)-3-isobutylgaba, activates the descending noradrenergic system to alleviate neuropathic pain in the mouse partial sciatic nerve ligation model, *Neuropharmacology*, 2007, **53**, 842–853.
9. N. Narita, *et al.*, Systemic pregabalin attenuates sensorimotor responses and medullary glutamate release in inflammatory tooth pain model, *Neuroscience (Amsterdam)*, 2012, **218**, 359–366.
10. C. P. Taylor, T. Angelotti and E. Fauman, Pharmacology and mechanism of action of pregabalin: The calcium channel $\alpha_2\delta$ (alpha$_2$-delta) subunit as a target for antiepileptic drug discovery, *Epilepsy Res.*, 2007, **73**, 137–150.
11. Pfizer Inc, Data on File, New York, NY.
12. Lyrica (pregabalin), Highlights of U.S. prescribing information, Pfizer, Inc.
13. C. P. Taylor, *et al.*, Potent and stereospecific anticonvulsant activity of 3-isobutyl GABA relates to in vitro binding at a novel site labeled by tritiated gabapentin, *Epilepsy Res.*, 1993, **14**, 11–15.
14. N. Klugbauer, E. Marais and F. Hofmann, Calcium channel $\alpha_2\delta$ subunits: differential expression, function, and drug binding, *J. Bioenerg. Biomembr.*, 2003, **35**, 639–647.
15. J. Arikkath and K. P. Campbell, Auxiliary subunits: essential components of the voltage-gated calcium channel complex, *Curr. Opin. Neurobiol.*, 2003, **13**, 298–307.
16. N. Qin, S. Yagel, M.-L. Momplaisir, E. E. Codd and M. R. D'Andrea, Molecular cloning and characterization of the human voltage-gated calcium channel $\alpha_2\delta$-4 subunit, *Mol. Pharmacol.*, 2002, **62**, 485–496.
17. N. Klugbauer, L. Lacinova, E. Marais, M. Hobom and F. Hofmann, Molecular diversity of the calcium channel $\alpha_2\delta$ subunit,, *J. Neurosci.*, 1999, **19**, 684–691.
18. A. J. Thorpe and J. Offord, The α_2-δ protein: an auxiliary subunit of voltage-dependent calcium channels as a recognized drug target, *Curr. Opin. Investig. Drugs*, 2010, **11**, 761–770.
19. A. J. Thorpe and C. P. Taylor, Calcium channel α_2-δ ligands: gabapentin and pregabalin,, *Compr. Med. Chem. II*, 2006, **8**, 227–246.

20. C. P. Taylor, Mechanisms of analgesia by gabapentin and pregabalin—calcium channel α2-δ [Cavα2-δ] ligands, *Pain*, 2009, **142**, 13–16.
21. F. Bian, *et al.*, Calcium channel alpha2-delta type 1 subunit is the major binding protein for pregabalin in neocortex, hippocampus, amygdala, and spinal cord: An ex vivo autoradiographic study in alpha2-delta type 1 genetically modified mice, *Brain. Res.*, 2006, **1075**, 68–80.
22. S. M. Lotarski, *et al.*, Anxiolytic-like activity of pregabalin in the Vogel conflict test in $\alpha_2\delta$-1 (R217A) and alpha2delta-2 (R279A) mouse mutants, *J. Pharmacol. Exp. Ther.*, 2011, **338**, 615–621.
23. M. J. Field, *et al.*, Identification of the α_2-δ-1 subunit of voltage-dependent calcium channels as a molecular target for pain mediating the analgesic actions of pregabalin, *Proc. Natl. Acad. Sci. U. S. A.*, 2006, **103**, 17537–17542.
24. T. R. Belliotti, *et al.*, Structure-activity relationships of pregabalin and analogues that target the α_2-δ protein, *J. Med. Chem.*, 2005, **48**, 2294–2307.
25. C. Y. Li, *et al.*, Calcium channel $\alpha_2\delta_1$ subunit mediates spinal hyperexcitability in pain modulation, *Pain*, 2006, **125**, 20–34.
26. H. L. Melrose, *et al.*, [3H]-Pregabalin binding is increased in ipsilateral dorsal horn following chronic constriction injury, *Neurosci. Lett.*, 2007, **417**, 187–192.
27. M. Narita, *et al.*, Role of spinal voltage-dependent calcium channel $\alpha_2\delta$-1 subunit in the expression of a neuropathic pain-like state in mice, *Life. Sci.*, 2007, **80**, 2015–2024.
28. W. Xiao, A. Boroujerdi, G. J. Bennett and Z. D. Luo, Chemotherapy-evoked painful peripheral neuropathy: Analgesic effects of gabapentin and effects on expression of the alpha-2-delta type-1 calcium channel subunit, *Neuroscience*, 2007, **144**, 714–720.
29. J. C. Fehrenbacher, C. P. Taylor and M. R. Vasko, Pregabalin and gabapentin reduce release of substance P and CGRP from rat spinal tissues only after inflammation or activation of protein kinase C, *Pain*, 2003, **105**, 133–141.
30. B. Brawek, M. Loffler, A. Weyerbrock and T. J. Feuerstein, Effects of gabapentin and pregabalin on K+-evoked 3H-GABA and 3H-glutamate release from human neocortical synaptosomes, *Naunyn Schmiedebergs Arch., Pharmacol.*, 2009, **379**, 361–369.
31. M. Kammerer, B. Brawek, T. M. Freiman, R. Jackisch and T. J. Feuerstein, Effects of antiepileptic drugs on glutamate release from rat and human neocortical synaptosomes, *Naunyn Schmiedebergs Arch. Pharmacol.*, 2011, **383**, 531–542.
32. N. Kumar, A. Laferriere, J. S. Yu, A. Leavitt and T. J. Coderre, Evidence that pregabalin reduces neuropathic pain by inhibiting the spinal release of glutamate, *J. Neurochem.*, 2010, **113**, 552–561.
33. J. E. Quintero, D. J. Dooley, F. Pomerleau, P. Huettl and G. A. Gerhardt, Amperometric measurement of glutamate release modulation by gabapentin and pregabalin in rat neocortical slices: role of voltage-sensitive Ca2+ α2δ-1 subunit, *J. Pharmacol. Exp. Ther.*, 2011, **338**, 240–245.
34. B. Brawek, M. Loffler, D. J. Dooley, A. Weyerbrock and T. J. Feuerstein, Differential modulation of K(+)-evoked (3)H-neurotransmitter release

from human neocortex by gabapentin and pregabalin, *Naunyn Schmiedebergs Arch. Pharmacol.*, 2008, **376**, 301–307.
35. D. J. Dooley, C. M. Donovan, W. P. Meder and S. Z. Whetzel, Preferential action of gabapentin and pregabalin at P/Q-type voltage-sensitive calcium channels: inhibition of K+-evoked [3H]-norepinephrine release from rat neocortical slices, *Synapse*, 2002, **45**, 171–190.
36. D. J. Dooley, C. M. Donovan and T. A. Pugsley, Stimulus-dependent modulation of [(3)H]norepinephrine release from rat neocortical slices by gabapentin and pregabalin, *J. Pharmacol. Exp. Ther.*, 2000, **295**, 1086–1093.
37. K. Fink, *et al.*, Inhibition of neuronal Ca(2+) influx by gabapentin and pregabalin in the human neocortex, *Neuropharmacology*, 2002, **42**, 229–236.
38. M. N. Di Guilmi, F. J. Urbano, C. G. Inchauspe and O. D. Uchitel, Pregabalin modulation of neurotransmitter release is mediated by change in intrinsic activation/inactivation properties of ca(v)2.1 calcium channels, *J. Pharmacol. Exp. Ther.*, 2011, **336**, 973–982.
39. C. S. Bauer, *et al.*, The increased trafficking of the calcium channel subunit $\alpha_2\delta$-1 to presynaptic terminals in neuropathic pain is inhibited by the $\alpha_2\delta$ ligand pregabalin, *J. Neurosci.*, 2009, **29**, 4076–4088.
40. L. A. Bee and A. H. Dickenson, Descending facilitation from the brainstem determines behavioural and neuronal hypersensitivity following nerve injury and efficacy of pregabalin, *Pain*, 2008, **140**, 209–223.
41. H. N. Bockbrader, *et al.*, Clinical pharmacokinetics of pregabalin in healthy volunteers, *J. Clin. Pharmacol.*, 2010, **50**, 941–950.
42. H. N. Bockbrader, P. Burger and L. Knapp, Pregabalin effect on steady-state pharmacokinetics of carbamazepine, lamotrigine, phenobarbital, phenytoin, topiramate, valproate, and tiagabine, *Epilepsia*, 2011, **52**, 405–409.
43. C. M. Hill, *et al.*, Pregabalin in patients with postoperative dental pain, *Eur. J. Pain (London)*, 2001, **5**, 119–124.
44. S. Arroyo, *et al.*, Pregabalin add-on treatment: a randomized, double-blind, placebo-controlled, dose-response study in adults with partial seizures, *Epilepsia*, 2004, **45**, 20–27.
45. H. Lesser, U. Sharma, L. LaMoreaux and R. M. Poole, Pregabalin relieves symptoms of painful diabetic neuropathy: a randomized controlled trial, *Neurology*, 2004, **63**, 2104–2110.
46. J. Rosenstock, M. Tuchman, L. LaMoreaux and U. Sharma, Pregabalin for the treatment of painful diabetic peripheral neuropathy: a double-blind, placebo-controlled trial, *Pain*, 2004, **110**, 628–638.
47. R. W. Richter, *et al.*, Relief of painful diabetic peripheral neuropathy with pregabalin: A randomized, placebo-controlled trial, *J. Pain*, 2005, **6**, 253–260.
48. R. van Seventer, *et al.*, Efficacy and tolerability of twice- daily pregabalin for treating pain and related sleep interference in posttherpetic neuralgia: a 13-week, randomized trial, *Curr. Med. Res. Opin.*, 2006, **22**, 375–384.
49. R. H. Dworkin, *et al.*, Pregabalin for the treatment of postherpetic neuralgia: a randomized, placebo-controlled trial, *Neurology*, 2003, **60**, 1274–1283.

50. R. Sabatowski, *et al.*, Pregabalin reduces pain and improves sleep and mood disturbances in patients with post-herpetic neuralgia: results of a randomised, placebo-controlled clinical trial, *Pain*, 2004, **109**, 26–35.
51. A. Beydoun, W. Nasreddine and S. Atweh, Efficacy and tolerability of pregabalin in partial epilepsy, *Expert Rev. Neurother.*, 2008, **8**, 1013–1024.
52. A. Beydoun, *et al.*, Safety and efficacy of two pregabalin regimens for add-on treatment of partial epilepsy, *Neurology*, 2005, **64**, 475–480.
53. J. A. French, A. R. Kugler, J. L. Robbins, L. E. Knapp and E. A. Garofalo, Dose-response trial of pregabalin adjunctive therapy in patients with partial seizures, *Neurology*, 2003, **60**, 1631–1637.
54. D. Feltner, *et al.*, Long-term efficacy of pregabalin in generalized anxiety disorder, *Int. Clin. Psychopharmacol*, 2008, **23**, 18–28.
55. D. E. Feltner, *et al.*, A randomized, double-blind, placebo-controlled, fixed-dose, multicenter study of pregabalin in patients with generalized anxiety disorder, *J. Clin. Psychopharmacol.*, 2003, **23**, 240–249.
56. A. C. Pande, *et al.*, Efficacy of the novel anxiolytic pregabalin in social anxiety disorder. A placebo-controlled, multicenter study, *J. Clin. Psychopharmacol.*, 2004, **24**, 141–149.
57. C. Pande Atul, *et al.*, Pregabalin in generalized anxiety disorder: a placebo-controlled trial, *Am. J. Psychiatry*, 2003, **160**, 533–540.
58. R. B. Pohl, D. E. Feltner, R. R Fieve and A. C Pande, Efficacy of pregabalin in the treatment of generalized anxiety disorder: double-blind, placebo-controlled comparison of BID versus TID dosing, *J. Clin. Psychopharmacol.*, 2005, **25**, 151–158.
59. K. A. Criswell, *et al.*, Mode of action associated with development of hemangiosarcoma in mice given pregabalin and assessment of human relevance, *Toxicol. Sci.*, 2012, **128**, 57–71.
60. Lyrica EPAR product information, Pfizer, Inc.
61. R. H. Dworkin, *et al.*, Considerations for extrapolating evidence of acute and chronic pain analgesic efficacy, *Pain*, 2011, **152**, 1705–1708.
62. P. J. Siddall, *et al.*, Pregabalin in central neuropathic pain associated with spinal cord injury. A placebo-controlled trial, *Neurology*, 2006, **67**, 1792–1800.
63. D. D. Cardenas, *et al.*, A randomized trial of pregabalin in patients with neuropathic pain due to spinal cord injury, *Neurology*, 2013, **80**, 533–539.
64. M. Arnold Lesley, *et al.*, A 14-week, randomized, double-blinded, placebo-controlled monotherapy trial of pregabalin in patients with fibromyalgia, *J. Pain*, 2008, **9**, 792–805.
65. R. J. Little, *et al.*, The prevention and treatment of missing data in clinical trials, *N. Engl. J. Med.*, 2012, **367**, 1355–1360.
66. L. J. Crofford, *et al.*, Pregabalin for the treatment of fibromyalgia syndrome: results of a randomized, double-blind, placebo-controlled trial, *Arthritis. Rheum.*, 2005, **52**, 1264–1273.
67. L. J. Crofford, *et al.*, Fibromyalgia relapse evaluation and efficacy for durability of meaningful relief (FREEDOM): A 6-month, double-blind, placebo-controlled trial with pregabalin, *Pain*, 2008, **136**, 419–431.

CHAPTER 13

Case Study: The Butrans® Patch

JOSEPH V. PERGOLIZZI JR,[a] ROBERT B. RAFFA[b] AND ROBERT TAYLOR JR*[c]

[a] Department of Medicine, Johns Hopkins University, Baltimore, MD 21287, USA; Department of Pharmacology, Temple University School of Medicine, Philadelphia, PA 19122, USA; Georgetown University School of Medicine, Washington, DC 20057, USA; [b] Department of Pharmaceutical Sciences, Temple University School of Pharmacy, Philadelphia, PA 19140, USA; [c] NEMA Research, Inc., Naples, FL 34108-1877, USA
*Email: robert.taylor.phd@gmail.com

13.1 Introduction

Providing adequate pain relief for chronic pain patients generally requires around-the-clock drug administration and long-term use. Conventional routes of administration (*e.g.* oral) have disadvantages regarding achieving adequate around-the-clock pain therapy and are not always suitable for some patient populations. Therefore, alternative routes of administration for the chronic pain patient have been sought. Transdermal drug delivery is an attractive route of drug delivery that offers several advantages over conventional routes of administration for treating chronic pain. In its simplest form, a transdermal drug delivery system (TDDS) utilises a skin 'patch', which is a device placed on the outermost layer of the skin, and usually through passive diffusion, delivers medication to the capillary bed located in the skin. Delivering compounds through the skin as an alternative to taking medicines by conventional routes has been demonstrated to be generally both safe and effective.

Many patients on long-term treatment with oral agents forget to take their medication or sometimes get tired of taking multiple pills a day. Thus patches

RSC Drug Discovery Series No. 35
Pain Therapeutics: Current and Future Treatment Paradigms
Edited by Charlotte Allerton

Published by the Royal Society of Chemistry, www.rsc.org

that last up to a week are an attractive alternative for this patient population. Furthermore, absorption through the skin bypasses first-pass liver metabolism and removes the risk of drug–drug interactions by this mechanism (which is commonly seen with oral medication). Most importantly, patches provide a means for steady absorption of drug, thus decreasing the risk of blood level spikes associated with oral medication. All transdermal products today offer these advantages.[1]

Modern designs of transdermal patches have been around since at least the 1970s. Some drugs marketed in patches include scopolamine (for motion sickness), nicotine (for quitting smoking), oestrogen (for menopause and to prevent osteoporosis after menopause), nitroglycerine (for angina), lidocaine (to relieve the pain of shingles (herpes zoster)), testosterone (for low testosterone) and opioid analgesics (chronic pain management). Most of these patches consist of a backing layer, drug, adhesive, and protecting/release liner. How the drug is suspended in the patch varies with patch design and the choice of materials in each layer is dictated mostly by their compatibility with the drug. There is no universal process of development; each patch must achieve a balance among several factors, including delivery, adhesion and skin tolerability. Most importantly, the drug chosen must be suitable to pass through the skin.

For chronic pain patients requiring around-the-clock analgesia, there are only two opioid analgesics marketed in the United States that are formulated into a transdermal patch: fentanyl and buprenorphine. Many other opioids are not suitable or do not possess all of the properties necessary to permeate the skin efficiently. A high octanol/water coefficient and semi-rapid skin flux are two properties that are essential for efficient permeation through the skin. Morphine, for example, which has a very low octanol/water coefficient of –0.1 (log *P*), has a very slow skin flux (0.006 $\mu g\,cm^{-2}\,h^{-1}$).[2] Buprenorphine, on the other hand, has an octanol/water coefficient of 4.98 (log *P*) and a skin flux of 1.4 $\mu g\,cm^{-2}\,h^{-1}$.[2] In this chapter we review some of the basic principles of transdermal drug delivery and patch development, using the buprenorphine transdermal patch, Butrans®, as a specific example.

13.2 Skin Structure

The skin is the largest organ of the body. On average, the skin covers a surface of approximately 2 m^2 and receives about one-third of the blood supply. This characteristic makes the skin an excellent entry point for delivering molecules to the body. It is a highly impermeable structure composed of multiple layers and serves many functions; its primary role is to act as boundary and first line of protection against harmful substances and the hazardous environment.[3]

13.2.1 Layers of Skin

The skin is composed of three main layers known as the epidermis, dermis and hypodermis. The composition of each of the layers is relevant for drug diffusion. The epidermis is the outermost layer of skin, the layer that plays the largest role in the skin's barrier function.[4] The epidermis thickness ranges from 0.06 to 0.8 mm, depending on the location on the body, being thickest on the palms and soles of the feet and thinnest on the eyelids.[5]

The epidermis has been further divided into specific layers, with the outermost layer known as the stratum corneum.[4] The stratum corneum is a flexible yet almost impermeable layer and provides the first obstacle for transdermal drug diffusion. The stratum corneum is composed of 10–25 layers of dead, keratinised cells known as corneocytes. Corneocytes are flat, anucleated and composed of keratin filaments. Derived from the differentiation of keratinocytes, corneocytes are embedded in a matrix of enriched non-polar lipids and organised as lamellar lipid layers.[6] In addition, this layer contains approximately 40–70 hair follicles and 200–250 sweat ducts per square centimetre. On average, it is composed of 20% water and is a very lipophilic membrane. The thickness of the stratum corneum varies depending on the state of hydration and can range from 10 to 20 μm. The hydrophobic environment of the epidermis is one of the main limiting factors in drug permeability.

Beneath the stratum corneum lies the epidermis. This layer consists of the stratum lucidum, stratum granulosum, stratum spinosum and stratum basale.[6] Stratum basale is the layer where keratinocyctes undergo mitosis and it provides a renewable source of cells for the upper layers of the epidermis. Cells that move from the lowest layer of the epidermis to the upper layers undergo a process known as keratinisation.

Below the epidermis lies a much thicker layer known as the dermis. Ranging from 3 to 5 mm in thickness, this layer is composed of connective tissue (blood vessels, lymph vessels and nerves). The blood vessels in this layer are essential for regulation of body temperature, provide nutrients and oxygen to the skin, and remove toxins and waste products. The blood supply keeps dermal concentration of permeates very low. The establishment of a concentration gradient across the epidermis and the dermis drives transdermal drug diffusion.

The innermost layer of the skin, the hypodermis, has a variety of functions. These include tissue support for dermis and epidermis, fat storage, temperature regulation, and nutritional support, among others.

13.2.2 Traversing the Layers of the Skin

For most molecules, passive diffusion is the major mechanism of absorption through the skin. Water-soluble substances can pass through or around the hair follicles and sweat ducts (*i.e.* follicular transport), but this contributes very minimally to the overall permeation of substances. The major driving force for transdermal diffusion is the drug concentration gradient established between the patch and the skin. The gradient is established due to the high concentration of drug in the patch and the lower concentration within the skin. The concentration gradient is maintained due to the continual absorption/uptake of the drug into the systemic circulation.

There are two types of possible diffusion routes. The appendageal/follicular route (a.k.a. shunt route) is transport through the skin's sweat glands and hair follicles.[7] This pathway avoids diffusion through the stratum corneum directly and is a minor route of permeation. The small surface area covered by hair follicles or sweat ducts contribute minimally to diffusion. The second and most common route, is known as the epidermal route. Two pathways compose the epidermal route. In the first, known as the paracellular pathway, the drug passes through the intercellular space of cells to the dermis (around or between the cells).[8] In the second route, known as the transcellular pathway, the drug passes through cell membranes and through the cytoplasm of the cells.[8] This route provides the shortest distance, but most resistance. Drugs must possess the physical properties that are compatible with crossing through a cell's highly lipophilic plasma membrane and the hydrophilic cellular contents in the cytoplasm. Active transport of ionic and polar compounds, endocytosis, and transcytosis of macromolecules also contribute to drug permeation.

13.2.3 Variations in Human Skin (Biological Factors)

Both the rate and the amount of diffusion that takes place at the skin layer is not the same for every individual. Many factors can affect the transport of drugs through the skin. These can be grouped based on the biology of the skin and on the physiochemical structure of the skin. Biological factors include skin state, skin age, blood flow, skin site, and local metabolism. Physiochemical factors include skin hydration, temperature and pH, among others.

13.2.3.1 Skin Conditions

Human skin is exposed to many potentially damaging environmental factors. Skin that is more intact or non-diseased provides a better barrier than does damaged skin.[9] Some examples of how alterations in skin conditions affect skin permeability include sunburn (decreased permeability), peeling skin (decreased permeability), thermal burns (increased permeability), eczema (increased permeability), psoriasis (decreased permeability) and chemical peels (increased permeability), among others.

13.2.3.2 Skin Age

Young skin is more permeable than older skin and children are more sensitive to toxins. The skin's main barrier function diminishes with increasing prematurity, which results in an increase of water loss and increased absorption of toxins and other chemicals.[10] Reasons for the increased absorption in children and infants include a thinner stratum corneum and higher surface area-to-volume ratio. A high surface area-to-volume ratio, immature drug metabolising systems and decreased subcutaneous fat stores effectively increase the absorptive area, while decreasing the volume of distribution, of a drug or toxin.[11,12]

13.2.3.3 Blood Supply

Molecules that pass into the dermis layer can be absorbed into the systemic blood circulation. This aids in developing the 'sink' necessary to maintain the concentration gradient required for transdermal drug delivery. Alteration in the rate of blood flow in the skin can alter the rate of absorption through the skin.[13] Disruption alters the concentration gradient and the rate of diffusion from the stratum corneum to the dermis.

13.2.3.4 Skin Site

The thickness of the skin can affect the absorption rate of drugs.[14] The thicker the layer, the lower the rate of permeation. The stratum corneum thickness varies throughout the body, with the soles of the feet and palms of hands being the thickest and that behind the ear and on the eyelids being the thinnest.[5]

13.2.3.5 Skin Metabolism

The skin has the ability to metabolise drugs such as steroids, hormones and chemical carcinogens, among others.[15]

13.2.4 Variations in Human Skin (Physicochemical Factors)

13.2.4.1 Skin Hydration

Skin hydration is one of the challenges of transdermal delivery. Fully hydrated stratum corneum has less barrier function than does its dehydrated counterpart. Humectants, substances that retain water, are commonly incorporated in transdermal delivery to increase the hydration of the skin.[16]

13.2.4.2 Temperature and pH

Skin permeation can vary up to tenfold with temperature variation. For example, the diffusion coefficient decreases as temperature decreases. Increasing skin temperature increases transdermal drug absorption.[17]

Skin pH also plays an important role in the skin's barrier function.[18] It can affect the dissociation of weak acids and weak bases. The proportion of unionised drug determines the drug concentration in the skin and is important for the establishment of the drug concentration gradient. For example, increase of pH may increase the permeability coefficients of some opioids.[19]

13.3 Transdermal Patches

13.3.1 History of Patches

The patch concept has been around since at least the ancient Greeks. The Greek physician Galen described an opium-based ointment that forms a 'patch' when it is applied externally.[20] Known as 'Olympic victor's dark ointment' (OVDO), it has been reformulated and shown to be effective when applied to abdominal mouse skin.[20] Modern transdermal patches began development in the 1970s and the first approval was granted by the U.S. Food and Drug Administration (FDA) in 1979. Known as the scopolamine patch, it was the first of its kind to deliver therapeutic doses of scopolamine for up to 3 days for symptoms of motion sickness. Since that time, transdermal patches have undergone significant changes in order to increase efficacy, safety and the number and types of drugs compatible for patches.

Patch 'generations' have been described by Prausnitz *et al.*[21] Many of the first-generation patches consisted of plastic coatings with a drug that was dissolved in alcohol and used conventional adhesives. The alcohol and poor adhesive generally created a high percentage of skin reactions and the patches did not adhere well to the skin. Improvements began in second-generation patches with the introduction of the 'drug in the adhesive' model. In these, the adhesive serves two roles: to keep the patch attached to the skin and as a suspension that contains the drug. However, these patches were not without problems and their major downfall was that the drug dictated the tackiness of the patch. Patch sizes and administration frequency varied depending on drug concentration and drug type. Delivery technology improved with the third-generation patches, when reservoir systems were incorporated to hold the drug. Many of these patches are currently on the market, accompanied by fourth-generation patches, which incorporate an enhancer mechanism to facilitate the delivery of the drug through the skin. These include chemical enhancers as well as mechanical enhancers such as iontophoresis, ultrasound, microneedles, sonophoresis, lasers and electrophoretic methods. These enhancements are now aimed at increasing the number of drugs that can be incorporated into a transdermal patch.

13.3.2 Advantages of a Patch

Transdermal patches provide a number of advantages compared to conventional oral dosing. These are summarised in Table 13.1.

13.3.3 Disadvantages of a Patch

The most common disadvantage of transdermal patches is the possibility of developing a skin irritation or reaction at the site of application. Erythema, itching and local oedema have been documented with the development of the first-generation patches. Irritation can be caused by the adhesive, the drug or the excipients in the drug formulation.[22]

The skin barrier function limits the type of molecules that can diffuse successfully to the dermis and thus only a limited number of drugs can be developed for transdermal application. Drugs that have hydrophilic structures cannot permeate well and thus drugs with lipophilic properties and of small molecular size are best suited for patch delivery. However, current development

Table 13.1 Advantages of transdermal patches.[55,56]

Pharmacokinetic	Continuous delivery of drug
	Sustained drug levels in blood (reduction in peaks and troughs)
	Decrease in systemic adverse events
	Increased probability of achieving optimal therapeutic dose
	Independent of food intake
	Avoids first-pass effect
	Reduced gastrointestinal upset
	Control dose and time
Convenience[57]	Small and discreet
	Suitable when oral route is not an option
	Non-invasive
	Minimal applications from once daily to once a week
	Easy to apply and monitor
	Comfortable to wear
	Empowers the care-giver
	Visual reminder to treat and reassurance that medication is being taken
	Avoids accidental overdosing
	Easily removed in the event of an emergency
	Increase patient compliance
	Site-specific dosing (*e.g.* arm, back, knee)
	Easily identified in emergencies because of physical presence and markings

Table 13.2 Disadvantages of transdermal patches.

May not adhere to certain skin types (*e.g.* oily, water, sweating)
May cause minor to severe skin irritation
Cannot deliver ionic drugs
Cannot achieve high drug levels in blood/plasma
Cannot develop for drugs of large molecular size
Cannot deliver drugs in a pulsatile fashion
Surplus of drug in disposed of patch can be a hazard to children and pets if not disposed of properly (*e.g.* used fentanyl patches can cause overdose in children)
If patch is damaged, poor control of release can occur

is under way to enhance the delivery of hydrophilic compounds. Other disadvantages are listed in Table 13.2.

13.3.4 Types of Transdermal Patches

Classification of transdermal patch type is generally based on how the drug is stored in the patch. Transdermal patches currently consist of three main types (with additional patches incorporating variations of these main types): the drug in the adhesive, the drug in the reservoir and the drug in the matrix. Each of these systems has advantages and disadvantages.

13.3.4.1 Adhesive Diffusion Controlled System

The adhesive diffusion controlled system is the simplest form of the transdermal patch.[23] The drug is directly dispersed in the adhesive polymer. The adhesive is then applied on to a drug-impermeable metallic plastic backing to form a thin drug reservoir layer. A non-medicated rate-controlling adhesive polymer of constant thickness is applied on the top of this layer. This system may be comprised of a single layer or may have a multilayer design (*e.g.* additional layer of drug-in-adhesive). Advantages of this type of patch include improved patient compliance, improved cosmetic acceptance and better adhesion. Disadvantages include possible drug-adhesive incompatibility or patch tackiness.

13.3.4.2 Membrane Moderated System

The reservoir system has the drug in a gel or solution and delivery is determined by a rate-controlling membrane between the drug reservoir and the skin.[23] The drug is homogeneously dispersed in a solid polymer matrix, suspended in an unleachable, viscous liquid medium to form a paste-like suspension, or is dissolved in a releasable solvent to form a clear drug solution. The rate-controlling membrane can be a microporous or polymeric membrane with specific drug permeability. The external surface can have pressure-sensitive adhesive polymer. Varying the drug reservoir formulation, the permeability coefficient and the thickness of the rate-controlling membrane controls the drug release rate. Advantages of the system consist of tighter control of delivery rate; disadvantages include possible initial burst of drug release or excess release due to damaged membrane.

13.3.4.3 Matrix

In the matrix system, the drug is incorporated in an adhesive polymer matrix.[23] The drug is continuously released into the skin. The dose depends on the drug amount in the matrix and the area of patch applied to the skin. Advantages include direct exposure of polymeric matrix to skin, no interference of adhesive and even distribution of the drug throughout the patch. The patch can be cut to the adjust dose, and it carries less risk of accidental overdose or abuse.

13.3.4.4 Other Patch Variants

Other patches are variants of the three major patch forms. These include the membrane matrix hybrid type patch and the microreservoir transdermal patch. The matrix hybrid is a modification of the reservoir-type transdermal patch. The liquid formulation of the drug reservoir is replaced with a solid polymer matrix (*e.g.* polyisobutylene) which is sandwiched between a rate-controlling membrane and backing laminate. The microreservoir patch is similar to the drug reservoir system in that drug is suspended in multiple tiny microscopic spheres or reservoirs. This allows the dose to be adjusted by cutting the patch; however, dose adjustment may not necessarily be proportional to size of patch.[23]

13.3.4.5 Future Patches

Current developmental focus is finding ways to increase the delivery of ionic and large-molecular-mass molecules through the skin. Therefore, patches are increasingly incorporating enhancer mechanisms into the patch design. Enhancers can either be chemical or mechanical. Chemical enhancers disrupt the structure of the skin barrier to allow for increased diffusion of molecules. Mechanical designs incorporate a number of technologies:

- **Iontophoresis**:[24] a few milliamperes of current is passed to a small area of skin through an electrode placed in contact with the formulation. There are active delivery systems which include iontophoresis.
- **Sonophoresis**:[25] ultrasonic energy is used to disrupt the stratum corneum barrier and enhance transport through the skin.
- **Microneedles**:[26] tiny microstructures disrupt the stratum corneum (create channels) and allow larger molecules and water-soluble drugs to be delivered through the skin.
- **Electroporation**:[1,27] short, high-voltage electrical pulses are applied to the skin. Skin permeability is disturbed and diffusion of drugs is increased by four orders of magnitude. Electrical pulses create aqueous pores in the stratum corneum that allow passage of drug.
- **Ultrasound**:[28] low-frequency ultrasound enhances transport of macromolecules through skin.

13.3.5 Components of Transdermal Patches

There are a number of components that make up transdermal patches, in addition to drug, regardless of the method of drug storage. Common components of transdermal systems may include polymer matrix/drug reservoir, permeation enhancers, pressure-sensitive adhesive, backing laminates, release liner and other excipients such as plasticisers and solvents.

13.3.5.1 Drug Reservoir—Polymer Matrix

Transdermal systems are composed of multiple layers, sandwiched together in order to deliver drug in one direction, through the skin. The drug polymer matrix is placed between an impervious outer backing layer and an inner layer composed of an adhesive. In the drug reservoir system, the outer layer consists of the impermeable backing layer and the inner layer (the rate-controlling membrane).

13.3.5.2 Rate Membrane

The rate-controlling membrane dictates the rate of diffusion of the drug from the drug reservoir to the skin. The membrane is generally made up of tiny micropores that allow passage of the drug to the skin. The design of the membrane makes sure that the rate of delivery to the skin is less than the rate of absorption from the skin. This offsets any variation in the rate of absorption according to the properties of different skin sites and helps maintain a continuous flow.

13.3.5.3 Enhancers

Enhancers may be chemical compounds or mechanical designs that increase permeability of the stratum corneum. Chemical enhancers interact with structural components of the skin, such as proteins or lipids, in the stratum corneum causing modification of the natural barrier function of the skin. Depending on the enhancer, they may disrupt lipid organisation or interact with the keratin structure. Common enhancers used today include alcohols, polyalcohols, pyrrolidones, amines, amides, fatty acids, sulphoxides, esters, terpenes, alkanes, surfactants and phospholipids. For example, alcohols are used to solubilise the lipid bilayer while fatty acids are used to induce lipid fluidisation as well as phase separation within the membrane.[29,30] Mechanical enhancers include ionotrophoresis, ultrasound, sonification, microneedles, *etc.*

13.3.5.4 Pressure-Sensitive Adhesives

One of the most important components of a transdermal patch is the adhesive.[31,32] The adhesive must remain stable for the length of the patch wear, which may be up to 1 week for some patches. Pressure-sensitive adhesives are used in transdermal preparation to help maintain contact between patch and skin. Desirable characteristics include easy adherence, sufficient duration of adherence and no residual residue after removal. Polyacrylates, polyisobutylene and silicon adhesives are commonly used. The adhesives should be physiochemically and biologically compatible with drug and not alter drug release.

13.3.5.5 Backing Laminate

A backing laminate protects the transdermal drug formulation when attached to the skin and prevents the drug from leaking out of the patch, as well as preventing absorption of other materials into the patch. The backing material can influence the delivery profile, adhesion, wear ability, occlusion of excipients and appearance of the patch. The patch environment can be extreme and thus has to be able to withstand varying conditions on the skin.

13.3.5.6 Protecting/Release Liner

A protective liner covers the patch while the patch is stored, and is removed before the patch is applied to skin.[33] The liner also protects the patch during its shelf life. Since the liner is in contact with the system, it should comply with specific requirements regarding chemical inertness and permeation to the drug, penetration enhancer and water.

13.3.6 Evaluation of Transdermal Patches

A new transdermal patch formulation has to undergo a multitude of tests (Table 13.3). Many of the initial tests determine the stability of the patch and the patch materials. From there, tests on patch durability and adhesion are

Table 13.3 Patch tests.

Test	*Description*
Physiochemical studies	
Thickness tests	Various instruments are used to determine thickness of the film at different patch sections. These may include traveling microscope, dial gauge or screw gauge
Uniformity of weight	Individual weights of patches are compared to average weight of X amount of patches
Drug content determination	Film containing drug is dissolved in a solvent and drug concentration is determined spectrophotmetrically
Content uniformity test	Test to determine content variability between patches. Normally 10 patches are selected and compared against each other
Moisture content	Films are weighed and kept in desiccators containing calcium chloride. Films are constantly weighed at specific intervals after 24 hours until constant weight is obtained. Moisture content is then calculated
Moisture uptake	Films are kept in desiccators and exposed to 84% relative humidity using Potassium chloride until a constant weight is achieved. % moisture is then calculated
Flatness	Tests to determine if patch will constrict with time
Folding endurance	Tests to determine the durability of the patch. Patch is folded repeatedly at same place until it breaks
Tensile strength	Polymeric films are stretched and elongated until the film breaks. The length of elongation is measured as well as weight required to break the films
Adhesive studies	
Thumb tack test	Test to determine amount of force required to remove thumb from the patch
Rolling ball test	A steel ball is rolled over the adhesive surface and distance is measured. The farther the ball rolls, the less tacky the adhesive
Quick stick (Peel tack) test	The adhesive tape is pulled away from experimental surface at a specific rate to determine the required force to remove adhesive
Probe tack test	Force required to pull a probe away from an adhesive at a fixed rate is recorded as tack
Pharmacological testing	
In vitro release studies	Tests to determine drug release rate include the paddle over disc, the cylinder modified USP basket, the reciprocating disc and diffusion cells
In vitro permeation studies	Usually permeation studies are performed by placing the fabricated transdermal patch with rat skin or synthetic membrane in between receptor and donor compartment in a vertical diffusion cell such as Franz diffusion cell or Keshary–Chien diffusion cell
In vivo studies	Animal models and human clinical trials

This list is not exhaustive.

conducted. Once confirmed stable and durable, a patch undergoes various *in vitro* and *in vivo* studies.

Studies of transdermal dosage forms often include:[34]

- physicochemical evaluation: thickness, weight, uniformity of drug content, moisture content and uptake, flatness, folding endurance and tensile strength
- water vapour transmission studies
- microscopic studies
- adhesive studies:[35] thumb tack tests, rolling ball tests, quick stick tests, probe tack test
- *in vitro* studies: drug release studies, permeation studies
- *in vivo* studies: animal, human

13.4 Pharmacokinetics of Transdermal Patch

13.4.1 Effects on Drugs (Physical/Chemical Properties of Compound)

Once a patch is applied, a drug concentration gradient becomes established. The drug moves down the concentration gradient from inside the patch to the dermis layer of the skin. During this process, a drug reservoir develops in the stratum corneum. Drug gets absorbed into the local capillary vasculature and enters the systemic circulation. The absorption process creates a delay between drug absorption and minimum effective concentration and varies between drugs. The time to reach steady-state concentration varies and may be achieved after 2–3 patch applications.

13.4.2 Drug Properties

Not all drugs are capable of being formulated for transdermal delivery. The key requirements necessary for a drug to be formulated for transdermal delivery[21,36,37] include:

- low molecular mass ($<$500 Da)
- high diffusion coefficient (high lipophilicity, 10 $<$ oil–water partition coefficient $<$1000)
- low required daily dose ($<$10 mg day^{-1})

Additional desirable properties include:

- half-life $\leq$10 hours
- pH of saturated aqueous solution 5–9
- high partition coefficient ($\sim$1–4)
- low oral bioavailability
- non-irritating and non-sensitising to the skin
- skin permeability coefficient should be $<0.5 \times 10^{-3}$ cm h^{-1}.

13.5 Analgesic Patches

13.5.1 Pain

Pain can be classified in several ways, such as by location (*e.g.* abdominal, colorectal, joint, dental), tissue type (*e.g.* muscle, bone), disease state (*e.g.* cancer, sickle-cell anaemia), sensation (*e.g.* sharp, dull, lancing) or physiology (*e.g.* nociceptive, neuropathic). Pain can also be classified by magnitude (*e.g.* mild, moderate or severe) or by duration (*e.g.* acute or chronic). At the present time there are no fully accepted mechanistic distinctions among pain classifications. For example, the difference between acute and chronic pain is currently defined by duration. However, mechanisms have been hypothesised. For example, it is thought that peripheral injury leads to excitability in peripheral nociceptors that leads to peripheral sensitisation, primary hyperalgesia,[38] and continuous stimulation of central nervous system neurons, which, in turn, can result in secondary hyperalgesia and central sensitisation. A phenomenon known as 'hyperalgesic priming' has been postulated—in which neuroplastic changes in primary afferent nociceptors trigger a protracted hypersensitivity to inflammatory cytokines—as a possible explanation for the transition from acute to chronic pain.[39,40] In this scenario, acute pain is the initiation phase of a cascade that establishes the biological foundation for chronic pain[41,42] and, if unchecked, progresses to chronic pain,[43] a process of 'chronification'.[44]

Transdermal delivery provides a continuous delivery of medication and is convenient for extended use. For example, the Butrans® transdermal system is '... indicated for the management of moderate to severe chronic pain when a continuous, around-the-clock opioid analgesic is needed for an extended period of time'.[45]

13.5.1.1 Prevalence

Pain is relatively common (http://pain.myhealthtrends.com/pain_statistics.do):

- 89% of Americans age ≥18 suffer from pain at least once a month
- Over a 3 year period, 76 million Americans reported pain lasting ≥24 hours
- 42% of adults age ≥20 who reported pain stated the pain lasted ≥1 year
- 40% of children and adolescents complain of pain that occurs at least once weekly
- Chronic pain affects at least 15% of children
- 50% of older persons living in nursing homes have pain that interferes with normal function
- 59% of nursing home residents experience persistent pain.

13.5.1.2 Beneficial and Deleterious Effects

Acute pain is often beneficial, preventing further injury if the source of the pain (tissue damage) is external. Pain that persists beyond this protective stage is generally not beneficial, and in fact is often deleterious. Unaddressed pain leads to decreased quality of life, lost work and recreation time, delayed recovery and extended hospital stays. It is one of the most common symptomatic causes for utilisation of healthcare resources (*e.g.* patients in the United States spend more than $100 billion annually to relieve pain).[46]

13.5.2 Opioid Analgesics

There are several analgesics available that are effective for the treatment of mild to moderate pain. For example, non-steroidal anti-inflammatory drugs (NSAIDs) and paracetamol (acetaminophen) are effective pain relievers with broad efficacy. However, they are often less effective against stronger pains. In addition, NSAIDs have been associated with dose-related side effects[47,48] and should not be given to certain patients,[49] and paracetamol should not be taken in excess of recommendations.[50] Opioids are generally safe and effective for stronger pains and there is evidence of an increase in opioid prescribing by U.S. primary care physicians in the past 10 years.[51]

13.5.3 Opioids in Patches

Not all opioids possess the physiochemical properties and other characteristics compatible for use in patch formulations. Several features of fentanyl and buprenorphine allow for such application. They have a low molecular mass, high lipid solubility, high analgesic potency and the requisite stability needed for use in patches.

13.6 Fentanyl Patch

13.6.1 Description

Fentanyl is an opioid receptor agonist that interacts predominately with the μ-opioid receptor subtype. The fentanyl patch (DURAGESIC®) is a rectangular transparent unit comprised of a protective liner and two functional layers: a backing layer consisting a of polyester/ethyl vinyl acetate film and a drug-in-adhesive layer (before use, a protective liner covering the adhesive layer is removed and discarded).

13.7 Buprenorphine

13.7.1 Pharmacology of Buprenorphine

Buprenorphine is a centrally acting analgesic that binds to opioid receptors with high affinity. It has particularly high affinity for μ-opioid receptors.[52] It also binds to other receptors with individual efficacy, ranging from agonist to antagonist, and has demonstrated efficacy against a variety of acute and chronic pain conditions.[53] Butrans® is indicated for the management of moderate to severe chronic pain when a continuous, around-the-clock opioid analgesic is needed for an extended period of time.[54]

13.7.2 Butrans®

13.7.2.1 Description

Butrans® is a rectangular or square, beige-coloured system (patch) consisting of a protective liner and functional layers. Proceeding from the outer surface toward the surface adhering to the skin, the layers are: a web backing layer; an adhesive rim; a separating layer over the buprenorphine-containing adhesive matrix; the buprenorphine-containing adhesive matrix; and a peel-off release liner (removed and discarded before use) (Figure 13.1).

13.7.2.2 Dosage and Administration

Three strengths of Butrans® are available: 5, 10 and 20 µg/hour (the proportion of buprenorphine mixed in the adhesive matrix is the same in each of the three strengths). The actual amount of buprenorphine released from each system per hour is proportional to the active surface area of the system. The skin is the limiting barrier to diffusion from the system into the bloodstream.

Each Butrans® patch is intended to be worn for up to 7 days. The initial dosing regimen should be individualised for each patient, taking into account their prior analgesic treatment experience and other factors (consult the prescribing information for this and other clinical instruction). Butrans® is applied to a hairless or nearly hairless skin site on the upper outer arm, upper chest, upper back or the side of the chest. These four sites (each present on both sides of the body) provide eight possible sites and applications should be rotated among the eight sites. After Butrans® removal from one site, a minimum of 21 days should elapse before reapplication to the same site. If problems with adhesion of Butrans® occur, the edges may be taped with first aid tape. If Butrans® falls off during the 7 days dosing interval, it should be properly disposed of and a new Butrans® applied at a different skin site. The patient should be monitored closely for respiratory depression, especially within the first 24–72 hours of initiating therapy. Butrans® should be individually titrated to a dose that provides adequate analgesia and minimises adverse reactions.

Butrans® should not be discontinued abruptly. When the patient no longer requires therapy, a gradual downward titration of the dose every 7 days should

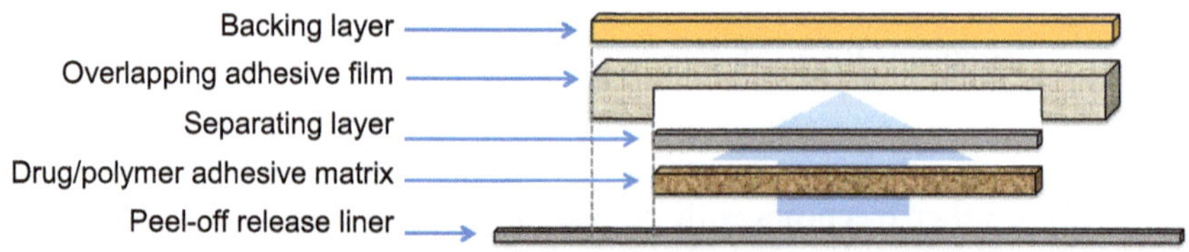

Figure 13.1 Essential components of the Butrans® transdermal system.

be used in order to prevent signs and symptoms of withdrawal in the physically dependent patient.

13.7.2.3 Pharmacokinetics

Based on the pharmacokinetic profile and time to reach steady state levels, the minimum titration interval is 72 hours. Because steady-state plasma concentrations are approximated within 72 hours, Butrans® dosage adjustments may be done every 3 days. Patients who experience breakthrough pain may require adjustment of dose or the administration of a rescue medication with appropriate dose of an immediate-release opioid or non-opioid drug.

13.8 Conclusion

Transdermal patches have been developed to improve clinical efficacy of the drug and to enhance patient compliance by delivering a smaller amount of drug at a predetermined rate. Transdermal administration of strong opioid analgesics provides many advantages (including scientific, sociological and economic) over other routes of drug administration. They are not without problems, but when used properly and in the appropriate patient population, they can provide a convenient and effective mechanism for continuous delivery of pain medication to patients who require a continuous, around-the-clock opioid analgesic for an extended period of time.

Dedication

The authors wish to dedicate this chapter to the memory of one of our beautiful nephews, Lawrence Langone Jr, who passed away while celebrating Thanksgiving with his loving family.

References

1. R. K. Subedi, S. Y. Oh, M. K. Chun and H. K. Choi, *Arch. Pharm. Res.*, 2010, **33**, 339–351.
2. L. Margetts and R. Sawyer, *Cont. Ed. Anaesth. Crit. Care. Pain*, 2007, **7**, 171–176.
3. C. M. Chuong, B. J. Nickoloff, P. M. Elias, L. A. Goldsmith, E. Macher, P. A. Maderson, J. P. Sundberg, H. Tagami, P. M. Plonka, K. Thestrup-Pederson, B. A. Bernard, J. M. Schroder, P. Dotto, C. M. Chang, M. L. Williams, K. R. Feingold, L. E. King, A. M. Kligman, J. L. Rees and E. Christophers, *Exp. Dermatol.*, 2002, **11**, 159–187.
4. A. Baroni, E. Buommino, V. De Gregorio, E. Ruocco, V. Ruocco and R. Wolf, *Clin. Dermatol.*, 2012, **30**, 257–262.
5. J. Sandby-Moller, T. Poulsen and H. C. Wulf, *Acta. Derm. Venereol.*, 2003, **83**, 410–413.

6. E. Proksch, J. M. Brandner and J. M. Jensen, *Exp. Dermatol.*, 2008, **17**, 1063–1072.
7. F. Knorr, J. Lademann, A. Patzelt, W. Sterry, U. Blume-Peytavi and A. Vogt, *Eur. J. Pharm. Biopharm.*, 2009, **71**, 173–180.
8. Y. Komarova and A. B. Malik, *Annu. Rev. Physiol.*, 2010, **72**, 463–493.
9. M. Fartasch, *Microsc. Res. Tech.*, 1997, **38**, 361–372.
10. T. Shwayder and T. Akland, *Dermatol. Ther.*, 2005, **18**, 87–103.
11. G. N. Stamatas, J. Nikolovski, M. A. Luedtke, N. Kollias and B. C. Wiegand, *Pediatr. Dermatol.*, 2010, **27**, 125–131.
12. F. S. Afsar, *Clin. Exp. Dermatol.*, 2009, **34**, 855–858.
13. K. Higaki, K. Nakayama, T. Suyama, C. Amnuaikit, K. Ogawara and T. Kimura, *Int. J. Pharm.*, 2005, **288**, 227–233.
14. H. Takeuchi, M. Ishida, A. Furuya, H. Todo, H. Urano and K. Sugibayashi, *Biol. Pharm. Bull.*, 2012, **35**, 192–202.
15. T. Hikima, K. Tojo and H. I. Maibach, *Skin Pharmacol. Physiol.*, 2005, **18**, 153–159.
16. K. S. Paudel, M. Milewski, C. L. Swadley, N. K. Brogden, P. Ghosh and A. L. Stinchcomb, *Ther. Deliv.*, 2010, **1**, 109–131.
17. K. Tominaga and K. Tojo, *Biol. Pharm. Bull.*, 2010, **33**, 1983–1987.
18. M. H. Schmid-Wendtner and H. C. Korting, *Skin. Pharmacol. Physiol.*, 2006, **19**, 296–302.
19. S. D. Roy and G. L. Flynn, *Pharm. Res.*, 1990, **7**, 842–847.
20. A. P. Harrison, S. H. Hansen and E. M. Bartels, *Pain. Pract.*, 2012, **12**, 620–625.
21. M. R. Prausnitz and R. Langer, *Nat. Biotechnol.*, 2008, **26**, 1261–1268.
22. I. Ale, J. M. Lachapelle and H. I. Maibach, *Adv. Ther.*, 2009, **26**, 920–935.
23. A. M. Ball and K. M. Smith, *Am. J. Health. Syst. Pharm.*, 2008, **65**, 1337–1346.
24. Y. Wang, R. Thakur, Q. Fan and B. Michniak, *Eur. J. Pharm. Biopharm.*, 2005, **60**, 179–191.
25. S. Mitragotri and J. Kost, *Biotechnol. Prog.*, 2000, **16**, 488–492.
26. M. G. Nava-Arzaluz, I. Calderon-Lojero, D. Quintanar-Guerrero, R. Villalobos-Garcia and A. Ganem-Quintanar, *Curr. Drug. Deliv.*, 2012, **9**, 57–73.
27. A. R. Denet, R. Vanbever and V. Preat, *Adv. Drug. Deliv. Rev.*, 2004, **56**, 659–674.
28. S. Mitragotri, D. A. Edwards, D. Blankschtein and R. Langer, *J. Pharm. Sci.*, 1995, **84**, 697–706.
29. T. Kai, V. H. W. Mak, R. O. Potts and R. H. Guy, *J. Controlled Release*, 1990, **12**, 103–112.
30. A. Naik, L. A. R. M. Pechtold, R. O. Potts and R. H. Guy, *J. Controlled Release*, 1995, **37**, 299–306.
31. H. S. Tan and W. R. Pfister, *Pharm. Sci. Technolol. Today*, 1999, **2**, 60–69.
32. S. Venkatraman and R. Gale, *Biomaterials*, 1998, **19**, 1119–1136.
33. A. M. Wokovich, M. Shen, W. H. Doub, S. G. Machado and L. F. Buhse, *J. Pharm. Sci.*, 2010, **99**, 3177–3187.

34. G. A. Van Buskirk, D. Arsulowicz, P. Basu, L. Block, B. Cai, G. W. Cleary, T. Ghosh, M. A. Gonzalez, D. Kanios, M. Marques, P. K. Noonan, T. Ocheltree, P. Schwarz, V. Shah, T. S. Spencer, L. Tavares, K. Ulman, R. Uppoor and T. Yeoh, *AAPS PharmSciTech*, 2012, **13**, 218–230.
35. F. Cilurzo, C. G. Gennari and P. Minghetti, *Expert Opin. Drug. Deliv.*, 2012, **9**, 33–45.
36. A. Naik, Y. N. Kalia and R. H. Guy, *Pharmaceut. Sci. Technol. Today*, 2000, **3**, 318–326.
37. N. S. Chandrashekar and R. H. Shobha Rani, *Indian J. Pharm. Sci*, 2008, **70**, 94–96.
38. C. Wang, J. Hah and I. Carroll, *Curr. Pain Headache Rep.*, 2009, **13**, 7–11.
39. L. Ferrari, O. Bogen and J. Levine, *Neuroscience*, 2010, **165**, 896–901.
40. D. Reichling and J. Levine, *Trends. Neurosci.*, 2009, **32**, 611–618.
41. D. Carr and L. Goudas, *Lancet*, 1999, **353**, 2051–2058.
42. D. Niv and M. Devor, eds., *Transition from Acute to Chronic Pain*, Williams & Wilkins, Baltimore, MD, 3rd edn., 1998.
43. J. Katz and Z. Seltzer, *Expert Rev. Neurother.*, 2009, **9**, 723–744.
44. J. V. Pergolizzi, R. B. Raffa and R. Taylor, *Pain Manage Nurs.*, 2012, doi:10.1016/j.pmn.2012.07.004.
45. Purdue Pharma L. P., Stamford, CT, July 2012.
46. K. Simpson, *Br. J. Anaesth.*, 2004, **92**, 327–328.
47. I. Bjarnason, *J. Roy. Soc. Med.*, 2007, **100**, 11–14.
48. T. Chiba, K. Sato and N. Kudara, *Inflammopharmacology*, 2008, **16**, 16–20.
49. G. Gisalson, J. Rasmussen, S. Abilstrom, T. Schramm, M. Hansen, E. Fosbol, R. Sorensen, F. Folke, P. Buch, N. Gadsboll, S. Rasmussen, H. Poulsen, L. Kober, M. Madsen and C. Torp-Pedersen, *Arch. Intern. Med.*, 2009, **169**, 141–149.
50. in *Ortho-McNeil Pharmaceuticals, Inc.*, Ortho-McNeil Pharmaceuticals, Inc., Raritan, 2000.
51. Y. Olsen, D. Daumit and D. Ford, *J. Pain*, 2006, **4**, 225–235.
52. P. Huang, G. B. Kehner, A. Cowan and L. Y. Liu-Chen, *J. Pharmacol. Exp. Ther.*, 2001, **297**, 688–695.
53. K. Budd and R. B. Raffa, *Buprenorphine—The Unique Opioid Analgesic*, Thieme, Stutgart, 2005.
54. Purdue Pharma L.P., Stamford, CT, 2012.
55. L. Priano, M. R. Gasco and A. Mauro, *Drugs Aging*, 2006, **23**, 357–375.
56. G. Small and B. Dubois, *Curr. Med. Res. Opin.*, 2007, **23**, 2705–2713.
57. L. Allan, H. Hays, N. H. Jensen, B. L. de Waroux, M. Bolt, R. Donald and E. Kalso, *BMJ*, 2001, **322**, 1154–1158.

Subject Index

References to tables and charts are in **bold** type